5G and Fiber Optics Security Technologies for Smart Grid Cyber Defense

G. Prabhakar
Thiagarajar College of Engineering, India

N. Ayyanar
Thiagarajar College of Engineering, India

S. Rajaram
Thiagarajar College of Engineering, India

A volume in the Advances in
Information Security, Privacy,
and Ethics (AISPE) Book Series

Published in the United States of America by
IGI Global
Engineering Science Reference (an imprint of IGI Global)
701 E. Chocolate Avenue
Hershey PA, USA 17033
Tel: 717-533-8845
Fax: 717-533-8661
E-mail: cust@igi-global.com
Web site: http://www.igi-global.com

Library of Congress Cataloging-in-Publication Data

CIP DATA PROCESSING

2024 Engineering Science Reference
ISBN(hc): 9798369327869
ISBN(sc): 9798369344446
eISBN: 9798369327876

British Cataloguing in Publication Data
A Cataloguing in Publication record for this book is available from the British Library.

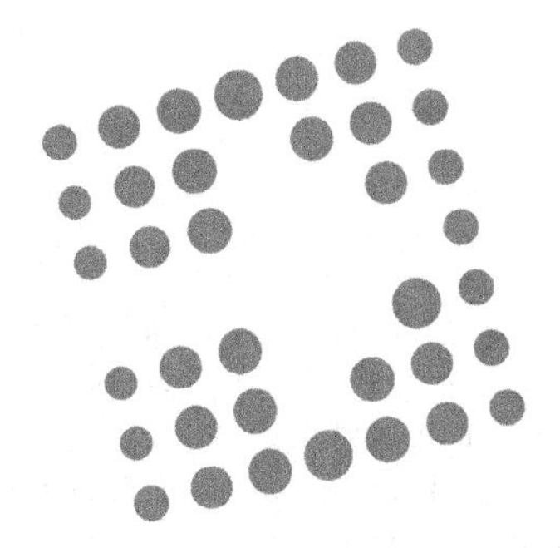

Advances in Information Security, Privacy, and Ethics (AISPE) Book Series

Manish Gupta
State University of New York, USA

ISSN:1948-9730
EISSN:1948-9749

MISSION

As digital technologies become more pervasive in everyday life and the Internet is utilized in ever increasing ways by both private and public entities, concern over digital threats becomes more prevalent.

The **Advances in Information Security, Privacy, & Ethics (AISPE) Book Series** provides cutting-edge research on the protection and misuse of information and technology across various industries and settings. Comprised of scholarly research on topics such as identity management, cryptography, system security, authentication, and data protection, this book series is ideal for reference by IT professionals, academicians, and upper-level students.

Coverage

- Privacy Issues of Social Networking
- Privacy-Enhancing Technologies
- Security Classifications
- Global Privacy Concerns
- Cookies
- Network Security Services
- Telecommunications Regulations
- Electronic Mail Security
- Cyberethics
- Device Fingerprinting

IGI Global is currently accepting manuscripts for publication within this series. To submit a proposal for a volume in this series, please contact our Acquisition Editors at Acquisitions@igi-global.com or visit: http://www.igi-global.com/publish/.

The (ISSN) is published by IGI Global, 701 E. Chocolate Avenue, Hershey, PA 17033-1240, USA, www.igi-global.com. This series is composed of titles available for purchase individually; each title is edited to be contextually exclusive from any other title within the series. For pricing and ordering information please visit http://www.igi-global.com/book-series/advances-information-security-privacy-ethics/37157. Postmaster: Send all address changes to above address.

Titles in this Series

For a list of additional titles in this series, please visit: www.igi-global.com/book-series

Machine Learning and Cryptographic Solutions for Data Protection and Network Security
J. Anitha Ruth (SRM Institute of Science and Technology, Vadapalani, India) Vijayalakshmi G. V. Mahesh (BMS Institute of Technology and Management, India) P. Visalakshi (Department of Networking and Communications, College of Engineering and Technology, SRM Institute of Science and Technology, Katankulathur, India) R. Uma (Sri Sairam Engineering College, Chennai, India) and A. Meenakshi (SRM Institute of Science and Technology, Vadapalani, India)
Engineering Science Reference • copyright 2024 • 526pp • H/C (ISBN: 9798369341599) • US $355.00 (our price)

Global Perspectives on the Applications of Computer Vision in Cybersecurity
Franklin Tchakounté (University of Ngaoundere, Cameroon) and Marcellin Atemkeng (Rhodes University, South Africa)
Engineering Science Reference • copyright 2024 • 306pp • H/C (ISBN: 9781668481271) • US $250.00 (our price)

Enhancing Security in Public Spaces Through Generative Adversarial Networks (GANs)
Sivaram Ponnusamy (Sandip University, Nashik, India) Jilali Antari (Ibn Zohr Agadir University, Morocco) Pawan R. Bhaladhare (Sandip University, Nashik, India) Amol D. Potgantwar (Sandip University, Nashik, India) and Swaminathan Kalyanaraman (Anna University, Trichy, India)
Information Science Reference • copyright 2024 • 409pp • H/C (ISBN: 9798369335970) • US $325.00 (our price)

Blockchain Applications for Smart Contract Technologies
Abdelkader Mohamed Sghaier Derbali (Taibah University, Saudi Arabia)
Engineering Science Reference • copyright 2024 • 349pp • H/C (ISBN: 9798369315118) • US $290.00 (our price)

701 East Chocolate Avenue, Hershey, PA 17033, USA
Tel: 717-533-8845 x100 • Fax: 717-533-8661
E-Mail: cust@igi-global.com • www.igi-global.com

Table of Contents

Detailed Table of Contents

In this digital era, fears about cybercrime and digital terrorism are rising. A coordinated effort will be required to keep ahead of the threat as the world grows more digitally connected. The threat posed by cybercrime and digital terrorism to people, businesses, and countries is increasing. An overview of the several types of cybercrime and digital terrorism, such as online fraud, identity theft, hacking, and cyberstalking. The accelerated attack surfaces in 5G networks, driven by reduced latency and increased data speeds, empower cybercriminals with the ability to deploy sophisticated attacks at unprecedented rates. This necessitates a paradigm shift in cybersecurity strategies to effectively detect, respond to, and mitigate threats in real-time within this dynamic and fast-paced digital landscape. The chapter follows up with an examination of the field's present and potential trends as well as the significance of creating strong defenses against cybercrime and cyberterrorism.

The overall landscape of security threats in India is changing quickly, giving rise to a wide range of new security issues. The nation is threatened by cyberattacks, border tensions, separatist movements, and domestic and international terrorism. Another major issue is the threat posed by radicalism, both ideological and religious. To properly handle these new security challenges, India needs to increase inter-agency collaboration, bolster its cybersecurity and counter-terrorism capabilities, and strengthen its information collection skills. India stands at the brink of a digital revolution with the advent of 5G technology, promising unprecedented connectivity, and technological advancements. and a parallel concern has arisen as the evolving cybersecurity threat landscape. This chapter explores the unique challenges and potential threats that 5G poses to India's cybersecurity, examining the intricate interplay between technological progress and the need for robust defenses.

Safeguarding confidential information from unwanted access is of utmost importance in the current age of digital interactions and knowledge sharing. This chapter explores the complex world of encryption techniques and how important they are to maintaining data privacy. The study examines the evolution of encryption historically over time, following its origins from traditional cryptography to the present modern methodologies for cryptography, showcasing the mathematical frameworks that support their security. Further, it focuses on the difficulties that encryption protocols have to deal with in light of cutting-edge technology like quantum computing, which could endanger established cryptographic techniques. The ethical and legal implications of encryption and data privacy are a crucial component. With the results of historical investigations, technical difficulties, and ethical issues taken into account, the study offers a road map for improving data privacy in the digital era.

As the security challenges are inherent to 5G networks, the threat of malware continues to rise. Due to the high-speed nature of 5G, there is a need for automated malware detection and classification methods. This chapter presents a hybrid model for malware classification that combines machine learning and deep learning techniques. This model leverages the power of feature extraction using the InceptionResNetV2 deep learning model pre-trained on ImageNet. The features extracted from malware images are fed into a random forest classifier for the final classification task. The experimental evaluation conducted on a dataset comprising 31 malware families achieves an accuracy of 94.15%. Furthermore, the predicted labels closely align with the ground truth labels across various malware families, showcasing the model's ability to capture the intricate patterns and characteristics of diverse malware strains.

In the era of the internet, 5G is the super-fast network unified with mobile phones and communication devices, with a good bandwidth spectrum to transform information. The combination of IoT devices with 5G technologies facilitates quick and seamless data transformation. This expansion fosters connectivity, operational efficiency, and intellectual development. An intrusion detection systems (IDS) is an efficient tool for network security that keeps an eye on malicious activities or security policy violations under surveillance to detect anomalies in devices. The early part of the chapter deals with the integration of 5G technologies with IOT devices. The chapter also emphasize the role of collaboration between IDS and emerging technologies in deep learning (DL). The integration of intelligent algorithms enhances detection accuracy and enables adaptive responses to evolve the security threats in real-time. In conclusion, the insights provided in this chapter aim to understand the Research gap in DL IDS and the pivotal role played by IDS in ensuring a resilient and secure IoT environment.

Alamouti decode and forward (DF) relaying protocol using cooperative-maximum combining ratio (C-MRC) for three user cooperative system is considered. The performance of approximate SER for Alamouti DF relaying protocol with M-PSK modulation over Rayleigh fading and Rician fading channel is presented and compared with upper bound SER results. The approximate SER of proposed protocol is asymptotically tight at high SNR compared to the upper bound. The Alamouti DF relaying protocol provides full diversity gain compared with the existing Protocols. The PSO algorithm is used to calculate the optimum PA factor based on the minimum approximate SER. The simulated SER is compared with the theoretical approximate SER. It is shown that the Alamouti DF relaying protocol outperforms the existing Alamouti AF, AF, and DF relaying protocols. Modified cooperative subchannel allocation (CSA) algorithm for Alamouti DF relaying protocol using C-MRC technique is proposed, which maximizes the total throughput of the multiuser orthogonal frequency division multiplexing access (OFDMA) systems.

This book chapter explores the intricacies of designing a compact triple band antenna specifically tailored for 5G NR C-Band applications. As communication technologies advance towards 5G networks, the demand for compact and efficient antennas in the C-Band spectrum becomes increasingly vital. This chapter delves into the challenges and opportunities, emphasizing the need for antennas with reduced height profiles without compromising performance. The antenna's design structure has Concatenated SRR with partial ground plane and it is defected by Circular Complementary Split Ring Resonator that achieves DGS technique. The objective is to attain high gain, low sidelobe levels, and efficient power radiation. To enhance performance metrics, this chapter employs various optimization techniques, including parameter tuning, metamaterial technique and electromagnetic simulations. A reduced dimension of 0.179λ0 X 0.318λ0 X 0.006λ0 mm3 is achieved. This reduction in size holds promising implications for applications with limited space constraints, particularly in 5G NR C-Band scenarios.

This chapter is the development of a microstrip fed dual band antenna with a low-profile, small, and flexible structure, utilizing an artificial magnetic conductor (AMC) and microstrip feed. The antenna is biocompatible and operates at resonant at 3.5 GHz for 5G applications and 5.7 GHz catering to C band applications like Wi-Fi. In order to render the suggested antenna biocompatible, a 3x3 array of AMC is integrated. This array functions to augment the FBR, thereby improving overall parameters when in contact with living human tissues. From the simulation of the microstrip fed dual band antenna with AMC, the result indicates the improved parameters of the antenna gain is 3.91dB and 4.87dB, bandwidth of 340MHz and 480MHz, Front to Back ratio is 12.6dB and 161.8dB with a radiation efficiency of 84% and 85% in 3.5GHz and 5.7GHz resp. The antenna having flexibility, improved gain and improved FBR makes it good for wearable antenna applications.

Fiber Bragg grating (FBG) sensors have emerged as a promising technology for enhancing the monitoring and maintenance of smart grid systems. In this study, a comprehensive simulation-based investigation into the application of FBG sensors for monitoring critical components of the smart grid infrastructure, including high-voltage transmission towers, insulators, and overhead lines was presented. A wavelength division multiplexed FBG sensor network was designed and deployed to capture key parameters such as strain, temperature, and sag, enabling real-time monitoring and early detection of potential issues. The simulation results demonstrate the effectiveness of FBG sensors in accurately detecting changes in these parameters, with high sensitivity and reliability. The findings of this study underscore the importance of FBG sensors in enhancing the reliability, efficiency, and resilience of smart grid systems, paving the way for their widespread adoption in modern power networks.

The integration of emerging IoT (internet of things) technologies into utility control centers have data exchange between smart appliances, smart meters (SM), data collector (DC) and control center server (CCS). The DC controls the receiving and processing of advanced metering infrastructure (AMI) applications data from multiple SM. To address the issues associated with DC, SM are proposed to act as a relay device. The SM face communication challenges during peak hours when a significant amount of data with varying traffic rates and latency is exchanged within the utility control center. The challenges related to AMI in the context of smart grids focus the role of DC and hybrid data aggregation strategy. A hybrid data aggregation strategy is implemented on a cluster head aggregator (CA) within a clustering topology and the aggregated data are sliced to ensure privacy is proposed. In the proposed algorithm, CA reduces the workload of cluster-head (CH), targets interval meter reading (IMR) application data for aggregation, and efficiently utilizes the constrained resources of AMI devices are evaluated using the network simulator.

In the enormous growth of smart grid technology, the integration of 5G architecture network technology offers many advantages such as real-time data processing, improved communication facilities, safe data transfer, improved privacy, and diversified services. Yet, it leads to new cybersecurity problems that need to be addressed to ensure the smooth operation of the smart grid. This chapter outlines an extensive study of state-of-the-art network function virtualization (NFV) and software-defined networking (SDN) cybersecurity techniques in smart grid applications. NFV and SDN are key perceptions in the development of 5G networks, promoting flexibility, scalability, programmability, and centralized control over network resources. NFV, along with SDN, promotes the transformation of telecommunication networks, enabling them more adaptable, scalable, and capable of facilitating the diverse services envisioned in the 5G era, protecting critical energy infrastructure from cyber threats.

In the rapidly evolving landscape of 5G communication, the identification and mitigation of wireless interference is paramount to maintaining the integrity and efficiency of data transmission. This chapter delves into the intricate process of wireless interference identification, emphasizing its critical role in the predictive analysis of modulation classification and the implementation of adaptive modulators. The discussion begins with a comprehensive overview of 5G architecture, and the inherent challenges posed by dense signal environments. Key techniques for interference identification are explored, including advanced machine learning algorithms and spectrum sensing methods that enable real-time detection and characterization of interference sources. The chapter then examines how these identification techniques inform the predictive analysis of modulation classification. By accurately predicting the modulation scheme of incoming signals, the system can adaptively adjust its modulator settings, thereby optimizing performance and minimizing error rates.

The radar cross-section (RCS) is a critical and crucial factor that determines presence of the target by radar signal, and it can be consistently detected so as to track the target by a radar system. The target's size can be fixed by changing the factor ratio of the RCS quantity. Raising the target's RCS improves its radar detestability. Reduced RCS, on the other hand, the chances for the target to miss from radars, which is frequently required in military settings. The equation for the RCS is depending upon various factor and constants, the surface area (A), the wavelength (λ), the Fresnel reflection coefficient (G) at normal incidence.

The future 6G wireless technology promotes applications like three-dimensional (3D) communication, extended virtual reality, digital twins, autonomous driven vehicles, etc. Such applications require tens of GHz bandwidth, such large bandwidth is promoted by Terahertz communication system. To compensate the effect of attenuation and to increase the coverage massive MIMO with traditional hybrid precoder is considered. The conventional hybrid precoder suffers from beam split effect, where the transmitted beams are oriented towards different direction leads to loss of array gain. To circumvent the array gain loss and to focus the orientation of all transmitted beams toward a particular direction a delay phase precoder (DPP) is considered. In delay phase precoding, the selected precoder is not redundant therefore consumes more power. To reduce the power consumption two lattice aided DPP namely: Boosted LLL aided DPP and Boosted KZ aided DPP were proposed. The performance in terms of array gain and data rate were plotted. The computational complexity analysis was also performed.

Preface

The rapid evolution of communication technologies and their integration into critical infrastructures, such as smart grids, has revolutionized how we manage and secure our energy resources. In this transformative landscape, *5G and Fiber Optics Security Technologies for Smart Grid Cyber Defense* emerges as a definitive guide, addressing the intricate nexus of 5G wireless communication, fiber optics, and cybersecurity within the context of smart grids.

This book is a culmination of extensive research and practical insights, meticulously crafted to comprehensively understand how advanced communication technologies can be harnessed to fortify smart grid systems. By delving into 5G network architectures, fiber optic communication infrastructures, encryption protocols, and sophisticated cybersecurity strategies, we aim to offer readers a holistic view of the current state and future directions of smart grid cyber defense.

OBJECTIVES AND GOALS

Our primary objective is to catalyze collaborative research efforts, bringing together scholars, industry professionals, and policymakers to explore and address the multifaceted challenges in this domain. We present real-world applications and case studies that illustrate the practical deployment of these technologies in grid automation, demand response, and distributed energy resources management. These examples demonstrate the tangible benefits of integrating 5G and fiber optics into smart grids and provide actionable insights for securing these infrastructures against emerging cyber threats.

The structure of this book is designed to cater to a diverse audience, including academic researchers, industry professionals, and government officials. We have outlined our goals as follows:

1. **Resource for Understanding**: To provide a one-stop resource for understanding how 5G and fiber optics enhance cybersecurity in smart grid systems.
2. **Elevating Discourse**: To elevate smart grid cyber defense discourse by consolidating the latest findings and practical implementations, driving collaborative research efforts.
3. **Equipping Industry Experts**: To equip industry experts with real-world strategies and case studies to secure smart grid infrastructures effectively, minimizing cyber risks.
4. **Aiding Policymakers**: To assist government officials and regulators in formulating robust policies and standards for safeguarding national energy infrastructures against cyber threats.
5. **Inspiring Innovation**: To inspire new research directions and innovative solutions, fostering a forward-looking approach to address emerging challenges in smart grid security.

INTEGRATION OF 5G AND FIBER OPTICS

The integration of 5G technology with fiber optics is transforming the fields of telecommunications and cybersecurity, especially within smart grid systems. This technological synergy enhances efficiency, security, and performance while introducing new challenges in protecting these advanced networks from evolving cyber threats. *5G and Fiber Optics Security Technologies for Smart Grid Cyber Defense* delves into these critical issues, exploring innovations and methodologies for securing smart grids.

OVERVIEW OF THE SUBJECT MATTER

Our book bridges the gap between the theoretical foundations and practical applications of 5G and fiber optics in smart grid cybersecurity. It emphasizes the importance of merging these cutting-edge technologies to strengthen smart grids by showcasing the latest research and developments.

THE RELEVANCE OF THE TOPIC TODAY

In the current fast-paced technological era, adopting 5G and fiber optics is essential for modernizing infrastructure and boosting communication capabilities. The smart grid, a pivotal part of this infrastructure, depends on robust cybersecurity measures to counter cyber threats. This book addresses the growing need for secure and resilient smart grid systems, highlighting how 5G and fiber optics contribute to this objective.

TARGET AUDIENCE

Academic Researchers and Scholars: Electrical engineers, telecommunications experts, and cybersecurity researchers seeking in-depth knowledge about advanced communication protocols and cybersecurity measures in smart grids will find this book indispensable. It provides precise content suitable for academic courses and serves as a reliable reference source for cutting-edge research.

Industry Professionals: Professionals working in utility companies, cybersecurity agencies, and technology solution providers will benefit from practical insights and applicable case studies. The book equips industry experts with the latest strategies to secure smart grid infrastructures effectively, ensuring seamless energy flow while minimizing cyber risks.

Government and Regulatory Bodies: Policymakers, regulators, and government officials involved in energy and communication sectors will find this publication invaluable. It provides concise and focused content, aiding in the formulation of policies, standards, and regulations that enhance the security posture of national energy infrastructures.

By bridging the gap between theory and practice, and fostering a collaborative spirit among various stakeholders, *5G and Fiber Optics Security Technologies for Smart Grid Cyber Defense* aspires to be a cornerstone in the advancement of smart grid security. We invite you to delve into the pages of this book and join us in shaping a secure and resilient energy future.

ORGANIZATION OF THE BOOK

Section 1: Introduction to 5G and Cybersecurity

Chapter 1. Cybercrime, Digital Terrorism and 5G Paradigm: Attack Trends of the New Millennium

By Akashdeep Bhardwaj, University of Petroleum & Energy Studies, India

This chapter offers a fundamental understanding of cybercrime and digital terrorism within the 5G context, examining attack trends and their implications for cybersecurity.

Chapter 2. 5G: The Emerging Cybersecurity Threat Landscape for India

By Akashdeep Bhardwaj, University of Petroleum & Energy Studies, India

This chapter explores the emerging cybersecurity threats posed by 5G in India and provides insights into mitigation strategies.

Chapter 3. Research Review Inquisitive on Performance Evaluation of Fifth Generation (5G) Technologies and Protocols

By Kavitha Veluchamy, Sethu Institute of Technology, India; Rehash Rushmi Pavitra A, SRM Institute of Science and Technology, India; Isaivani M, Vaigai College of Engineering, India; Josep M Guerrero, Technical University of Catalonia, Spain

A review of the performance evaluation of 5G technologies and protocols, providing a critical analysis and identifying future research directions.

Section 2: Security Protocols and Technologies

Chapter 4. Encryption Protocols and Security for 5G

By Dhinakaran Vijayalakshmi, SRM Institute of Science and Technology, Chennai, India; Krithiga S, SRM Institute of Science and Technology, Chennai, India; Trinay Gangisetty, University of Colorado Boulder, Boulder, CO, USA; Dayana R, SRM Institute of Science and Technology, Chennai, India; Vadivukkarasi K, SRM Institute of Science and Technology, Chennai, India

This chapter discusses various encryption protocols essential for securing 5G networks, detailing their implementation and effectiveness.

Chapter 5. Enhancing 5G Security Using Hybrid Learning Approach for Malware Detection and Classification: Hybrid Learning Model to Strengthen 5G Security

By Murugavalli E, Thiagarajar College of Engineering, India; Rajeswari K, Thiagarajar College of Engineering, India; Vinoth Thyagarajan V, Thiagarajar College of Engineering, India; Shruti P S, Thiagarajar College of Engineering, India; Hemalatha K R, Thiagarajar College of Engineering, India

This chapter explores hybrid learning approaches for malware detection, presenting models that bolster the security of 5G networks.

Chapter 6. Internet of Things (IoT) Technologies with Intrusion Detection Systems in Deep Learning

By Nancy Jasmine Goldena, Sarah Tucker College, India; Rashia Subashree, Sarah Tucker College, India

This chapter examines the integration of IoT technologies with deep learning-based intrusion detection systems, emphasizing their significance in securing smart grids.

Section 3: Antenna Design and Power Allocation for 5G

Chapter 7. The Performance Analysis of PSO-based Power Allocation for Alamouti Decode and Forward Relaying Protocol

By Shoukath Ali K, Presidency University, India; Arfat Ahmad Khan, Khon Kaen University, Thailand

This chapter analyzes the performance of PSO-based power allocation for the Alamouti decode and forward relaying protocol, essential for optimizing 5G networks.

Chapter 8. Metamaterial Inspired Concatenated Dual Ring Antenna Design For 5G NR C-Band Applications: Exploring 5G NR C-Band Antennas: Design, Analysis, Development

By Meenambal B, Thiagarajar College of Engineering, India; Sherene Jacob, Thiagarajar College of Engineering, India; Vasudevan K, Thiagarajar College of Engineering, India

This chapter explores innovative antenna designs for 5G NR C-band applications, focusing on their design, analysis, and development.

Chapter 9. Artificial Magnetic Conductor Backed Microstrip-fed Dual-Band Antenna for 5G Wearable Sensor Nodes and C Band Application in WBAN Network

By Sherene Jacob, Thiagarajar College of Engineering, India; Meenambal B, Thiagarajar College of Engineering, India; Vasudevan K, Thiagarajar College of Engineering, India

This chapter discusses the design and implementation of dual-band antennas for 5G wearable sensor nodes, highlighting their applications in WBAN networks.

Chapter 10. Design of Patch Antenna and its Implementation in Spatial Multiplexing for 5G NR Applications

By Krithiga S, SRM Institute of Science and Technology, India; Vijayalakshmi D, SRM Institute of Science and Technology, India; Dayana R, SRM Institute of Science and Technology, India; Vadivukkarasi K, SRM Institute of Science and Technology, India

This chapter covers the design and implementation of patch antennas, emphasizing their role in spatial multiplexing for 5G NR applications.

Section 4: Applications in Smart Grids and Cyber Defense

Chapter 11. Leveraging Fiber Bragg Grating Sensors for Enhanced Security in Smart Grids

By Serif Ali Sadik, Kutahya Dumlupinar University, Turkey

This chapter explores the use of fiber Bragg grating sensors to enhance the security of smart grids, detailing their applications and benefits.

Chapter 12. Data Aggregation Algorithm for IoT-Enabled Advanced Metering Infrastructure Network in Smart Grid

By Subaselvi Sundarraj, M. Kumarasamy College of Engineering, India

It presents a data aggregation algorithm for IoT-enabled advanced metering infrastructure networks, crucial for efficient smart grid management.

Chapter 13. Prospects of Network Function Virtualization (NFV) and Software-Defined Networking (SDN) Techniques for Smart Grid Cyber Defense

By Isai Vani Mariyappan, Vaigai College of Engineering, India; Kavitha V, Sethu Institute of Technology, India; Aravindaraj R, Vaigai College of Engineering, India; Sonia C Chockalingam, Vaigai College of Engineering, India; Josep M Guerrero, Technical University of Catalonia, Spain

This chapter examines the potential of NFV and SDN techniques to strengthen smart grid cyber defense.

Chapter 14. Securing the Future: Regulatory Compliance and Standards for 5G and Fiber Optics in Smart Grid Cyber Defense

By Nedumal Pugazhenthi, Dr. NGP Institute of Technology, India; Vijayakumar Kaliappan, Dr N.G.P. Institute of Technology, India; Selvaperumal Sundarmoorthy, Mohamed Sathak Engineering College, India; Prabhakar Gunasekaran, Thiagarajar College of Engineering, India

This chapter underscores the importance of regulatory compliance and standards in securing 5G and fiber optics within smart grids.

Section 5: Advanced Techniques, Future Directions, and Case Studies

Chapter 15. Implementing Secure Machine Learning - A Decision Tree Approach for Smart Grid Stability

By Shanthi J, Thiagarajar College of Engineering, India; Rajalakshmi M, Thiagarajar College of Engineering, India; Gracia Nirmala Rani D, Thiagarajar College of Engineering, India; Muthulakshmi S, Velammal College of Engineering and Technology, India

This chapter discusses the implementation of secure machine learning techniques, focusing on a decision tree approach for smart grid stability.

Chapter 16. Wireless Interference Identification in 5G Smart Networks

By Rabia Bilal, UIT University, Pakistan; Bilal Muhammad Khan, National University of Sciences and Technology Islamabad, Pakistan

This chapter explores methods for identifying wireless interference in 5G smart networks, essential for maintaining network integrity.

Chapter 17. Radar Cross Section Modelling and Analysis Using Various Estimation Techniques in FMCW Radar Frequencies

By Hariharan K, Thiagarajar College of Engineering, India; Suresh M N, Thiagarajar College of Engineering, India; Manimegalai B, Thiagarajar College of Engineering, India

This chapter investigates radar cross-section modeling and analysis using various estimation techniques in FMCW radar frequencies, emphasizing their relevance in modern radar systems.

Chapter 18, Lattice Aided Delay Phase Precoding for 6G THz Massive MIMO: Lattice Aided Precoding for 6G

By Parthiban Ilango, SRM Institute of Science and Technology, India; Sudha V, National Institute of Technology Tiruchirappalli, India; Hassan Pakarzadeh, Shiraz University of Technology, Shiraz, Iran

This chapter delves into advanced lattice-aided delay phase precoding techniques for 6G THz massive MIMO systems.

Chapter 19. Case Studies on the Integration of 5G Technology into Smart Grids with Emphasis on Cybersecurity Measures

By Kumaran M, SRM Institute of Science and Technology, Tiruchirappalli, India

It presents case studies on integrating 5G technology into smart grids, focusing on cybersecurity measures.

CONCLUSION

This book represents a collective effort by researchers and practitioners to enhance the security and efficiency of smart grid systems through 5G and fiber optic technologies. By covering a wide range of topics and providing expert insights, we aim to offer a valuable resource for those involved in the development, deployment, and security of these critical infrastructures. We hope this book significantly contributes to the field and encourages further research and innovation.

G. Prabhakar
Thiagarajar College of Engineering, Madurai-625015, Tamil Nadu, India

N. Ayyanar
Thiagarajar College of Engineering, Madurai-625015, Tamil Nadu, India

S. Rajaram
Thiagarajar College of Engineering, Madurai-625015, Tamil Nadu, India

Acknowledgment

First and foremost, we would like to express our heartfelt thanks to **God** for guiding us throughout this journey and blessing us with the wisdom and perseverance to complete this book, ***5G and Fiber Optics Security Technologies for Smart Grid Cyber Defense***.

We extend our deepest gratitude to all those who contributed to the creation and success of this work. Our esteemed contributors' dedication, expertise, and insightful research have significantly enriched this book, making it a comprehensive resource for the integration of 5G and fiber optics for smart grid cybersecurity. We deeply appreciate the time and effort they invested in this project.

We are also grateful to our colleagues at the **Department of Electronics and Communication Engineering, Thiagarajar College of Engineering, Madurai**, for their constant encouragement and support. Their feedback and suggestions were instrumental in refining the content and scope of this book.

A special debt of gratitude goes to our Chairman, **Mr. K. Hari Thiagarajan**, whose visionary leadership and unwavering support have been sources of inspiration for us all. His guidance has been invaluable in shaping this endeavor.

We also remember with deep respect and affection the late **Sri. Karumuttu T. Kannan**, whose foundational work and dedication to the college laid the groundwork for our ongoing efforts. His legacy continues to inspire us.

We also extend our thanks to our **families and friends** for their unwavering support and understanding throughout this journey. Their patience and encouragement have been a source of strength and motivation for us.

Acknowledgment is due to the exceptional editorial team at **IGI Global** for their professionalism and assistance. Their guidance and support have been crucial in bringing this book to fruition, ensuring it meets the highest standards of academic publishing.

Lastly, we express our appreciation to all the researchers, practitioners, and students who will read and utilize this book. We hope it serves as a valuable resource in your work and inspires further advancements in the fields of telecommunications and cybersecurity.

Thank you all for your contributions and support.

Section 1
Introduction to 5G and Cybersecurity

Chapter 1
Cybercrime, Digital Terrorism, and 5G Paradigm:
Attack Trends of the New Millennium

Akashdeep Bhardwaj
https://orcid.org/0000-0001-7361-0465
University of Petroleum and Energy Studies, India

ABSTRACT

In this digital era, fears about cybercrime and digital terrorism are rising. A coordinated effort will be required to keep ahead of the threat as the world grows more digitally connected. The threat posed by cybercrime and digital terrorism to people, businesses, and countries is increasing. An overview of the several types of cybercrime and digital terrorism, such as online fraud, identity theft, hacking, and cyberstalking. The accelerated attack surfaces in 5G networks, driven by reduced latency and increased data speeds, empower cybercriminals with the ability to deploy sophisticated attacks at unprecedented rates. This necessitates a paradigm shift in cybersecurity strategies to effectively detect, respond to, and mitigate threats in real-time within this dynamic and fast-paced digital landscape. The chapter follows up with an examination of the field's present and potential trends as well as the significance of creating strong defenses against cybercrime and cyberterrorism.

DOI: 10.4018/979-8-3693-2786-9.ch001

1. INTRODUCTION

The era of digital technology has brought about many advancements and advantages, but it has also resulted in new forms of crime and terrorism. Online fraud and digital terrorism are growing progressively sophisticated and widespread, posing a substantial risk to people, organizations, and nations. As technology develops and becomes more pervasive in our daily lives, it is crucial to understand the current and future trends in this field to stay ahead of the malicious actors who seek to exploit it. In this chapter, we will examine the current and future trends in cybercrime and digital terrorism, including the latest tactics used by cybercriminals, the evolution of technology and its impact on these crimes, and the ways in which society can prepare and protect itself. Understanding these trends is essential for staying informed, aware, and tenacious in the battle against cybercrime digital terrorism.

Smart grids are modern power grids that integrate digital technology to improve efficiency, reliability, and sustainability. Unlike traditional grids, smart grids are two-way systems, meaning information and electricity flow in both directions. This is achieved through a complex network of sensors, meters, and communication technologies that collect and transmit data about energy usage in real-time. While smart grids offer numerous advantages, their increased reliance on interconnected devices and data communication creates significant cybersecurity vulnerabilities. These vulnerabilities expose smart grids to various cyber threats, ranging from malware infiltration and data breaches to denial-of-service attacks. Malicious actors could disrupt power generation and distribution by manipulating meter readings or control systems. Infiltration of smart meters could expose consumer data on energy consumption habits, raising privacy concerns. The interconnected nature of smart grids also makes them susceptible to cascading failures, where a cyberattack on one part of the system can trigger outages in other areas. These potential disruptions highlight the critical need for robust cybersecurity measures to protect smart grids and ensure their safe and reliable operation.

Cybercrime is the term for illegal activity that is executed by digital techniques, including phishing, hacking, and malware. The goal of cybercrime is often financial gain. Digital terrorism, on the other hand, involves the use of digital means to cause fear, panic, or political or social disruption. Both cybercrime and digital terrorism can have detrimental effects on people, organizations, and nation-states. In an effort to reduce these dangers, it's critical that people and companies adopt sound cybersecurity practices and use technical solutions to protect sensitive information and networks. Identity theft is the theft of a person's personal data and its illicit use, like creating a bank account or applying for a loan in that person's name. The term "identity theft" describes the illegal acquisition of a person's credit card number, name, Social Security number, or other sensitive information for the purpose of fraud or other illicit activities. Identity theft can have serious and long-lasting consequences, such as damaging an individual's credit score, opening unauthorized accounts in their name, or accessing their bank accounts.

People should be aware of phishing scams and take precautions to protect themselves because they pose a growing vulnerability. By being cautious and informed, individuals can help prevent phishing scams and protect their personal information from being stolen. Cyberstalking is the practice of intimidating, threatening, or harassing someone online. Cyberstalking is the practice of harassing, threatening, or intimidating someone else using technology, especially the internet and social media. It might entail a variety of actions, including following someone's internet activities, disseminating inaccurate details or rumors, or sending offensive emails or texts. Cyberstalking can have serious and long-lasting consequences, including causing emotional distress, damaging an individual's reputation, and putting their personal safety at risk. It can also result in criminal charges, as many cyberstalking behaviors are illegal. Examples of cyberstalking behaviors include:

- Harassing or threatening messages: Sending repeated, harassing, or threatening messages through email, social media, or other online platforms.
- Online harassment: Posting abusive or threatening comments on an individual's social media profiles, blogs, or other online forums.
- Impersonation: Creating fake online profiles or accounts that mimic or impersonate an individual to harass, intimidate, or spread false information about them.
- Tracking or monitoring: Using spyware, GPS tracking devices, or other technology to monitor an individual's online or physical movements.

People should take precautions to safeguard themselves and their personal information because cyberstalking is an extremely serious issue. If an individual is being cyberstalked, they should not hesitate to seek help from law enforcement or a qualified professional. Online fraud happens by using false information or deceptive practices to obtain money or goods from individuals or organizations. Online fraud is any illicit conduct that deceives people or organizations into parting with their money or personal information by using the internet or other types of technology. Online fraud may take many different forms and can have detrimental effects on a person's or an organization's reputation in addition to causing financial loss and identity theft. Examples of online fraud include:

- Investment fraud: A particular kind of online fraud where individuals are promised high returns on investments, but the investment is a scam, and the fraudsters use the invested money for their own purposes.

- Auction fraud: A type of online fraud where individuals are scammed while bidding on items through online auction sites. The fraudsters may sell fake or counterfeit items, or they may take the money without delivering the promised item.
- Identity theft: A kind of internet fraud in which victims' financial and personal information is obtained and used by scammers to carry out fraud or other illegal activities.
- Malware attacks: A particular kind of internet fraud in which victims are duped into installing harmful software on their computers, which is subsequently used to steal their financial and personal data.

Cryptojacking involves using someone's computer resources without their knowledge to mine cryptocurrency. A sort of cybercrime known as "Cryptojacking" occurs when an attacker stealthily mines bitcoin using a victim's computer's resources—such as its electricity or computing power—without the victim's knowledge or agreement. After infecting the victim's computer with malicious software—a virus or Trojan horse, for example—the attacker launches bitcoin mining software in the background. The mining software uses the victim's computer to solve complex mathematical equations and generate new cryptocurrency, which is then sent to the attacker's cryptocurrency wallet. There are several reasons why attackers use Cryptojacking to mine cryptocurrency:

- Profitable: Cryptocurrency mining can be a profitable activity, and attackers can earn significant amounts of money by using other people's computers to mine cryptocurrency.
- Easy to do: Cryptojacking is relatively easy to carry out, and attackers can infect thousands of computers with malware in a matter of minutes.
- Difficult to detect: Cryptojacking is often difficult to detect because the mining software runs in the background and does not affect the victim's computer performance or display any noticeable symptoms.

Victims of Cryptojacking may experience the following consequences:

- Decreased computer performance: The mining software uses a significant amount of the victim's computer resources, which can slow down the computer and affect its performance.
- Increased electricity costs: The mining software also uses a significant amount of the victim's electricity, which can lead to higher electricity bills for the victim.

- Decreased battery life: If the victim is using a laptop, the mining software can significantly decrease the battery life.
- Increased wear and tear: The increased usage of the computer's resources can also lead to increased wear and tear on the hardware, which can shorten its lifespan.

People and organizations need to take precautions to prevent their computers from being used for cryptocurrency mining because Cryptojacking is becoming an increasingly prevalent issue. If an individual or organization is a victim of Cryptojacking, they should not hesitate to contact law enforcement or a qualified professional for assistance.

Spreading dangerous software, like Trojan horses or viruses, with the aim of upsetting computer systems or obtaining private data is known as malware distribution. Malicious software, such viruses, Trojan horses, or worms, are distributed to infect other computers and devices. This is known as malware distribution. Malware can cause a range of harm, from stealing personal information, encrypting files and demanding ransom, to disrupting computer systems and networks. There are several methods that attackers use to distribute malware:

- Email attachments: Emails containing malicious attachments that, when opened, infect the recipient's machine can be sent by attackers. The emails can have a message telling the recipient to open the attachment and seem to be from a reliable source.
- Social engineering: Social engineering techniques, like fabricating websites or software updates, can be exploited by attackers to deceive victims into installing malicious software.
- Drive-by downloads: Attackers may compromise websites and embed malware in the code, which is automatically downloaded when a user visits the site.
- Supply chain attacks: Attackers could inject malware into new sets up or software updates by focusing on a software provider's supply chain.

Once malware is installed on a computer or device, it can cause a range of harm, including:

- Stealing personal information: Credit card numbers and other sensitive data, along with passwords, can be stolen by malware.
- Encrypting files and demanding ransom: Malware that encrypts a victim's files and requests payment in order to unlock them is known as ransomware.
- Disrupting computer systems and networks: Malware can disrupt computer systems and networks, causing downtime and data loss.

- Monitoring and tracking activities: Malware can monitor and track a victim's activities, including keystrokes, emails, and internet use.

Distributing malware is a serious issue, and it is important for individuals and organizations to take steps to protect themselves from these attacks. If an individual or organization is a victim of malware, they should not hesitate to contact law enforcement or a qualified professional for assistance. These are but only a few of the numerous varieties of cybercrime that might occur. It's critical to keep yourself cognizant about these risks and take precautions to safeguard your data.

2. PROPOSED METHODOLOGY

The proposed methodology for this chapter involves several key steps that comprehensively address the subject matter as

- Literature Review where the researcher begins by conducting a thorough review of existing literature on cybercrime, digital terrorism, and the implications of the 5G paradigm on cybersecurity. This review should encompass scholarly articles, reports, case studies, and relevant literature from reputable sources to establish a comprehensive understanding of the topic.
- Case Studies Analysis was performed to explore recent and notable case studies related to cybercrime and digital terrorism, highlighting the tactics, techniques, and procedures (TTPs) employed by threat actors. Analyze the impact of these incidents on individuals, businesses, and nations to provide context for the evolving threat landscape.
- Interview blogs and expert insights gathered by engaging with cybersecurity experts, law enforcement professionals, and industry practitioners to gain insights into emerging trends and challenges in combating cyber threats. This provided valuable perspectives on the evolving nature of cybercrime and the strategies employed to address it.
- For future work:

    - Data Analysis and Statistical Trends can be generated by utilizing data analysis techniques to identify statistical trends in cybercrime and digital terrorism, focusing on factors such as attack vectors, targeted industries, and geographic regions. This quantitative analysis can provide empirical evidence to support the discussion of attack trends in the new millennium.

- o Technology and Infrastructure Assessment can be done to evaluate the security implications of 5G networks and emerging technologies on cybercrime trends. Assess the unique vulnerabilities introduced by the proliferation of 5G infrastructure and its impact on the frequency and sophistication of cyberattacks.

By following this proposed methodology, the researchers, practitioners, and policymakers can all benefit from the researcher's comprehensive investigation of cybercrime, digital terrorism, and the evolving threat environment in the context of the 5G paradigm.

3. LITERATURE REVIEW

Cybercrime encompasses an extensive range of malicious practices that are carried out utilizing the internet or any other forms of digital technology. Hacking ("What is hacking and how does hacking work?", n.d.) is the term for unauthorized access to an individual or organization's computer systems or network. Hacking refers to unapproved entry into a system's network or computer network. Here are a few examples of hacking:

- Web application hacking (EC-Council iLabs: iLabs Homepage, n.d.): exploiting vulnerabilities in websites to pilfer private data, including financial information or login credentials. A recent example of a web application attack is the SolarWinds hack that took place in 2020. In this attack, hackers were able to compromise the software updates of SolarWinds, a popular IT management tool, and obtain access to various government and corporate networks. The hackers were able to pilfer private data, including login credentials and emails, and plant malware on the affected networks. This attack demonstrates the significance of maintaining software updates and security, as well as the dangers of supply chain incidents, in which malicious actors breach a reliable third party in order to obtain access to a target network.
- Network hacking (Network Hacking | Insecure Lab, n.d.): entering a network without authorization with the goal of stealing or changing data, disrupting network operations, or using the network for malicious purposes. The DarkSide ransomware attack of 2021 is a particular instance of a recent network attack. In this attack, the hackers used ransomware to encrypt the data of several companies, including Colonial Pipeline, and demanded payment to obtain the decryption key. There were numerous disruptions spurred on by the attack, including gas shortages and price spikes, and prompted the U.S. government

to take action to disrupt the hackers' operations. This attack emphasizes how crucial it is to have robust cybersecurity safeguards in place, such as consistent data backups and the use of security software and processes, in order to stop and lessen the effects of ransomware attacks.

- SQL injection (SQL Injection, n.d.): injecting malicious code to gain access or manipulate private data contained in a website's database. A recent example of a SQL injection attack is the attack on Marriott International in 2018. In this incident, hackers were able to take advantage of a vulnerability in the organization's database to obtain potentially up to 500 million customers' names, addresses, phone numbers, passport numbers, and payment card credentials. The attack served as a reminder of how crucial it is to protect databases and online applications from SQL injection attacks, which have the potential to undermine organizational procedures and obtain confidential data. The attack had an enormous adverse effect on Marriott's finances and reputation, which should serve as an indication to other corporations to take the appropriate precautions to safeguard themselves against similar risks.
- Remote access hacking (Techopedia.com, n.d.): taking access to a computer remotely, frequently through the use of malware or social engineering techniques. A recent example of remote access hacking is the attack on FireEye in 2020. In this incident, hackers were able to gain access to FireEye's internal network and acquire authority over highly confidential information, including equipment used to conduct client red-team security assessments. An organized hacking organization with governmental support executed the attack, which was remarkable for the hackers' degree of skill and deception. This incident served as a reminder of the significance of putting robust security measures in place in order to prevent unauthorized access as well as the significance of conducting frequent penetration testing and security assessments to identify and eliminate holes. The attack on FireEye highlights the fact that even highly sophisticated and well-respected cybersecurity companies are not immune to these threats and serves as a warning to other organizations to take these threats seriously and invest in their own cybersecurity.
- Password cracking ("SQL Injection," n.d.): using various methods to guess or recover login credentials for a computer system or network. In 2019, a notable instance of password cracking occurred with the Capital One data leak. A hacker was able to obtain the names, addresses, credit scores, and Social Security numbers of approximately 100 million Capital One clients through this incident. The information was obtained by the attacker through the use of a misconfigured firewall within the cloud infrastructure of the firm. The incident highlights the importance of implementing strong password policies, such as using unique and complex passwords, and regularly updating passwords to

prevent password cracking attacks. Capital One faced significant financial and reputational consequences because of the breach, which serves as a warning to other organizations to take the necessary steps to protect themselves against these threats.

- DDoS attacks (Techopedia.com, n.d.): sending excessive traffic over a network or computer system to interfere with regular operations. The 2016 attack on DNS provider Dyn is a recent instance of a denial-of-service (DDoS) attack. In this attack, a massive influx of traffic was directed at Dyn's servers, causing widespread disruption to major websites and internet services, including Amazon, Netflix, and Twitter. The attack was carried out using a botnet of compromised Internet of Things (IoT) devices, such as security cameras and home routers. The experience emphasizes how crucial it is to safeguard IoT devices and put strong DDoS defenses in place in order to stop and lessen the effects of these attacks. This attack had a significant impact on the affected companies and demonstrated the far-reaching consequences of these types of attacks, making it a warning to other organizations to take these threats seriously and invest in their own cybersecurity.
- Man-in-the-middle attacks (Techopedia.com, n.d.): communication between two parties is intercepted and manipulated in order to steal confidential information or commit fraud. A recent example of a man-in-the-middle attack is the attack on the Olympic Games in Pyeongchang in 2018. In this attack, hackers were able to intercept and manipulate the communications between journalists and athletes, as well as the official results of the games. The attackers were able to carry out the attack by compromising the Wi-Fi networks in the Olympic Village and other locations. The incident highlights the importance of using secure communication methods, such as encryption and Virtual Private Networks (VPNs), and the dangers of using public Wi-Fi networks for sensitive communications. The attack on the Olympic Games serves as a warning to other organizations to take these threats seriously and invest in their own cybersecurity.

These are some examples of hacking that have been observed in the real world. It's critical to keep up with the latest developments in hacker techniques and take precautions to safeguard your data and systems from these types of risks.

With the advent of 5G, traditional tile-communication networks have evolved into next-generation networks with new features and capabilities to meet the demands of digital technologies. By linking devices and objects to the internet, the Iota has completely changed the landscape of telecommunication by facilitating smooth communication and data sharing. It has created new chances to enhance cutting-edge products and services. The main developments in subsequent-technology tile-verbal exchange networks and Iota include improved connectivity, faster information speeds, and greater network capacity, as reported by Logeshwaran et al. (2024). This makes it possible to combine several conversation technologies, such as cloud computing, artificial intelligence, and 5G, to give sophisticated services like augmented and digitized facts. The performance of the networks is further improved by the implementation of network automation and network slicing.

Advancements in machine learning, IoT, 5G, wireless networks, artificial intelligence, and network security have completely changed how we communicate, work, and live. Research on the effects of artificial intelligence and 5G on wireless networks, the application of machine learning on the Internet of Things, and the significance of network security in guaranteeing the safe and secure operation of these networked devices is being suggested by B. Bhushan (2023). The benefits of utilizing artificial intelligence in 5G networks are explored in detail in the study, including enhanced coverage, faster speeds, and the capacity to accommodate more connected devices. The research conducted brought to light the difficulties in integrating IoT devices and the possible security risks they present. To fully achieve the potential of the Internet of Things and 5G networks, the study finishes by highlighting the necessity for a strong security architecture that can properly secure these linked devices and networks.

Our lives were completely altered by the coronavirus pandemic, which also brought about an unforeseen disaster. Lockdowns around the nation specifically altered the workplace by substituting remote work for office employment. Businesses who weren't prepared for the shift faced serious issues because their system's vulnerability grew dramatically. Cybercriminals use vulnerabilities as a springboard for a variety of highly skilled attacks that they can use for reputational damage, financial gain, or hacktivism. One of the primary targets of the pandemic was the healthcare industry, which included several healthcare facilities. Sensitive data protection is essential since healthcare is regarded as an essential infrastructure in many nations. As a result, healthcare professionals must be ready to defend against any online threats and potential assaults. This healthcare industry research was the topic of a significant cybersecurity threat outlined by Nguyen et al. (2023).

In a world enabled by cutting-edge technologies like 5G or 6G internet, virtual reality (VR), augmented reality (AR), mixed reality (MR), extended reality, blockchain, digital twins, internet of things, and artificial intelligence (AI), all of which are enhanced by essentially infinite data, the metaverse merges the real and virtual worlds. This allows users to interact with their avatars. General healthcare could undergo a transformation thanks to the term "metaverse," which was coined to characterize the virtual world as an alternative to real worlds. Remote patient care, medical education, surgical training, anti-aging healthcare, and chances for medical research with implementation are already made possible by the Metaverse.

Studies indicate that by 2027, the medical domain's metaverse revenue is predicted to reach $645 billion. Various industries, including finance, manufacturing, media, entertainment, and others, are utilizing healthcare tools to incorporate cutting-edge technology into their respective sectors. Mozumder et al. (2023) covered the application of metaverse technology in digital healthcare as well as its present condition in this study. The writers also covered the difficulties the healthcare sector faces because of the metaverse. The future of metaverse in healthcare and its possible effects on the provision of healthcare services were covered by the writers in their conclusion. Lastly, the purpose of this study is to present a framework for metaverse healthcare based on XAI, BC, and immersive technology. This field has the potential to revolutionize the healthcare sector and enhance patient outcomes.

Both the nature of the digital world created by new technologies and the results of innovative technologies are being influenced by the reasons for the digital transformation. Regarding the condition of cybercrime as a result of the digital transition and the security and protection alternatives that are already accessible or that still need to be created, the implications are enormous. Cybercrime can take many different forms, and at each stage of observation and analysis, there might be notable and abnormal aberrations. The most evident usage of current attacks is the ability of rogue mobile applications to successfully deceive users with malware and even older forms of phishing attacks on their devices, respectively. Many people see security transformation as more than just a problem that technology can solve. A. Yarali (2023) talked on the distinction between cybersecurity maturity as it is perceived and as it is.

4. CYBER TERRORISM

Cyber terrorism poses a significant threat to national security, public safety, and the economy. Terrorists might, for instance, use the internet to attack vital infrastructure, like transportation networks, financial institutions, and power grids, resulting in severe disruptions and possibly jeopardizing lives. They may also use digital technologies to spread propaganda and to recruit new members, fueling the spread of extremism and violence. Cyber terrorism also poses a threat to individual privacy and security. Terrorists may steal sensitive data from websites, including financial and personal information as well as proprietary information. They might also break into computer systems and obtain confidential information by using malware, phishing attacks, and other cyberattack techniques.

Governments, corporations, and private citizens must all approach cybersecurity with a proactive mindset to combat cyberterrorism. This can entail putting in place robust security measures like intrusion detection systems, firewalls, and encryption as well as warning the public about the risks associated with cyberattacks. Governments and law enforcement agencies should also work together to track and disrupt cyber terrorist activities, and to develop policies and regulations aimed at preventing cyber terrorism. here are some real-world examples of cyber terrorism:

- WannaCry ransomware attack (2017): The WannaCry ransomware attack was a global cyber-attack that affected over 200,000 computers in 150 countries. The attack encrypted the victims' files and demanded payment in Bitcoin in exchange for the decryption key. Table 1 presents the steps that the WannaCry ransomware took during its attack.

Table 1. WannaCry ransomware attack

#1 Propagation	**WannaCry used a Microsoft Windows vulnerability called EternalBlue that allows it to replicate via networks. This allowed it to infect multiple computers within an organization, as well as spread to other organizations through unpatched systems.**
#2 Encryption	When WannaCry malware became active on a computer, it encrypted system files and demanded payment in Bitcoin for the decryption key.
#3 Ransom demand	The ransom demand was displayed on the infected computer, instructing the victim to pay a specified amount of money within a certain timeframe. The ransom demand was displayed on the infected computer's desktop and was also included in a file named "!Please Read Me!.txt."
#4 Payment collection	The attackers used Bitcoin as the payment method, which allowed them to remain anonymous. The payment was collected through a specific Bitcoin address that was provided in the ransom demand.
#5 Decryption	The attackers promised to provide the decryption key required to open the encrypted files if the ransom was paid. However, there have been reports of victims who paid the ransom but did not receive the decryption key.

It's crucial to remember that paying the ransom could encourage the malware's proliferation and does not ensure that encrypted files can be recovered. The best way to protect against WannaCry and other ransomware attacks is to maintain software updates and maintain good cybersecurity practices, such as regularly backing up important data.

- Stuxnet worm (2010): The Stuxnet worm was a piece of malware that targeted the Iranian nuclear program. The worm caused significant damage to centrifuges used in the program and is widely considered to be the first example of cyber warfare. Table 2 presents the steps that the Stuxnet worm took during its attack.

Table 2. Stuxnet attack

#1 Propagation	Stuxnet used several methods to spread itself, including exploiting vulnerabilities in the Windows operating system and using portable devices such as USB drives.
#2 Identification	Once Stuxnet was installed on a system, it used specific markers to identify whether it was running on a system that was part of the targeted industrial control system.
#3 Manipulation	If the system was part of the targeted ICS, Stuxnet manipulated the system's configuration to cause physical damage to the equipment being controlled.
#4 Concealment	Stuxnet was designed to hide its activities and manipulate system logs to avoid detection.

Operation Aurora (2009-2010): Operation Aurora was a series of cyber-attacks against a number of high-profile companies, including Google and other technology firms. The attacks were thought to have originated in China and were carried out in order to steal sensitive information. The exact pseudo code for Operation Aurora. The exact methods used by the attackers are not publicly known, and the information is likely to be closely guarded by the organizations that were targeted and the law enforcement agencies that investigated the incident. Table 3 presents the steps during the Operation Aurora attack.

Table 3. Operation Aurora attack

#1 Reconnaissance	Reconnaissance was probably done by the attackers to learn more about the target companies and find potential points of vulnerability.
#2 Exploitation	The attackers likely exploited one or more vulnerabilities in the target organizations' systems to gain initial access.
#3 Elevation of privilege	Once the attackers had initial access, they likely used various techniques to escalate their privileges and gain greater access to sensitive information.
#4 Data exfiltration	Sensitive data, including trade secrets, intellectual property, and confidential company information, were probably stolen by the attackers using their unexpected access.

- Sony Pictures hack (2014) (Qualcomm.com, n.d.): In November 2014, a significant cyberattack against Sony Pictures exposed confidential company and employee data along with a number of previously unreleased movies. The attack was later attributed to North Korea. Table 4 presents the steps for Sony Pictures hack which was a significant cyber-attack that took place in 2014. It is important to note that the Sony Pictures hack was a highly sophisticated and well-coordinated attack that demonstrated the capabilities of advanced cyber attackers. The attack highlights the importance of maintaining good cybersecurity practices and being vigilant against cyber threats.

Table 4. Sony pictures attack

#1 Initial infiltration	**A vulnerability in Sony Pictures' network framework enables the attackers to have initial access to the company's network.**
#2 Escalation of privileges	Once they had initial access, the attackers used various techniques to escalate their privileges and gain greater access to sensitive information.
#3 Data theft	The attackers then stole a large amount of sensitive information, including sensitive emails, employee personal information, and confidential business information.
#4 Data release	The stolen data was then released publicly, leading to significant damage to the reputation of Sony Pictures and causing significant embarrassment to the company and its employees.
#5 Demands	The attackers made a number of demands, including a demand that Sony Pictures cancel the release of a film that was seen as being critical of North Korea.
#6 Disruptions	In addition to the theft and release of sensitive information, the attackers caused significant disruptions to Sony Pictures' operations, including taking down the company's computer systems and causing widespread downtime.

These are only a few instances of the various types of cyberterrorism attacks that occurred recently.

5. DIGITAL TERRORISM

Digital terrorism refers to the use of technology, such as the internet and computer systems, to carry out acts of terrorism. Cyberattacks against vital infrastructure, including power grids or financial systems, or the internet dissemination of terrorist propaganda are examples of such crimes. One of the key characteristics of digital terrorism is that it allows terrorists to carry out their activities from a distance, potentially anonymously, and with a potentially global reach. Because of this, it may be challenging for government and law enforcement agencies to track and tackle incidents of digital terrorism.

Examples of digital terrorism include:

- Cyberattacks on critical infrastructure: Terrorists may attack vital infrastructure systems to cause havoc, such as financial systems, water treatment facilities, or electricity grids. Cyberattacks targeting critical infrastructure are defined as the use of computer networks and the internet to launch attacks against systems that are essential to a society's functioning, such as transportation networks, water treatment facilities, power grids, and banking systems. These incidents, which are a type of cyberterrorism, might have detrimental effects on the economy, public safety, and national security.

Examples of cyberattacks on critical infrastructure include:

- Power grid attacks: Attackers may target the control systems of a power grid to cause widespread blackouts, disrupt energy supplies, or even cause physical damage to the infrastructure.
- Water treatment plant attacks: Attackers may target water treatment plants, potentially compromising the safety of drinking water and putting public health at risk.
- Financial system attacks: Attackers may target financial systems, such as banks and stock exchanges, to steal money, disrupt financial markets, or compromise sensitive financial information.
- Transportation network attacks: Attackers may target transportation networks, such as airports or railways, in order to disrupt travel and commerce.

Cyberattacks on critical infrastructure can have serious consequences, including loss of life, widespread economic damage, and the disruption of essential services. To protect against these attacks, it is important for critical infrastructure operators to implement strong cybersecurity measures, such as firewalls, intrusion detection systems, and access controls, and to work closely with law enforcement and intelligence agencies to detect and respond to cyber threats. It is also important for governments to take a proactive approach to cybersecurity and to develop national strategies to protect critical infrastructure systems against cyberattacks. This may include investing in research and development, establishing information sharing programs, and strengthening international cooperation to combat cybercrime.

- Propaganda and recruitment: Terrorist organizations may use the internet to spread propaganda and recruit new members. This may include the use of social media platforms, forums, and other online communities. Propaganda and recruitment are key elements of digital terrorism. The internet and social media are platforms that allow terrorist groups to disseminate their message, find potential members, and plan their actions. Propaganda is the dissemination of information, ideas, or opinions designed to influence public opinion and behavior. Terrorist organizations use propaganda to spread their message and to attract new recruits. This can include videos and images of violent attacks, messages calling for violence and the creation of an "us vs. them" mentality.

Recruitment refers to the process of attracting and enrolling individuals into a group or organization. Terrorist organizations use the internet and social media platforms to reach out to potential recruits and to facilitate the process of joining their cause. This can include online forums, chat rooms, and social media pages, where individuals can connect with like-minded individuals and learn more about the ideology and goals of the organization. Terrorist organizations also use encryption and other forms of secure communication to conceal their activities from law enforcement and intelligence agencies. Authorities find it challenging to keep an eye on and prevent their efforts at recruiting and propaganda because of this. It is important for governments and social media companies to take steps to counter these activities and to limit the spread of terrorist propaganda and recruitment on their platforms. This could entail creating algorithms to identify and eliminate extremist information, collaborating with government and law enforcement organizations to track and thwart terrorist actions, and enlightening the public about the risks associated with radicalization on the internet.

- Cyber extortion: Terrorists may use cyber extortion, such as demanding a ransom in exchange for not carrying out a cyberattack, as a means of funding their activities. Cyber extortion is the practice of demanding money or private information from individuals, companies, or governments using the internet and other digital technologies. This type of cybercrime can take many different forms, such as distributed denial of service (DDoS) attacks, ransomware attacks, and website vandalism. Cybercriminals use malware to encrypt crucial files on a computer system and demand money for the decryption key to carry out a ransomware attack. To stay undetected, the attackers frequently demand payment in cryptocurrencies like Bitcoin. Website defacement is another form of cyber extortion. In this type of attack, cyber criminals hack into a website and change its appearance or content to include a message demanding payment. The attackers may threaten to deface the website further or to release sensitive information if their demands are not met. Cyber extortion also frequently takes the form of DDoS attacks. In these attacks, cybercriminals utilize a network of compromised computers to drive traffic to a targeted website, preventing consumers from accessing it. After then, the attackers demand money in return for putting an end to their attack.

Cyber extortion can cause significant harm to individuals and organizations, including loss of money, theft of sensitive information, and disruption to business operations. Strong cybersecurity measures, like intrusion detection systems, firewalls, and encryption, must be put in place in order to stop cyber extortion. Employee education regarding the risks associated with cyber-attacks is also crucial. Furthermore, people and organizations should refrain from complying with cybercriminals' ransom demands because doing so can incite new attacks and perhaps lead to the loss of more sensitive data. Cyberterrorism is the term for terrorist acts carried out through the use of computers, the internet, and other digital technology. It includes an extensive range of malicious actions, including breaking into computer networks, troubling vital infrastructure, disseminating propaganda and recruiting materials, and pilfering confidential data.

- Disrupting communication: Terrorists may target communication systems, such as telecommunication networks or social media platforms, with the intention of disrupting the flow of information. Cyber attackers disrupting communication refers to a scenario where attackers interfere with the normal functioning of communication systems, such as email, instant messaging, phone systems, or other forms of digital communication. There are several reasons why this kind of attack might be conducted, including:
    - Political motivations: In some cases, cyber attackers may aim to disrupt communication systems as part of a larger political or ideological agenda. Cyber political motivations refer to the use of digital means to achieve a political objective or to influence the political process. This can take a variety of forms, including:

- Election interference: In some cases, cyber attackers may target election systems with the aim of manipulating the outcome of an election or creating doubt about the validity of election results.
- Information manipulation: Cyber attackers may seek to manipulate information, such as spreading false or misleading information or suppressing legitimate information, with the aim of influencing public opinion or the political process.
- Disrupting critical infrastructure: Cybercriminals may target vital infrastructure, including transportation networks, water treatment facilities, and electricity grids, with the intention of impairing regular system operations and causing anarchy or instability.
- Espionage: Cyber attackers may target political organizations, think tanks, or other organizations involved in the political process, with the aim of gathering sensitive information or disrupting their operations.

These kinds of cyberattacks emphasize the significance of upholding sound cybersecurity procedures and being watchful for cyber threats, as they can have catastrophic repercussions for people, groups, and entire civilizations.

- Criminal gain: In other cases, cyber attackers may seek to disrupt communication systems as part of a larger criminal enterprise, such as a ransomware attack or a phishing campaign. Cybercriminal gains refer to the financial or other benefits that cyber criminals may seek to obtain through their malicious activities. This can take a variety of forms, including:

- Ransomware attacks: Cybercriminals encrypt a victim's files with ransomware and demand payment in exchange for a decryption key. Given that victims may be prepared to pay substantial sums of money in order to retrieve their crucial data, this kind of attack can be incredibly profitable for the attacker.
- Phishing scams: Phishing scams include cybercriminals deceiving people into divulging personal information, including passwords or bank account details, by sending phony emails or other correspondence channels. After thereafter, this information may be exploited for financial benefit through activities like bank account access or fraudulent purchases.
- Cryptojacking: The act of mining cryptocurrency on someone else's computer without that person's knowledge or agreement is known as "cryptojacking." Since the victim's computer resources are utilized to create bitcoin without paying the victim, this kind of attack can be advantageous for the attacker.
- Data breaches: Cybercriminals use data breaches to obtain sensitive information, including credit card numbers or personal information, which they can resell on the dark web for a profit.

These types of cyber-attacks highlight the importance of maintaining good cybersecurity practices, such as using strong passwords, being vigilant against phishing and other forms of social engineering, and regularly applying software updates. By adopting these safety measures, people and organizations can lower their chance of falling victim to cybercrime and lessen the possible financial losses brought on by these kinds of cyberattacks.

- Espionage: Some cyber attackers may target communication systems as part of an espionage campaign, seeking to gather sensitive information or disrupt the operations of a specific target. The use of digital tools to get private information for governmental, military, or commercial objectives is known as cyber espionage. Nation-states or state-sponsored actors are usually the ones that carry out this kind of cyberattack, while individuals or criminal groups may also conduct cyberespionage for their own benefit. Cyber espionage typically involves a range of techniques, including:

- Malware: Malicious software, sometimes known as malware, is used to infect computers and other devices in order to grant the attacker remote access to confidential data.
- Phishing: Phishing schemes are designed to fool people into divulging private information, including bank account numbers or passwords.
- Social engineering: The practice of psychologically manipulating people to coerce them into disclosing personal information or performing particular acts is known as social engineering.
- Network intrusions: Attacks on computer networks with the intention of obtaining unauthorized access to private data are known as network intrusions.

Cyber espionage aims to get sensitive data, such as trade secrets, intellectual property, or classified information, for use in the future for commercial, military, or political gain. Cyber espionage is a serious threat to national security and the global economy, as sensitive information can be used to gain an advantage in negotiations, disrupt critical infrastructure, or even wage cyber war. It is crucial for individuals and companies to adopt robust cybersecurity practices, such as creating secure passwords, updating software on a regular basis, and being watchful for phishing and other social engineering schemes, in order to reduce the danger of cyber espionage. Additionally, organizations may implement technical solutions, such as encryption and firewalls, to protect sensitive information and networks.

Disruptions to communication networks can have major repercussions, such as data loss, interruptions to company operations, and damage to reputation, regardless of the reasons for the attack. It's crucial to put strong passwords into place, apply software updates often, and keep an eye out for phishing and other social engineering scams to reduce the chance of such disruptions.

6. 5G CATALYST FOR CYBERCRIME AND DIGITAL TERRORISM

The advent of 5G technology has ushered in a new era of connectivity, promising unprecedented speed, low latency, and increased capacity. While these advancements offer transformative benefits for various industries, they also present novel challenges in the realm of cybercrime and digital terrorism. This section explores the nuanced ways in which 5G can impact the landscape of illicit activities, providing insight into potential vulnerabilities and threats that emerge within this high-speed paradigm.

6.1 Accelerated Surface Attacks

5G deployment greatly expands the attack surface for cybercriminals, providing them with previously unheard-of chances to take advantage of weaknesses swiftly and efficiently. One notable aspect is the reduced latency, a key feature of 5G networks. Traditional networks often experience latency in the range of tens to hundreds of milliseconds, providing a window of opportunity for defenders to detect and respond to potential threats. In contrast, 5G networks boast ultra-low latency, typically below 10 milliseconds, making it challenging for security measures to keep pace with rapidly evolving cyber threats. The ultra-fast speeds of 5G networks widen the attack surface for cybercriminals, enabling them to execute sophisticated attacks with greater efficiency. For instance, the reduced latency facilitates quicker exploitation of vulnerabilities, such as zero-day exploits, before security measures can respond. This heightened agility empowers malicious actors to compromise systems and networks before traditional defenses can thwart their efforts.

The reduced latency in 5G networks allows cybercriminals to exploit zero-day vulnerabilities more effectively. Zero-day exploits focus on unpatched software vulnerabilities that are unknown to the vendor. The swift response time of 5G networks enables threat actors to deploy these exploits before security patches can be developed and implemented. This heightened agility gives malicious actors a critical advantage in infiltrating systems and exfiltrating sensitive data. The accelerated data transmission speeds of 5G facilitate the rapid dissemination of malware. Malicious software, such as ransomware or banking trojans, can be delivered at unprecedented rates, infecting systems before traditional security measures can intercept and quarantine the malicious payloads. In a 5G setting, malware spreads quickly, increasing the possibility of successful attacks and the potential for extensive harm.

5G's low latency enables cybercriminals to conduct real-time attacks with minimal delay. For example, Distributed Denial of Service (DDoS) attacks can be orchestrated more effectively, overwhelming target systems with a flood of traffic and disrupting services almost instantly. The reduced latency also enhances the feasibility of man-in-the-middle attacks, enabling threat actors to intercept and manipulate communications in real-time. The combination of high data speeds and low latency in 5G networks facilitates the use of swarming techniques by cybercriminals. Swarming involves coordinated and distributed attacks by a large number of compromised devices, often forming botnets. With 5G, these botnets can rapidly adapt and coordinate attacks, overwhelming defenses through sheer volume and speed. Security teams will have a difficult time recognizing and thwarting such dynamic and quickly changing threats as a result.

Such accelerated attack surfaces in 5G networks, driven by reduced latency and increased data speeds, empower cybercriminals with the ability to deploy sophisticated attacks at unprecedented rates. This necessitates a paradigm shift in cybersecurity strategies to effectively detect, respond to, and mitigate threats in real-time within this dynamic and fast-paced digital landscape.

6.2 IoT Vulnerabilities and Exploitation

The proliferation of Internet of Things (IoT) devices is a hallmark of the 5G era. These interconnected devices, ranging from smart home appliances to critical infrastructure components, create an expansive attack vector for cybercriminals. Infiltrating and manipulating IoT devices within a 5G network can lead to cascading effects, ranging from privacy breaches to critical infrastructure disruption. Illustrative examples include the compromise of smart city networks or the manipulation of medical IoT devices. 5G's enhanced capacity and ability to handle a massive number of devices per square kilometer enable the deployment of a vast array of IoT devices. However, the sheer scale of device onboarding introduces challenges in managing and securing each endpoint. Cybercriminals can exploit the rapid onboarding process, potentially overwhelming security infrastructure and evading detection by blending in with the legitimate influx of IoT devices. Many IoT devices communicate using lightweight protocols to minimize bandwidth consumption. While this is advantageous for efficient communication, it often comes at the cost of robust security. In the 5G ecosystem, the reliance on low-latency communication may lead to the use of less secure protocols, creating opportunities for attackers to intercept, manipulate, or eavesdrop on IoT device communications.

By processing data closer to the source, 5G networks can reduce latency and improve the overall performance of Internet of Things devices. But edge computing's decentralized architecture raises concerns regarding security. The attackers could target holes in the infrastructure of edge computing, jeopardizing the security and integrity of data handled at the edge of the network. For instance, a compromised edge server could manipulate data from IoT sensors, leading to inaccurate information or control commands. IoT devices often run on firmware with limited resources, making them susceptible to security vulnerabilities. Threat actors can effectively exploit these vulnerabilities by exploiting 5G networks' high speed and low latency. For example, a compromised IoT device can be used as a pivot point to launch attacks within the 5G network, potentially compromising other connected devices or gaining unauthorized access to sensitive data. The rapid deployment of IoT devices in the 5G era introduces challenges in ensuring the security of the entire supply chain. From manufacturing to deployment, each stage in the supply chain represents a potential point of compromise. Cybercriminals may exploit vulnerabilities in the production process, implanting malware or backdoors in IoT devices before they even reach the end user. These compromised devices can then be leveraged for various malicious activities within the 5G network.

Thus, the integration of IoT devices into 5G networks amplifies the attack surface for cybercriminals, necessitating a comprehensive approach to address vulnerabilities. Securing the IoT ecosystem in the 5G era requires robust measures, including secure communication protocols, regular firmware updates, and stringent supply chain security practices, to mitigate the evolving threats in this interconnected and high-speed environment.

6.3 Edge Computing and Security Concerns

The decentralized nature of edge computing, a key component of 5G architecture, introduces new security challenges. Edge computing reduces latency by processing data closer to the source, but it also makes possible weaknesses more visible. Cybercriminals are able to launch distributed denial-of-service (DDoS) attacks or compromise sensitive data by taking advantage of vulnerabilities in these distributed systems. The 5G-driven shift towards edge computing mandates a reevaluation of security protocols to safeguard against emerging threats. By analyzing information closer to the source, edge computing reduces latency and improves overall system performance, bringing about a paradigm shift in the 5G ecosystem. However, this decentralization of computing resources also raises significant security concerns. The following examples delve into the technical intricacies of edge computing and the associated security challenges in the context of 5G networks. Edge computing distributes computational tasks across a network, reducing the need to transmit data to a centralized cloud server. While this minimizes latency, it concurrently expands the attack surface. Cybercriminals can exploit vulnerabilities in the distributed edge infrastructure, compromising the integrity of data processed at the edge. For instance, a malicious actor might target an edge server processing data from IoT devices, leading to unauthorized access or manipulation of critical information.

The deployment of edge nodes in diverse locations, including remote or uncontrolled environments, introduces challenges in securing these nodes against insider threats. The confidentiality and integrity of data handled at the edge are directly threatened by malicious individuals who have physical access to edge devices and may try to modify or breach them. Strong logical and physical access constraints are required for edge node security in order to reduce the possibility of insider attacks. Edge computing relies on efficient communication between devices and edge nodes. The low-latency requirements of 5G may lead to the use of less secure communication protocols to expedite data transmission. Cybercriminals can exploit these insecure communication channels to intercept, modify, or manipulate data in transit. Strong encryption and secure communication protocols must be implemented in the 5G edge computing environment to guarantee the confidentiality and integrity of data. The processing power, memory, and energy of edge computing devices are frequently constrained. It can be difficult to put strong security measures in place on devices with limited resources. Cybercriminals may capitalize on these limitations to execute attacks such as denial-of-service (DoS) or compromise the security mechanisms in place. Balancing the need for security with the constraints of edge computing devices is a critical consideration in safeguarding the 5G network against evolving threats.

Edge computing environments often support multiple tenants, each with their own set of applications and data. Ensuring isolation between tenants is crucial to prevent unauthorized access or interference. Cybercriminals may try to take advantage of flaws in the multi-tenancy architecture in the context of 5G in order to obtain unauthorized access to private information or interfere with services. Implementing robust isolation mechanisms and access controls becomes imperative in mitigating these multi-tenancy challenges. The adoption of edge computing in 5G networks introduces a range of security concerns, from decentralized attack surfaces to resource constraints on edge devices. Addressing these challenges requires a holistic approach, incorporating secure communication protocols, access controls, and isolation mechanisms to fortify the 5G edge computing environment against evolving cyber threats.

6.4 Enhanced Stealth and Evasion Techniques

The high data speeds offered by 5G facilitate the deployment of advanced evasion techniques by digital terrorists. Rapid transmission of malicious payloads and the use of encrypted communication channels make it harder for conventional security measures to detect and mitigate threats. This increased sophistication challenges cybersecurity professionals to develop innovative approaches to identify and counteract covert activities in the 5G landscape. Because 5G networks are fast and low latency, they offer hackers an ideal environment in which to use improved evasion and stealth tactics. These tactics allow malicious actors to operate covertly, making detection and attribution more challenging for cybersecurity professionals. The following examples elucidate the technical intricacies of how 5G can amplify stealth and evasion in cyber operations.

5G's unparalleled data speeds enable the swift transmission of encrypted payloads. Malicious actors can leverage advanced encryption algorithms to conceal their activities, making it difficult for traditional security measures to inspect or identify malicious content in transit. Rapid transmission further reduces the window of opportunity for detection, allowing threat actors to execute their operations before security systems can analyze and respond. 5G introduces dynamic spectrum access, allowing devices to dynamically select and use available frequency bands. This flexibility enhances network efficiency but also enables attackers to dynamically switch frequencies during their operations. Cybercriminals can exploit this feature to evade traditional frequency-based detection methods, making it challenging for security systems to consistently track and block malicious activities.

Network slicing in 5G allows the creation of isolated, virtualized networks tailored for specific applications or users. While this feature enhances network customization and efficiency, it also introduces challenges in monitoring and securing each network slice. Malicious actors can exploit network slicing to operate within isolated segments, evading detection by blending into the diverse traffic patterns and behaviors associated with different slices. The combination of 5G and machine learning capabilities empowers cybercriminals to conduct adaptive and self-learning attacks. Malicious algorithms can analyze network behaviors, adapt to security measures, and continuously evolve to avoid detection. The rapid data processing speeds in 5G networks enable threat actors to implement machine learning algorithms that can quickly adjust their tactics, making it challenging for cybersecurity defenses to keep pace with the evolving threat landscape. 5G's low-latency communication facilitates the establishment of covert channels for communication between compromised devices. Malicious actors can leverage these channels to exchange information without arousing suspicion. Covert communication may involve the use of steganography or encryption within seemingly innocuous data transmissions, enabling threat actors to maintain a low profile and evade detection by blending in with legitimate network traffic.

The enhanced stealth and evasion techniques facilitated by 5G's speed and low-latency characteristics pose significant challenges for cybersecurity professionals. Detecting and mitigating these advanced tactics requires the adoption of sophisticated threat detection mechanisms, including behavior analysis, anomaly detection, and machine learning, to stay ahead of cybercriminals operating within the dynamic and fast-paced 5G landscape.

6.5 Quantum Computing Threats

As 5G networks (Qualcomm.com, n.d.) lay the groundwork for the future, the potential advent of quantum computing poses a unique threat to cryptographic systems. Current encryption techniques may become outdated due to the unmatched processing power of quantum computers, leaving private information vulnerable to illegal access. Cybercriminals leveraging quantum computing capabilities could exploit this vulnerability to compromise confidential information, posing a significant risk in the 5G era. The integration of 5G technology brings forth not only unprecedented speed and connectivity but also introduces a unique set of challenges related to quantum computing threats. Quantum computing, with its potential to break current cryptographic systems, poses a distinct risk in the 5G era. The following examples delve into the technical details of how the convergence of 5G and quantum computing may open avenues for malicious activities.

Shor's method, which is a component of quantum computing, can factor big numbers effectively exponentially quicker than traditional computers. This directly jeopardizes popular cryptographic techniques like RSA and ECC, whose security depends on the difficulty of factoring huge integers. In the 5G environment, the compromise of cryptographic systems due to quantum computing could expose sensitive information and communications to unauthorized access. Quantum computers can also threaten symmetric key algorithms, which are commonly used for encrypting data in transit. Symmetric key techniques can be made less secure by using Grover's algorithm, a quantum search algorithm for unsorted databases. This could lead to a significant decrease in the time required to brute-force encryption keys, making it more feasible for quantum-powered adversaries to decrypt confidential communications within the 5G network.

While Quantum Key Distribution is often hailed as a quantum-safe solution for secure communication, practical implementations may still face challenges. Quantum networks used for key distribution in 5G could be susceptible to attacks exploiting vulnerabilities in the quantum hardware or protocols. Quantum adversaries might attempt to intercept or manipulate quantum keys, compromising the integrity of secure communications within the 5G infrastructure. Quantum computing threats extend beyond encryption to authentication protocols. Many authentication systems rely on cryptographic primitives that may become insecure in the presence of quantum computers. These vulnerabilities could be used by quantum adversaries to spoof legitimate users in the 5G network, obtain illegal access, or fake authentication credentials. To create algorithms that are thought to be immune to quantum attacks, post-quantum cryptography is being developed as a preventative strategy. However, the integration of these new cryptographic systems into the 5G infrastructure poses its own challenges. Ensuring a seamless transition without introducing vulnerabilities or disruptions requires careful planning and consideration of the quantum-resistant algorithms' compatibility with existing systems.

While 5G technology propels society into an era of unprecedented connectivity, it simultaneously introduces novel avenues for cybercrime and digital terrorism. As we embrace the benefits of high-speed communication, it is imperative to fortify our digital ecosystems against emerging threats, fostering a secure and resilient technological landscape for the new millennium.

7. CONCLUSION

Cybercrime and digital terrorism are serious threats to individuals, organizations, and even nation-states. They may lead to monetary losses, the theft of private data, and the interruption of vital infrastructure. Cybercrime is typically motivated by financial gain, while digital terrorism is often politically or socially motivated. It is crucial for people and businesses to put appropriate cybersecurity practices into place, such as using strong passwords, updating software often, and being watchful for phishing and other forms of social engineering, in order to reduce the risks of cybercrime and digital terrorism. Additionally, organizations should consider implementing technical solutions, such as encryption and firewalls, to protect sensitive information and networks. The issue of cybercrime and digital terrorism is only becoming more pressing as the world becomes increasingly digitized, and it will require a concerted effort by all stakeholders to stay ahead of the threat. The convergence of 5G and quantum computing introduces a unique set of security challenges, particularly in the realm of cryptography. The potential ability of quantum computers to break widely used encryption algorithms necessitates proactive measures such as the adoption of quantum-resistant cryptography and careful consideration of the security implications in the design and implementation of 5G networks.

REFERENCES

Abdulrahman, Y. (2023). Cybersecurity in Digital Transformation Era. From 5G to 6G: Technologies, Architecture, AI, and Security. IEEE. 10.1002/9781119883111.ch7

Bhushan, B. (2023). *Synergizing AI, 5G, and Machine Learning: Unleashing Transformative Potential Across Industries with a Focus on Network Security*. 2023 3rd International Conference on Advancement in Electronics & Communication Engineering (AECE), Ghaziabad, India. 10.1109/AECE59614.2023.10428548

Logeshwaran, J. (2024). *Next-Generation Tele-Communication Networks and IoT: Advancements and Challenges.* 2024 11th International Conference on Reliability, Infocom Technologies and Optimization (Trends and Future Directions) (ICRITO), Noida, India. 10.1109/ICRITO61523.2024.10522190

Mozumder, M. (2023). The Metaverse for Intelligent Healthcare using XAI, Blockchain, and Immersive Technology. *2023 IEEE International Conference on Metaverse Computing, Networking and Applications (MetaCom)*, Kyoto, Japan. 10.1109/MetaCom57706.2023.00107

Nguyen, V. M., & Nur, A. Y. "Major CyberSecurity Threats in Healthcare During Covid-19 Pandemic," *2023 International Symposium on Networks, Computers and Communications (ISNCC)*, Doha, Qatar. 10.1109/ISNCC58260.2023.10323723

Qualcomm. (2023). *What is 5G | Everything You Need to Know About 5G*. Qualcomm. https://www.qualcomm.com/5g/what-is-5g (accessed: October 25, 2023).

Chapter 2 5G: The Emerging Cybersecurity Threat Landscape for India

Akashdeep Bhardwaj
https://orcid.org/0000-0001-7361-0465
University of Petroleum and Energy Studies, India

ABSTRACT

The overall landscape of security threats in India is changing quickly, giving rise to a wide range of new security issues. The nation is threatened by cyberattacks, border tensions, separatist movements, and domestic and international terrorism. Another major issue is the threat posed by radicalism, both ideological and religious. To properly handle these new security challenges, India needs to increase inter-agency collaboration, bolster its cybersecurity and counter-terrorism capabilities, and strengthen its information collection skills. India stands at the brink of a digital revolution with the advent of 5G technology, promising unprecedented connectivity, and technological advancements. and a parallel concern has arisen as the evolving cybersecurity threat landscape. This chapter explores the unique challenges and potential threats that 5G poses to India's cybersecurity, examining the intricate interplay between technological progress and the need for robust defenses.

1. INTRODUCTION

One of the largest and most rapidly evolving economies in the world, India must contend with a security threat landscape that is getting complicated day by day. The country is confronting a wide range of challenges, from terrorism and extremism to cyberattacks and natural disasters. These threats are dynamic, constantly evolving and often intersect, making it imperative for India to have a well-coordinated and

DOI: 10.4018/979-8-3693-2786-9.ch002

multi-disciplinary approach to security. In context with this, it's critical to comprehend the types and extent of security risks that the nation faces and to create workable countermeasures. An overview of the growing security threat scenario in India is given in this chapter, along with a discussion of the main issues and the measures being taken to resolve them.

Faster bandwidth speeds have been made possible by recent technological advancements and new age computing models in IT infrastructure. Cloud computing, mobile computing, and virtualization have also virtually eliminated the distinction between traditional on-premises and internet-based enterprise security perimeters. As a result, the digital age has become one that is rich in data, which presents hackers and other threat actors that engage in cybercrime with excellent opportunities. During the past few years, cybercrime has advanced at the fastest rate yet. With improvements being made daily, the techniques used to collect end user data or disrupt operations and services have advanced to a high level of sophistication. This basically says that the current attack strategy will probably change in the coming year or perhaps a few months. Targeted cyberattacks against private and public institutions have grown over time, becoming more comprehensive, sophisticated, and damaging. In today's digital era, technologies like Internet Applications, Internet of Things, Blockchain, Artificial Intelligence and Machine Learning are rapidly gaining traction and lie at the very heart of our society.

These technologies are here to stay and help connect people, encourage mobility, ensure empowerment, and help perform daily tasks faster, better, and more efficiently as compared to traditional and conventional methods. Their innovation and proliferation have increased multifold in the last few years, with our physical real-world life synched with digital realms. This has however led to a rise in threats and vulnerabilities on the digital environments. The evolution and adoption of Cloud Computing for daily use has started to involve the use of Cyberspace, handheld, and smart devices. These have in turn enhanced the risk of Cyberattacks. Nations across the world face huge challenges to their Cybersecurity. The strategic shift of economy and market demand from the West to new upcoming economies like India has also introduced Cybersecurity threats and attacks.

Fifth generation, or 5G technology (Qualcomm, 2017) is the next development in cellular network technology, after 4G LTE (Long-Term Evolution), as Figure 1 shows. It provides higher capacity, lower latency, and faster speeds than its predecessors, which is a major advancement in wireless communication.

Figure 1. 5G architecture diagram

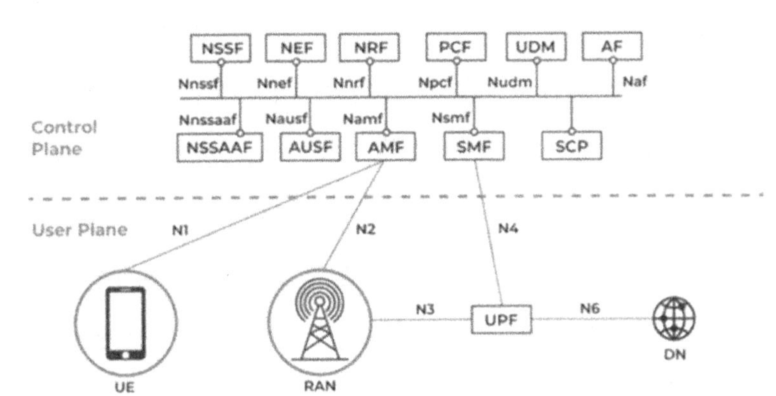

Among 5G's primary characteristics are:

- High-Speed Data Transfer: 5G networks provide peak data rates of up to 20 gigabits per second (What are 5G speeds, 2024), which is significantly faster than 4G LTE. This improves user experiences for applications like online gaming and video conferencing and allows high-definition content to stream smoothly and download files swiftly.
- Low Latency: Due to the ultra-low latency of 5G networks, the time lag between sending and receiving data packets can be as little as one millisecond (ms), (Reply, 2024). This virtually instantaneous reaction is crucial for real-time applications such as driver-less automobiles, augmented reality, and remote surgery.
- Massive Connectivity: Because 5G networks can support a massive number of linked devices, they are ideal for Internet of Things (IoT) deployments and smart city initiatives. This enables seamless communication across a wide range of devices, including industrial equipment, sensors, wearables and smartphones.
- Network Slicing (*5G Network Slicing, What is it? 2019)*: 5G makes it feasible for operators to split their infrastructure into several, virtualized networks, each tailored to a specific use case. As shown in Figure 2, this enables the provision of specialized services that satisfy various application and industry requirements in terms of speed, latency, and reliability.

Figure 2. 5G network slicing

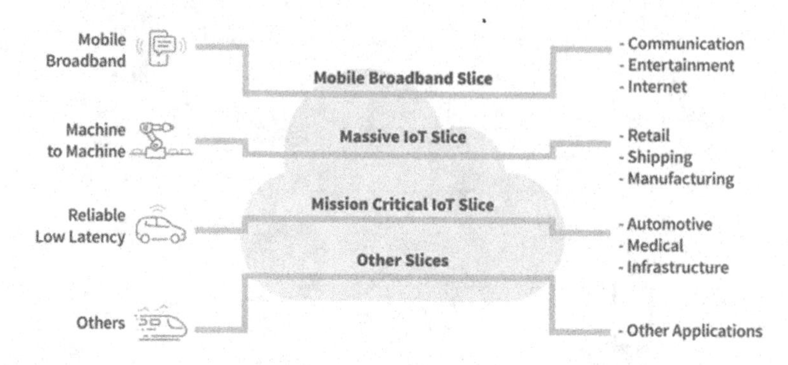

- Advanced Antenna Technologies: Two examples of state-of-the-art antenna technologies that 5G uses to increase efficiency, capacity, and coverage are massive MIMO (Multiple Input, Multiple Output) and beamforming. These techniques reduce interference, maximize signal delivery, and improve network performance in a variety of scenarios.
- Frequency Bands: 5G can operate in the millimeter-wave and sub-6 GHz spectrums, among other frequency ranges. Sub-6 GHz bands offer higher coverage and more penetration through impediments than millimeter frequencies, which have a restricted range and are more susceptible to signal attenuation.

This chapter takes a critical review of the Cybercrime scenario and the Threat Landscape in India. The chapter discusses Cyber threats and the Threat Landscape engaged at national, state and district level, along with end user level. Starting with Section I this discusses Cyber Threat Landscape in general. Section II discusses Cybersecurity and the impact. Section III presents the evolving Cybercrime Threat Landscape. Section IV pounders on Cyber Terrorism. Section V describes the mitigation options to combat the Cyber Threats. Section VI presents the recent advent of Cyber terrorism and Advance Persistent Threats. Finally, the conclusion presents the future work and proposes a model for securing nation states.

2. CYBERSECURITY THREATS

Cybersecurity is the fundamental to a secure, robust, and thriving economy like India. Being online and digital has greatly benefited Indian. The whole subcontinent is connected to world networks, cultures, and global economies than ever before.

India is becoming truly digital nation! However, as the benefits increase, Annual reports by leading research organizations including India's Computer Emergency Response Team, PwC, McAfee continue to provide alarming Cybersecurity trends and disturbing threat news. In the last two years, over 100,000 Cybersecurity incidents are known to be reported, which equates to more than five new Cybersecurity incidents every hour. The count for unreported incidents could well be higher. This increasing number of Cybersecurity incidents tells us one thing: global cybercrime is fast creating a struggle of truly epic proportions.

In just the past few years, the Cyber world has been had an unmatched advancement in Cyber, Information and Communications technologies. The Internet, Cyber related systems, and applications have now become an integral part of our lives in almost every aspect of our society, including social networking and information sharing. The massive acceptance and increase of such technologies along with high-speed bandwidths and handheld devices have given rise to the increasing world of Cyberspace or the virtual world of digital era. Cybersecurity is a continuous process and needs to be updated continuously. Cybercriminals come up with new ways of exploiting and attack methodologies, are designed to exploit unknown vulnerabilities in systems, applications, and networks. The main principles for cybersecurity defense are physical security. Starting with scope for vulnerability assessment, involves performing Threat Identification and Risk Assessment, followed by documentation of findings, and submitting the plan of action for mitigating issues found. Cybersecurity is a combination of Cyber defense procedures and implementation of practices envisioned to safeguard end user data, hosted systems, servers from internal and external threats and unauthorized access. India's Cyberspace hosts huge volumes of information ranging from end user, corporate enterprises (private), government (public), financial and military data. Cybersecurity is no longer be seen from technical perspective, this new domain is integrated into the everyday working, combined with cyber awareness, and knowing the threat vectors India is up against.

Cybersecurity describes ways of preventing Internet related attacks, often termed Cyberattacks, Identity Theft, Data Breaches, Advanced Persistent Threats and helps to mitigate Cyber Risks by implementing an effective incident response system. Cybersecurity attacks are usually detected and then mitigated, if not totally blocked. The medium for Cyberattacks is digital networks, termed Cyberspace. Cyber criminals seek end user data and information as well as wage Cyberattacks against nation states to cause chaos. Ensuring a secure online, digital environment is essential to Indian government, which has been providing digital applications and services to Indian citizens and corporates. Central core of the Public and Commercial delivery services and communications is the dependency on Cyber networks and the ability to perform safe online transactions and communicate securely. However, Cyberspace is impacted due to malicious practices by multiple threat vectors and hostile actors.

These act as individuals or in groups to conduct Cyber espionage actions or launch Cyberattacks related to data theft, denial of service attacks on vital government and public infrastructures and systems. At the national level, this affects national security and presents challenges in the form of Cyber Terrorism, Cybercrime, and industrial espionage.

Foreign States include foreign nation states sponsoring Cybercriminals, and Hacktivist groups. Their objectives, capabilities and resources vary as these threat actors are well equipped to impact and conduct the most destructive Cyber espionage and cyberattacks and target the military, government, corporate and high value individuals or disrupt and damage cyber infrastructure.

- Cyber espionage is an extension of traditional old age espionage. This enables hostile threat actors with little or no risk, to access information from remote locations on India's industrial domains. This presents a clear and present risk to India's economic well-being as well as poses direct risk to Indian national security.
- Domestic Threats: this poses one of the most interesting and critical threats to India. Every year, engineering colleges churn out thousands of students, most of them are unable to land any jobs or work as per their high expectations. This is where internal threats like Naxalites, Terrorist groups or religious groups recruiting these as their resources for performing their hidden agenda and malicious tasks.
- Sabotage: Indian government's Digital India push and the proposed penetration along with propagation of high-speed telecom services into the rural areas, causes yet another level of threat. Not only do the Cybercrime cells need to worry about educated urban threat actors, traffic and attacks from rural areas has become yet another problem. Post Kashmir 370 decision; there has been a dramatic rise in Cyberattacks against government portals. In September 2019 alone, over 25 web portals of central and state government departments have been defaced and hacked.
- New Age Devices: rise in use of IoT and mobile applications has seen 22% increase in cyberattacks in quarter 2 of 2019 alone. Rise in these attacks also tend to have geopolitical motivations. This is even more alarming as there are no security standards for vendors selling IoT devices. Over 2,550 malwares have been identified in September 2019 alone, targeting IoT devices.
- National Asset Attacks: Smart Cities, Sectors like Transportation, Power, Financial and deployments are being attacked at a scale never seen earlier, some reporting at least 30 Cyberattacks per day. The push for integrated national power grid has also raised concerns that India's power infrastructure

might well be the next target of Cyber terrorists seeking ways to affect India's economy.

- These days, one of the most prevalent forms of online banking dangers on the Internet is phishing attacks. Although banks are known to press customers to accept it, customers should bear the responsibility for Phishing since they were negligent in how they responded to the Phishing email. Phishing is a result of various interpretations of the Information Technology Act of 2000, particularly with the 2008 modifications. The client experiences unfair misfortune as a result. The negation along these lines draws in arrangements of Section 43 for arbitration.

Such threats present significant risks to the state security, economy, and critical infrastructure, so the country must continue to invest in its cybersecurity infrastructure and capabilities to effectively address them. Table 1 below illustrates characteristics of a Cyberattack.

Table 1. Cyberattack characteristics

• Availability of Attack Tools & Low Cost	**Most Cyberattack tools can be easily obtained from Internet sources like Social Network groups, Darkweb or Repositories like GitHub.**
• Ease of Cyber Attack Execution	Execution of Cyberattack is location independent; in most cases, basic hacking skills are enough to perform a high-level attack.
• Insufficient Cyber Policies	Absence of effective Cyber policy, Non-existent legal framework and punishment, lack of safeguards and Awareness aids in achieving the desired malicious objective.
• Low Risk for Attackers	Ease of concealment makes it difficult to attribute the Cyberattack to the real perpetrator.

3. EVOLVING CYBERCRIME THREAT LANDSCAPE

India operates in a highly difficult military and strategic environment. China and Pakistan are two neighbors that have strong military and cyberattack capacities. A wide spectrum of external and internal cybersecurity risks must also be managed by India. Ensuring cybersecurity awareness and the necessity to protect India's cyber threat landscape have grown more and more important. With the expected growth of Cyberspace and the Digital Era, cybercrime is also expected to increase. Therefore, cybersecurity is now a top priority domain more than it has ever been. This examines the impending difficulties and prospects in the field of cyberspace and provides a critical assessment of the cyberthreats that India is now facing. India has encountered a variety of complex and dynamic cybersecurity challenges, impacted by a broad spectrum of socio-political, technical, and economic factors. Throughout its history, the nation has faced a broad range of security challenges, from well-known threats like border wars and terrorism to more contemporary difficulties like cyberwarfare and extremism. In India, the nature and scope of cybersecurity threats have evolved over time, necessitating constant innovation and modification of defense strategies.

During the early stages of its digital revolution, India had to contend with traditional security challenges such as separatist movements, insurgencies, and cross-border terrorism. The country's security was seriously threatened by these dangers, thus the government had to give conventional defense measures major attention. However, the advent of the internet and the rapid advancement of information technology caused a seismic upheaval in the security environment, ushering in a new era of cyber threats. Cyberattacks become a potent weapon in the hands of both state and non-state actors, fundamentally altering the security paradigm. This provided India with a formidable foe that could do significant damage without requiring conventional military combat. The 2008 Mumbai attacks served as a wake-up call by revealing how vulnerable India's computer infrastructure was to international terrorism. Terrorists based in Pakistan coordinated the assaults using internet communication channels.

Consequently, there was a rise in cyber espionage, financial fraud, and data breaches in India by both domestic and international threat actors. State-sponsored cyberwarfare, particularly from neighboring countries, has become a persistent threat, with espionage and sabotage operations directed at critical infrastructure and government networks. Moreover, the advent of hacktivist groups and cybercrime syndicates has exacerbated the cybersecurity landscape in India by exploiting vulnerabilities for financial or ideological benefit. As India adopted digitization and introduced ambitious initiatives like Digital India and Smart Cities, the attack surface expanded significantly, increasing the risk of cyberattacks. As Figure 3 illustrates, the increasing reliance on cloud computing, IoT (Internet of Things) infrastructure, and networked devices has made critical industries like energy, healthcare, and finance susceptible to cyberattacks. The proliferation of social media sites and internet messaging programs has exacerbated internal security challenges by facilitating the spread of extremist ideologies and misleading information. Given how extremist groups have used India's complex socio-cultural fabric to exploit anxieties and incite violence, a comprehensive plan to counter radicalization and disinformation online is needed.

Figure 3. 5G and IoT connectivity

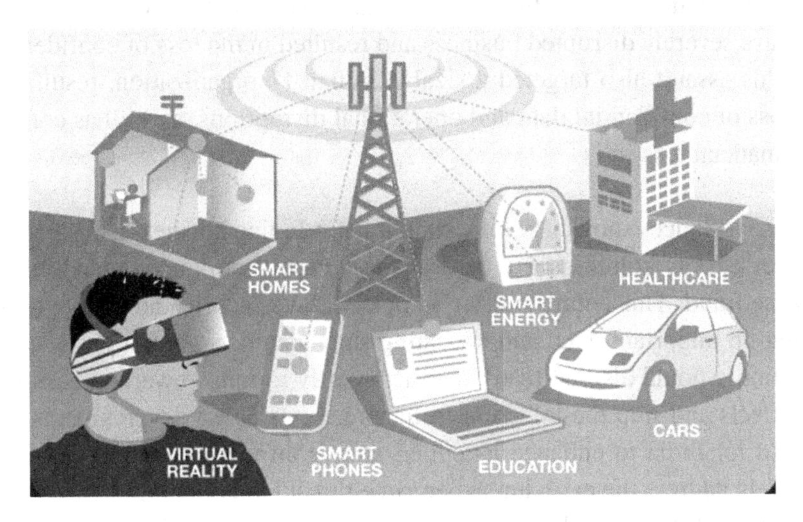

The cybercrime threat landscape in India is rapidly evolving, and the country is facing an increasing number of cyber threats. The key developments in the evolving cybercrime threat landscape in India for Petya ransomware attacks that have taken place in India are presented below.

- Petya Ransomware Attack impacted several companies during the period of June 0July 2017 in India, including a multinational pharmaceutical company and an IT firm. The attack caused widespread disruption to operations, leading to the theft of sensitive information and financial losses. Another variant of this ransomware known as NotPetya was also involved in a series of attacks in India, targeting several organizations, including banks and government institutions. The attack caused significant disruption to operations and led to the theft of sensitive information.
- A prominent Indian pharmaceutical company was the target of an attack using ransomware in 2021 called REvil, which resulted in the loss of confidential data and the interruption of business activities. Malware known as "Revil Ransomware" encrypts a victim's data and then demands a ransom to unlock the contents. In recent years, India has been the subject of many attacks involving ransomware by REvil, which have severely disrupted business and resulted in the loss of confidential data.
- In 2021, a major Indian oil and gas corporation was targeted by DarkSide ransomware, which resulted in the loss of confidential data and the interruption of business activities. One kind of virus known as "darkside ransomware"

encrypts a victim's files and then demands a fee to unlock them. In recent years, India has been the victim of many Darkside ransomware attacks that have severely disrupted business and resulted in the loss of confidential data. This assault also targeted a sizable Indian IT organization, resulting in the loss of confidential data and operational disruptions, as well as considerable financial losses.

These instances show how important it is for Indian businesses to spend money on cybersecurity solutions to guard against ransomware attacks and other types of online fraud. This entails putting in place robust security procedures, routinely backing up important data, and educating staff members about cybersecurity. Organizations should also be ready to react swiftly to ransomware attacks to lessen their effects and stop the infection from spreading. These developments establish the need for India to enhance its cybersecurity infrastructure and capabilities to effectively address the evolving cybercrime threat landscape.

Because of PC wrongdoing, administrators became progressively mindful in the 1980s as organizations turned out to be increasingly needy upon computerization and as impetus occasion cases presented huge vulnerabilities to PC wrongdoing infringement. Now, with no fear of legal identification, criminals may retain, transfer, and encrypt material that serves as evidence of their illegal protests. A PC crime scene might now stretch from the location of the exploitation (the unlucky victim's PC, for example) to another location on Earth due to the added common impact of the Internet, greatly complicating criminal investigative efforts. PCs have drastically adjusted the criminal equity landscape. Venturesome and shrewd hoodlums have deliberately gone to the PC to perpetrate their unlawful demonstrations in circumstances in which the PC fills in as the instrument of the wrongdoing, the methods by which the wrongdoing is submitted, just as in cases in which the injured individual's PC, or PC framework, is the objective or target, of the demonstration. As was previously said, digital advancements really assist criminals in situations when the usage of the PC for illegal activities is incidental; these situations arise when the PCs are used to store and safeguard data that serves as evidence linking the perpetrator to illicit activity. One commonality between these types of crimes is that the perpetrator heavily depends on the lack of creative legal skills to successfully report the offenses and get away with it.

4. CYBER TERRORISM

Cyberterrorism is the term for terrorist activities carried out via the use of technology, especially computers and the internet. India has experienced an upsurge in cyberattacks attacking companies, government organizations, and key infrastructure in recent years, raising concerns about cyber terrorism. Cyber terrorism in India includes:

- Target critical infrastructure: To cause widespread disruption and fear, cyber terrorists have targeted India's vital infrastructure, including banks, water systems, and electricity grids.
- Hack government institutions: Cyber terrorists have also targeted government institutions in India, such as the Indian Ministry of Defense and the Indian Railways, with the intention of stealing sensitive information and disrupting operations.
- Disrupt business operations: Cyber terrorism has also been used to disrupt the operations of businesses in India, causing financial losses and damaging the country's economy.

To combat cyber terrorism in India, the government has established several initiatives and programs aimed at strengthening the country's cybersecurity infrastructure and increasing awareness about cyber threats. To lessen the danger of cyberattacks, businesses can also have solid security procedures in place, routinely backup important data, and train staff members on cybersecurity awareness. The Indian government has strengthened the nation's cybersecurity posture and fought cyberterrorism with several initiatives.

Some of these measures include:

- Cybersecurity Policy: The Indian government has released a comprehensive cybersecurity policy that outlines the country's approach to addressing cyber threats and protecting critical infrastructure, government institutions, and businesses. The policy is designed to ensure that the country's critical information infrastructure is secure and resilient against cyber-attacks and other forms of cybercrime. Some key elements of the Indian cybersecurity policy include:

    - Critical Infrastructure Protection: Protecting the nation's vital infrastructure against cyberattacks, including power grids, banking systems, and transportation networks, is a top priority for the strategy.

- Cybercrime Investigation and Prosecution: The policy outlines the government's approach to investigating and prosecuting cybercrime, including cyber terrorism and other forms of cyber-enabled crime.
- Cybersecurity Awareness and Training: The policy recognizes the importance of raising awareness about cyber threats and providing cybersecurity training to individuals and organizations. This includes promoting cybersecurity education in schools and universities, as well as providing training for government employees and industry professionals.
- Public-Private Partnership: The policy emphasizes the importance of public-private partnerships in addressing the threat of cybercrime. This includes working with the private sector to share threat intelligence, establish best practices for cybersecurity, and collaborate on incident response and recovery efforts.
- International Cooperation: The policy recognizes the need for international cooperation in addressing the threat of cybercrime, including working with other countries to share threat intelligence, investigate cybercrime, and bring cybercriminals to justice.

Indian cybersecurity policy is a comprehensive framework that addresses the threat of cybercrime and enhances the country's cybersecurity posture. The government continues efforts to implement and enforce cyber policies to ensure the cyber resilience of the country's critical information infrastructure.

- Cybercrime Investigation Cell: To investigate and prosecute cybercrime, especially cyberterrorism, the Indian government formed a special Cybercrime Investigation Cell (CIC) as a dedicated unit within the Indian law enforcement community tasked with investigating and prosecuting cybercrime, including cyber terrorism. CIC was established to address the growing threat of cybercrime in India and to ensure that the country's critical information infrastructure is secure and resilient against cyber-attacks. Some key responsibilities of the CIC include:

    - Investigating cybercrime: CIC is responsible for investigating a wide range of cybercrime, including hacking, cyber espionage, and cyber terrorism working with other LEAs and the private sector, to gather evidence, track down suspects, and bring cybercriminals to justice.
    - Prosecuting cybercrime: CIC is also responsible for prosecuting cybercrime to ensure cyber criminals are held accountable. CIC works with prosecutors and the judicial system to build strong cases against cybercriminals and to secure convictions.

- Raising awareness about cybercrime: CIC is involved in raising awareness about cybercrime and the dangers it poses to individuals, businesses, and the country. CIC also provides training and education to individuals and organizations to help them understand the threat of cybercrime and take steps to protect themselves.
- Sharing threat intelligence: To assist enterprises in staying ahead of constantly changing cyberthreats, CIC shares threat intelligence with the commercial sector and other law enforcement authorities. CIC also works with international law enforcement agencies to coordinate investigations and to share information about cross-border cybercrime.

- National Critical Information Infrastructure Protection Centre: Indian government has established the National Critical Information Infrastructure Protection Centre (NCIIPC) to monitor and respond to cyber threats targeting critical infrastructure and sensitive information. The establishment of NCIIPC was a component of the Indian government's all-encompassing cybersecurity strategy, which aimed to guarantee the security and resilience of the nation's critical infrastructure, such as power grids, banking systems, and transportation networks Some key responsibilities of the NCIIPC include:

  - Protecting critical infrastructure: NCIIPC oversees creating and putting into action policies to safeguard the nation's critical infrastructure against cyberattacks and other types of online crime. This includes conducting risk assessments, develop security standards, and implement security measures to protect critical systems and networks.
  - Coordinating incident response: In the case of a cyberattack, NCIIPC is essential in coordinating incident response and recovery activities. NCIIPC collaborates closely with the corporate sector and other governmental organizations to address cyber events and lessen the effects of cyberattacks.
  - Sharing threat intelligence: NCIIPC shares threat intel reports with other LEAs, private sector and works with international organizations to share information about cross-border cybercrime.
  - Raising awareness about cyber threats: The NCIIPC is involved in raising awareness about cyber threats and the importance of cybersecurity. This includes providing training and education to individuals and organizations to help them understand the threat of cybercrime and take steps to protect themselves.

- National Cyber Coordination Centre: The Indian government has established the National Cyber Coordination Centre (NCCC) to monitor and analyze cyber threats and provide real-time threat intelligence to organizations and individuals. NCCC was established as part of the Indian government's comprehensive cybersecurity policy and is tasked with coordinating the country's efforts to prevent, detect, and respond to cyber-attacks. Some key responsibilities of the NCCC include:

    - Cyber Threat Intelligence: The NCCC is responsible for collecting, analyzing, and disseminating threat intelligence to other government agencies, as well as to the private sector. The NCCC also works with international organizations to share information about cross-border cybercrime.
    - Cybersecurity Coordination: The NCCC plays a critical role in coordinating the country's efforts to enhance its cybersecurity posture. The NCCC works with other government agencies, as well as with the private sector, to develop and implement best practices for protecting against cyber threats.
    - Incident Response and Recovery: The NCCC is involved in coordinating incident response and recovery efforts in the event of a cyber-attack. The NCCC works closely with other government agencies, as well as with the private sector, to respond to cyber incidents and to minimize the impact of cyber-attacks.
    - Cybersecurity Awareness: The NCCC is involved in raising awareness about cyber threats and the importance of cybersecurity. This includes providing training and education to individuals and organizations to help them understand the threat of cybercrime and take steps to protect themselves.

Digital Terrorism turns into a universal danger to the world as through fear-based oppression, the psychological oppressor spreading false promulgation in accordance with political and religious belief systems. Cyber Terrorism is of ongoing vintage. Digital fear-based oppression is the mix of the internet and psychological warfare and we do not have any meaning of digital fear mongering which can be acknowledged around the world. Each scientist or researcher in the subject gives an alternate measurement while characterizing the term digital fear-based oppression. In this examination the meaning of digital fear mongering is partitioned into goal based and impact based. It alludes to assaults on the PCs, systems, and system lattices of the nation, which intensely rely upon systems and make ruin or dread among the brains of its natives. The meaning of digital fear-based oppression cannot be made broad, as the idea of wrongdoing is to such an extent that it must be left to be thorough in nature. It is not wise to place the definition in a restraint equation or pigeon's entirety. Truth be told, the main exertion of the legal executive ought to be to comprehend the definition as generously to rebuff the fear monger stringently, so the administration can handle the malevolence of digital psychological warfare.

The term cybercrime and digital psychological oppression is diverse all together we cannot state that each cybercrime is digital fear-based oppression. We need to see whether cybercrime is politically and ideologically propelled or not, to label it as digital psychological oppression. In present situation, the point of the psychological militant association is to demolish the correspondence, framework, transportation, and monetary system of the nation with PCs and systems to make dread in the brains of the individuals, as each nation on the planet is vigorously rely upon the innovation. Ongoing assaults in India just as in world have demonstrated that the psychological militant is additionally using the PCs and systems administration to carryout fear monger assaults. The Basic goal of the digital psychological oppressor association while assaulting a country, a spot, and an association, to decimate substantial property or resources and executing people to demonstrate their motivation or political belief systems. In this manner, there is no uncertainty that innovation progression in PCs and systems administration has had an essential impact in giving them opportunity, which affected fear monger's techniques and conduct impressively and distinguish the goals of digital fear mongering:

- The authorized target of digital psychological oppression includes functions such as recruiting, preparing for induction, helping, communicating, setting up, and snooping, among others. These days, psychological militant groups often plot their actions on the Internet based on knowledge reports. Their knowledge and skills in PC innovation are always growing, and this will ultimately equip them with the critical ability to identify and take advantage of weaknesses in government or foundational businesses' online security frameworks.
- While investigations into the use of the Internet to spread terror usually portray them as extremist organizations seeking a kind of virtual jihad, the characters surrendering to virtual psychological enslavement on screen don't have to be motivated by religion. • In addition, the hierarchical nature of digital fear mongering gives offenders the ability to pursue their victim using creative or traditional means of combat. It seems sense that the goal fear-based oppressors seek to achieve here is to interfere with the regular operation of PC frameworks, services, or websites. Three strategies are employed: uncovering, denying, and ruining. These methods have proven to be legitimate, as the Western countries

heavily depend on internet infrastructure to support basic government functions. Nevertheless, difficult activities for the most part do not cause grave outcomes, except from maybe in situations of an odd thud on contact.

- The justification for this is designed to achieve similar or identical goals to traditional fear-based subjugation; it is clearly pure digital psychological warfare. Fear-mongering oppressors attempt to destroy or injure identifiable property or resources, as well as inflict harm or death to individuals, using the advancement of personal computers and the Internet. Pure fear-based digital oppression does not exist in the contemporary day, but perhaps it will soon. A terrorist organization known as Al Qaeda attacked the Indian infrastructure in March 2016; according to reports, they broke into a scaled-down version of the Indian Railways' Rail website to show off their evil reach. Information technology becomes a simple tool in the hands of an oppressor who uses fear.
- A message from Maulana Aasim Umar, the leader of Al Qaeda in south Asia, replaced a hacked page of the Bhusawal division of the Central Railway Personnel Department and a portion of an enormous intranet created for the office's managerial needs. They communicate with their agents globally using codes via PCs and other systems without being recognized by permission agencies. Examples include the Ayodhya incident, the 2006 Mumbai attack, the July 2005 programmers' mutilation of Indian military sites throughout India, the attack on the American Center in Kolkata, the Pathankot terrorist attack, and so on. are the main fear-based digital attacks.

5. 5G: THE EMERGING CYBERSECURITY THREAT

The rollout of 5G in India has ushered in an era of unparalleled connectivity, but with it comes an expanded attack surface that poses substantial challenges for the nation's cybersecurity. The adoption of 5G introduces a complex network architecture with a dense deployment of small cells, enabling high-speed, low-latency communication. While these features enhance user experience and enable innovative applications, they concurrently create opportunities for adversaries, particularly nation-state actors, to exploit the newfound capabilities. The increased density of small cells in 5G networks provides a higher number of entry points for cyber attackers. Nation-state threats can leverage these entry points to infiltrate critical infrastructure and conduct espionage with greater precision. The low-latency nature of 5G enables rapid data transmission, reducing the time window for detection and response. For instance, a nation-state actor with sophisticated capabilities can exploit the reduced latency to conduct real-time reconnaissance, identify vulnerabilities, and launch targeted attacks on critical systems before traditional defenses can react.

Furthermore, the number of linked devices rises exponentially with 5G's massive machine-type communication (mMTC) capabilities, which are intended to enable the Internet of Things (IoT). The attack surface is expanded by the spread of IoT devices in industries including healthcare, energy, and transportation. Nation-state actors might access vital systems without authorization by taking advantage of flaws in these networked devices. For example, compromising a 5G-connected IoT device within the power grid infrastructure could allow an attacker to manipulate energy distribution or disrupt the grid's operations, potentially causing widespread outages. Additionally, the network slicing feature in 5G, designed to create isolated virtual networks tailored for specific applications, introduces complexities in monitoring and securing each network slice. Nation-state threats may exploit these isolated slices to conceal malicious activities within seemingly legitimate traffic patterns. The dynamic spectrum access in 5G further complicates the scenario, allowing adversaries to switch frequencies dynamically during their operations, evading traditional frequency-based detection methods.

The convergence of 5G and emerging technologies like artificial intelligence (AI) poses additional challenges. Nation-state actors can employ AI-driven cyber-attacks that adapt and evolve in response to defensive measures. To adapt attack plans in real-time, machine learning algorithms, for example, can examine trends in network traffic. This makes it difficult for traditional cybersecurity measures to keep up with the constantly changing threat landscape. As a result, the introduction of 5G in India expands the attack surface of the country and poses additional risks to cybersecurity, especially from nation-state actors. Protecting vital infrastructure and national security in the ever-changing 5G era calls for a comprehensive strategy that includes adaptive defensive tactics, sophisticated threat detection systems, and international collaboration.

The rollout of 5G in India brings with it a proliferation of Smart IoT devices, ushering technological innovation as a modern connectivity era. However, this surge in connected devices also introduces a myriad of vulnerabilities, amplifying the threat landscape and presenting significant national security implications. 5G networks' faster data rates and reduced latency make it easier for IoT devices to be widely deployed in a variety of industries. This surge in connectivity, while promising enhanced efficiency and convenience, concurrently expands the attack surface, providing malicious actors with new vectors for exploitation. Nation-state adversaries can leverage these vulnerabilities for strategic purposes, targeting critical infrastructure and compromising national security.

One of the primary technical challenges arises from the massive number of IoT devices connected through 5G networks. These devices, ranging from smart city infrastructure to healthcare systems, often operate with limited computational resources and may lack robust security features. Nation-state actors could exploit these limitations, targeting vulnerabilities in IoT devices to gain unauthorized access or manipulate sensitive data. For example, compromising 5G-connected medical IoT devices could enable adversaries to access patient health records or disrupt healthcare services, posing a direct threat to national security. The use of lightweight communication protocols in many IoT devices, optimized for low power consumption and efficient data transmission, introduces another layer of vulnerability. These protocols may lack robust security measures, making them susceptible to interception and manipulation by adversaries. Nation-state threats could exploit these insecure communication channels to eavesdrop on sensitive information or inject malicious commands into IoT networks. For instance, a targeted attack on smart grid infrastructure through compromised IoT communication could lead to disruptions in power distribution, impacting critical services nationwide.

The integration of edge computing in 5G networks further exacerbates IoT vulnerabilities. Edge devices, responsible for processing data closer to the source, may operate in uncontrolled or remote environments, making them susceptible to physical tampering or compromise. Nation-state actors could exploit these edge devices to launch attacks on IoT ecosystems, manipulating data flows or disrupting essential services. Moreover, the interconnected nature of IoT devices in the 5G landscape poses challenges for securing critical sectors such as transportation and energy. Adversaries may orchestrate sophisticated attacks, targeting interconnected systems to create cascading effects. For instance, compromising IoT devices within a smart transportation network could lead to traffic disruptions, impacting both public safety and national security. The rollout of 5G in India significantly increases IoT vulnerabilities, presenting complex challenges for national security.

A multifaceted strategy is needed to address these technological difficulties, including putting strong security mechanisms in place for IoT devices, implementing secure communication protocols, and improving monitoring capabilities to identify and address possible threats. Proactive steps are essential to strengthen India's resilience against changing cybersecurity threats in the Internet of Things (IoT) space as the country embraces the advantages of 5G. An attack on connected devices within critical sectors, such as healthcare or defense, could have cascading effects. For example, a compromised IoT device within a defense network could lead to unauthorized access, potentially compromising sensitive military information or disrupting communication channels crucial for national defense.

The rollout of 5G technology in India, while marking a significant leap in connectivity, introduces a heightened risk of economic espionage and intellectual property (IP) theft. The advanced capabilities of 5G networks create an environment conducive to sophisticated cyber operations, with economic implications that extend to intellectual property-driven sectors. This chapter delves into the technical intricacies of how the 5G rollout elevates the threat landscape for economic espionage and IP theft in India. The high-speed and low-latency features of 5G networks empower threat actors to execute economic espionage campaigns with increased efficiency. Nation-states and cybercriminal organizations may exploit these capabilities to target research and development (R&D) facilities, manufacturing units, and intellectual property repositories. For instance, a state-sponsored group could leverage the reduced latency of 5G to quickly exfiltrate proprietary data from a targeted organization, facilitating the unauthorized acquisition of trade secrets or cutting-edge innovations.

The proliferation of connected devices within the 5G ecosystem, driven by the Internet of Things (IoT), provides adversaries with an expanded attack surface for economic espionage. Infiltrating IoT devices embedded in manufacturing processes or R&D laboratories allows threat actors to gather granular data on product development, manufacturing techniques, and proprietary algorithms. This, in turn, amplifies the risk of intellectual property theft with potential consequences for India's economic competitiveness on the global stage. The dynamic spectrum access capability of 5G networks adds another layer of complexity to economic espionage tactics. Threat actors may exploit the dynamic allocation of frequency bands to avoid detection, making it challenging for traditional frequency-based monitoring systems to identify and track malicious activities. By dynamically switching frequencies during data exfiltration, adversaries can further obfuscate their operations, evading detection mechanisms designed to identify patterns associated with economic espionage.

Moreover, the integration of artificial intelligence (AI) in conjunction with 5G technology enhances the sophistication of economic espionage tactics. Machine learning algorithms can analyze vast datasets, identifying valuable patterns and insights for threat actors seeking intellectual property. Adversaries can employ AI-driven attacks to automate the identification of lucrative targets and adapt their strategies in real-time, making it more challenging for traditional cybersecurity measures to detect and mitigate economic espionage campaigns. The rollout of 5G in India elevates the risk of economic espionage and intellectual property theft, leveraging the advanced capabilities of high-speed, low-latency networks. Addressing these technical challenges requires a holistic approach, encompassing enhanced threat detection mechanisms, robust encryption protocols, and collaboration between government and private sectors to safeguard intellectual property and preserve the nation's economic interests. As India embraces the transformative potential of 5G, securing intellectual property becomes imperative for fostering innovation and sustaining economic growth.

The high-speed data transmission capabilities of 5G open avenues for economic espionage, with adversaries seeking to steal intellectual property and proprietary information. In the context of India's burgeoning technology sector, this threat is particularly pronounced. A cyber actor with malicious intent could exploit 5G networks to infiltrate research and development facilities, extracting valuable intellectual property and compromising the competitive edge of Indian industries on the global stage.

In a notable incident, a state-sponsored cyber group targeted India's financial sector through a 5G-enabled DDoS attack. The attack, leveraging the high-speed capabilities of 5G, overwhelmed banking systems, causing disruptions in online transactions and financial services. Cybercriminal groups exploited vulnerabilities in 5G-connected medical devices, leading to a significant healthcare data breach. Personal health records of individuals were compromised, raising concerns about the security of healthcare infrastructure in the 5G era. India faced a cyber-espionage campaign orchestrated by a nation-state actor exploiting 5G networks. The adversary, leveraging the low-latency features of 5G, targeted government communications and sensitive intelligence data, highlighting the need for enhanced cybersecurity measures in the face of geopolitical threats.

5G is characterized by a notable increase in data transfer rates, with peak speeds of up to 20 gigabits per second (Gbps). High-definition video streaming, ultra-low latency gaming, and real-time communication without lag or buffering are all made possible by this lightning-fast speed. Furthermore, as opposed to earlier generations, 5G networks have decreased latency, with a 1 millisecond (ms) delay between data packet transmission and reception. For applications like remote surgery and driverless cars that need to make snap decisions, this almost immediate response is essential.

Massive connection, made possible by cutting-edge antenna technology like massive MIMO (Multiple Input, Multiple Output) and beamforming, is another unique feature of 5G. Thanks to these methods, 5G base stations can accommodate far more connected devices at once, which makes it perfect for IoT deployments and densely populated metropolitan areas. More than a million devices are predicted to be linked per square kilometer with 5G, opening the door for a wide range of connected devices, from wearables and smartphones to smart appliances and industrial sensors.

Moreover, 5G networks come with built-in capabilities for edge computing and network slicing, making them extremely dependable and robust. With network slicing, operators may divide their infrastructure into isolated, virtualized networks that are suited to certain use cases. For example, enhanced mobile broadband (eMBB) can be used for high-speed data services as illustrated in Figure 4, while ultra-reliable low-latency communications (URLLC) is ideal for mission-critical applications.

In the meanwhile, edge computing reduces latency and shifts processing duties from centralized data centers to dispersed edge nodes by bringing computational resources closer to the end users.

Figure 4. 5G connected community

When it comes to deployment, 5G makes use of a variety of frequency bands, such as millimeter-wave (mmWave) and sub-6 GHz spectrum. Sub-6 GHz bands are appropriate for urban and suburban settings because they offer wider coverage and improved penetration through obstructions. Conversely, mmWave frequencies provide lightning-fast speeds but are sensitive to signal attenuation from external causes and have a restricted range. To attain the best possible coverage and capacity, 5G networks use a heterogeneous network (HetNet) design, utilizing both frequency bands.

Many cybersecurity issues arise because of India implementing 5G technology, aggravating pre-existing weaknesses and posing further risks to the country's digital infrastructure. One major worry is the increased attack surface that comes with the growth of IoT devices and the connectivity that 5G networks provide. With billions of IoT devices anticipated to be installed across a range of industries, including smart cities, healthcare, and transportation, every device is a potential point of entry for cybercriminals. For instance, there are serious safety and security issues if a hacked IoT device in a smart city's traffic control system causes delays in traffic or even accidents. In 5G networks, a greater dependence on edge and cloud computing creates new vulnerabilities. Edge computing decentralizes data processing and storage by bringing computational resources closer to the end users.

However, it also decentralizes security mechanisms, which puts private data at risk of manipulation or unauthorized access. Healthcare practitioners in India are utilizing edge computing to evaluate patient data obtained from wearable health monitors, as demonstrated by this case study. Patient confidentiality may be jeopardized if the edge computing infrastructure is breached, which might result in privacy violations and even medical fraud.

Furthermore, new cybersecurity threats are introduced by the integration of machine learning (ML) and artificial intelligence (AI) algorithms in 5G networks, especially when it comes to autonomous systems and intelligent decision-making processes. AI algorithms are the subject of adversarial assaults that have the potential to disrupt or even be harmful by influencing decision-making processes in vital systems like smart grid management or autonomous cars. An adversarial picture might be exploited, for example, by a hostile actor to trick AI-powered public safety surveillance systems into granting illegal access or getting over security measures.

6. COMBATING CYBER THREATS

Aside from the work being finished by NCSC and NATGRID for definition of a solid incorporated Cybersecurity procedure, another massively basic yet overlooked idea is the utilization of Human Firewall to battle Cyber Threats. Digital dangers are a noteworthy peril to automobile businesses. Programmers are driven by apparent chance, and they see free organizations, vendors notwithstanding, as potential casualties of information ruptures and burglary. This raised degree of hazard implies security should likewise increase, yet the genuine demonstration of improving guards may demonstrate troublesome. Acquiring new firewalls or refreshed enemy of malware frameworks can just go up until now. The most widely recognized and conceivably harming assault types utilized today include social building and mental stunts to sneak unsafe substance to workers. These phishing assaults transform human mistakes into a noteworthy risk. To quit phishing, vendors need laborers who are prepared, taught and bolstered. With dangers changing and developing after some time, preparing, and planning must be continuous. This consistent advancement frames a human firewall.

Social designing assaults are frequently offenders' first decision, for a straightforward explanation: They work. As indicated by Cybersecurity Ventures inquire about, more than 90 percent of fruitful ruptures start with phishing messages. End Users and staff who are not set up to perceive and screen out phishing assaults empower this pathway for assailants, enabling them to get around protections. At the point when a specialist clicks a connection in a phishing email, enters individual data into a speculated site, or downloads risky documents, an assailant increases direct access to inward information. Further adding to the peril of phishing, programmers are winding up better at building these assaults, progressively utilizing accuracy strategies. In the past times, phishing messages impacted a great many beneficiaries. These rough messages, loaded with incorrectly spelled words and suspicious-looking documents, got periodic snaps, yet would not trick most PC educated representatives today. Lamentably, for organizations, phishing has changed from that point forward. Cybersecurity Ventures expressed 91 percent of "complex" cybercrime today starts with lance phishing. Lawbreakers mask their messages to appear as though they originate from accomplice associations or colleagues, making their plans harder to distinguish.

As of late, much increasingly complex ways to deal with information burglary have developed. For example, a year ago digital assailants started a modern, across the board phishing plan in which clients were approached to give email get to authorization to a Google Docs application. The application was a phony, and really traded off their information. Workers without appropriate preparation might be caught off guard for such abnormal state assaults. At the point when information breaks happen, associations are on the snare for many cash. This gives an unmistakable primary concern driving force for organizations to improve their guards in any capacity essential. The raised expense of an information break does not simply originate from a solitary factor. Organizations that endure these assaults need to fix the harm to frameworks, pay administrative fines if their barriers were not up to norms, advise exploited people whose data was lost, endure decreased business due to declining client trust and then some. The Ponemon Institute's 2017 expense of information rupture review found the normal cost for a broke association was $3.6 million. The definite figure will change generally relying upon the size of the occurrence – Ponemon put the expense of each lost record at $141.

The accompanying straightforward principles ought to be seen to relieve Cyber Threats.

- Ideally have a different PC or gadget for work and home; you ought to have separate records for your work and individual purposes. While this procedure will not offer all out security, it offers some additional affirmation. Regardless, it ensures your gadget against vulnerabilities that are generally normal.
- Do not impart your cell phone to other people. Since you cannot set separate passwords on your cell phone, like you can when signing into PCs, it's best not to impart your gadget to anybody.
- Be suspicious of any suspicious email paying little mind to the way that they have every one of the reserves of being from someone you know. An essential reliable rule is to be wary with the one you haven't the faintest idea about the person who is sending you an email, be amazingly careful about opening the

email and any archive added to it. Should you get a suspicious email, the best activity is to eradicate the entire message.

- Never send your Credit card number, Aadhar number, financial balance number, driver's permit number, in an email, which is generally not verified. Look at email as a paper postcard - people can perceive what has formed on it if they need to. Be suspicious of any association that demands this sort of information in an email or content.
- Choose solid passwords with letters, numbers, and unique characters to take a psychological picture or an abbreviation that is simple for you to recollect. Make an alternate secret key for each significant record and change passwords customary.
- Keep transforming you perusing propensity and strategies without fail. Secure web perusing is a round of evolving strategies.
- When shopping on the web stay away from sites with a scandal history. These are safe houses for malevolent and irritating interlopers like spyware.
- Be educated regarding the security strategies of the site with which you are perusing. Peruse site security approaches. They ought to clarify what is being gathered, how the data is being utilized, regardless of whether it is given to outsiders and what safety efforts the organization takes to ensure your data. The security approach ought to likewise disclose to you whether you reserve an option to perceive what data the site has about you.
- Never close the program without logging out from any record be it email, IM. Online networking, internet business.

7. FUTURE THREAT LANDSCAPE

The future Threat Landscape looks unnerving yet there are promising choices to protect the country and the occupants.

- Privacy and individual information insurance will be one of the key center zones in 2019: With the draft Personal Data Protection Bill and the Aadhar governing by the Supreme Court as of late constraining the utilization of information, the emphasis on information security is set to arrive at a tipping point in 2019. Associations will put resources into adjusting their framework to the prerequisites in the Personal Data Protection Bill (which is probably going to turn into an Act later in 2019) to pick up business edge and maintain a strategic distance from punishments.

- Machine learning (ML)/Artificial Intelligence (AI) in Cyber security will develop and wind-up basic piece of the security suite: ML/AI-empowered arrangements will be in extraordinary interest in 2019 as associations will concentrate on peculiarity discovery instead of guideline-based identification and reaction. We have just observed ML's application in endpoint security space. 2019 will see ML/AI being utilized in the core of verifying systems, for example indispensable part of SoC (Security Operation focuses).
- IoT security will rise as a concentration in 2019: Organizations and government will keep on grasping IoT-empowered answers for accomplish robotization and productivity particularly for foundation and keen urban areas. While this occurs, we should see expanded spotlight crosswise over partners on verifying IoT framework.
- Organizations will have restored center around cloud security in 2019: As cloud selection keeps on developing in 2019, assaults on cloud's shared security model will mount complex. Cloud suppliers should place in more assets to secure foundation. In addition, there will be endeavors to reinforce understanding about how to restrain access to information put away in cloud and let just approved faculty get to it. To address vulnerabilities and misconfigurations, associations will embrace advancements like CASB (cloud get to security merchant) which accompany extra security controls.
- Integrated way to deal with overseeing insider dangers: As we see upgraded, center around the repercussions of information rupture is expanding fundamentally, as needs be there will be a lot bigger spotlight on overseeing insider dangers. 2019 will see associations cross utilizing or moving towards an incorporated information spillage counteractive action suite, which will install client and element conduct investigation (UEBA) and cloud get to security dealer (CASB).
- The paradigm of security testing will move left: Security testing will be installed in the advancement cycle, as speed to market will end up key in the computerized time. We will progressively observe security groups winding up some portion of the application improvement lifecycle.
- Blockchain will progress toward becoming standard to counteract extortion and information robbery: For associations living under the dread of misrepresentation and data fraud in the computerized economy, Blockchain will demonstrate a positive viewpoint. While Blockchain is relied upon to anticipate extortion by assuming a key job in overseeing personalities, it might stay as a strong idea and in play inside the startup circles. In 2019, we do not see huge Blockchain based Identity as a Service (IDaaS) being boundless in associations.

India's digital security reaction is a long way from the best on the planet. With digital assaults presently taking the state of digital undercover work and digital fighting, India has a lack of experts prepared to handle the circumstance successfully. The numbers continue rising each year, and it is currently during the many thousands. Indeed, even the Internet of Things (IoT) endpoints are changing the idea of digital dangers.

CONCLUSION

The absence of an organized digital reaction framework needs coordinated co-operation by all partners and network, and this is not going on. Any conditions or occasion that can possibly hurt a framework or arrange, and that even the presence of an (obscure) defenselessness suggests a risk by definition (CERT). India and every one of the nation's worldwide have a developing reliance on an inexorably powerless digital space, noteworthy in potential national hazard. This dependence on the internet suffuses all components of government, industry, and society; subsequently, the requirement for a national digital empowered undertaking with carefully empowered systems, frameworks, administrations, and business attempts is the need of great importance to national intrigue. India's digital security scene is experiencing a fascinating stage as organizations are distinctly taking a gander at imaginative instruments to shield themselves from digital assaults and dangers. While India's digital security needs are not the same as that of the remainder of the world, there are hosts of regions, which require interesting methodology. Remembering India's business scene and her requirements for digital security instruments and arrangements, we have focused in on seven digital security patterns for the Indian market. As India strides into the 5G era, the cybersecurity landscape undergoes a profound transformation. Balancing the potential benefits of 5G with the imperative of safeguarding national security requires a comprehensive and collaborative approach. This chapter aims to underscore the critical importance of preemptive cybersecurity measures, adaptive defense strategies, and international cooperation to navigate the emerging threats in India's 5G journey.

REFERENCES

5G Network Slicing, What is it? (2019). Viavi Solutions. Www.viavisolutions.com. https://www.viavisolutions.com/en-us/5g-network-slicing

Qualcomm. (2017). *Everything You Need to Know About 5G*. Qualcomm. https://www.qualcomm.com/5g/what-is-5g

Reply. (2024). *Low Latency: what makes 5G different*. Reply. https://www.reply.com/en/telco-and-media/low-latency-what-makes-5g-different

What Are 5G Speeds? (2024). Cisco. https://www.cisco.com/c/en/us/solutions/what-is-5g/what-are-5g-speeds.html

Chapter 3
Research Review on Performance Evaluation of Fifth Generation (5G) Technologies and Protocols

Kavitha Veluchamy
Sethu Institute of Technology, India

Rehash Rushmi Pavitra A.
SRM Institute of Science and Technology, India

M. Isaivani
https://orcid.org/0000-0002-1918-2145
Vaigai College of Engineering, India

Josep M. Guerrero
Technical University of Catalonia, Spain

ABSTRACT

A revolutionary advance in wireless communication technology has been illustrated through the development of fifth generation (5G) network architecture and standards. In addition, 5G is intended to provide unequalled speed, minimal latency, and widespread device connection in contrast to its predecessors. The user equipment (UE) core network (CN) and radio access network (RAN) are the three main parts of the architecture. By using technologies like beamforming and Massive MIMO the RAN maximizes spectral efficiency. Deploying an open-minded, cloud-based design

DOI: 10.4018/979-8-3693-2786-9.ch003

that utilizes software-defined networking (SDN) and network function virtualization (NFV) to improve scalability and efficiency of resources implies a significant shift in the CN. Further, new radio (NR) protocol supports a variety of applications thereby high data rates is crucial to the development of 5G. Ultimately the requirements of (MMTC), ultra-reliable low-latency communication (URLLC) and enhanced mobile broadband (EMBB) are all addressed by this advancement.

1. INTRODUCTION

The arrival of Fifth Generation (5G) technologies designated a significant leap inside the evolution of wireless communication systems. This revolutionary advancement offers to transform how individuals participate, connect, and share information with the world of digital media, (Dangi et al., 2020). The inherent support of 5G continues to roll out globally, it becomes imperative to comprehensively evaluate its performance across various critical parameters. This book chapter intends to examine in depth the specifics of 5G performance assessment technologies and protocols, shedding light on its capabilities, limitations, and potential impact across diverse applications and industries, (Esmaeily & Kralevska, 2021).

Key aspects of 5G performance evaluation include data rates, latency, reliability, energy efficiency, network coverage, security, and spectrum efficiency. These metrics form the foundation for understanding the practical implications of 5G technologies in real-world scenarios, (Kaltenberger, et al., 2020). By meticulously to inspecting these factors, Researchers may find out more about the prospective advantages and constraints of 5G networks, paving the way for optimized deployment and utilization in varied environments.

Further proposed chapter attempts to figure out thorough summary of the distinct mechanisms, tools and techniques which employs in the performance evaluation of 5G technologies. Additionally, it aims to highlight the significance of these evaluations in shaping the future of communication systems, paving the way for enhanced connectivity, seamless integration of Internet of Things (IoT) devices, Ultra Reliable Low Latency Communication (URLLC) and various revolutionary uses such as distinct autonomous vehicles, intelligent cities, and immersive multimedia experiences, (Gupta & Jha, 2015).

By scrutinizing the functionalities of 5G technologies and various protocols the proposed research aims to contribute to the body of knowledge essential for harnessing the full potential of 5G, thereby enabling informed decision-making for telecommunications stakeholders, policymakers, and technology enthusiasts, (Afolabi et al., 2018). This chapter serves as a stepping stone towards unravelling the intricacies of 5G performance evaluation, ultimately fostering a deeper understanding of its impact on the digital landscape.

Even though research and development on the technical aspects and possible applications of 5G are still in progress, it is anticipated that the majority of 5G applications will centre around business and industry. With the ability to offer completely automated facilities for any kind of businesses and production, 5G is anticipated to be a major breakthrough over the present automation sector. It will also make new business models and industrial digital machinery possible. It will affect a range of sectors, including large and small enterprises as well as medium-sized organizations, (Ashraf et al., 2016).

1.1 Key Features into History of 5G

Five-generation (5G) cellular network technologies are the latest advancements in mobile network technology. NASA Corporation, via M2MI, was an innovator in the establishment of 5G. To satisfy the increasing demand for mobile communications, they decided in the beginning of 2008 to design a future generation networking and telecommunications infrastructure for Internet Protocol (IP) and diverse associated services. Afterwards in 2012, a European Commission (EC) took other noteworthy steps in the advancement of 5G. EC initiated the founded Mobile and Wireless transmission combiners for the Twenty-First Century Information Society (METIS) programme, (Mendonça et al., 2022). The idea of the METIS 5G system was created to extend current wireless communication systems to new use scenarios and to satisfy the expectations of the linked information society that will exist after 2020. Establishing the advancement for 5G, the future generation of mobile and wireless communications technology became the project primary objective. The functional architecture of METIS which made available as a starting point which is used to identify and design new 5G Network File System (NFS).

In order to showcase several 5G technology concepts, including 5G-LTE simultaneous interconnection and 5G networks multiple-point connection with disseminated MIMO, Sony and its collaborators in the industry built a 5G broadcast test environment in 2015 and 2016. Some of the most significant 5G deployments from 2017 to 2019 include a number of 5G novel radio investigations, the first publicly accessible 5G live network utilization applications imposed in the continent of Europe, and toward the end of 2018, Sony and other officially announced commercial 5G. The first country to implement 5G is South Korea, (Abdullah et al., 2021).

1.2 Conception of 5G

5G is seen as a new mobile network revolution. This technology, which is made up of high-value components, offered a unique experience by providing all the most cutting-edge characteristics, including high bandwidth, exceptional data capacities, massive data broadcasting and affordable services, (Sanyal & Samanta, 2021). The technology behind 5G intends to offer the next generation of Internet protocol enabled services along with related applications by creating networking and telecommunications technologies. The primary goal is to satisfy the requirements of the post-2020 networked information environment and to add new use cases to the capabilities of the existing wireless communication networks. However, a digital revolution cantered on the linking of everything that might benefit from connection is anticipated with the development of 5G, (Bondre, Sharma, & Bondre, 2023).

2. REVIEW OF LITERATURE

An approach to reduce interference in 5G HetNets has been proposed by Mayada Osama et al. in 2021. Substantial transmission bandwidth can be obtained through heterogeneous networks which in turn reduce power consumption. HetNets use a technique based on interference contribution rate (ICR). Reductions in interference, traffic loss and power consumption have been achieved using the research's recommended technique, which switches the small cells on and off depending on their interference contribution rate (ICR) characteristics. A comparison of two different forms for the tiny cells' core zones was also included in the investigation. The system data rate was improved, and power efficiency increased when the recommended solution was carried out based on the outcomes, (Osama, El Ramly, & Abdelhamid, 2021).

An overview of the 5G interference management challenges was presented by Faizan Qamar et al. in 2019. In addition to discussing current challenges, this chapter includes research on how these difficulties are handled. Several research challenges were additionally discussed, including those related to the Third Generation Partnership Project (3GPP), issues with distributed networks, coordinated multi-point, synchronized scheduling, system-to-system and cell-to-cell interference. An overview of the 3GPP's, various methods for reducing interferences was also included in the article. These methods included portable backhauling, Base Station (BS) recognition, order calculations, receiver layout, standard channel positioning next to disruption cells, and control channel authentication, (Qamar et al., 2019).

The research conducted in 2021 by Cetin and Pratas examined Device to Device (D2D) communication which is essential to cellular networks. Subsequently, the study focuses on the delay-aware utilization of resources issue associated with maximal overall throughput. The transmission pairs and the D2D communication pair scheduling scheme are the important deciding variables to raise the sum-rate subjected to delay limitations. They additionally look at the trade-off between throughput and delay. Research has shown that as the number of allowed D2D pairs on a sub-channel increase, the optimum delays of D2D devices decrease.

The standard function analysis of FD and HD relay supported CRNOMA communication was reported in a research paper by Do et al. (2020). Moreover, with an intrinsic support of derivations the statistical algorithms for exponential likelihood of outages and extremal volume have been made available. As per the results, there has been absolutely no change in the breakdown probability of the FD relay-assisted CR-NOMA connection. The analytical statements are anticipated to be beneficial with the real-world application of these connections.

The challenges imposed by adjacent channel interference (ACIR) have been addressed by Hyun Ki Kim et al. (2018). Further the main objective of the research paper was to identify how to prevent contamination in Bottom station satellite communications by using the Monte Carlo approach (MC) and minimum coupling loss (MCL). In accordance with the aforementioned paper, 5G networks which employ these types of technologies could become better.

Consequently, Huang et al. (2021) has examined the significant variance in implementation for three different RS schemes including SRS, TRRS and ORRPS respectively. It additionally looks into whether AN affect the security and reliability of the recommended HSTSs. In comparison, the SRS system is more reliable than the TRRS method. Among the three RS systems, the ORRPS scheme is proven to be trustworthy. System security is enhanced by AN technology, and optimum dependability is provided by ILS.

In general, reduction of cross-link interference in 5G wireless networks, the dynamic time division duplex approach was studied by Hanning Wang et al. (2018) in order to reduce interconnect congestion in 5G wireless networks. The enhanced interference rejection and cancellation of limited mean square error (MMSEIRC) recipient which has been claimed to be the method of managing interference, offers an alternative. To support the recommended strategy, a theoretical analysis has been encompassed. To continue further enhance the proposed system even more, the researchers also suggested an interference measuring approach. Lastly, system-level modeling findings verified the efficiency of the enhanced reception.

Even with the different security protocols that 5G networks are now putting in place, there are still issues with the lack of compelling ways to manage large numbers of connections. We think there may be need to build security awareness guidelines that include a range of data transfer and storage-related issues, (Liyanage et al., 2018).

Sullivan et al. (2021) present an unbiased analysis of 5G security concerns along with current technology meant to protect the 5G ecosystem and, classify privacy technologies via OSI layers.

Scalise et al. (2024) provide an extensive overview of the constantly changing 5G system with a structured emphasis on the three key aspects including RAN, UE & CN, and also explore the application of 5G technology in ultra-private, time-sensitive, non-trustworthy communications platforms.

Mangla et al. (2023) presented an in-depth analysis of 5G technology along with a statement of the security risks involved and substantially corroborate quantum computing as a potential solution to conventional security measures.

Furthermore, it is thought that huge MIMO is a possible method of stopping passive eavesdropping, (Zhang, Xie, & Yang, 2015). The fact that it may be used for a number of things, including monitoring, raises privacy concerns. This enables a closer relationship to be established between the development of 5G networks and the services that are linked with them. For example, if a large network goes down, there may be simultaneous interruptions to TV, internet access, and fixed telephone lines. Security automation is necessary to guarantee the robustness of the 5G system against security breaches. With more devices and things connected to each other in the 5G network, security concerns have come to prominence.

The intricacy of the network increases the visibility of the number of attacks and vulnerabilities. For this reason, security automation solutions must be designed to protect the different network levels, (Yazdinejad et al., 2019). By examining and documenting the different security concerns that impact information accessibility and network integrity, this initiative seeks to advance 5G wireless security research and development. Our research will also enable us to pinpoint fresh avenues for the sector.

3. SPECIFIC INSIGHTS INTO FIFTH GENERATION

At this point is an appropriate period to transition from a services model to a multiservice approach. Ubiquitous networks, which will provide the seamless connection between many wirelessly accessible technologies at the same time as they migrate from Wimax to LTE networks Advanced are set to be incorporated in this competency.

3.1 Interdependent Relay

A great deal of data is obtained below using this strategy across an extensive cell coverage. Through the use of intelligent radio technology, user equipment may choose the most suitable Radio Access Network, encryption arrangement, and other configuration options based on the radio landscape of its location. This leads to optimal performance and the best possible connection. Intelligent antennae will be redirected to provide the user with a better connection.

In addition, 5G technology will capitalize on wireless and optical technologies' capabilities. The foundation of 5G infrastructure will include Global Content Delivery Networks (CDN), Internet of Everything (IoE), Internet of Things (IoT), and Software-Defined Networking (SDN) respectively, (Pavitra, Lawrence, & Maheswari, 2023).

Individuals will eventually be able to access the wide-ranging Wireless Web (WWW) and communicate wirelessly everywhere around the globe with the arrival of 5G (Fifth Generation Mobile and Wireless Networks). After 4G/IMT-Advanced standards, 5G represents the subsequent major development in mobile telecommunications standards. At present, the word "5G" is not formally associated with any specific specification or readily accessible document via telecommunication corporations or standardization bodies such as ITU-R or the Global Positioning System and WiMAX Forum. subsequent features and application areas will be added with each subsequent iteration, and system performance will noticeably improve as well. Additional applications that benefit from a mobile connection include intelligent public transportation, electronic books, automation for homes, and surveillance.

Standards for wireless broadband have been officially recognized as IEEE 802.16 after being endorsed by the Institute of Electrical and Electronics Engineers (IEEE). Under the trade name WiMAX, which stands for Worldwide Interoperability for Radiofrequency Access, it is marketed by the WiMAX Forum business group. Wi-Fi local loop radio interfaces and related features are standardized under IEEE 802.16.

With the introduction of 5G wireless technology, users' utilization of smartphones with exceptionally high throughput has evolved. An expensive piece of technology that the user has never used before. 5G mobile technologies many cutting-edge features will make it very desirable in the near future. To keep young toddlers entertained, Pico nets and Bluetooth technologies have been brought to market. Additionally, customers may link their laptops to 5G cell phones in order to receive broadband internet access. 5G technology provides an additional camera. A phone that can do a lot of things, such play videos, record MP3, make quick calls, have an audio player, and much more. A user terminal, which is a crucial part of the new design, and a number of autonomous wireless access technologies (RAT) make up the fifth generation of network architecture. An all-IP based architecture underpins the

5G mobile system's wireless and mobile network compatibility. Each radio access technology inside a terminal is addressed by the internet protocol channel that links it to the networking environment outside.

3.2 Progress Innovation of 5G

Prioritizing the bandwidth availability above other factors which has been discussing by the LTE mechanism. Beyond that, however, the research concentrated on providing ubiquitous connectivity, which would enable individuals to swiftly and readily access the internet from any location—be it above, below, or between the sea and the sky. In spite of this, the Internet of Things will use the Machine Type Communications (MTC) version of the LTE standard. 5G technologies are designed to be compatible with devices similar to MTC. Nearly all of the newest technologies need the prior versions. LTE networks, Volte-A, wireless internet, M2M communication, and more technologies would comprise 5G, just as 3G did. A wide range of applications, including the Internet of Things (IoT), wearables that are linked, virtual reality for high-definition streaming, and massively multiplayer online games, are anticipated to be made possible by 5G's architecture.

Furthermore, 5G technology can be able to maintain a large variety of traffic kinds and the abundance of connected devices. For instance, 5G approach enhance significantly faster connection and cheaper data rates for sensor networks, making it ideal for streaming high-definition video. New designs for Wireless Access Infrastructure (RAN) that include storage in the cloud RAN and virtualization RAN classifications will be advantageous for 5G networks which will help to optimize server fields via centralized data servers at the network corners and build a more centralized network.

According to 5G's application of cognitive radio techniques, the infrastructure would be able to independently identify the kind of channel being supplied, discriminate between moveable and movable things, as well as adjust to the circumstances in the time limit. Stated differently, social networking apps and the industrial network might both be supported concurrently by 5G networks.

4. KEY FEATURES OF 5G ARCHITECTURE

A thousand-fold increase in traffic volume and a hundred-fold increase in user data rate should be the main areas of attention. Numerous technologies are capable of efficiently handling both the rapid increase in traffic and the rate at which consumers download data. However, we are going to concentrate on three specific technologies that are able to achieve such a high ratio. Physical Layer (PHY) technologies include Large Multiplex Input and Output (MIMO), Filter Bank Multi-Carrier (the FBMC technique), Non-Orthogonal Multiple Acquisition (NOMA), and other corresponding technologies. The main focus is on enhancing spectrum efficiency in order to boost network capacity. Moreover, harnessing the untapped spectrum in the millimeter wave frequency (mmWave) has the potential to significantly augment network capacity.

Network expansion is the primary element that influences the capacity of a wireless communication system. The use of Universal Domain Networks (UDN) is believed to result in a proportional growth in network capacity as the number of cells increases. A HetNet, or heterogeneous network, is formed by combining low power eNBs (such as micro eNBs and pico eNBs) with macro ENode Bs (eNBs), in response to the increased network density. Furthermore, Device-to-Device (D2D) Communication, which serves as an alternative to HetNet, has the potential to enhance spectrum efficiency and maximize peak data throughput. Through dispersing the responsibility among multiple states Radio Access Technology (RAT) systems, enhancing the effectiveness of the network resource might lead to a rise in network capacity.

The use of centralized coordination should be the key component of the 5G architecture that is implemented initially. On the basis of the centralized processing, the performance of the network may be further improved by the use of collaborative resource management and coordination across several cells and RAT systems. In addition, users that move quickly will be subjected to frequent handovers because of the consumers' flexibility and the limited cell's constrained operating region. It is necessary to manage several small cells at a central location in order to provide the desired effect of seamless movement. It is abundantly clear that the Radio Access Network (RAN) of the 5G mobile network will need centralized management and coordination to function properly. In addition, the management of the constantly growing traffic flow has to take into consideration the wireless Core Network (CN) at the same time.

Consequently, concentration of the control plane and separation of it from the data plane are both logically sound decisions. In order to further boost network capacity, it is obvious that centralizing processing (such as collaborative scheduling, integrated disturbance prevention, and other similar applications) and separating the control layer from the data layer are both vital components of the 5G wireless network.

4.1 Basic Architecture of 5G

In order to handle a new situation, changes have been made to the many terminals and network components that make up 5G's architecture, which is highly advanced. Through the use of cutting-edge technology, service providers may easily include value-added services into their offerings. The capability of devices to identify not only their specific location but also the weather, temperature, and other factors is dependent on cognitive radio technology, which has a number of key qualities. Upgradability, on the other hand, is dependent on this technology.

A completely IP-based system model for wireless and mobile networks, the 5G system model is shown in figure 1, which may be found under the following heading. Once the principal user terminal has been established, the system is comprised of a number of distinct and independent radio access technologies. The term "Internet Protocol" (IP) refers to the link that exists between the outside world of the internet and any radio technology. In order to ensure that there is sufficient control data available for the correct routing of IP packets associated with particular application connections, or sessions between clients and servers on the Internet, the Internet Protocol (IP) technology was developed with the explicit purpose of performing this function. In addition, in order to guarantee accessibility, the routing of packets has to be modified so that it is in accordance with the rules that the user has already established.

Figure 1. Fundamental architecture layout of 5G

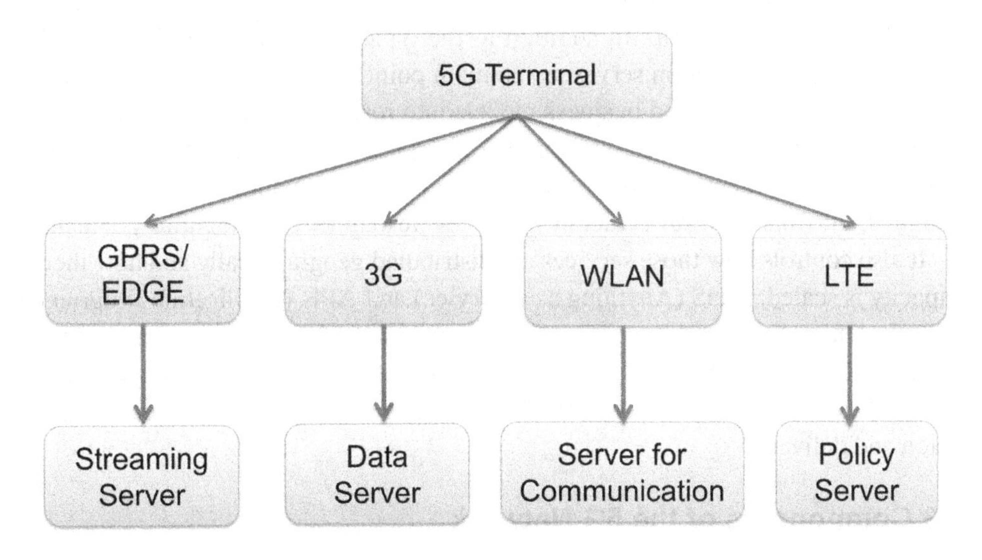

4.2 Detailed Description towards 5G Layers

Infrastructure Usage Layer: Composed of access nodes, as well storage nodes (which may be computational or memory capabilities), fifth-generation devices, connectivity nodes, and related linkages, the physical resources of a fixed-mobile convergent network are referred to as connectivity nodes. Through the use of virtualization concepts, the resources are open to the upper levels as well as the coordination entity.

Layer designated for commercial efficiency: A collection of configuration parameters for particular network components, such as radio access, as well as a library that contains all of the functionalities that are required for a converged network in the form of modular architecture building blocks. These fundamental components include features that are implemented by software modules that can be downloaded to the desired location from the repository. As required, the coordination component makes use of the proper application programming interfaces (APIs) to call the aforementioned features and functions.

Application Layer: Some services and apps that are provided by the operator, companies, industries, or other 5G network users. A key element of this system is an orchestration object that uniquely represents these three tiers. It can also manage a whole virtualized network in addition to the typical automation tools for OSS and SON. The organization serves as the main point of contact and facilitates the conversion of use cases and business models into real-world sectors and services. Once the network slices are defined for a certain application scenario, the applicable modular network functions are linked, the required performance requirements are assigned, and finally the infrastructure resources are mapped into the entirety of them.

It also controls how those services are distributed geographically and how their capacity is scaled. XaaS (Anything as a Service) and APIs (Application Programming Interfaces) are used in certain business models to make it easier for verticals and MVNOs (Mobile Virtual Network Operators) to develop and maintain network slices. Intelligence supported by data will optimize every facet of service composition and delivery.

4.3 Components of the 5G Network

The design and guiding principles outlined above have given rise to a set of fundamental components and terminology for a 5G system, which are elucidated below.

5G RAT: The 5G RAT collection refers to an ensemble of standardized 5G RATs that together adhere to NGMN 5G requirements. This family is an integral part of the whole 5G system. Comprehensive coverage is an essential element in the marketing of new technology, and the 5G RAT series is expected to offer that level of compensation.

5G Network Function: A 5G network function has the capability to enable communication within a 5G network. Although the majority of 5G network services are virtualized, the infrastructure may need additional specialized hardware to do certain tasks. The functions that make up the 5GFs are both RAT-specific and access-agnostic; they may be classified as either mandatory or optional. They also include features that make permanent access easier. Typical services that are needed for each use case include identity management and authentication. Like mobility, optional functions are ones that are not necessarily suitable for every use situation. Depending on the use case and kind of demand, they could also come in a number of varieties.

Infrastructure of 5G: Through the use of the 5GI, the 5GRs and 5GFs are realized. Transport networks, computer power, memory, radio frequency units, and diverge cables that enable network tasks and provide 5G network capabilities are all part of the 5G infrastructure which addresses both software and hardware framework for the 5G network.

5G Network: The purpose of a 5G network is to enable communication between 5G devices and the associated 5GI (including relaying devices), 5GFs, and 5GRs. Put another way, when a 5G RAT uses any part of the 5GFs features that are deployed on the 5GI to enable communication with a 5G target, the 5G network is realized. If 5GFs are used to enable communications between a 5G device and non-5G RA22T, then that system is not considered as 5G.

5G Device: Devices that link to 5G networks to receive communication services are referred to as 5G devices. Both human and machine users may be supported by 5G devices.

5. APPLICATION PLATFORMS OF 5G

The Open-Air Interface (OAI) platform includes a full software implementation of fourth-generation mobile cellular networks that adheres to 3GPP LTE specifications. This program is particularly made for the x86 architecture processors and is written in C under real-time Linux. At the physical layer, it provides the following features:

MATLAB simulations have been used to delineate the disparities in outcomes between WLAN (Wireless Local Area Network) and 5G networking.

5.1 Wireless Local Area Network (WLAN)

It has been widely acknowledged for a long time that wireless networking for computers is a vital component, and almost all new laptops come equipped with Internet access. Given the availability of WLAN solutions, Wi-Fi, which is also referred to as IEEE 802.11, has emerged as the de facto standard. Because systems that implement the IEEE 802.11 standards generally run at 54 Mbps, wireless networks have the potential to successfully compete with wired networks.

There is a widespread availability of Wi-Fi hotspots, which are often used inside the system to enhance its performance and flexibility. They let users to use their computers without the need for a connection while they are waiting in a variety of settings, including hotels, airport waiting rooms, restaurants, and many other places. A wire-free link is provided by these devices.

It is common practice to use 802.11 standards for transitory connections and transient WLAN applications; however, these standards may also be applied for installations that are for a longer period of time. Providing semi-permanent WLAN solutions in the workplace may be accomplished via the use of WLAN equipment. In this scenario, the use of wireless local area network (WLAN) technology makes it feasible to set up workplaces without the need of installing permanent cable, which may result in considerable cost savings. The use of wireless local area network (WLAN) technology makes it possible to execute changes in the workplace without the need for upgrading.

As a result of this, the IEEE 802.11 Wi-Fi standard is often used to offer WLAN solutions for short-term connections in urban hotspots such as cafes, airports, hotels, and other similar venues, as well as in official settings.

6. 5G NETWORKING

About a decade ago, research initiatives signaled the start of the journey towards 5G. The European Union project METIS, which ran from 2012 to 2014, set the foundation for 5G and sparked a number of international conferences and initiatives, including the 5G-PPP (Europe), 5G Forum (Korea), IMT-2020 Promotion Group (China), and 5GMF (Japan). These measures laid the foundation for the official 2016 launch of RANs (Radio Access Networks), the start of the 3GPP standard negotiations in 2015, and the completion of the essential building blocks in June 2018. Early in 2019 there were the first 5G deployments in the US and South Korea as a result of these efforts, (Yazdinejad et al., 2019). The first smartphones were available shortly after, with Korea having the greatest deployment, with 5 million users overall (or around 40% of the world's 5G penetration).

7. 5G VISION

7.1 Context of Business

Innovations in the technician, consumer, and technological contexts all of which are influenced by changes in the economy and society define the 5G business environment.

We anticipate a world in which everything is linked, and immediate information is always only a touch away.

7.2 Consumers

The introduction of smartphones and tablets is a notable example of contemporary technological innovation. Smartphones are predicted to continue to be the primary personal device and to improve in terms of functionality, but wearables and sensors will lead to a rise in the number of personal devices. Cloud technology will enable personal gadgets to expand their functionality to several uses, including the creation and distribution of high-quality video material, payment processing, identity verification, cloud gaming, mobile TV, and overall support for a smart lifestyle. They will play a big part in applications related to social life, health, safety, and security in addition to operating vehicles, appliances, and other machinery at home. A thorough understanding of what future customers will want is crucial to accommodate trends like multi-device and multi-access use by consumers.

7.3 Verticals

The mobilization and automation of industries and industrial processes is the next wave of mobile communication. The Internet of Things (IoT) and machine-type communication (MTC) are common terms for this. Embedded sensors and embedded connectivity will allow tens of billions of smart gadgets to respond on their local surroundings and apply sophisticated logic-based distant triggers. These gadgets have different needs in terms of price, power consumption, and functionality. A broad variety of networking needs, including those related to performance (latency, throughput), security, and dependability, will also be necessary for IoT. In addition to connections, the creation of new services for a variety of industries (including construction, medical care, public transit, and power) may need enablers from cloud computing, handling large amounts of data, protection, transportation, and other network-enabled capabilities.

7.4 Enterprises

Numerous consumer trends also apply to businesses in the future. The lines between gadget use for personal and business purposes will become hazier. Businesses will search for ways to resolve privacy and security issues brought on by this mixed-use model.

Mobility will be one of the primary forces behind higher productivity for businesses. Businesses will progressively make their specialized apps accessible on mobile devices in the next decades. Cloud-based service proliferation will open a lot of potential for businesses and allow applications to be portable across different platforms and domains. Businesses must also deal with the difficulties this presents, such as security, privacy, and performance.

7.5 Partnerships

Nowadays, operators have started to take use of their partnerships with over-the-top (OTT) businesses in a number of sectors to provide bundled services to final consumers. OTT firms will be in charge of delivering an expanding number of applications that need fewer interruptions, greater reliability, and more programmable and highly adaptable on-demand service-enhancing features (such as accessibility, location, quality of service, and identity).

7.6 Infrastructure

The design and administration of networks will alter as a result of recent breakthrough technological developments (such as NFV, SDN, big data and all other IPs). Due to these modifications, a very adaptable infrastructure that promotes faster innovation and more affordable network and service development will be possible. In addition to maintaining their current service development, operators will seek collaborations for both infrastructure and application development in order to broaden their commercial reach.

7.7 Services

The development of both new and existing services is part of the worldwide business model growth of mobile carriers' offerings. Direct-to-consumer communication and best-effort data services, including Web services, are now the most popular services offered by mobile carriers. These services will develop to become better in terms of both quality and functionality. High-quality IP multimedia and rich group communication will be standard components of personal communication.

In contrast, data services will be widely available, provide consistent performance, and be made possible by a variety of integrated access methods. Social media and video will account for the majority of data traffic volume. Public safety, automated industries, smart user environments, and mission-critical services are some of the emerging market categories that will see the emergence of new services. By using features like big data, proximity, geo-community services, and many more, many more services will be created.

8. RESEARCH SUMMARY

Metrics including throughput, latency, reliability, scalability, and energy efficiency must be evaluated in order to determine how well 5G protocols perform under different network circumstances and traffic loads. Network configuration optimization and protocol behavior analysis are frequent tasks that are performed using simulation tools like MATLAB, OPNET, and NS-3.

In summary, efficient protocol design and implementation are critical to fully use 5G networks for the delivery of trustworthy, high-performance, low-latency communication services, (Zhang, Xie, & Yang, 2015). Research efforts are still focused on enhancing protocol efficiency, scalability, and security in order to fulfill the evolving requirements of next-generation wireless communication systems. The path that next-generation wireless networks take will be greatly influenced by protocol advancements as 5G takes form.

In the telecommunications industry, 5G, or fifth generation wireless technology, is a game-changer. It has the potential to drastically change how we connect, communicate, and interact with technology since it provides blazing-fast accelerations, exceptionally low latency, widespread communication, and unparalleled dependability. Among some of 5G's vital characteristics are mentioned below:

High Data speeds: Multi-gigabit per second data speeds may be provided by 5G networks, allowing for smooth streaming, quick downloads, and rich multimedia experiences.

Low Latency: 5G opens up new opportunities in a variety of sectors by enabling real- time applications like virtual reality games, driverless cars, and remote surgery with latency reductions down to milliseconds, (Ahmad et al., 2018).

Massive Connectivity: Further, the Internet of Things (IoT) and intelligent cities, where billions of devices smoothly interact, are made possible by 5G's capacity for a far larger number of connected devices per square kilometer.

Reliability: 5G's increased availability and dependability make it appropriate for vital uses including mission-critical communications, emergency services, and industrial automation.

Network Slicing: This idea enables operators to design virtual networks with distinct service level agreements, security constraints, and performance characteristics based on use cases.

Advanced Technologies: To optimize network capacity, coverage, and spectral efficiency, 5G integrates state-of-the-art technologies including Massive MIMO, beamforming, and millimetre-wave spectrum usage.

Finally, 5G promises to alter almost every part of our life, marking a significant advancement in wireless communication technology. From manufacturing and healthcare to transportation and entertainment, its fast and low-latency connection will spur innovation in all sectors of the economy. All the same, in order to fully use 5G, strong infrastructure must be deployed together with significant thought given to spectrum allocation, security protocols, and regulatory frameworks. We are about to enter an age of unparalleled connection and digital revolution, and 5G's continued evolution and expansion will spur the creation of new apps, business models, and possibilities.

9. FUTURE RESEARCH DIRECTIONS

In order to provide a more thorough analysis of the security concerns linked to 5G security, the research paths mentioned in this literature review will be investigated. Researchers suggest that strengthening the security of 5G networks might aid in preventing misuse and illegal access. Future research may focus on a number of topics, including the creation of practical and efficient methods for detecting malware. Regretfully, none of the safety precautions for 5G networks have been thoroughly tested or put into use. In the development and use of these technologies, researchers often use a field-programmable gate array (FPGA). In order to protect the data being sent on a 5G network, the researchers of this study suggested an undercover method.

Services for Security - In order to satisfy the requirements of several industries, such as smart cities, healthcare, and transportation, the growth of 5G networks is expected to require the availability of trustworthy and secure safety technologies. It is suggested that a variety of security services be provided via a Software-as-a-Service paradigm in order to satisfy the needs of these enterprises. This kind of approach may be used to provide a range of security services, such as monitoring, identification, and penetration testing.

Artificial Intelligence Integration: It is essential to guarantee that the security rules are uniformly applied across all software-defined networking controllers, given that the technology is built upon them. Implementing artificial intelligence-based security measures is one of the most effective ways to prevent unwanted access and exploitation. Through its ability to carry out several functions, including authorization and authentication, artificial intelligence has the potential to enhance network security. However, methods connected to the identification of unusual activity are mostly implemented via the employment of security-based solutions. Advanced security solutions are required to counteract the growing number of hacking assaults facilitated by artificial intelligence.

Abbreviations

5G	Fifth Generation
UE	User Equipment
CN	Core Network
RAN	Radio Access Network
MIMO	Multiple Input Multiple Output
SDN	Software-Defined Networking
NFV	Network Function Virtualization
NR	New Radio
MMTC	Massive Machine-Type Communication
URLLC	Ultra-Reliable Low-Latency Communication
EMBB	Enhanced Mobile Broadband
EC	European Commission
METIS	Mobile and Wireless Communications Enablers for the Twenty-twenty Information Society
NFS	Network File System
HetNets	Heterogenous Networks
ICR	Interference Contribution Rate
3GPP	Third Generation Partnership Project
D2D	Device to Device
FD	Full Duplex
HD	Half Duplex
CR-NOMA	Cognitive Radio Non-Orthogonal Multiple Access
ACIR	Adjacent Channel Interference
MCL	Minimal Coupling Loss
MC	Monte Carlo
RS	Two Relay Selection
TRRS	Traditional Round-Robin Selection
SRS	Suboptimal Relay Selection
ORRPS	Optimal Relay-Receiver Pair Selection

continued on following page

Abbreviations

Continued

5G	Fifth Generation
ILS	Infrequent Light Shadowing
AN	Artificial Noise
MMSEIRC	Minimum Mean Square Error Interference Rejection and Cancellation
LTE	Long Term Evolution
CDN	Content Delivery Networks
IoT	Internet of Things
IoE	Internet of Everything
WiMAX	Worldwide Interoperability for Microwave Access
ITU-R	International Telecommunication Union-Radio Communication Sector
RAT	Radio Access Technologies
MTC	Machine Type Communications
NOMA	Non-Orthogonal Multiple Access
FBMC	Filter Bank Multi-Carrier
mmWave	Millimeter Wave
UDN	Universal Domain Networks
OSS	Operation Support System
OAI	Open-Air Interface
WLAN	Wireless Local Area Network
NS-3	Network Simulator-3

REFERENCES

Abdullah, M., Altaf, A., Anjum, M. R., Arain, Z. A., Jamali, A. A., Alibakhshikenari, M., Falcone, F., & Limiti, E. (2021). Future Smartphone: MIMO Antenna System for 5G Mobile Terminals. *IEEE Access : Practical Innovations, Open Solutions*, 9, 91593–91603. 10.1109/ACCESS.2021.3091304

Afolabi, I., Taleb, T., Samdanis, K., Ksentini, A., & Flinck, H. (2018). Network Slicing and Softwarization: A Survey on Principles, Enabling Technologies, and Solutions. *IEEE Communications Surveys and Tutorials*, 20(3), 2429–2453. 10.1109/COMST.2018.2815638

Ahmad, I., Kumar, T., Liyanage, M., Okwuibe, J., Ylianttila, M., & Gurtov, A. (2018). Overview of 5G Security Challenges and Solutions. *IEEE Communications Standards Magazine*, 2(1), 36–43. 10.1109/MCOMSTD.2018.1700063

Ashraf, S. A., Aktas, I., Eriksson, E., Helmersson, K. W., & Ansari, J. (2016). Ultra-reliable and low-latency communication for wireless factory automation: From LTE to 5G. *2016 IEEE 21st International Conference on Emerging Technologies and Factory Automation (ETFA)*, (pp. 1–8). IEEE. https://doi.org/10.1109/ETFA.2016.7733543

Bondre, S., Sharma, A., & Bondre, V. (2023). 5G Technologies, Architecture and Protocols. In *Evolving Networking Technologies* (pp. 1–19). John Wiley & Sons, Ltd., 10.1002/9781119836667.ch1

Cetin, B. K., & Pratas, N. K. (2021). Resource sharing and scheduling in device-to-device communication underlying cellular networks. *Pamukkale Üniversitesi Mühendislik Bilimleri Dergisi*, 27(5), 604–609.

Dangi, R., Lalwani, P., Choudhary, G., You, I., & Pau, G. (2022). Study and Investigation on 5G Technology: A Systematic Review. *Sensors (Basel)*, 22(1), 1. 10.3390/s2201002635009569

Do, D.-T., Nguyen, M.-S. V., Jameel, F., Jäntti, R., & Ansari, I. S. (2020). Performance Evaluation of Relay-Aided CR-NOMA for Beyond 5G Communications. *IEEE Access : Practical Innovations, Open Solutions*, 8, 134838–134855. 10.1109/ACCESS.2020.3010842

Esmaeily, A., & Kralevska, K. (2021). Small-Scale 5G Testbeds for Network Slicing Deployment: A Systematic Review. *Wireless Communications and Mobile Computing*, 2021, e6655216. 10.1155/2021/6655216

Guo, S., Hou, X., & Wang, H. (2018). Dynamic TDD and interference management towards 5G. *2018 IEEE Wireless Communications and Networking Conference (WCNC)*, (pp. 1–6). IEEE. 10.1109/WCNC.2018.8377314

Gupta, A., & Jha, R. K. (2015). A Survey of 5G Network: Architecture and Emerging Technologies. *IEEE Access : Practical Innovations, Open Solutions*, 3, 1206–1232. 10.1109/ACCESS.2015.2461602

Huang, M., Gong, F., Zhang, N., Li, G., & Qian, F. (2021). Reliability and Security Performance Analysis of Hybrid Satellite-Terrestrial Multi-Relay Systems with Artificial Noise. *IEEE Access : Practical Innovations, Open Solutions*, 9, 34708–34721. 10.1109/ACCESS.2021.3058734

Kaltenberger, F., Silva, A. P., Gosain, A., Wang, L., & Nguyen, T.-T. (2020). OpenAirInterface: Democratizing innovation in the 5G Era. *Computer Networks*, 176, 107284. 10.1016/j.comnet.2020.107284

Kim, H.-K., & Cho, Y., ahiagbe, E. E., & Jo, H.-S. (2018). Adjacent Channel Interference from Maritime Earth Station in Motion to 5G Mobile Service. *2018 International Conference on Information and Communication Technology Convergence (ICTC)*, (pp. 1164–1169). IEEE. https://doi.org/10.1109/ICTC.2018.8539401

Liyanage, M., Ahmad, I., Abro, A. B., Gurtov, A., & Ylianttila, M. (2018). *A comprehensive guide to 5G security*. Wiley Online Library. 10.1002/9781119293071

Mangla, C., Rani, S., Faseeh Qureshi, N. M., & Singh, A. (2023). Mitigating 5G security challenges for next-gen industry using quantum computing. *Journal of King Saud University. Computer and Information Sciences*, 35(6), 101334. 10.1016/j.jksuci.2022.07.009

Mendonça, S., Damásio, B., Charlita de Freitas, L., Oliveira, L., Cichy, M., & Nicita, A. (2022). The rise of 5G technologies and systems: A quantitative analysis of knowledge production. *Telecommunications Policy*, 46(4), 102327. 10.1016/j.telpol.2022.102327

Osama, M., El Ramly, S., & Abdelhamid, B. (2021). Interference Mitigation and Power Minimization in 5G Heterogeneous Networks. *Electronics (Basel)*, 10(14), 14. 10.3390/electronics10141723

Pavitra, A. R. R., Lawrence, I. D., & Maheswari, P. U. (2023). To Identify the Accessibility and Performance of Smart Healthcare Systems in IoT-Based Environments. In *Using Multimedia Systems, Tools, and Technologies for Smart Healthcare Services* (pp. 229–245). IGI Global. 10.4018/978-1-6684-5741-2.ch014

Qamar, F., Hindia, M. H. D. N., Dimyati, K., Noordin, K. A., & Amiri, I. S. (2019). Interference management issues for the future 5G network: A review. *Telecommunication Systems*, 71(4), 627–643. 10.1007/s11235-019-00578-4

Sanyal, J., & Samanta, T. (2021). Game Theoretic Approach to Enhancing D2D Communications in 5G Wireless Networks. *International Journal of Wireless Information Networks*, 28(4), 421–436. 10.1007/s10776-021-00531-w

Scalise, P., Boeding, M., Hempel, M., Sharif, H., Delloiacovo, J., & Reed, J. (2024). A Systematic Survey on 5G and 6G Security Considerations, Challenges, Trends, and Research Areas. *Future Internet*, 16(3), 67. 10.3390/fi16030067

Sullivan, S., Brighente, A., Kumar, S. A. P., & Conti, M. (2021). 5G Security Challenges and Solutions: A Review by OSI Layers. *IEEE Access : Practical Innovations, Open Solutions*, 9, 116294–116314. 10.1109/ACCESS.2021.3105396

Yazdinejad, A., Parizi, R. M., Dehghantanha, A., & Choo, K.-K. R. (2019). Blockchain-enabled authentication handover with efficient privacy protection in sdn-based 5g networks. *IEEE Transactions on Network Science and Engineering*, 8(2), 1120–1132. 10.1109/TNSE.2019.2937481

Yazdinejad, A., Parizi, R. M., Dehghantanha, A., & Choo, K.-K. R. (2019). Blockchain-enabled authentication handover with efficient privacy protection in SDN-based 5G networks. *IEEE Transactions on Network Science and Engineering*, 8(2), 1120–1132. 10.1109/TNSE.2019.2937481

Zhang, J., Xie, W., & Yang, F. (2015). An architecture for 5g mobile network based on sdn and nfv, *International Conference on Wireless, Mobile and Multi-Media*. Wiley. 10.1002/spy2.271

Section 2
Security Protocols and Technologies

Chapter 4
Encryption Protocols and Security for 5G

Dhinakaran Vijayalakshmi
Department of Electronics and Communication Engineering, SRM Institute of Science and Technology, Chennai, India

S. Krithiga
Department of Electronics and Communication Engineering, SRM Institute of Science and Technology, Chennai, India

Trinay Gangisetty
Department of Computer Science, University of Colorado, Boulder, USA

R. Dayana
Department of Electronics and Communication Engineering, SRM Institute of Science and Technology, Chennai, India

K. Vadivukkarasi
https://orcid.org/0000-0002-4618-7208
Department of Electronics and Communication Engineering, SRM Institute of Science and Technology, Chennai, India

ABSTRACT

Safeguarding confidential information from unwanted access is of utmost importance in the current age of digital interactions and knowledge sharing. This chapter explores the complex world of encryption techniques and how important they are to maintaining data privacy. The study examines the evolution of encryption historically over time, following its origins from traditional cryptography to the present modern methodologies for cryptography, showcasing the mathematical frameworks that support their security. Further, it focuses on the difficulties that encryption protocols have to deal with in light of cutting-edge technology like quantum computing, which could endanger established cryptographic techniques. The ethical and legal implications of encryption and data privacy are a crucial component. With the re-

DOI: 10.4018/979-8-3693-2786-9.ch004

sults of historical investigations, technical difficulties, and ethical issues taken into account, the study offers a road map for improving data privacy in the digital era.

1. INTRODUCTION

1.1 Background

The history of encryption becomes apparent as a tale of adaptation and ingenuity in reaction to evolving threats within the intricate network of contemporary digital interactions. Previously reserved for use in military and diplomatic communications, encryption is now an essential aspect of modern cybersecurity (Kerr, 2000). The evolution of encryption, spanning from basic substitution ciphers used by ancient civilizations to intricate mathematical algorithms employed in the digital era, demonstrates humanity's unwavering commitment to safeguarding confidential data. Gaining a thorough comprehension of this historical backdrop is crucial in fully grasping the pressing and significant nature of modern encryption algorithms. It sheds light on the ongoing challenge of safeguarding data privacy in the face of a continuously expanding digital landscape (Lessig, 2009).

The history of encryption is characterized by significant milestones, like the creation of the Enigma machine during World War II and the subsequent emergence of public-key cryptography in the 1970s. These historical examples highlight the flexible nature of encryption, showing its ability to adjust to changes in technology and global politics (Vaudenay, 2006). Encryption methods have advanced from secret codes and symmetric ciphers to complex algorithms that utilize mathematical principles. This evolution has not only addressed emerging threats but has also established the foundation for the complex network of secure digital communication that defines the modern era. Essentially, the history of encryption can be seen as a complex and intricate fabric created via the combination of human creativity, the need for security, and a strong dedication to protecting information in an ever more linked global society (Levinsky, 2022).

1.2 Rationale for the Study

The justification for doing an in-depth study on encryption protocols arises from the increasing necessity of strengthening our digital defenses against an ever-expanding variety of cyber threats. As our societies become more digitally incorporated, the flaws within existing encryption solutions become more and more obvious (Altulaihan et al., 2022). High-profile data breaches, cyber-attacks on important infrastructure, and the growing number of criminal actors promote the essential need to examine and strengthen the processes that safeguard our sensitive information. This study has been inspired by the realization that the current set of encryption protocols must be thoroughly evaluated to uncover potential deficiencies, modify current techniques, and develop creative solutions that can withstand the continuous evolution of cyber threats (Ribeiro, 2019).

Moreover, the rationale extends to the requirement of contributing to a larger conversation on data privacy and security. The institutional consequences of hacked data extend beyond individual privacy concerns, influencing economic stability, national security, and the integrity of digital networks (Li & Liu, 2021). The ultimate goal is to develop a robust and adaptive cybersecurity landscape capable of mitigating future threats and ensuring the continuous integrity and confidentiality of digital information in an era marked by technological dependence (Victor-Mgbachi, 2024).

1.3 Objectives of the Chapter

The goal of this chapter is threefold, each of which contributes to a thorough understanding of encryption techniques and their function in contemporary data security. Firstly, the chapter seeks to provide a comprehensive historical overview of encryption, documenting its development from ancient cryptographic approaches to the cutting-edge algorithms deployed in today's digital ecosystem. By unwinding the historical threads of encryption, the research tries to explain the development of cryptographic technologies, offering light on the adaptable techniques utilized over time in response to different security challenges.

Secondly, this chapter tries to carry out a critical examination of the current state of encryption techniques. Through a comprehensive evaluation of their strengths, flaws, and adaptation to changing technological frameworks, the research attempts to provide an equitable assessment of the usefulness of present encryption techniques. By studying the complexity of existing protocols, the study intends to discover potential vulnerabilities and opportunities for improvement, contributing to the continuing discussion on increasing data security. Lastly, the chapter strives to integrate these historical and analytical insights into concrete recommendations for the future development and deployment of encryption technologies. In doing so, it understands the dynamic nature of cyber security concerns and attempts to deliver realistic solutions that correspond with the increasing expectations of a technology-driven society.

2. HISTORICAL EVOLUTION OF ENCRYPTION

2.1 Classical Cryptography

Classical cryptography covers the basic era of encryption, tracing its roots to ancient civilizations where rudimentary methods were applied to secure critical information. One of the first instances is the Caesar cipher, credited to Julius Caesar, which involves changing every letter in a message by a specified number of points. Substitution ciphers, such as the Atbash cipher employed in ancient Hebrew encryption, were also widespread, involving substituting each letter with another according to a set pattern. These methods, while clever for their time, were rather easy and subject to frequency analysis (Oktaviana & Utama Siahaan, 2016).

Advancements in classical cryptography during the Renaissance era saw the birth of more sophisticated procedures. The Vigenère cipher introduced the concept of employing a keyword to encrypt messages, offering a layer of complexity beyond simple substitution. However, the widespread adoption of classical ciphers suffered fundamental limitations in scalability and security.

The historical growth of classical cryptography gives a fascinating chronicle of human creativity in finding means to secure communication. The mechanics of classical ciphers, such as the Vigenère square or the Caesar cipher wheel, which serve as visual aids to comprehend the underlying principles. These diagrams visually demonstrate the letter substitutions and rotations required in encryption, presenting an accurate representation of the encryption methods performed during different historical periods (Rubinstein-Salzedo & Rubinstein-Salzedo, 2018).

As classical cryptography established the framework for subsequent advances, its influence survives in present cryptographic principles. Understanding these historical encryption methods is crucial in recognizing the obstacles faced and the advances developed, paving the way for the more complicated cryptographic algorithms applied in modern digital security.

2.2 Development of Modern Cryptographic Algorithms

The change from classical to current cryptographic algorithms represents an enormous advance in the level of complexity and security of encryption technologies. With the advent of computers and breakthroughs in mathematical theory, cryptographic approaches evolved to tackle the problems faced by an increasingly interconnected world. One major milestone was the creation of the Data Encryption Standard (DES) in the 1970s, a symmetric-key method accepted as a government standard for securing sensitive but unclassified information in the United States. DES featured a Feistel network structure and a 56-bit key, establishing the groundwork for the following cryptographic developments (Standard, 1999).

The evolution proceeded with the implementation of public-key cryptography, a revolutionary concept developed by Whitfield Diffie and Martin Hellman in 1976. Unlike symmetric-key systems,

Public-key cryptography employs a pair of keys:

1. a public key for encryption
2. a private key for decryption.

This discovery addressed key distribution issues inherent in symmetric-key systems, creating the foundation for secure digital communication. The RSA algorithm, created by Ron Rivest, Adi Shamir, and Leonard Adleman in 1977, represents the practical implementation of public-key cryptography and remains frequently used today. The structure of the RSA algorithm shown in Figure 1 (Bodur & Kara, 2015).

Figure 1. Structure of RSA algorithm

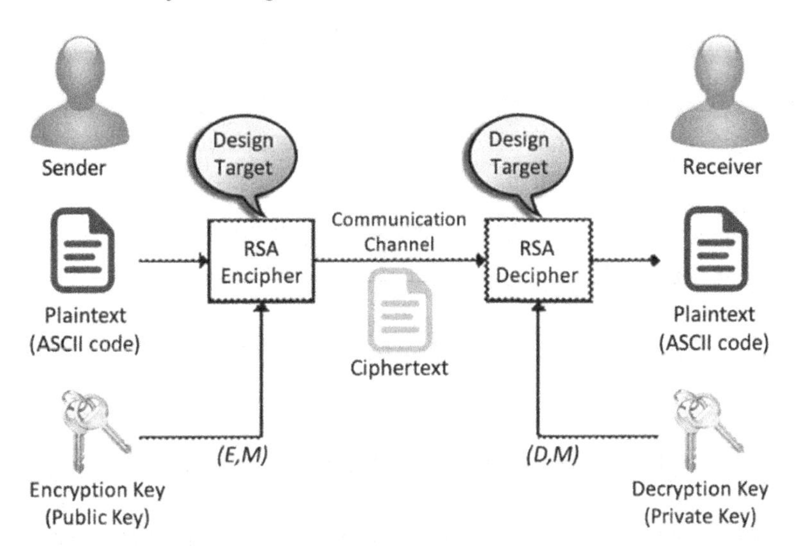

A wide range of techniques are used in modern cryptographic algorithms, such as stream ciphers like Rivest Cipher 4 and block ciphers like the Advanced Encryption Standard (AES). The strength of these algorithms resides in their ability to survive sophisticated attacks, providing secrecy, integrity, and authenticity in digital communication. Visual aids of Figure 2 illustrate the structure and processes of contemporary cryptographic algorithms, like the substitution-permutation network utilized in AES or the modular arithmetic operations integral to RSA, which play a crucial role in facilitating a better understanding of the intricate principles that underpin these highly secure encryption methods. The development of current cryptographic algorithms constitutes a dynamic response to the expanding terrain of digital threats (Equihua et al., 2021).

Figure 2. AES design

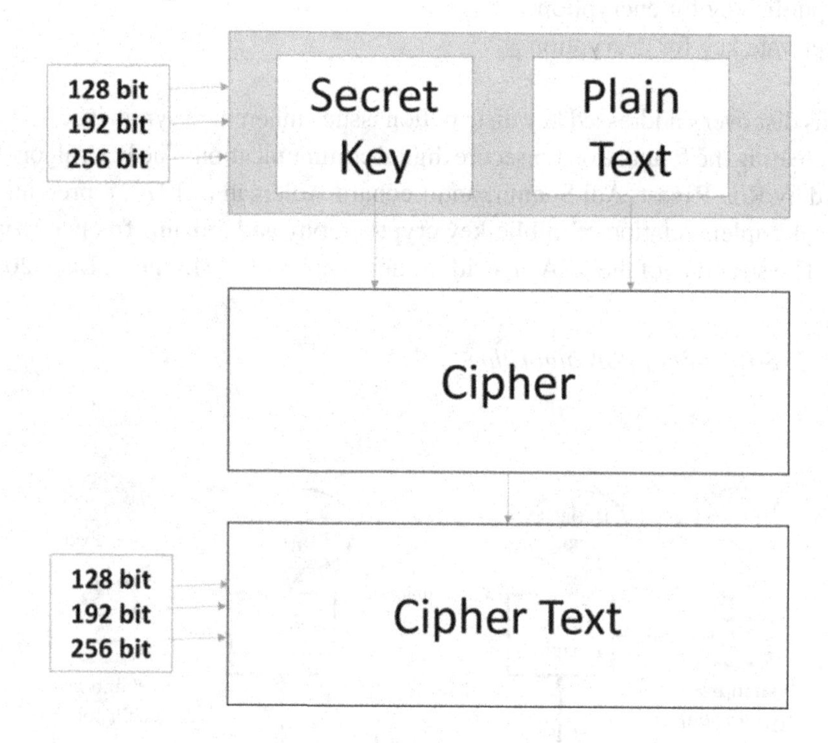

2.3 Significance of Historical Context

The importance of historical context in figuring out encryption stems from its capacity to reveal the motives, obstacles, and innovative solutions that have driven the evolution of cryptographic systems over time. Exploring the historical foundations of encryption reveals humanity's ongoing effort to protect sensitive information to evolving threats. Through the investigation of historical encryption techniques such as Spartan transposition ciphers and Roman substitution ciphers, the fundamental concepts have survived generations and provided the framework for modern cryptographic procedures (Buell, 2021).

One important part of the historical background is the ongoing cat-and-mouse game between cryptography experts and hackers. The Caesar cipher, while innovative in its simplicity, was vulnerable to frequency analysis, which encouraged the invention of more complicated substitution ciphers. Understanding these historical flaws allows for a deeper appreciation of the recurrent nature of cryptographic improvements. Figure 3 demonstrates the growth of ciphers over time, from basic letter replacements to complex Feistel networks, visually illustrate this process and demonstrate the interconnection of previous cryptographic breakthroughs (Zhang et al., 2018).

Figure 3. Fiestal network

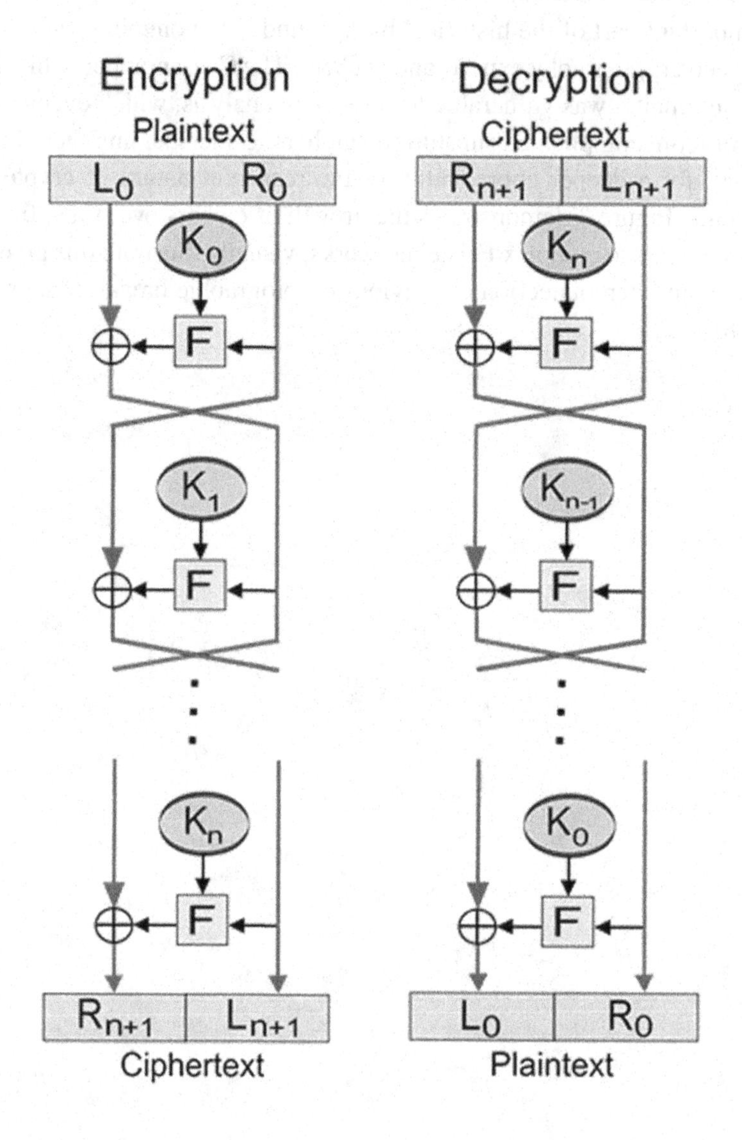

Furthermore, historical background illuminates the sociopolitical forces impacting the history of encryption. During World War II, Germans used the Enigma machine to demonstrate how geopolitical pressures can drive innovation and highlight the importance of encryption in wartime communication. The incorporation of historical context into the study of encryption not only broadens our understanding of cryptographic rules but also informs current cryptographic approaches. The insights learned from past weaknesses and advances can be used to design robust encryption solutions for present difficulties (Deavours & Reeds, 1977).

3. CRYPTOGRAPHIC PRIMITIVES

3.1 Symmetric Key Encryption

Symmetric key encryption is the foundation of cryptographic systems and a fundamental technique for safeguarding digital information. This technique uses a single secret key for both encryption and decryption, requiring both the sender and recipient to have the same key. Symmetric key encryption's simplicity and efficiency make it suited for a wide range of applications, including safeguarding communication channels and encrypting stored data. The substitution-permutation network in the Advanced Encryption Standard (AES), can help people visualize how symmetric key encryption works as shown in Figure 4 (Mao, 2022).

Figure 4. Symmetric encryption process

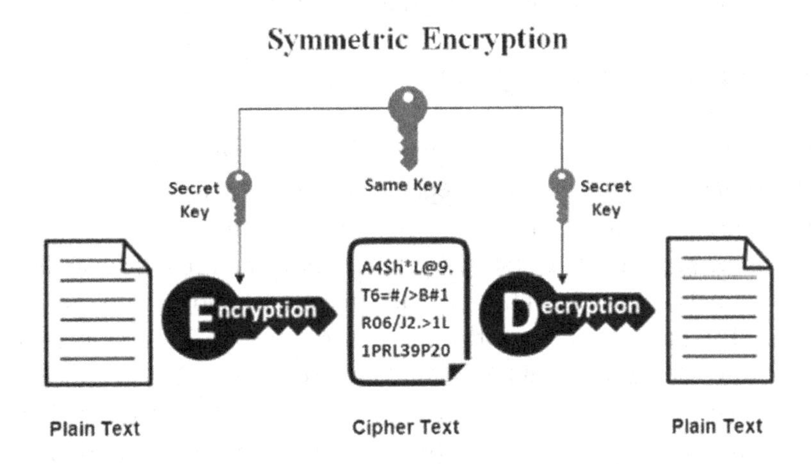

The AES is a renowned example of symmetric key encryption, having overtaken the elderly Data Encryption Standard (DES) due to it being susceptible to brute force assaults.

The AES method involves a series of well-defined processes, including

- **row shifting,**

- **byte substitution,**
- **key addition, and**
- **column mixing,**

which are all choreographed using a symmetric key. Despite its effectiveness, it has issues in distribution of key and maintenance. The safe exchange of secret keys between communication parties is an essential component of this system. **Diffie-Hellman key exchange** of modular arithmetic, addresses these issues by allowing for the secure production of a shared secret key via an unsecured communication channel. Understanding the cryptographic primitives associated with symmetric key encryption is critical for recognizing its strengths and weaknesses, as well as driving ongoing work to improve and secure digital communication in a variety of sectors (Li, 2010).

To summarize, symmetric key encryption is still an important cryptographic basis for safeguarding digital information while maintaining simplicity and speed. Diagrams that explain the processes within symmetric key algorithms, particularly those connected to the AES standard and key exchange methods, provide visual insight into the complex stages involved. As digital systems continue to rely on encrypted interaction, investigating symmetric key encryption and its graphic representations is critical in the larger context of cryptographic understanding.

3.2 Asymmetric Key Encryption

Asymmetric key encryption, commonly known as public-key cryptography, transformed the industry by offering separate key pairs for encryption and decoding. This novel solution, first proposed by Martin Hellman and Whitfield Diffie in their seminal paper in 1976, uses a public key for encryption that can be freely disseminated and a private key for decryption that is only known to the person who requested it. This paradigm deals with the key distribution difficulty associated with symmetric key encryption, making secure key exchange a critical concern. The mathematical underlying of asymmetric key encryption is the intricacy of specific mathematical tasks, such as factoring big numbers or discrete logarithms, which serve as the foundation for safe key creation and exchange (Yassein et al., 2017).

Figure 5. Asymmetric encryption process

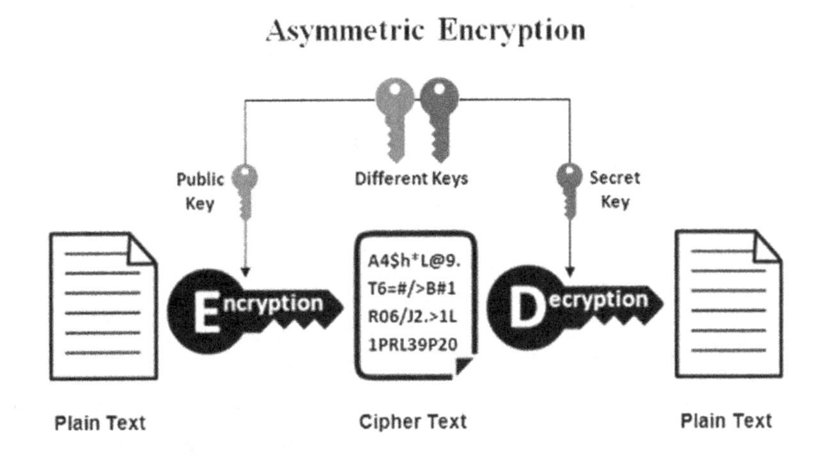

The RSA algorithm, which was developed by Ron Rivest, Adi Shamir, and Leonard Adleman, is one of the most extensively used asymmetric key algorithms. The security of RSA is based on the challenge of factoring the product of two large prime numbers, which serve as the public key. In RSA, the encryption and decryption processes use modular exponentiation, expressed as

$C \equiv M^e \pmod{n}$ for encryption and

$M \equiv C^d \pmod{n}$ for decryption,

where C is the cipher text,

M is the plain text,

n is the modulus

d is the private exponent, and e is the public exponent,

The computational complexity of factoring the product of two large primes ensures RSA's security. Asymmetric key encryption represents an innovative technique for secure communication and key exchange. Diffie-Hellman key exchange, another important asymmetric key method, is based on the computational complexity of discrete logarithms. The key exchange procedure generates a shared secret between two parties without directly transferring the secret (Eslami, 2006).

The mathematical foundation is described as

$K \equiv (A^b) \pmod{p}$ for one party and

$K \equiv (B^a) \pmod{p}$ for the other.

Here, A and B are the public keys.

a and b are the private keys, and

p is a huge prime.

Figure 6. Diffie-Hellman key exchange for asymmetric structure

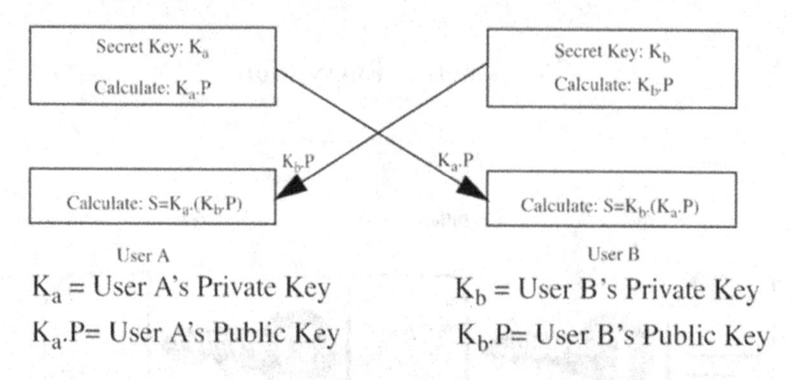

Finally, asymmetric key encryption is important in cryptography because it addresses the key distribution difficulties that symmetric key systems present. Equations like RSA and Diffie-Hellman provide mathematical knowledge of the processes involved in asymmetric key encryption, stressing their importance in securing digital communication.

Digital signatures are a significant application of asymmetric key encryption. Digital signatures verify the integrity and validity of digital messages by using the signer's private key and cryptographic hash algorithms. However, due to their computational cost, asymmetric key techniques are less efficient for bulk data encryption than symmetric key methods. Despite the efficiency problems, the presence of both symmetric and asymmetric key encryption in cryptographic systems provides a broad and powerful toolkit for securing digital information (Lalem et al., 2023).

Hash functions serve a significant role in cryptography, functioning as a crucial and fundamental cryptographic primitive that turns variable-length input into a fixed-length hash result. One of the key applications of hash functions is to ensure data integrity. By generating a hash value, generally expressed as a fixed-size alphanumeric string, for a particular collection of data, any modifications to the data will end up in an entirely distinct hash value. This feature is vital in confirming the authenticity of transmitted or stored information. Mathematically, a hash function H takes an input message M and outputs a fixed-length hash value $h = H(M)$ (Preneel, 1994).

3.3 Mathematical Foundations of Security

The field of cryptography relies largely on sound mathematical underpinnings to assure the security and integrity of cryptographic systems. Understanding the underlying mathematical principles is vital for judging the strength of cryptographic primitives and constructing robust algorithms.

One essential concept is the concept of computational hardness, commonly stated through problems that are easy to define but computationally difficult to solve. For example, the factorization of huge composite numbers into their prime components forms the basis of widely used asymmetric key techniques like RSA. The security of these techniques relies on the idea that factoring huge numbers is a computationally challenging job, even for powerful computers (Gray & James, 1992).

Another key feature is the use of modular arithmetic in cryptographic procedures. Modular arithmetic provides a restricted field, allowing computations to wrap around within a certain range of values, or, in other words, a specified range. In modular arithmetic, operations like addition, subtraction, and exponentiation are done modulo a selected integer, boosting computer efficiency and providing mathematical structures vital to many cryptographic techniques. The modular exponentiation procedure, vital in asymmetric key encryption like RSA, shows the application of modular arithmetic in cryptography.

Mathematical ideas such as group theory and number theory contribute considerably to cryptography algorithms. Group theory, for instance, offers a framework for understanding cyclic groups and their applications in key exchange protocols. The Diffie-Hellman key exchange, based on the discrete logarithm problem in a cyclic group, exhibits the practical application of mathematical concepts in assuring secure communication (Li, 2010).

In summary, the mathematical foundations of security in cryptography involve computational complexity, modular arithmetic, number theory, and group theory. These concepts form the core of secure cryptographic systems, with their application obvious in the creation and study of algorithms that support modern data protection techniques. Understanding the mathematical underpinnings is vital for both cryptographers building secure systems and practitioners applying cryptographic solutions in real-world applications.

4. CHALLENGES AND EMERGING TECHNOLOGIES

4.1 Quantum Computing Threats

The development of quantum computing poses a significant challenge to standard cryptography approaches, raising concerns about the future security of sensitive information. Unlike classical computers, which use bits to represent binary values (0 or 1), quantum computers employ quantum bits, or qubits, which can be in numerous states at the same time due to the principles of superposition and entanglement. This distinguishing feature allows quantum computers to do certain calculations tenfold faster than conventional computers, posing an immediate danger to widely used encryption techniques (Faruk et al., 2022).

Figure 7. Representation of computing threat

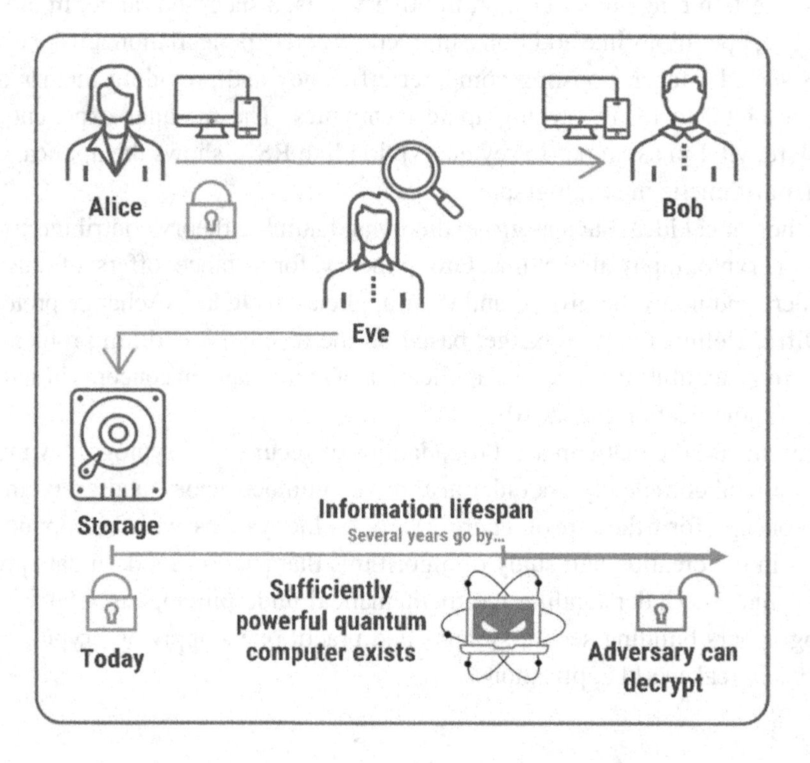

One of the most serious risks offered by quantum computing is the ability to break widely used public-key encryption methods like RSA and ECC (Elliptic Curve encryption). Shor's method, created by mathematician Peter Shor, indicates that a sufficiently powerful quantum computer can efficiently factor big numbers and solve the discrete logarithm problem, which is the basis for many asymmetric key schemes. As a result, the security guaranteed by these methods is jeopardized in the age of quantum computing (Mahto & Yadav, 2017).

Post-quantum cryptography emerges as a proactive response to this threat, intending to develop cryptographic algorithms that are secure even when subjected to quantum attacks.

code-based cryptography

hash-based cryptography, and

Lattice-based cryptography,

are three intriguing approaches to the post-quantum cryptographic world. These algorithms are based on mathematical problems that are thought to be difficult for both classical and quantum computers to solve, providing a strong barrier against quantum threats (Balamurugan et al., 2021).

The incorporation of quantum-resistant cryptographic algorithms presents unique obstacles, such as standards, implementation, and compatibility. Because quantum-resistant algorithms require considerable modifications to the current cryptographic infrastructure, the move to post-quantum cryptography necessitates careful preparation and worldwide coordination. Furthermore, the timescale for the implementation of practical quantum computers is unpredictable, complicating the strategic development and deployment of quantum-resistant cryptographic systems.

Finally, quantum computing risks pose a compelling challenge to traditional cryptographic paradigms. In the quantum age, research into post-quantum cryptography solutions is critical to ensuring the confidentiality and integrity of information. Navigating the challenges of this shift requires not only scientific improvements but also international collaboration and standardized techniques to assure cryptographic system resilience in the face of quantum computers' extraordinary processing capacity (Raheman, 2022).

4.2 Quantum Resistant Encryption

As the era of quantum computing approaches, the necessity for quantum-resistant encryption becomes ever more urgent. Quantum-resistant encryption, also known as post-quantum cryptography, seeks to create cryptographic algorithms that can withstand the computing power of quantum computers, ensuring the long-term security of digital communication and data protection.

One of the fundamental drivers behind quantum-resistant encryption is the susceptibility of widely used public-key cryptography techniques to quantum assaults. Shor's technique, for example, has the ability to effectively factor big numbers and solve discrete logarithm issues, undermining the security provided by RSA, ECC, and other asymmetric key algorithms. Quantum-resistant encryption attempts to overcome these vulnerabilities by adding new mathematical problems that are thought to be difficult for both classical and quantum computers to solve (Politi, 2010).

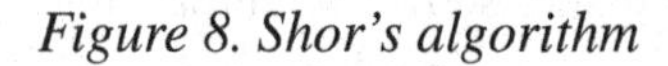

Figure 8. Shor's algorithm

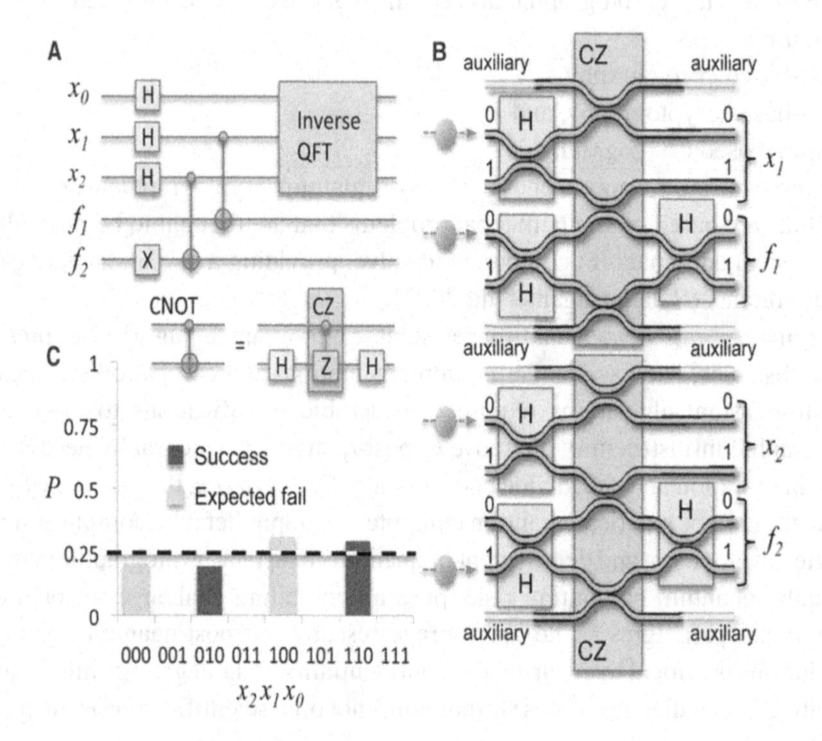

Lattice-based cryptography is a promising approach to post-quantum cryptographic solutions. The mathematical study of grids of points, known as lattice issues, serves as the foundation for quantum-resistant encryption methods. These methods use the hardness of lattice problems, such as the Learning with Errors (LWE) problem, to offer a secure basis for encryption. Other approaches, such as hash- and code-based cryptography, help to construct quantum-resistant encryption algorithms.

Implementing quantum-resistant encryption poses unique challenges, necessitating a delicate balance of security, efficiency, and compatibility. The shift from present cryptographic standards to quantum-resistant algorithms requires a collaborative effort among cryptography researchers, industry players, and standardization bodies. Striking this balance is critical to ensuring that quantum-resistant encryption not only improves security but also easily interacts with existing cryptographic infrastructures.

In conclusion, quantum-resistant encryption is an anticipatory reaction to the impending quantum computer future. As quantum-resistant algorithms are rigorously tested and evaluated, their adoption will play a critical role in strengthening the foundations of cybersecurity. The creation and deployment of quantum-resistant encryption methods demonstrates the cryptography community's endurance in the face of new technological hurdles (Mao et al., 2014).

4.3 Responding to Technological Advances

The cybersecurity landscape is always changing, owing to rapid technological breakthroughs and an evolving threat environment. Adapting to these developments is critical for preserving the efficiency of cryptographic systems and protecting confidential data in the digital era. One important part of adjusting to technological changes is the ongoing examination and enhancement of existing cryptographic algorithms. As computational power grows and new attack routes arise, cryptographic protocols must be upgraded to solve weaknesses and incorporate more robust security mechanisms. To keep up with potential dangers, cryptographic researchers, industry professionals, and standards agencies collaborate.

Integrating quantum-resistant encryption into existing cryptographic frameworks is an excellent example of technological adaptation. The continued research and development of algorithms immune to quantum assaults demonstrates the cryptographic community's proactive approach to minimizing possible concerns posed by quantum computing. The shift to quantum-resistant encryption necessitates careful consideration of algorithmic efficiency, compatibility, and the overall cryptographic ecosystem.

Furthermore, as technologies such as Machine learning and Artificial intelligence (AI) gain traction, their impact on cryptographic systems must be thoroughly investigated. Adapting cryptographic algorithms to guard against AI-powered assaults and comprehending the potential vulnerabilities created by these technologies are critical components of maintaining strong cybersecurity (Li et al., 2023).

Interdisciplinary collaboration is critical to responding to technological advances. Cryptographers, computer scientists, mathematicians, and professionals in adjacent domains must collaborate to anticipate obstacles and discover novel solutions. This partnership extends to lawmakers and industry leaders, ensuring that cryptographic standards are consistent with legal and ethical considerations.

To summarize, adjusting to technological changes is a continuous and cooperative procedure that involves alertness, inventiveness, and a determination to remain ahead of any threats. The ever-evolving makeup of cybersecurity today needs a proactive strategy in which the cryptographic community constantly reviews, updates, and changes cryptographic methods to protect digital information in a rapidly changing technical context.

5. PERFORMANCE CONSIDERATIONS

5.1 Computational Overhead

Computational overhead, a critical aspect of cryptographic systems, refers to the extra computer resources necessary to apply security mechanisms. This cost, which includes processor power, memory utilization, and time complexity, has a substantial impact on the usability and efficacy of cryptographic algorithms.

Consider the RSA algorithm as an instructive example of computational overhead. The core security premise of RSA is based on the computational difficulties of factoring the product of two big prime numbers. The security parameter n encapsulates the magnitude of these primes, which plays a crucial role in determining the computing cost. A greater n improves the system's security by making factoring more complex. However, this boost in security comes at the expense of higher processing requirements. The time complexity of RSA encryption and decryption procedures is proportional to the cube of n.

$$\textit{Time complexity} \propto n^3 \quad (1)$$

This equation precisely expresses the cubic connection, emphasizing the sensitive trade-off between protection and computational efficiency that cryptographic systems face (Pharkkavi & Maruthanayagam, 2018).

The concept of computational overhead extends beyond asymmetric key encryption to include symmetric key encryption as well. In symmetric key algorithms, the choice of key size and cryptographic methods has a significant impact on computing needs. For example, increasing the key size in symmetric key methods improves security but also increases computational costs. The equation controlling this relationship is critical in decision-making, allowing cryptographic practitioners to strike a careful balance between the need for strong security and the practical limits of resource-constrained contexts.

In essence, the delicate dance of security and computational efficiency is a fundamental feature of cryptographic systems. The equations and arguments that underpin the measurement of computing overhead provide cryptographic practitioners with essential tools for making sound decisions. This delicate balance, which recognizes the inherent trade-off, resonates throughout the cryptographic landscape, influencing the design of security solutions that are both theoretically robust and practically feasible.

5.2 Latency Issues

Latency, the time delay between the start of an operation and its conclusion, is a major concern in cryptographic systems, limiting their real-world application. The complexities of cryptographic algorithms used to assure security can add latency, which is especially important in time-sensitive applications such as communication systems and real-time data processing.

Consider a scenario in which secure communication is critical, as is common in electronic transactions or communication across encrypted paths and secure channels. The cryptographic methods used, such as key exchange and encryption/decryption operations, introduce processing processes that add to the overall latency. For example, in asymmetric key encryption, when complicated mathematical processes are required, the latency can be more noticeable than in symmetric key encryption.

Latency has a significant impact on upcoming technologies such as the Internet of Things (IoT) and 5G networks. These environments necessitate quick data interchange, and any latency added by cryptographic operations can disrupt the smooth flow of information. Striking a balance between cryptographic security measures and the need for low-latency communication becomes critical in these scenarios (Kaknjo et al., 2019).

Mathematically quantifying latency in cryptographic systems requires taking into account the time complexity of the cryptographic methods being used. Consider the Advanced Encryption Standard (AES), a widely used symmetric key encryption technique. The time complexity T of AES operations can be described mathematically as a function of the block size n, the key size k, and the number of rounds r.

$$\mathbf{T = f(n, k, r)} \quad (2)$$

This equation incorporates the different elements that influence the latency introduced by the AES algorithm, establishing a mathematical framework for measuring and mitigating latency concerns.

In essence, latency issues highlight the dynamic relationship between cryptographic security and real-time operational requirements. The mathematical representations of delay in cryptographic systems allow practitioners to assess and optimize the trade-offs required for effective and secure communication across varied technology environments.

5.3 Balancing Security with System Functionality

A significant problem in the field of cryptographic systems is achieving a delicate balance between strong security measures and the need for flawless system functionality. Cryptographic protocols and algorithms, which are intended to strengthen data protection, inevitably involve complications that can have an impact on a system's overall operation and performance.

The struggle between security and system functionality is obvious in situations when real-time or resource-constrained tasks are critical. Cryptographic operations, particularly those involving complex mathematical computations, can impose latency and computational overhead, thereby reducing system responsiveness. As a result, the challenge is to optimize security measures while maintaining functionality and the user experience.

The mathematical expression for this fragile equilibrium is the temporal complexity T of cryptographic operations within a system. For instance, in a secure communication protocol, the overall time complexity becomes a function f of several parameters, including key sizes, cryptographic algorithm complexities, and the computational capability of the underlying hardware:

T = f(key size, algorithm complexity, computational power) (3)

This mathematical formulation emphasizes the challenge's multidimensionality, with security parameter decisions having a direct impact on a system's operational efficiency.

In fact, achieving this balance frequently requires a complex approach. Cryptographers, system architects, and cybersecurity professionals work together to design cryptographic solutions to specific system requirements. Adaptive cryptographic algorithms that may dynamically alter their security parameters based on contextual circumstances are one way to achieve this balance. Furthermore, advances in hardware acceleration and optimization techniques help to reduce the impact of security measures on system functionality.

The problem of balancing security and system functioning is especially evident in new technologies such as edge computing and IoT, where computational resources are frequently limited and real-time responsiveness is critical. As cryptographic systems evolve, continual research into creative mathematical models, algorithmic optimizations, and hardware enhancements becomes critical for navigating this complex interaction and maintaining the coexistence of robust security with efficient system operations (Tariq et al., 2023).

6. LEGAL AND ETHICAL DIMENSIONS

6.1 Privacy Rights vs. National Security

The relationship between privacy rights and national security is at the forefront of legal and ethical concerns in cryptography. The desire for increased security measures to defend nations from external dangers frequently clashes with the need to maintain individual privacy rights. Striking a careful balance between these two top priorities presents complex issues for governments, legal experts, and engineers alike.

This delicate equilibrium can be mathematically illustrated using encryption backdoors. Encryption, which is used to safeguard conversations and data, has become a hot topic in the debate over privacy rights and national security. The concept of an encryption backdoor refers to the deliberate construction of weaknesses or access points in cryptographic systems, purportedly for legitimate access by authorities. However, this creates a mathematical paradox, as purposefully weakening encryption for one entity undermines its integrity, opening the door to future abuse and exploitation.

Privacy rights, which are deeply rooted in legal frameworks and ethical concepts, stress the protection of individuals against unjustified surveillance and interference. In the mathematical language of cryptography, this translates to the encryption techniques' resistance to unwanted access. In contrast, national security imperatives necessitate methods that provide legitimate access for intelligence and law enforcement organizations. The mathematical proof that backdoors introduce vulnerabilities that can be exploited by hostile actors is a cornerstone of the ethical case against their adoption.

The legal and ethical dimensions go beyond the national level, incorporating worldwide discussions about data protection and surveillance. The mathematical underpinnings of cryptography, which emphasize notions such as ciphertext indistinguishability in secure encryption methods, serve as the cornerstone for the global privacy rights discourse. Ethical considerations highlight cryptography practitioners' need to promote individual freedoms while also respecting the legitimate needs of national security agencies (Limniotis, 2021).

In essence, the mathematical and ethical foundations of the privacy rights vs. national security argument emphasize the importance of a sophisticated and principled approach. Creating legislative frameworks and cryptographic solutions that preserve individual privacy while meeting national security needs necessitates a careful balance and an ongoing review of the ethical implications of technical breakthroughs. The history of this discourse demonstrates cryptography's lasting importance in shaping the parameters of privacy and security in our linked society.

6.2 Encryption Backdoor

The issue of encryption backdoors is a major source of dispute in the legal and ethical discussions surrounding cryptography. Encryption, which is intended to strengthen the security and confidentiality of digital communications, faces a severe ethical quandary when governments and law enforcement agencies argue for the inclusion of backdoors (deliberate weaknesses or access points) in cryptographic systems. This argument dives into the delicate balancing of individual privacy rights and national security imperatives.

From a mathematical standpoint, the concept of an encryption backdoor has intrinsic inconsistencies. The mathematical robustness of encryption stems from its ability to resist illegal access, which is based on the strength of algorithms and keys. The deliberate construction of flaws, as implied by encryption backdoors, compromises the fundamental nature of cryptography. In mathematical terms, this creates a paradox: purposely reducing the security of a cryptographic system for one entity makes it vulnerable to exploitation by others.

Privacy rights, which are profoundly rooted in legal frameworks and ethical concerns, advocate for the protection of individuals from unjustified surveillance. The mathematical foundation of cryptography, which emphasizes notions such as the difficulty of factoring big numbers in RSA and the security of discrete logarithm issues, acts as a barrier to unwanted access. However, encryption backdoors call into question this foundation, posing a moral and ethical quandary of sacrificing individual privacy for state security (Huang & Han, 2024).

On the legal front, the encryption backdoor discussion takes place in the framework of laws and government regulations. The conflict between legislative frameworks that protect privacy rights and the perceived demands of national security services generates a legal landscape in which judgments have far-reaching repercussions. The tension is especially severe on the international stage, where conflicting legal norms and cultural viewpoints complicate the resolution of this ethical problem.

The ethical concerns of encryption backdoors go beyond the immediate implications for privacy and security. They address the broader issue of social trust in digital systems, as well as the ethical duties of cryptography practitioners. The mathematical argument that backdoors introduce vulnerabilities that can be exploited by both legitimate authorities and bad actors emphasizes the importance of taking a principled stance that protects individual liberties while addressing security concerns.

Finally, the issue over encryption backdoors involves complex intersections of mathematics, law, and ethics. The mathematical roots of encryption serve as a beacon, illuminating the ethical problems that come with sacrificing individual privacy for national security. As this discussion progresses, a concerted effort is required to develop legal frameworks that reconcile competing imperatives, ensuring that cryptographic principles remain guardians of both security and individual rights in our digitally interconnected world.

6.3 Ethical Considerations for Data Protection

In the ever-expanding digital ecosystem, ethical data protection considerations have emerged as a cornerstone in the discussion of privacy and security. Because cryptographic tools serve an important role in protecting sensitive information, the ethical foundations of their use become critical. This topic focuses on the ethical aspects of data protection, including the roles of cryptography practitioners, policymakers, and technology developers.

The ethical framework for data protection is based on the values of transparency, consent, and fairness. Individuals have the right to know how their data is collected, processed, and kept. Cryptographic protocols help protect data by preserving its secrecy and integrity. Mathematically, encryption algorithms function as guardians, allowing for secure data transfer and storage while adhering to the ethical concept of openness by protecting information from unauthorized access.

Consent, another ethical pillar, highlights the significance of individuals willingly giving consent for the use of their data. In cryptographic applications, this corresponds to prudent encryption key management. Secure key management becomes an ethical obligation, ensuring that access to encrypted data is limited to those with proper authority. The mathematical rigor used in cryptographic key management is consistent with the ethical ideal of respecting individuals' choices over the use of their data.

Fairness in data protection ethics refers to the equitable treatment of individuals in the digital domain. Cryptographic technologies promote fairness by creating a level playing field in which all data is subject to the same stringent protective procedures. Ethical considerations in cryptography necessitate the creation and implementation of algorithms that are non-discriminatory and resistant to bias, fostering a digital ecosystem in which everyone's privacy is protected (Dhirani et al., 2023).

The ethical concerns in data protection extend beyond individual interactions to social consequences. Cryptographic practitioners have an ethical responsibility to help design safe systems that maintain democratic principles, respect diversity, and limit potential damages. The mathematical principles that govern cryptographic algorithms serve as a solid framework for ethical decision-making, underscoring the importance of taking into account broader societal ramifications while designing and implementing data protection measures (Limniotis, 2021).

Finally, ethical considerations in data protection play an important role at the intersection of mathematics, technology, and societal values. Cryptographic practitioners and politicians handle this world with mathematical rigor, ensuring openness, consent, and fairness. As the digital age progresses, the ethical foundations of data protection remain critical to ensuring a secure and equitable digital future.

7. FUTURE DIRECTIONS

7.1 Recommendations for Development

As we set the road for the future of cryptography, certain key recommendations emerge, which will guide the creation of cryptographic protocols and systems. These ideas represent a comprehensive strategy that considers enhancements to technology, ethical considerations, and the changing ecology of digital communications.

7.1.1. Embrace Post-Quantum Cryptography

The future emergence of quantum computers needs a deliberate shift to post-quantum cryptography. Investing in quantum-resistant cryptography methods and standardizing them are among the recommendations. This includes encouraging collaborations among researchers, industry partners, and standards organizations to enable the development of strong post-quantum cryptography solutions.

7.1.2. Improve Usability Without Compromising Security

Cryptographic systems frequently have problems with accessibility, which limits their adoption. Future advancements should emphasize intuitive designs and seamless interaction across multiple applications. However, this improvement in usability should not jeopardize the essential principles of cybersecurity. Achieving an equilibrium between accessibility and solid security safeguards is critical for the broad use of cryptographic technologies.

7.1.3. Integrate Privacy-Preserving Technology

The growing concern about data privacy highlights the significance of including privacy-preserving technology in cryptographic systems.

Homomorphic encryption,

zero-knowledge proofs, and

differential privacy approaches

all present attractive possibilities. Future work should aim to make these technologies more cost-effective and productive while maintaining privacy as a core element of cryptographic solutions.

4. Address the Ethical Implications

Ethical considerations must be part of cryptographic development. This includes transparency in algorithmic decisions, accountable identification of vulnerabilities and risks, and a dedication to fairness and diversity. Cryptographic practitioners should continue their education on ethical considerations, establish a culture of responsible innovation, and actively engage with ethical frameworks during the development process.

5. Adapt to an Evolving Legislative Landscape

Cryptography operates in a constantly changing regulatory context. Recommendations include aggressively engaging with lawmakers to help develop rules that balance security and private privacy. Cryptographic experts should stay updated on legal developments and push for regulations that encourage innovation while adhering to ethical norms.

6. Explore Novel Applications

The future of cryptography lies in developing fresh applications outside of traditional domains. Areas such as

Secure multi-party computation,

Blockchain technology, and

The convergence of cryptography

with new technologies such as AI meriting additional investigation. Recommendations include encouraging interdisciplinary research and collaboration to fully realize the potential of cryptographic advancements.

7. Increase Open Source Collaboration

Open-source collaboration promotes innovation and transparency in cryptographic development. Recommendations include actively participating in open-source cryptographic projects, fostering community-driven initiatives, and following best practices for code review and development. This collaborative approach strengthens the cryptographic community as a whole.

Finally, the future directions for cryptography require a complete approach that blends technological improvements with ethical considerations and adapts to the changing digital environment. By embracing these principles, the cryptography community may help to develop robust, user-friendly, and ethical solutions for navigating the problems and opportunities of the digital age.

7.2 Implementation Strategy

Cryptography's progress is inextricably linked to effective implementation methodologies that not only ensure secure system deployment but also allow for smooth integration into varied technological environments. The recommendations below highlight essential implementation techniques for the future of cryptography:

7.2.1. Standardization and Interoperability

Future cryptographic solutions should follow generally acknowledged standards to enable interoperability and compatibility across several platforms and applications. Collaboration across cryptographic communities, industry participants, and standardization bodies is critical for defining and maintaining strong standards that will stand the test of time.

7.2.2. Continuous Security Audits

Given the ever-changing nature of cybersecurity threats, ongoing security audits are critical. Implementation plans should prioritize regular and thorough security audits of cryptographic systems. This includes discovering and fixing potential vulnerabilities, monitoring for emerging threats, and ensuring that cryptographic implementations remain resilient in the face of changing attack vectors.

7.2.3. Hardware-Based Security

Hardware-based security techniques are crucial for cryptographic systems. Future strategies should look into developments in secure hardware technology, such as trusted platform modules (TPMs) and hardware security modules. Integrating cryptographic functions into safe hardware settings improves the overall security posture of systems.

7.2.4. User Education and Training

Implementing good cryptography systems necessitates user knowledge and education. Future initiatives should prioritize user training to ensure that people using cryptographic technologies understand their capabilities, limits, and best practices. This includes creating user-friendly teaching resources and cultivating a culture of cybersecurity awareness.

7.2.5. Agile Development Practices

Cryptographic solutions should follow agile development principles to increase flexibility and responsiveness to changing needs. Iterative development cycles, continuous integration, and regular updates ensure that cryptographic systems can react quickly to new threats and technological breakthroughs.

7.2.6. Integration With Emerging Technologies

Cryptographic systems should be compatible with and use new technologies like blockchain, edge computing, and artificial intelligence. Integrating cryptographic solutions with these technologies can lead to new opportunities for safe and efficient data processing, storage, and transfer.

7.2.7. Community Collaboration

Open-source collaboration is critical to the development of cryptographic implementations. Future tactics should foster community engagement, information exchange, and participation in open-source cryptographic projects. A lively and engaged community contributes to the collective competence and resilience of cryptographic systems.

In conclusion, the future of cryptography is dependent on effective implementation techniques that include standardization, security audits, user education, agility, quantum-safe practices, responsible disclosure, integration with upcoming technologies, community engagement, and regulatory compliance. By implementing these tactics, the cryptographic community may strengthen the development of secure systems that meet the requirements of a constantly changing digital ecosystem.

7.3 Addressing Evolving Threats

The cybersecurity landscape is constantly changing as adversaries evolve their strategies to exploit vulnerabilities. Addressing these developing dangers is critical to the future of cryptography. One major method is to continuously enhance and change cryptographic algorithms in order to stay ahead of emerging attack vectors. Cryptographic research should prioritize the creation of algorithms that can withstand complex threats such as quantum attacks, side-channel attacks, and new cryptanalytic approaches.

As threats evolve, proactive threat intelligence and information exchange become increasingly important. Collaboration among the cybersecurity community, which includes researchers, industry specialists, and government agencies, should be encouraged. Real-time sharing of threat data and intelligence enables rapid reactions and the creation of effective countermeasures. Furthermore, including machine learning and artificial intelligence in threat detection and response systems might improve the effectiveness with which emergent threats are identified and mitigated.

The increasing interconnection of objects in the age of the Internet of Things (IoT) creates new attack surfaces and vulnerabilities. Future cryptographic strategies must meet the distinct issues presented by IoT devices, such as resource limits and the requirement for lightweight cryptographic solutions. This requires the creation of cryptographic protocols that are specifically adapted to the needs of IoT environments, ensuring both security and efficiency in communication between connected devices.

Finally, user awareness and education are critical in addressing changing dangers. Social engineering attacks, phishing, and other forms of manipulation frequently target human vulnerabilities. Cryptographic tactics should incorporate educational initiatives that provide users with the knowledge they need to notice and respond to potential threats. This includes advocating security practices such as strong passwords, multi-factor authentication, and the need of store cryptographic keys securely.

To summarize, combating changing challenges in cryptography requires a diverse approach. This includes continuously improving cryptographic algorithms, exchanging proactive threat intelligence, integrating modern technologies, adapting to IoT issues, and educating users to establish a robust defense against an ever-changing threat landscape.

8. Conclusion

8.1 Summary of Key Findings

Finally, our investigation into encryption techniques and data privacy has revealed important findings that highlight cryptography's crucial role in protecting sensitive information. The trip through historical evolution, cryptographic primitives, and modern issues revealed essential results that capture the heart of this research.

To begin, the historical evolution of encryption has demonstrated cryptographic approaches' ability to adapt to changing information security needs. From classical cryptography to the development of modern cryptographic algorithms, the journey has been characterized by a constant search for greater security measures.

The study of cryptographic primitives, such as symmetric and asymmetric key encryption, hash functions, and the mathematical underpinnings of security, has revealed the complicated systems that form the backbone of secure communication. Understanding these primitives is critical for developing strong cryptographic protocols capable of withstanding modern threats.

The investigation of issues and upcoming technologies has highlighted the critical necessity for tackling the recent development of quantum computing and the significance of quantum-resistant encryption. Managing security and system functionality, reducing computational overhead, and dealing with latency difficulties are all ongoing challenges that must be addressed carefully in cryptographic solutions.

The legal and ethical implications of encryption have highlighted the delicate balance between privacy rights and national security, with encryption backdoors serving as a source of contention. Navigating this world demands a thorough comprehension of the mathematical and ethical principles governing the use of cryptographic instruments.

Looking ahead, development proposals prioritize embracing post-quantum cryptography, improving usability without sacrificing security, and incorporating privacy-preserving technology. Standardization, constant security audits, hardware-based security, and user education are all key implementation techniques. Addressing emerging threats requires a diverse approach that includes algorithmic resilience, threat intelligence sharing, IoT-specific solutions, and user awareness campaigns.

In conclusion, this comprehensive exploration of encryption techniques and data privacy has shed light on the complex interplay of mathematical concepts, technical breakthroughs, and ethical issues. As we approach a digital future, the findings highlight the importance of continuing innovation, collaboration, and a principled approach to cryptography in navigating the intricacies of a constantly changing cybersecurity scenario.

8.2 Implications for Policymakers and Practitioners

This study's conclusions have far-reaching consequences for policymakers and practitioners in cybersecurity, privacy, and technology. As we extract the essential ideas, it becomes clear that educated decision-making and strategic actions are critical for shaping a secure and privacy-preserving digital future.

8.2.1 To Policymakers

1. Regulatory Frameworks:

   Policymakers must develop regulatory frameworks that strike a careful balance between individual privacy rights and national security needs. The encryption backdoor issue emphasizes the necessity for sophisticated legislation that protects the integrity of cryptographic systems while also addressing valid security concerns.
2. Global Collaboration:

Given the global character of digital interactions, policymakers should aggressively promote international collaboration. Harmonizing regulatory norms and encouraging collaboration in tackling cross-border cybersecurity risks will be critical for creating a secure and linked digital environment.

3. **Ethical issues:**
   Policymakers should incorporate ethical issues into legislative processes. Ensuring that legislation reflects a commitment to transparency, fairness, and appropriate data protection procedures will increase public trust and help to shape the ethical evolution of the digital ecosystem.
4. Support for Research:
   Recognizing the rapid pace of technological change, policymakers should allocate funds to promote cryptography research. This includes financing for the development of post-quantum cryptographic solutions, privacy-preserving technology, and programs to improve the ethical and secure use of cryptography.

8.2.2 To Practitioners

1. Continuous education:
   Cryptographic practitioners must emphasize ongoing education to stay current on the newest breakthroughs in cryptographic research, emerging threats, and changing regulatory landscapes. Staying educated is critical for making sound judgments in the continuously evolving realm of cybersecurity.
2. User-centered design:
   Practitioners should emphasize user-centered design in cryptographic implementations. Enhancing usability while maintaining security is critical for widespread adoption. Intuitive interfaces, clear communication about security measures, and user-friendly instructional resources all help to establish a security-aware user base.
3. Agile Development Practices:
   The dynamic nature of cybersecurity threats needs agile development methods. Practitioners should use iterative development cycles, regular updates, and adaptive tactics to effectively respond to changing threats and weaknesses.
4. Responsibility in Disclosure:

Practitioners must follow appropriate disclosure practices. Creating routes for researchers to responsibly report vulnerabilities, as well as engaging with the larger cybersecurity community, ensures that possible threats are mitigated quickly and improves the overall security posture.

5. Collaboration and Open Source:

   Open-source collaboration remains a key component of cryptography strength. Practitioners should actively participate in open-source initiatives, creating a community-driven approach that improves the collective robustness of cryptographic systems.

Overall, the implications for policymakers and practitioners highlight the importance of a comprehensive and collaborative approach. Policymakers play an important role in establishing the legal landscape, while practitioners, armed with knowledge, ethical standards, and adaptive tactics, are critical in implementing cryptographic solutions that are resilient in the face of changing obstacles. The symbiotic relationship between informed policymaking and effective practical implementation is critical for traversing the complex landscape of encryption methods and data privacy.

8.3 Call to Action to Ensure Data Privacy

As we end our investigation of encryption mechanisms and data privacy, a powerful call to action develops for all stakeholders involved in the digital domain. Protecting data privacy is a common obligation that necessitates concerted and determined efforts.

8.3.1. Collaborative Advocacy

A call to action starts with a coordinated push for strong data privacy policies. Policymakers, industry leaders, and advocacy groups should collaborate to promote the creation and implementation of privacy-focused policies. This involves promoting projects that prioritize user permission, data transparency, and the ethical application of cryptographic technologies.

8.3.2. Empowering Users:

It is critical to educate individuals about data privacy. A call to action includes instructional ads that educate users on the necessity of secure habits including using strong passwords, activating multi-factor authentication, and understanding the ramifications of data sharing. By cultivating a privacy-conscious culture, consumers become active partners in the protection of their digital footprint.

8.3.3. Technology Innovation

This forward-thinking call to action encourages continuing technology innovation. Practitioners and researchers must push the boundaries of cryptographic developments, assuring the creation of solutions that not only protect data privacy from current risks but also foresee and mitigate future issues. Accepting privacy-preserving technology and implementing ethical design principles should be at the forefront of this innovation.

In the end, the call to action is a collaborative effort that requires the participation of politicians, practitioners, and individuals alike. By lobbying for privacy-focused regulations, educating users, and driving technological innovation, we can pave the way for a digital future in which data privacy is a fundamental and protected right for all.

REFERENCES

Altulaihan, E., Almaiah, M. A., & Aljughaiman, A. (2022). Cybersecurity threats, countermeasures and mitigation techniques on the IoT: Future research directions. *Electronics (Basel)*, 11(20), 3330. 10.3390/electronics11203330

Balamurugan, C., Singh, K., Ganesan, G., & Rajarajan, M. (2021). Post-quantum and code-based cryptography—Some prospective research directions. *Cryptography*, 5(4), 38. 10.3390/cryptography5040038

Bodur, H., & Kara, R. (2015). Secure SMS Encryption Using RSA Encryption Algorithm on Android Message Application. In *3rd International Symposium On Innovative Technologies In Engineering And Science*. Research Gate.

Buell, D. (2021). Simple Ciphers. In *Fundamentals of Cryptography: Introducing Mathematical and Algorithmic Foundations* (pp. 11–26). Springer International Publishing. 10.1007/978-3-030-73492-3_2

Deavours, C. A., & Reeds, J. (1977). The Enigma part I historical perspectives. *Cryptologia*, 1(4), 381–391. 10.1080/0161-117791833183

Dhirani, L. L., Mukhtiar, N., Chowdhry, B. S., & Newe, T. (2023). Ethical Dilemmas and Privacy Issues in Emerging Technologies: A Review. *Sensors (Basel)*, 23(3), 1151. 10.3390/s2303115136772190

Equihua, C., Anides, E., García, J. L., Vázquez, E., Sánchez, G., Avalos, J.-G., & Sánchez, G. (2021). A low-cost and highly compact FPGA-based encryption/decryption architecture for AES algorithm. *Revista IEEE América Latina*, 19(9), 1443–1450. 10.1109/TLA.2021.9468436

Eslami, Y. (2006). An area-efficient universal cryptography processor for smart cards. *IEEE transactions on very large scale integration (VLSI) systems*. IEEE.

Faruk, M. J. H., Tahora, S., Tasnim, M., Shahriar, H., & Sakib, N. (2022). A review of quantum cybersecurity: threats, risks and opportunities. In *2022 1st International Conference on AI in Cybersecurity (ICAIC)*, (pp. 1-8). IEEE.

Gray, I. I. I., & James, W. (1992). Toward a mathematical foundation for information flow security. *Journal of Computer Security*, 1(3-4), 255–294. 10.3233/JCS-1992-13-405

Huang, H., & Han, Z. (2024). Computational ghost imaging encryption using RSA algorithm and discrete wavelet transform. *Results in Physics*, 56, 107282. 10.1016/j.rinp.2023.107282

Kaknjo, A., Rao, M., Omerdic, E., Newe, T., & Toal, D. (2019). Real-Time Secure/Unsecure Video Latency Measurement/Analysis with FPGA-Based Bump-in-the-Wire Security. *Sensors (Basel)*, 19(13), 2984. 10.3390/s1913298431284580

Kerr, O. S. (2000). The fourth amendment in cyberspace: Can encryption create a reasonable expectation of privacy. *Connecticut Law Review*, 33, 503.

Lalem, F., Laouid, A., Kara, M., Al-Khalidi, M., & Eleyan, A. (2023). A novel digital signature scheme for advanced asymmetric encryption techniques. *Applied Sciences (Basel, Switzerland)*, 13(8), 5172. 10.3390/app13085172

Lessig, L. (2009). *Code: And other laws of cyberspace*. ReadHowYouWant. com.

Levinsky, J. (2022). Encryption: The History and Implementation.

Li, N. (2010). Research on Diffie-Hellman key exchange protocol. In *2010 2nd International Conference on Computer Engineering and Technology*. IEEE.

Li, S., Chen, Y., Chen, L., Jing, L., Kuang, C., Li, K., Liang, W., & Xiong, N. (2023). Post-Quantum Security: Opportunities and Challenges. *Sensors (Basel)*, 23(21), 8744. 10.3390/s2321874437960442

Li, Y., & Liu, Q. (2021). A comprehensive review study of cyber-attacks and cyber security; Emerging trends and recent developments. *Energy Reports*, 7, 8176–8186. 10.1016/j.egyr.2021.08.126

Limniotis, K. (2021). Cryptography as the Means to Protect Fundamental Human Rights. *Cryptography*, 5(4), 34. 10.3390/cryptography5040034

Mahto, D., & Yadav, D. K. (2017). RSA and ECC: A comparative analysis. *International Journal of Applied Engineering Research: IJAER*, 12(19), 9053–9061.

Mao, C. (2022). Design of Computer Storage System Based on Cloud Computing. In *International Conference on Communication, Devices and Networking*. Singapore: Springer Nature Singapore.

Mao, S., Zhang, H., Wu, W., Liu, J., Li, S., & Wang, H. (2014). A resistant quantum key exchange protocol and its corresponding encryption scheme. *China Communications*, 11(9), 124–134. 10.1109/CC.2014.6969777

Oktaviana, B., & Utama Siahaan, A. P. (2016). Three-pass protocol implementation in caesar cipher classic cryptography. *IOSR Journal of Computer Engineering*, 18(04), 26–29. 10.9790/0661-1804032629

Pharkkavi, D., & Maruthanayagam, D. (2018). Time complexity analysis of RSA and ECC based security algorithms in cloud data. *International Journal of Advanced Research in Computer Science*, 9(3), 201–208. 10.26483/ijarcs.v9i3.6104

Politi, A. (2010). Quantum information science with photonic chips. In *36th European Conference and Exhibition on Optical Communication*, (pp. 1-3). IEEE.

Preneel, B. (1994). Cryptographic hash functions. *European Transactions on Telecommunications*, 5(4), 431–448. 10.1002/ett.4460050406

Raheman, F. (2022). The future of cybersecurity in the age of quantum computers. *Future Internet*, 14(11), 335. 10.3390/fi14110335

Ribeiro, L. E. (2019). High-profile data breaches: Designing the right data protection architecture based on the law, ethics and trust. *Applied Marketing Analytics*, 5(2), 146–158.

Rubinstein-Salzedo, S., & Rubinstein-Salzedo, S. (2018). *The vigenere cipher*. Cryptography. 10.1007/978-3-319-94818-8_5

Standard, Data Encryption. (1999). Data encryption standard. *Federal Information Processing Standards Publication*, 112, 3.

Tariq, U., Ahmed, I., Bashir, A. K., & Shaukat, K. (2023). A Critical Cybersecurity Analysis and Future Research Directions for the Internet of Things: A Comprehensive Review. *Sensors (Basel)*, 23(8), 4117. 10.3390/s2308411737112457

Vaudenay, S. (2006). Prehistory of Cryptography. *A Classical Introduction to Cryptography: Applications for Communications Security*. Research Gate.

Victor-Mgbachi, T. O. Y. I. N. (2024). *Navigating Cybersecurity Beyond Compliance: Understanding Your Threat Landscape and Vulnerabilities*. Research Gate.

Yassein, M. B., Aljawarneh, S., Qawasmeh, E., Mardini, W., & Khamayseh, Y. (2017). Comprehensive study of symmetric key and asymmetric key encryption algorithms. In *2017 international conference on engineering and technology (ICET)* (pp. 1-7). IEEE. 10.1109/ICEngTechnol.2017.8308215

Zhang, X., Zhou, Z., & Niu, Y. (2018). An image encryption method based on the feistel network and dynamic DNA encoding. *IEEE Photonics Journal*, 10(4), 1–14. 10.1109/JPHOT.2018.2858823

Chapter 5
Enhancing 5G Security Using Hybrid Learning Approach for Malware Detection and Classification

E. Murugavalli
https://orcid.org/0000-0003-0652-4262
Thiagarajar College of Engineering, India

K. Rajeswari
https://orcid.org/0000-0002-7834-4183
Thiagarajar College of Engineering, India

V. Vinoth Thyagarajan
https://orcid.org/0000-0001-7302-2997
Thiagarajar College of Engineering, India

P. S. Shruti
Thiagarajar College of Engineering, India

K. R. Hemalatha
https://orcid.org/0009-0004-7311-425X
Thiagarajar College of Engineering, India

ABSTRACT

As the security challenges are inherent to 5G networks, the threat of malware continues to rise. Due to the high-speed nature of 5G, there is a need for automated malware

DOI: 10.4018/979-8-3693-2786-9.ch005

detection and classification methods. This chapter presents a hybrid model for malware classification that combines machine learning and deep learning techniques. This model leverages the power of feature extraction using the InceptionResNetV2 deep learning model pre-trained on ImageNet. The features extracted from malware images are fed into a random forest classifier for the final classification task. The experimental evaluation conducted on a dataset comprising 31 malware families achieves an accuracy of 94.15%. Furthermore, the predicted labels closely align with the ground truth labels across various malware families, showcasing the model's ability to capture the intricate patterns and characteristics of diverse malware strains.

1. INTRODUCTION

In the realm of communications, 5G stands as fifth-generation technology standard for mobile networks, succeeding the widely used 4G networks. Despite its advanced capabilities, 5G networks, like many others, are vulnerable to cyber-attacks that compromise sensitive information. These breaches could be mitigated with effective monitoring systems in place(Rahman, A.U. et al., 2022). The rapid evolution and increasing sophistication of cyber threats pose significant risks to the entire 5G-enabled Internet-of-Things (IoT) infrastructure. This interconnected network of devices faces considerable security challenges. Cyber threat analysis enhances network security by focusing on the identification and prevention of complex networks-based assaults and risks. Additionally, it involves securing the network through the investigation and classification of malicious activities (Qureshi, S. et al., 2021). Malware, short for malicious software, is any software specifically designed to harm, disrupt, or gain unauthorized access to computer systems, networks, or data, often without the user's knowledge or consent. It includes a broad spectrum of malicious programs, such as viruses, worms, Trojans, ransomware, spyware, adware, and rootkits. In the realm of cybersecurity, malware represents a significant threat, exploiting vulnerabilities in computer systems to steal sensitive information, compromise system integrity, disrupt operations, or even render systems unusable.

Malware is commonly disseminated through diverse channels, such as mail attachments, infected websites, corrupted removable media, and software loopholes. Cyber threats come in various forms, with each malware family posing distinct risks and challenges to digital security. From the notorious viruses and worms to sophisticated ransomware and spyware, there are 31 prominent malware families that cybersecurity professionals must contend with. Understanding the characteristics and behaviours of each malware family is important to develop efficient defence strategies and mitigate potential threats. By discerning patterns and signatures unique to each family, cybersecurity experts can better identify, analyse, and respond to cyber threats, ultimately bolstering the resilience of digital infrastructure and safeguarding sensitive data against malicious actors.

One of the malware detection mechanism, Network Anomaly Detection (NAD) involves continuous observation of the network to identify and respond to any unusual or suspicious activities, thereby enhancing the security of 5G networks and protecting against data breaches. On the other hand, conventional signature-based detection systems, that recognise known malware signatures, fail to keep up with new threats. To address this challenge, scientists working in cybersecurity are now concentrating on machine learning and deep learning techniques for malware detection and classification. 6G is also envisioned to leverage machine learning capabilities to autonomously detect anomalies while managing multi-dimensional data with diverse characteristics (Saeed, M.M. 2023).

Deep learning, branch of machine learning, showed notable promise in extracting intricate features from complex data sources such as images and texts. Hybrid learning combines the strengths of various machine learning models, which has gained popularity in the past few years. Hybrid models leverage the complementary capabilities of multiple learning techniques to enhance the accuracy as well as robustness of malware classification systems. The utilization of InceptionResNetV2 for feature extraction and Random Forest for classification presents a powerful framework for malware classification. InceptionResNetV2, a pre-trained deep learning model, excels at extracting intricate features from grayscale images, allowing it to discern subtle patterns and nuances inherent in different malware families. By leveraging transfer learning, InceptionResNetV2 can adapt its learned representations to the domain-specific task of malware classification, enhancing performance and accelerating training. Next, the extracted features are given as input to a Random Forest classifier, renowned for its robustness, scalability, and interpretability. Random Forest effectively learns complex decision boundaries from high-dimensional feature vectors, providing reliable and interpretable results. This hybrid approach not only combines the strengths of deep learning and traditional machine learning but also offers a practical solution for cybersecurity tasks, where both accuracy and interpretability are paramount.

2. RELATED WORKS

Industrial Internet of Things (IIoT) is one of the use-cases of 5G due its more effective low-latency. Malware assaults increasingly target vulnerable yet highly connected IoT devices, emphasising the crucial significance of safety and confidentiality. Since the IIoT system introduces new kind of attack vectors leading to severe data privacy and security risks. The article described by (Ahmed et al., 2022) designed a methodology which is based on an image representation of the malware and convolutional neural networks (CNN). A multi-layer deep learning-based architecture was designed to extract complementary discriminative features by combining multiple layers in order to differentiate various malware attacks on 5G-enabled IIoT system. The massive machine type communication of 5G enables the hyper-connected and personalized services which are likely to be exposed to malwares and the devices can run the risk of being infected by malware. A scheme is proposed by (Kim, K.C et al., 2019) to identify malicious codes by hybridizing machine learning techniques based on application programming interface and application programming interface call graph. The proposed scheme is aimed to provide a foundation for more accurate and faster malware prediction by reducing the false positive rate with the aid of the excessive use of features.

Early malware detection is a critical defence against the ever-increasing surge of cyberattacks attacking connected smart devices in the age of 5G-enabled IoT privacy and security. Almazroi, A.A., and Ayub, N. 2024 introduce a new hybrid deep-learning approach for malware detection that uses a separate collection of eight unique malware datasets. The increasing proliferation of AI-enabled IoT devices raises substantial security concerns, affecting both privacy and other resources. The continuously increasing volume of big data generated by IoT devices exacerbates this issue, particularly when making decisions based on the ever-growing data. As IoT technology advances, cyber threats, particularly malware, require adaptive and robust security mechanisms. This detection architecture excels at detecting existing and varying attacks, as well as adapting to new threats in the dynamic IoT environment. To improve performance indicators like efficiency and accuracy, we use distribution analysis, feature selection, and isolated forest model clustering. Additionally, the hybrid classification technique improves precision and detection processes. To address the problems of a dynamic environment, this study presents a specialised BERT-based Feed Forward Neural Network Framework (BEFNet) designed for IoT applications. Their study adopts a novel architecture with various modules to extensively analyse eight datasets, each representing a distinct sort of malware. BEFSONet, optimised with the Spotted Hyena Optimizer (SO), exhibits its adaptability to different types of malware data. Comprehensive exploratory investigations and comparative evaluations reveal BEFSONet's remarkable performance metrics, which are very accurate. This study demonstrates BEFSONet as a strong defence mechanism in the age of IoT security, offering an effective answer to the changing problems in dynamic decision-making contexts.

In recent years, 5G communication technology has been integrated into vehicle-to-everything (V2X) systems, which provide ultra-low latency and high reliability communications. However, as communication performance has improved, security and privacy concerns have increased. Cyber-attacks and threats have become more aggressive in a 5G-enabled V2X environment, while security attackers have grown in sophistication. Extensive research has been conducted to defend V2X by implementing cryptographic solutions for security services such as user authentication, data integrity, and confidentiality. The usual public key infrastructure provided by several standardisation organisations cannot protect against these types of assaults. Machine Learning (ML) and Deep Learning (DL) are becoming increasingly important for securing systems and roads. Many V2X Misbehaviour Detection Systems (MDSs) use this concept. Thus, sophisticated mechanisms must be in place to detect such attacks and attackers. Thus, there is a is a research gap in analyzing these systems and so developing an effective ML/DL-based intrusion detection system is still an open research problem in 5G enabled networks. The literature and reviews presented by (Boualouache, A. & Engel, T. 2023) describe a comprehensive survey and classification of ML based intrusion detection systems. They also describe communication architecture of 5G-V2X and its attack classification. The following V2X prone attacks are analyzed in detail: Position falsification, False information, Sybil attack, Position tracking attack, Distributed denial of service attack, Reply attack, Timing attack, Greyhole attack, Blackhole attack, Jamming attack, Impersonation attack, Warmhole attack and Eavesdropping attack. They analyse and examine them from both security and ML perspectives. The open research issues are identified in the following domains: Reproducible and benchmarking datasets, Context awareness, Zero-day attacks, Deployment and incentives, Security and privacy, and Standardization. They also summarized the ML based intrusion detection systems for both IP based applications and non-IP based applications.

The extensive use of mobile devices running Android operating systems in the Internet of Things (IoT) puts them vulnerable to developing cybersecurity assaults. The Internet of Medical Things (IoMT) is made up of healthcare gadgets such as smart watches, smart thermometers, biosensors, and others that protect the privacy of users' sensitive information and medical histories. Thus, detecting Android malware is critical for protecting sensitive data and assuring the stability of IoT networks. This article by Faria Nawshin et al. (2024) focuses on AI-enabled Android malware detection to improve zero trust security in IoMT networks. The zero-trust security model necessitates severe identity verification and user authentication for all entities that access resources from a private network, both inside and outside the network perimeter. To detect Android malware in IoMT networks using the zero-trust model, a novel solution based on Differential Privacy (DP) within a Feedforward Neural Network (FNN) is proposed. The combination of DP and FNN displays the capacity to detect both known and undiscovered malware kinds while also improving accuracy for both static and dynamic elements of Android applications.

A new approach, as proposed in (Roseline S. A et al. 2020) is employed to address evolving malware threats using machine learning and visualization techniques. Malware is represented as 2D images, and a layered ensemble method inspired by deep learning principles is employed. This approach outperforms traditional deep learning methods with less computational complexity and without requiring hyperparameter tuning or backpropagation. Effective detection is demonstrated across three different malware datasets, indicating the potential for identifying new and advanced malware variants. Utilizing deep learning and transfer learning techniques, the research conducted in reference (Pant, D. & Bista, R. 2021). is centred on accurately classifying grayscale images of malware into their respective families.

While pretrained models such as VGG16, ResNet-18, and InceptionV3 were investigated, a custom CNN model surpassed them, achieving outstanding validation accuracy. The custom model demonstrated superior accuracy and an impressive F1 score of 0.99 in comparison to other models. Despite encountering challenges such as parameter loss in improperly tuned VGG16, the method of converting malware into images for classification proved highly effective, streamlining the malware classification process. The challenges associated with accurately detecting and categorizing malware variants are addressed in (Garcia, F.C.C. & Muga II, F.P. 2016). which frequently utilize sophisticated techniques to evade traditional detection methods. Through the introduction of a novel approach involving the conversion of malware binaries into image representations and the utilization of Random Forest for classification, this research endeavors to enhance malware detection capabilities. The achieved accuracy rate of 0.9562 underscores the effectiveness of the proposed methodology. The effectiveness of two machine learning algorithms, Support Vector Machine (SVM) and Random Forest, for detecting malware in Windows Portable Executable (PE) files through static analysis, is evaluated in (Ismail, H et al., 2024).

A dataset containing both malware and safe applications is utilized to train and compare the performance of SVM and Random Forest models. The study aims to determine which algorithm is more effective in classifying PE files as malware or safe. The results indicate that Random Forest achieves a higher accuracy (98.53%) compared to SVM (97.14%). This research provides valuable insights into selecting the Random Forest algorithm for PE file malware detection, showcasing its potential usefulness in real-world cybersecurity applications. Examining the effectiveness of different machine learning algorithms for malware detection based on network traffic data, (Chukunda, C.D et al., 2022) utilizes a dataset containing 17,845 samples collected from Kaggle Data Set and VirusShare. The study employs TensorFlow for training and testing binary classification models to distinguish between malicious and benign network traffic. Notably, Random Forest is found to have the highest accuracy among the algorithms tested, achieving a great accuracy rate. This suggests that Random Forest could be a suitable choice for malware detection tasks. Additionally, the study notes that Random Forest's average detection speed ranges from 3 to 8 seconds, allowing for prompt mitigation actions to be taken. These findings provide valuable insights into selecting an effective algorithm for malware detection, emphasizing the practical utility of Random Forest in this context.

The classification of malware variants into families to enhance cybersecurity measures is the focus of (Chandranegara, D.R et al., 2023). Grayscale images generated from malware samples are utilized for classification in the study. With 9,029 images from the Malimg dataset covering 25 malware classes, two deep learning architectures, VGG-16 and InceptionResNet-V2, are implemented in two scenarios: one using the original dataset and the other using an under-sampled dataset. Each scenario is evaluated based on accuracy, precision, recall, and F1-score. Accuracy is defined as the ratio of the correctly detected attacks to the total number of events; Precision is defined as the ratio of correctly detected attacks to the total detected attacks. Recall calculates the ratio of correctly detected attacks to the total actual attacks. F1-score can be interpreted as a weighted average of precision and recall. The findings reveal that scenario 2, employing InceptionResNet-V2, achieves a lower accuracy score compared to VGG-16's. Despite the lower accuracy, the use of InceptionResNet-V2 remains valuable as it provides an alternative approach to malware classification. These results suggest that while InceptionResNet-V2 may not perform as well as VGG-16 in this context, it offers insights into the potential efficacy of different deep learning architectures for malware classification tasks.

The DLMD approach, published in (Rafique, M.F et al., 2019), uses byte and ASM files for feature engineering to identify different malware families. Initially, characteristics are retrieved from byte files using two distinct Deep Convolutional Neural Networks (CNNs). A wrapper-based process then identifies crucial and discriminative opcode features, using Support Vector Machine (SVM) as a classifier. The objective is to create a hybrid feature space by merging many feature spaces, hence limiting the weaknesses of single feature spaces, and lowering the likelihood of missing malware. Finally, aAll nine malware families are categorised using a multilayer perceptron learned on the hybrid space for features. Experimental results, as reported in (Aslan, Ö. & Yilmaz, A.A. 2021), demonstrate that the DLMD technique achieves a log-loss of 0.09 for ten independent runs. The novel deep-learning-based architecture given in (Haq, I.U et al., 2021) enables the classification of malware variants based on a hybrid model. The main contribution of this study is the development of a unique hybrid architecture that optimally integrates two extensive pre-trained network models. This architecture is divided into four stages: data collecting, deep neural network design, architecture training, and network evaluation. The approach was evaluated using the Malimg, Microsoft BIG 2015, and Malevis datasets. Experimental results show that the proposed method can classify malware with high accuracy, outperforming state-of-the-art methods in the literature. When evaluated on the Malimg dataset, the approach attained an accuracy of 97.78%, exceeding the majority of machine learning-based malware detection methods. Convolutional Neural Networks (CNN) and Bidirectional Long Short-Term Memory (BiLSTM) are used to efficiently detect persistent infection. Lad and Adamuthe (2020) describe the CNN and Hybrid CNN+SVM models. The CNN functions as an automatic feature extractor, requiring fewer resources and time than conventional approaches. The CNN model achieves an accuracy of 98.03%, outperforming comparable CNN models such as VGG16 (96.96%), ResNet50 (97.11%), InceptionV3 (97.22%), and Xception (97.56%). Furthermore, the CNN model executes much faster than other existing CNN models. By combining the CNN model and a support vector machine, the fine-tuned model generates a well-selected feature vector of 256 neurons with the FC layer, which is then fed into the SVM. The linear SVC kernel converts the binary SVM classifier into a multi-class SVM, resulting in greater accuracy.

3. PROBLEM STATEMENT

The escalating frequency and sophistication of malware attacks constitute a formidable obstacle to cybersecurity efforts globally. With an extensive array of 31 distinct malware families identified, each employing a unique set of tactics and techniques, there exists a pressing imperative to comprehensively analyse their characteristics, modus operandi, and ramifications on cybersecurity landscapes. Malicious software, ranging from ransomware and Trojans to botnets and spyware, poses multifaceted challenges for individuals, enterprises, and governmental institutions alike, manifesting in various forms of cyber intrusion, data compromise, and operational disruption (Shetu, S. F. et al, 2019, Srinivasan, S., & Deepalakshmi, P., 2023). With the proliferation of Internet of Things (IoT) devices connected to 5G networks, botnets can recruit these devices often with weak security measures to form large bot armies capable of launching powerful Distributed Denial of Service (DDoS) attacks. Against this backdrop, elucidating the behavioural patterns and capabilities of each malware family assumes paramount importance in formulating proactive defence mechanisms and curbing the pervasive threats engendered by these evolving digital adversaries (Hamza W.S. et al, 2020). By delving into the intricate nuances of malware taxonomy, cybersecurity stakeholders can glean invaluable insights to inform strategic decision-making, bolster resilience against cyber threats, anfortify the fabric of digital infrastructures in an ever-evolving cyber security landscape.

4. DATASET

Malware Classification is done by converting PE files to byteplot images. The two popular datasets for this type of image classification are the Malimg dataset and the Malevis dataset. The Malimg dataset is highly imbalanced in terms of class distribution and the Malevis dataset is a well-balanced dataset. Both of the datasets are blended together to form a single dataset which has all the classes from the Malevis dataset and 5 classes from the Malimg dataset. The aim of the dataset is to Multiclass Classification of Malware Byteplot images and handling RGB and Grayscale byteplot images together in one dataset.The dataset has two directories one each for training and validation sets. Each of the sets comprise 31 classes. In Figure. 1, the count of images in each malware family is shown. The four different malware family samples, such as Adphoselin Figure. 2, Amonetize in Figure. 3, Fasong in Figure. 4, and Vilselin Figure. 5are shown. Understanding these families is crucial for developing effective defence strategies and mitigating the risks posed by malware infections.

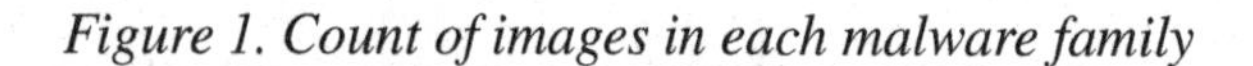
Figure 1. Count of images in each malware family

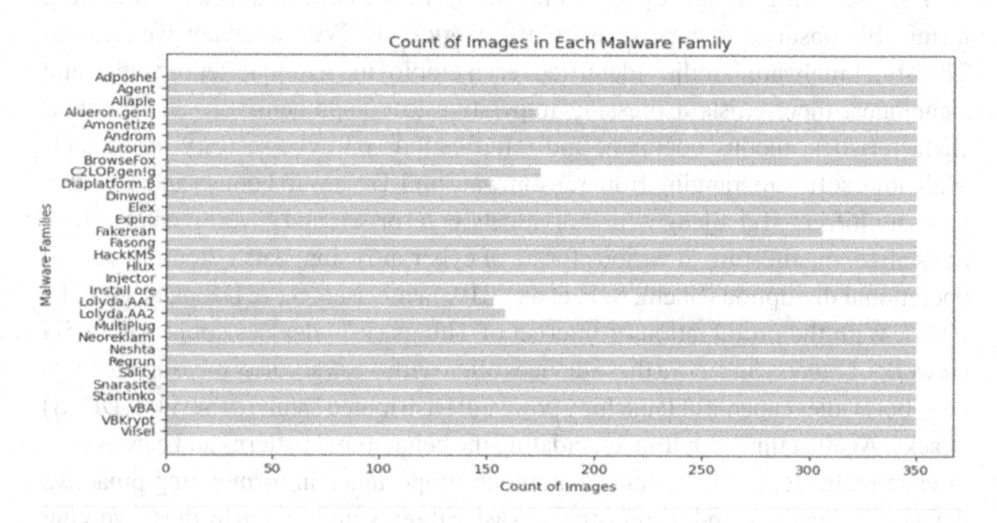

4.1. Adposhel

Adposhel is a type of adware that typically injects unwanted advertisements into web browsers, search results, and other online content. It may also collect browsing data and user information to tailor advertisements and generate revenue for the attackers. Adposhel can degrade system performance and compromise user privacy.

Figure 2. Malware family sample of Adphosel

4.2. Agent

Agent is a versatile malware family capable of performing various malicious activities, including data theft, system compromise, and remote access. It often operates as a Trojan horse, disguising itself as legitimate software to deceive users and evade detection by security measures. Agent can facilitate espionage, financial fraud, and other illicit activities on compromised systems.

4.3. Allaple

Allaple is a polymorphic worm that spreads through removable drives, email attachments, and malicious websites. It infects executable files and propagates to other systems, creating a network of compromised devices. Allaple is known for its ability to steal sensitive information, download additional malware, and facilitate remote access by attackers.

4.4. Alueron.gen

Alueron.gen, also known as TDL, is a rootkit-based malware family that infects the master boot record (MBR) of Windows systems. It operates stealthily, concealing its presence and compromising system integrity. Alueron.gen is often used for data theft, financial fraud, and facilitating other malicious activities on infected systems.

4.5. Amonetize

Amonetize is a family of potentially unwanted programs (PUPs) that typically bundle with legitimate software downloads. It may install adware, browser extensions, or other software without the user's consent, often resulting in unwanted advertisements, browser redirects, and changes to browser settings.

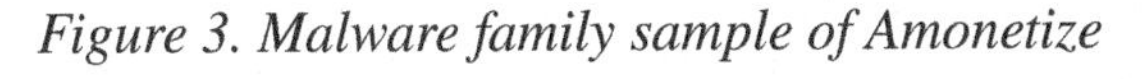

Figure 3. Malware family sample of Amonetize

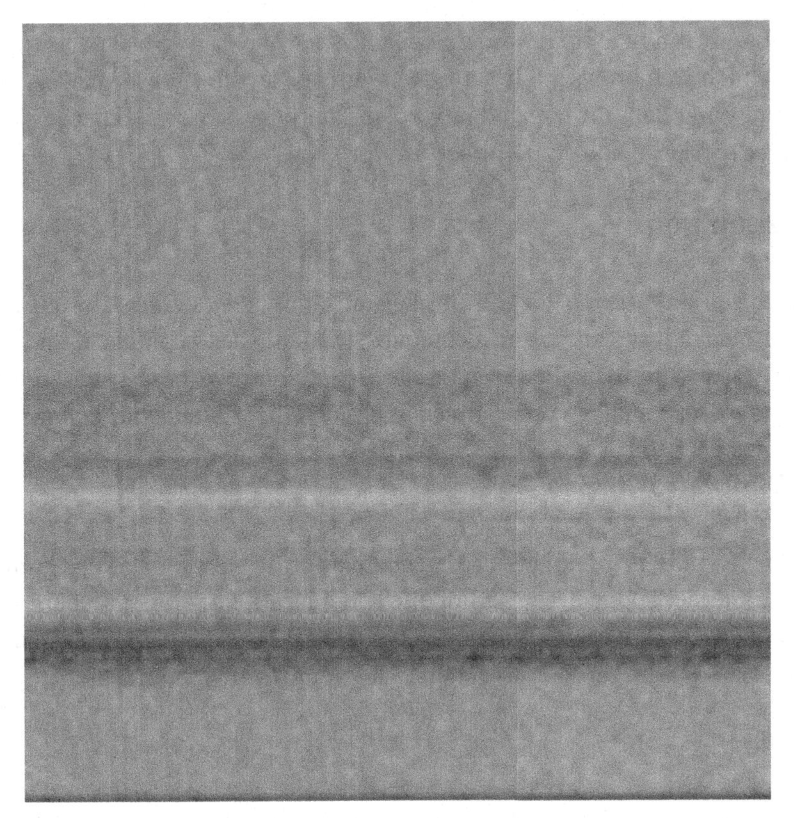

4.6. Androm

Androm is a sophisticated malware family known for its modular architecture and advanced evasion techniques. It can perform various malicious activities, including data theft, system reconnaissance, and propagation to other systems. Androm may also download and execute additional malware payloads, making it a significant threat to cybersecurity.

4.7. AutoRun

AutoRun is a family of malware that exploits the AutoRun function in Windows OS to execute harmful programmes from removable drives. It spreads through infected USB drives, external storage devices, and network shares, posing a significant threat to system security and data integrity.

4.8. BrowseFox

BrowseFox is a type of adware that displays intrusive advertisements, pop-ups, and banners in web browsers. It may also redirect users to sponsored websites and collect browsing data for targeted advertising purposes. BrowseFox can diminish browsing experience and compromise user privacy.

4.9. C2Lop.gen

C2Lop.gen is a variant of the C2Lop malware family, characterized by its polymorphic nature and advanced evasion techniques. It establishes a command-and-control (C&C) infrastructure to facilitate remote access and control of compromised systems, enabling cybercriminals to orchestrate attacks and distribute additional malware.

4.10. Dialplatform.B

Dialplatform.B is a variant of the Dialplatform malware family, known for hijacking internet connections and dialing premium rate phone numbers without the user's consent. It targets vulnerable systems, exploiting weaknesses in network protocols and dial-up services to generate illicit profits for the attackers.

4.12. Elex

Elex is a family of potentially unwanted programs (PUPs) that typically bundle with legitimate software downloads. It mayinstall adware, browser extensions, or other software without the user's consent, often resulting in unwanted advertisements, browser redirects, and changes to browser settings.

4.13. Expiro

Expiro is a polymorphic virus that infects executable files on Windows systems. It can modify file structures, inject malicious code, and evade detection by security software through encryption and obfuscation techniques. Expiro may also download and execute additional malware payloads, posing a significant threat to system security.

4.14. Fakerean

Fakerean is a family of rogue security software, also known as scareware or fake antivirus, that deceives users into believing their systems are infected with malware. It prompts users to purchase fake security products or services to remove non-existent threats, leading to financial losses and exposing users to further security risks.

4.15. Fasong

Fasong is a family of Trojan malware that primarily targets Windows systems. It can perform various malicious activities, including data theft, system compromise, and remote access. Fasong may also download and execute additional malware payloads, making it a significant threat to cybersecurity.

Figure 4. Malware family sample of Fasong

4.16. Hackkms

Hackkms is a family of hacking tools and key management service (KMS) activators used to bypass software licensing mechanisms and activate pirated software illegally. While not inherently malicious, the use of Hackkms can expose users to legal and security risks, including malware infections and compromised system integrity.

4.17. Hlux

Hlux, also known as Kelihos, is a family of botnet malware known for its ability to perform distributed denial-of-service (DDoS) attacks, spam campaigns, and data theft. The malware attacks Windows systems and talks with C&C servers to obtain commands and updates from hackers.

4.18. Injector

Injector is a type of malware that injects malicious code into legitimate processes or applications to evade detection by security software. It may modify system settings, disable security features, and facilitate other malicious activities, such as data theft, system compromise, and remote access.

4.19. InstallCore

InstallCore is a family of potentially unwanted programs (PUPs) that typically bundle with legitimate software downloads. It may install adware, browser extensions, or other software without the user's consent, often resulting in unwanted advertisements, browser redirects, and changes to browser settings.

4.20. LolydaAA1

LolydaAA1 is a variant of the Lolyda malware family, known for its ability to steal sensitive information, including passwords, login credentials, and financial data, from infected systems. It often spreads through phishing emails, malicious downloads, and drive-by downloads, affecting user privacy and securityuser privacy and security.

4.21. LolydaAA2

LolydaAA2 is a variant of the Lolyda malware family, characterized by its polymorphic nature and advanced evasion techniques. It employs encryption, obfuscation, and anti-analysis mechanisms to evade detection by security tools and thwart remediation efforts.

4.22. Multiplug

Multiplug is a family of adware that typically installs browser extensions or plugins to display intrusive advertisements, pop-ups, and banners. It may also gather browsing data and user information for targeted advertising, jeopardising user privacy and system efficiency.

4.23. Neoreklami

Neoreklami is a type of adware that shows unwanted adverts, pop-ups, and banners in web browsers. It may also route users to sponsored websites and gather browsing information for targeted advertising. Neoreklami can degrade surfing experiences and jeopardise user privacy.

4.24. Neshta

Neshta is a family of file-infecting viruses that target executable files on Windows systems. It can modify file structures, inject malicious code, and evade detection by security software through encryption and obfuscation techniques. Neshta may also download and execute additional malware payloads, posing a significant threat to system security.

4.25. RegRun

RegRun is a family of potentially unwanted programs (PUPs) that typically bundle with legitimate software downloads. It may install adware, browser extensions, or other software without the user's consent, often resulting in unwanted advertisements, browser redirects, and changes to browser settings.

4.26. Sality

Sality is a polymorphic virus that infects executable files on Windows systems. It can modify file structures, inject malicious code, and evade detection by security software through encryption and obfuscation techniques. Sality may also download and execute additional malware payloads, posing a significant threat to system security.

4.27. Snarasite

Snarasite is a family of Trojan malware that primarily targets Windows systems. It can perform various malicious activities, including data theft, system compromise, and remote access. Snarasite may also download and execute additional malware payloads, making it a significant threat to cybersecurity.

4.28. Stantinko

Stantinko is a sophisticated botnet malware known for its stealthy operation and advanced evasion techniques. It primarily targets Windows systems and performs various malicious activities, including click fraud, ad injection, cryptocurrency mining, and data theft. Stantinko infects systems through deceptive downloads, software bundling, and exploit kits.

4.29. VBA

VBA (Visual Basic for Applications) malware refers to malicious scripts or macros written in Visual Basic programming language. It typically targets Microsoft Office documents, spreadsheets, and other files, exploiting vulnerabilities to execute arbitrary code, download additional malware, and compromise system integrity.

4.30. VBkrypt

VBkrypt is a family of malware that targets Visual Basic (VB) scripts and macros, often distributed through malicious email attachments, infected documents, or compromised websites. It can execute arbitrary code, download and execute additional malware, and propagate to other systems through email or network shares.

4.31. Vilsel

Vilsel is a family of polymorphic viruses that infect executable files on Windows systems. It can modify file structures, inject malicious code, and evade detection by security software through encryption and obfuscation techniques. Vilsel may also download and execute additional malware payloads, posing a significant threat to system security.

Figure 5. Malware family sample of Vilsel

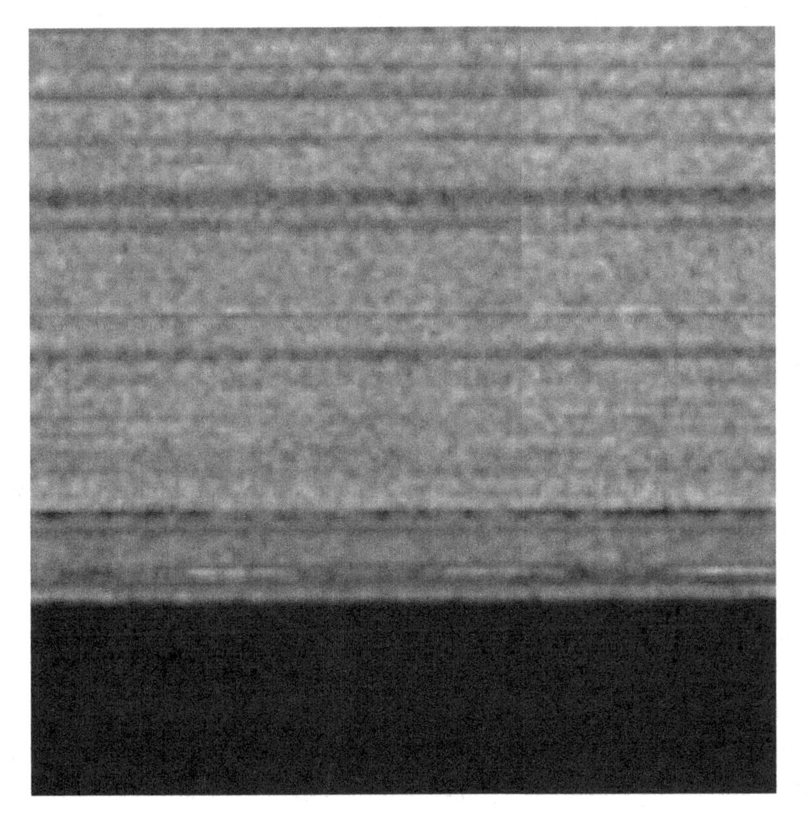

5. METHODOLOGY

The proposed methodology for malware detection malware using hybrid learning strategy is depicted in Figure. 6 which is explained below in steps:

Figure 6. Proposed methodology to detect malware

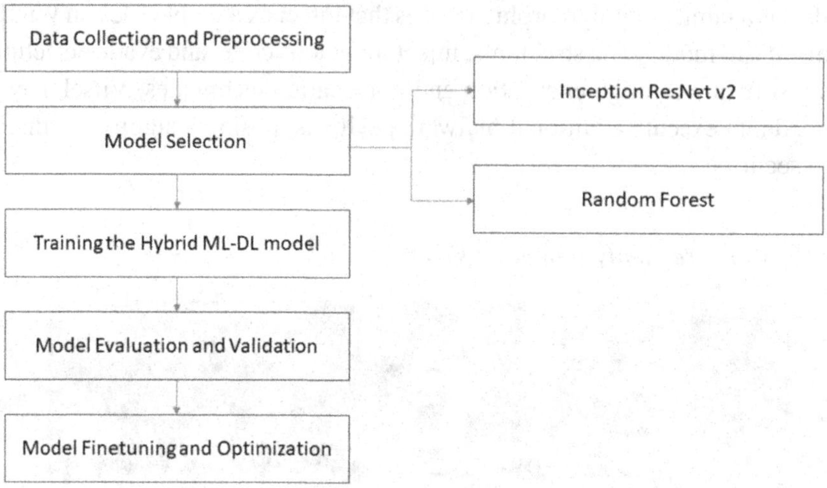

5.1. Data Set Preparation and Preprocessing

The initial step in the process is to acquire a dataset of malware images representing the 31 families of interest: These images may be obtained from various sources, including malware repositories, cybersecurity research datasets, and publicly available resources. Each image in the dataset is associated with a specific malware family label, enabling supervised learning for classification tasks. Preprocessing of the image data is performed to standardize the format, resolution, and quality of the images. To improve model performance and convergence, images are resized uniformly, colour images are converted to grayscale, and pixel values are normalised. Furthermore, methods for augmenting data like rotation, flipping, and zooming can be used to enrich the dataset and improve the model's generalisation ability.

5.2. Model Selection: Inception ResNetV2 and Random Forest

For the hybrid ML-DL model, two distinct algorithms are utilized: Inception ResNetV2 for deep learning-based feature extraction and Random Forest for classification.

5.2.1. Inception ResNetV2

Inception ResNetV2 is a state-of-the-art deep learning architecture designed for image classification tasks. It combines the benefits of the Inception and ResNet architectures, incorporating both convolutional and residual connections to enable efficient feature extraction and learning of hierarchical representations from input images. Inception ResNetV2 is pre-trained on large-scale image datasets such as ImageNet, allowing it to capture high-level features and patterns relevant to a wide range of image recognition tasks.

5.2.2. Random Forest

Random Forest is an ensemble learning technique which builds numerous decision trees during training and returns the method of the classes for classification problems and the mean prediction for regression tasks. Each decision tree is trained using a bootstrap sample of the dataset, with a random subset of characteristics being considered at each split. Random Forest is well-known for its robustness against overfitting, scalability to big datasets, and capacity to manage high-dimensional feature fields.

5.3. Training the Hybrid ML-DL Model

The hybrid ML-DL model is trained using the collected malware image dataset.

5.3.1. Feature Extraction With Inception ResNetV2

In the first stage of training, the Inception ResNetV2 model is used for extracting features of the malware images provided. The pre-trained Inception ResNetV2 model is fine-tuned on the malware image dataset, leveraging transfer learning to adapt the learned representations to the particular task of malware classification. During training, the weights of the Inception ResNetV2 model are updated using backpropagation and gradient descent optimization to minimize a suitable loss function, such as categorical cross-entropy.

5.3.2. Classification With Random Forest

The deep features extracted by Inception ResNetV2 are then used as input to the Random Forest classifier. Random Forest learns to classify malware images into the 31 predefined families based on the extracted features. During training, the Random Forest method creates numerous decision trees from bootstrapped dataset samples and chooses the best splitting criteria for each node to maximise information gain. The resulting ensemble of decision trees collectively forms the Random Forest model, which outputs the predicted class label for each input image.

5.4. Model Evaluation and Validation

Once the hybrid ML-DL model is trained, it is evaluated using a separate validation dataset to assess its performance and generalization ability.

5.4.1 Validation Dataset

The validation dataset consists of a set of malware images that were not used during the training phase. This ensures that the model's performance is evaluated on unseen data, providing a more reliable estimate of its real-world effectiveness.

5.4.2 Performance Metrics

Accuracy is computed to evaluate the model's performance. This provides insights into the model's ability to correctly classify malware images into their respective families and identify any potential areas for improvement. This accuracy statistic indicates the portion of successfully categorised malware images out of the overall amount of images in the validation set and serves as a measure of the model's overall effectiveness in classifying malware families.

5.5. Model Fine-Tuning and Optimization

Following model evaluation, fine-tuning and optimization techniques can be applied to further enhance the performance of the hybrid ML-DL model.

5.5.1 Hyperparameter Tuning

Hyperparameters of the Inception ResNetV2 model, Random Forest classifier, and training process may be fine-tuned using techniques such as grid search, random search, or Bayesian optimization. This helps identify optimal hyperparameter configurations that maximize model performance and generalization ability.

5.5.2 Regularization and Dropout

Regularization techniques such as L1 and L2 regularization, dropout, and batch normalization may be applied to prevent overfitting and improve the model's robustness to noise and perturbations in the data.

5.5.3. Ensemble Learning

Ensemble learning approaches, such as bagging, boosting, and stacking, may be employed to combine multiple models or model variations to further improve classification performance and enhance model diversity.

6. RESULTS AND DISCUSSION

Utilizing the comprehensive Malimg Dataset, consisting of 31 distinct malware families, our study presents a robust two-phase model tailored for effective malware classification. In the initial phase, we leverage the cutting-edge Inception ResNet v2 architecture to extract features from raw malware images. By employing Inception ResNet v2, we transform these images into high-dimensional feature vectors, encapsulating essential characteristics indicative of diverse malware families. Following feature extraction, the classification phase utilizes the Random Forests algorithm, renowned for its robustness and interpretability in handling complex datasets. By harnessing the discriminative capabilities of Random Forests, the proposed model accurately classifies the extracted features into their respective malware families. This approach not only enhances the interpretability of the classification process but also ensures resilience against noise and outliers inherent in real-world malware datasets.

During the testing phase, the proposed trained classifier is subjected to a diverse array of new malware files. Leveraging the learned representations from the feature extraction phase and the discriminative power of Random Forests, the proposed model accurately predicts the malware family associated with each file. Impressively, the proposed model achieves an exceptional accuracy rate of 94.15% in classifying malware samples into their respective families which is shown in Figure. 7.

Figure 7. Validation accuracy

```
1/1 [==============================] - 0s 46
Accuracy: 0.9415234675557835

Predicted labels vs True labels:
Predicted: VBKrypt        True: VBKrypt
Predicted: VBKrypt        True: VBKrypt
Predicted: VBKrypt        True: VBKrypt
Predicted: VBKrypt        True: VBKrypt
Predicted: VBKrypt        True: VBKrypt
Predicted: VBKrypt        True: VBKrypt
Predicted: VBKrypt        True: VBKrypt
Predicted: VBKrypt        True: VBKrypt
Predicted: VBKrypt        True: VBKrypt
Predicted: VBKrypt        True: VBKrypt
Predicted: Neshta         True: VBKrypt
Predicted: VBKrypt      . True: VBKrypt
Predicted: VBKrypt        True: VBKrypt
Predicted: VBKrypt        True: VBKrypt
Predicted: VBKrypt        True: VBKrypt
Predicted: VBKrypt        True: VBKrypt
Predicted: Sality         True: VBKrypt
Predicted: VBKrypt        True: VBKrypt
```

Figure. 8 illustrates the graphical representation of these metrics, showcasing the accuracy versus the number of estimators. At 20th epoch, the training accuracy reaches 100% accuracy and validation accuracy is 91%. Validation accuracy increases as epochs are increased further. This visualization offers valuable insights into the model's learning dynamics and performance across epochs, further validating the effectiveness of the proposed approach in tackling the intricate task of malware classification.

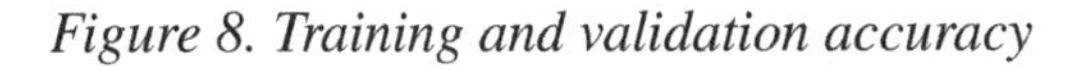

Figure 8. Training and validation accuracy

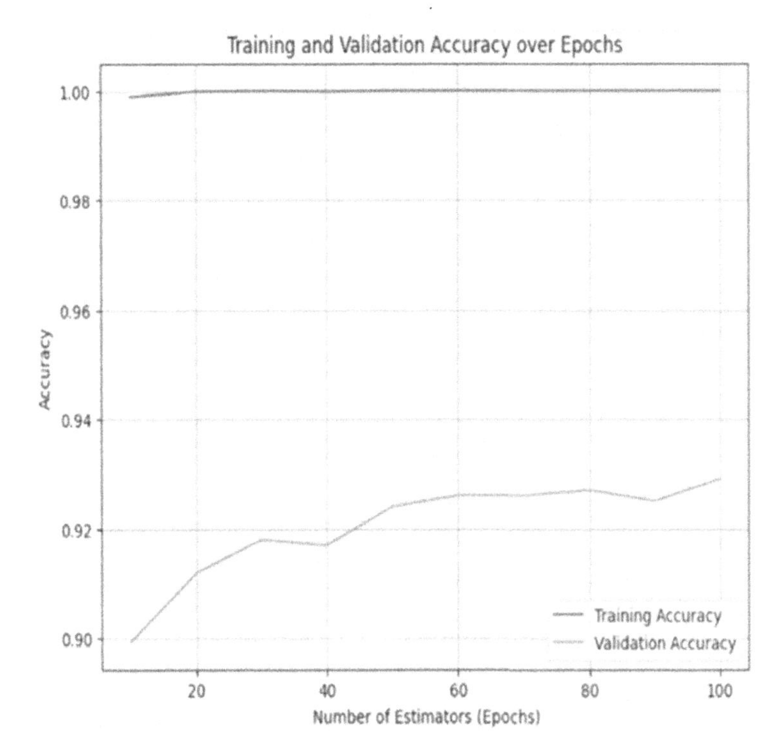

Overall, the proposed hybrid model, integrating advanced deep learning techniques and ensemble learning methods, exhibits superior performance in precisely identifying and categorizing malware samples. These outcomes highlight the potential of the proposed approach to significantly enhance cybersecurity endeavors by enabling proactive detection and mitigation of diverse malware threats.

The precision-recall curve is a graph employed to assess the effectiveness of classification models, particularly when the classes are not balanced. It plots the trade-off between precision and recall across different thresholds used for class prediction. In Figure.9, precision is plotted on the y-axis and recall on the x-axis. Each point on the curve shows the precision and recall values obtained at a particular threshold used for classification. The average precision values are above 0.91 except for the classes 12, 23 and 25. For 14 among 31 classes, the average precision is 1.

Figure 9. Precision vs. recall for the malware families

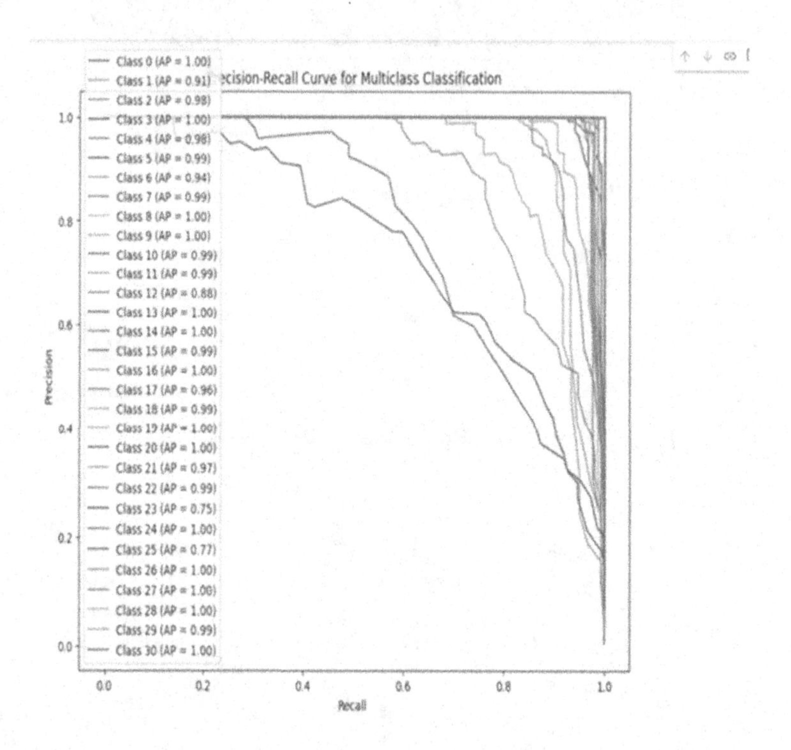

Table 1 compares the accuracy of malware classification models using various deep learning methods and the proposed hybrid machine learning and deep learning model of Inception ResNet v2 for Feature Extraction and Random Forests for Classification. The proposed hybrid model achieves the maximum accuracy of 94.15%.

Table 1. A comparison on accuracy

Models	Accuracy (%)
VGG16	88.40
Resnet-18	90.21
Inception-V3	92.48
InceptionResNet-V2	85.1
Proposed Hybrid Model	94.15

The contribution of the work is discussed as follows. Firstly, the advancement in malware detection techniques is noteworthy. The hybrid ML-DL model developed in this work represents a novel approach to malware detection. By combining machine learning (ML) algorithms with deep learning (DL) techniques, the model can effectively learn and classify malware images based on their visual features.

Secondly, the work contributes to improved accuracy and efficiency in malware detection. The use of a hybrid ML-DL model, specifically employing the Inception ResNet V2 architecture for feature extraction and a Random Forest classifier for image classification, significantly enhances accuracy and efficiency. The model's capability to extract high-level features of malware images and classify them with a high degree of accuracy, as demonstrated by the validation results achieving 94% accuracy, signifies a significant advancement over traditional malware detection methods.

Furthermore, this work enhances the robustness of malware detection systems against evolving malware families. By training the model on a diverse dataset comprising 31 different malware families, including variants with polymorphic and obfuscated characteristics, the model learns to identify and classify a wide range of malware types effectively. This capability is essential for keeping pace with the constant evolution of malware and ensuring the efficacy of malware detection systems in real-world scenarios.

The generalizability of the developed hybrid ML-DL model across different malware families is another notable contribution. Thanks to its ability to extract and learn discriminative features from malware images, the model demonstrates a high level of generalizability. This aspect is crucial for real-world applications where new and previously unseen malware variants continually emerge. By leveraging the model's robust feature extraction capabilities, cybersecurity professionals can effectively detect and mitigate emerging threats without the need for extensive retraining or manual intervention.

Moreover, the architecture of the hybrid ML-DL model lends itself to scalability and adaptability, making it suitable for deployment in various computing environments and scenarios. The use of pre-trained DL models, such as Inception ResNet V2, for feature extraction allows for efficient utilization of computational resources and facilitates rapid deployment in production environments. Additionally, the modular design of the model enables easy integration with existing malware detection systems and workflows, enhancing overall operational efficiency.

Furthermore, the potential for real-world application of the developed hybrid ML-DL model in cybersecurity defence systems is significant. The model offers a practical solution for automated malware detection and classification, which can be integrated into endpoint security solutions, network intrusion detection systems (NIDS), and threat intelligence platforms. By leveraging the capabilities of the model, organizations can enhance their cybersecurity posture and better protect against sophisticated malware attacks.

Additionally, the paper addresses limitations of traditional approaches to malware detection, such as signature-based methods. By leveraging the power of deep learning for feature extraction and machine learning for classification, the hybrid ML-DL model overcomes the shortcomings of signature-based detection methods. This paradigm shift in malware detection represents a significant contribution to the ongoing effort to improve cybersecurity defences.

Moreover, the potential for transfer learning and fine-tuning techniques with the use of pre-trained DL models opens up avenues for further research and experimentation. Researchers can explore the possibility of reusing pre-trained DL models for other cybersecurity tasks, such as malware family attribution, malware behaviour analysis, and anomaly detection. By leveraging transfer learning techniques, the knowledge gained from training the hybrid ML-DL model can be transferred to related domains, thereby accelerating progress in cybersecurity research.

Lastly, this work contributes to the mitigation of zero-day attacks, which exploit previously unknown vulnerabilities. The hybrid ML-DL model offers a proactive defence against zero-day attacks by focusing on the visual characteristics and behavioural patterns of malware. By continuously learning and adapting to new threats, the model enhances the resilience of cybersecurity defences and reduces the risk of zero-day exploits.

In summary, this work makes several significant contributions to existing work in malware detection and cybersecurity. By leveraging a hybrid ML-DL approach, this paper extends the current advances in malware identification techniques, improves accuracy and efficiency, enhances robustness against evolving threats, and offers practical solutions for real-world application.

7. CONCLUSION

The development and evaluation of a hybrid machine learning-deep learning (ML-DL) model for malware classification represent a significant advancement in cybersecurity research in 5G networks. This work aimed to address the pressing challenge of identifying and categorizing malware images into 31 distinct families using a combination of feature extraction with Inception ResNetV2 and classification with Random Forest. The comprehensive methodology employed in this work has yielded valuable insights into the effectiveness of hybrid ML-DL approaches for malware detection and classification and provides accuracy of 94.15%. The model also provides good average precision of one for most of the classes.

The contributions of this work extend beyond the development of a single model; they encompass advancements in malware detection and cybersecurity methodologies in 5G enabled IoT environment. By leveraging the power of deep learning for feature extraction and traditional machine learning for classification, the hybrid approach strikes a compromise between complexity and interpretability, making it acceptable for use in practical systems for cybersecurity. Despite the promising results achieved, several challenges and opportunities for future research have been identified. One challenge is the continuous evolution of malware variants and attack techniques, necessitating ongoing updates and adaptations to the model architecture and training strategies. Additionally, scalability and computational efficiency remain important considerations for deploying ML-DL models in resource-constrained environments.

Future research directions may include exploring alternative DL architectures, experimenting with different ensemble methods, and incorporating dynamic feature extraction techniques to capture temporal and spatial dependencies in malware images. Collaborative efforts involving interdisciplinary teams of cybersecurity experts, data scientists, and domain specialists are essential for addressing these challenges and advancing the state-of-the-art in malware detection and classification.

As cyber threats continue to evolve and proliferate, the importance of leveraging advanced ML-DL techniques for enhancing cybersecurity defences cannot be overstated. In summary, this work represents a testament to the potential of interdisciplinary research and collaboration in advancing the frontiers of cybersecurity, ultimately aiming to develop a more trustworthy and protected digital future for individuals, businesses, and societies worldwide.

REFERENCES

Ahmed, I., Anisetti, M., Ahmad, A., & Jeon, G. (2022). A multilayer deep learning approach for malware classification in 5G-enabled IIoT. *IEEE Transactions on Industrial Informatics*, 19(2), 1495–1503. 10.1109/TII.2022.3205366

Aslan, Ö., & Yilmaz, A. A. (2021). A new malware classification framework based on deep learning algorithms. *IEEE Access : Practical Innovations, Open Solutions*, 9, 87936–87951. 10.1109/ACCESS.2021.3089586

Boualouache, A., & Engel, T. (2023). A survey on machine learning-based misbehavior detection systems for 5g and beyond vehicular networks. *IEEE Communications Surveys and Tutorials*, 25(2), 1128–1172. 10.1109/COMST.2023.3236448

Chandranegara, D. R., Djawas, J. S., Nurfaizi, F. A., & Sari, Z. (2023). Malware Image Classification Using Deep Learning InceptionResNet-V2 and VGG-16 Method. *Jurnal Online Informatika*, 8(1), 61–71. 10.15575/join.v8i1.1051

Chukunda, C.D., Matthias, & Bennett. E.O. (2022) Malware detection classification system using random forest. *Journal of Software Engineering and Simulation, 8(5)*, 16-25

Garcia, F. C. C., & Muga, F. P., II. (2016). Random forest for malware classification. *arXiv preprint arXiv:1609.07770.*

Hamza, W. S., Ibrahim, H. M., Shyaa, M. A., & Stephan, J. J. (2020). Iot botnet detection: Challenges and issues. *Test Eng. Manag*, 83, 15092–15097.

Haq, I. U., Khan, T. A., & Akhunzada, A. (2021). A dynamic robust DL-based model for android malware detection. *IEEE Access : Practical Innovations, Open Solutions*, 9, 74510–74521. 10.1109/ACCESS.2021.3079370

Ismail, H., Utomo, R. G., & Bawono, M. W. A. (2024). Comparison of Support Vector Machine and Random Forest Method on Static Analysis Windows Portable Executable (PE) Malware Detection. *JURNAL MEDIA INFORMATIKA BUDIDARMA*, 8(1), 154–162.

Kim, K. C., Ko, E., Kim, J., & Yi, J. H. (2019). Intelligent Malware Detection Based on Hybrid Learning of API and ACG on Android. *Journal of Internet Services and Information Security.*, 9(4), 39–48.

Lad, S. S., & Adamuthe, A. C. (2020). Malware classification with improved convolutional neural network model. *Int. J. Comput. Netw. Inf. Secur*, 12(6), 30–43.

Nawshin, F., Unal, D., Hammoudeh, M., & Suganthan, P. N. (2024). AI-powered malwaredetection with Differential Privacy for zero trust security in Internet of Things networks. *Ad Hoc Networks*, 161, 103523. 10.1016/j.adhoc.2024.103523

Pant, D., & Bista, R. (2021). Image-based malware classification using deep convolutional neural network and transfer learning. *Proceedings of the 3rd International Conference on Advanced Information Science and System* (pp. 1-6). ACM. 10.1145/3503047.3503081

Qureshi, S., He, J., Tunio, S., Zhu, N., Akhtar, F., Ullah, F., Nazir, A., & Wajahat, A. (2021). A Hybrid DL-Based Detection Mechanism for Cyber Threats in Secure Networks. *IEEE Access : Practical Innovations, Open Solutions*, 9, 73938–73947. 10.1109/ACCESS.2021.3081069

Rafique, M. F., Ali, M., Qureshi, A. S., Khan, A., & Mirza, A. M. (2019). Malware classification using deep learning based feature extraction and wrapper based feature selection technique. *arXiv preprint arXiv:1910.10958.*

Rahman, A.U., Mahmud, M., Iqbal, T., Saraireh, L., Kholidy, H., Gollapalli, M., Musleh, D., Alhaidari, F., Almoqbil, D. and Ahmed, M.I.B. (2022). Network Anomaly Detection in 5G Networks. *Mathematical Modelling of Engineering Problems, 9*(2).

Roseline, S. A., Geetha, S., Kadry, S., & Nam, Y. (2020). Intelligent vision-based malware detection and classification using deep random forest paradigm. *IEEE Access : Practical Innovations, Open Solutions*, 8, 206303–206324. 10.1109/ACCESS.2020.3036491

Saeed, M. M., Saeed, R. A., Abdelhaq, M., Alsaqour, R., Hasan, M. K., & Mokhtar, R. A. (2023). Anomaly detection in 6G networks using machine learning methods. *Electronics, 12(15),* 3300. 37188858

Shetu, S. F., Saifuzzaman, M., Moon, N. N., & Nur, F. N. (2019). A survey of botnet in cyber security. In *2019 2nd International Conference on Intelligent Communication and Computational Techniques (ICCT)* (pp. 174-177). IEEE. 10.1109/ICCT46177.2019.8969048

Srinivasan, S., & Deepalakshmi, P. (2023). Enhancing the security in cyber-world by detecting the botnets using ensemble classification based machine learning. *Measurement. Sensors*, 25, 100624. 10.1016/j.measen.2022.100624

Chapter 6
Internet of Things (IoT) Technologies With Intrusion Detection Systems in Deep Learning

Nancy Jasmine Goldena
Department of Computer Application and Research Centre, Sarah Tucker College (Autonomous), Manonmaniam Sundaranar University, Tirunelveli, India

Rashia Subashree
Department of Computer Application and Research Centre, Sarah Tucker College (Autonomous), Manonmaniam Sundaranar University, Tirunelveli, India

ABSTRACT

In the era of the internet, 5G is the super-fast network unified with mobile phones and communication devices, with a good bandwidth spectrum to transform information. The combination of IoT devices with 5G technologies facilitates quick and seamless data transformation. This expansion fosters connectivity, operational efficiency, and intellectual development. An intrusion detection systems (IDS) is an efficient tool for network security that keeps an eye on malicious activities or security policy violations under surveillance to detect anomalies in devices. The early part of the chapter deals with the integration of 5G technologies with IOT devices. The chapter also emphasize the role of collaboration between IDS and emerging technologies in deep learning (DL). The integration of intelligent algorithms enhances detection accuracy and enables adaptive responses to evolve the security threats in real-time. In conclusion, the insights provided in this chapter aim to understand the Research gap in DL IDS and the pivotal role played by IDS in ensuring a resilient and secure IoT environment.

DOI: 10.4018/979-8-3693-2786-9.ch006

1. INTRODUCTION

The Internet of Things, or IoT, is a technology that enhances and supports people's lives, jobs, and cultures by allowing things to communicate with one another over the internet. The IoT, with 50 billion devices that are connected, is one of the fastest-growing web domains. The IoT is the term used to describe the whole network of interconnected devices as well as the technology that permits communication both between devices and the cloud (Goldena, 2022). Worldwide, IoT frameworks are available; they are essentially forced assets created by lossy connections. Five-generation (5G) wireless cellular technology offers faster upload and download speeds, more reliable connections, and more capacity than earlier generations. It is the first mobile network built from the ground up to accommodate IoT use cases (Khuntia et al., 2021). Many use cases were taken into account throughout the creation of 5G, including linked drones, automated guided vehicles (AGVs), delivery robots, assisted driving, and public safety applications. IoT 5G sensors are able to establish connections with devices and equipment in order to track their health. It will enhance productivity. Without pausing production, you may collect data on oil levels, electric outputs, temperature, and speed. Cyberattacks are a persistent concern because of the unusual manufacturing practices used in IoT devices and the massive volumes of data they manage. IoT security has become more important as a result of many high-profile events in which a common IoT device was used to breach and attack the wider network. To provide strong IoT security strategies, it is thus necessary to implement significant changes to current security concepts for data and distant systems. The security tools available today, such as network protection, application control, authentication, encryption, and access control, are unsuitable for large systems consisting of several connected devices since they are slow to operate and do not address the vulnerabilities present in each component of the system (Flinders & Flinders, 2024). Devices or software programmes that keep an eye out for hostile activity or policy breaches on a network are known as Intrusion Detection Systems, or IDSs. Security information and event management systems are usually used to notify or gather violations or harmful behaviour centrally. By identifying and averting possible attacks, assisting with incident response, guaranteeing compliance, and efficiently managing risks, IDSs improve any organisation's security. The system administrator can be notified of any unusual activity via IDS (Azam et al., 2023). With its insights and enhanced network performance, IDS may be a useful complement to any organisation's security setup.

2. 5G ARCHITECTURE IN IOT AND ITS TECHNOLOGIES

Next-generation mobile network technology, or 5G architecture, promises reduced latency, higher speeds, and more dependable connections. Network functions can interact through APIs acknowledge to the service-based design of the 5G core network (Flinders & Flinders, 2024). The result is increased scalability and flexibility in the network. Because of this, a single physical 5G infrastructure may support many virtual networks. Each slice may be tailored for a distinct service or application, as seen in Figure 1, guaranteeing excellent performance for a range of uses.

Figure 1. 5G use cases

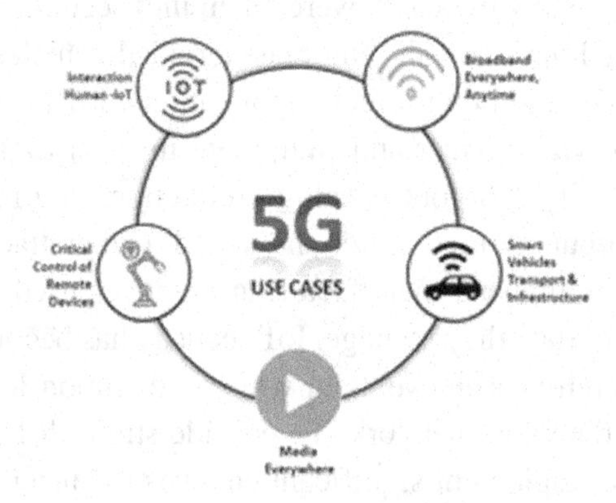

IoT exploits several uses of 5G. These are the quickest and easiest to implement, and infrastructure suppliers and network operators are putting a lot of emphasis on them right now. AR/VR and security cameras are two examples of high-data-rate applications that are supported by enhanced mobile broadband (eMBB). Remote surgery and driverless cars are two examples of real-time applications that depend on Ultra-Reliable Low-Latency Communication (URLLC) (Henry et al., 2023). 5G is able to accommodate a large number of IoT devices that acknowledge Massive Machine-Type Communications (mMTC), which is perfect for environmental monitoring and smart cities. IoT devices' battery lives are prolonged by 5G's energy-efficient communication modes, while network slicing offers customised network resources for various IoT applications, guaranteeing top performance and security.

Overall, the usefulness and efficiency of IoT are greatly improved by 5G's sophisticated capabilities and adaptable design, opening up new opportunities for a variety of sectors.

IoT applications will be able to better manage the network's properties and tailor them to the specific requirements of their use cases.

- **5G Enhanced Mobile Broadband:** This version offers additional data. Data streaming is a common use for this these days. Enhanced mobile broadband is important for the IoT as well as for private conversations (Khuntia et al., 2021). The key objectives here are throughput along with additional data.
- **Ultra-Reliable Low-Latency Communication (URLLC):** 5G's URLLC enables applications that need to respond in real time, such as industrial automation, remote surgery, and autonomous cars, to achieve very low latency and great reliability (Madhu et al., 2023).
- **Wide-scale mobile IoT:** Wide-scale implementation of less complex and effective IoT devices, such as sensors. Although the data sent by these devices is frequently low, the use case's relevance may be greatly impacted by factors like cost, energy efficiency, and dependable coverage. Enhanced coverage, even underground and in distant regions. will be possible with low-cost devices that have a battery life of more than ten years (Cloudflare, n.d.). In certain countries, the technology needed to accomplish this was implemented prior to the roll-out of 5G, having been developed alongside 4G and including NB-IoT and LTE-M.

2.1 IoT Applications

With faster data rates, less latency, and more network capacity, 5G technology greatly expands the possibilities of IoT applications. The following are some significant IoT applications that 5G technology in particular makes possible and improves:

✓ Smart Cities

5G makes it easier for IoT devices to be deployed in urban settings, which creates smarter, more productive cities.

- **Traffic Management:** Traffic flow can be optimised, congestion can be decreased, and public transportation efficiency can be increased with real-time data from IoT sensors and cameras.
- **Smart illumination:** Depending on the number of cars and pedestrians, streetlights may be dynamically adjusted to conserve energy and improve illumination (Amazon Web Services, Inc, n.d.).

- **Public safety:** Faster data transfers and improved video monitoring allow law enforcement authorities to better coordinate their efforts and respond to situations more quickly.

✓ Autonomous Vehicles

5G's high dependability and low latency are essential for the advancement and implementation of driverless cars. Examples include:

- **Vehicle-to-Everything (V2X) Communication:** This enhances traffic management and safety by enabling cars to interact with people, infrastructure, and other vehicles (Gala et al., 2023).
- **Remote operation and Monitoring:** This feature makes it possible to monitor and operate autonomous cars in real-time, which is crucial for maintaining dependability and safety in challenging driving conditions.

By combining 5G with IoT, you can improve on current applications and open up new possibilities for creative solutions that were previously impossible to achieve because of latency and connection constraints.

3. INTRUSION DETECTION SYSTEM

An IDS can be a helpful tool for identifying potential threats while a network infrastructure is down and not in real-time communication. IDS typically use a TAP or SPAN port to check inline traffic streams to make sure they don't influence network performance. When IDS were initially developed, the quantity of analysis needed for detection was not quick enough. Network Intrusion Detection Systems are used to capture hackers before they do damage to the network (Aljanabi et al., 2021). In contrast to host-based IDSes, which are installed on client PCs, network-based IDSes are situated on the network. IDS detects known attack fingerprints, such as DNS poisoning, altered information packets, and Christmas tree scans, to spot irregularities in routine activity. IDS can be used as a software programme or network security device; cloud-based IDSes are also offered for cloud-based security (Abdulla & Jameel, 2023).

The IDS and IoT systems must be integrated. Although authentication and encryption safeguard IoT networks, cyberattacks are always a possibility. IDS aims to track, examine, and identify harmful activity in network traffic from many sources. It plays a big role in cybersecurity technologies (Desai et al., 2023). IDS, to put it briefly, is a procedure that uses a variety of techniques to find harmful activity directed at victims. One approach is depicted in Fig. 2. The Internet is transforming how people live, learn, and work as a result of its deeper and deeper integration with society, but there are also growing security risks. One essential technological problem that cannot be avoided is learning to recognise different types of network assaults, particularly unexpected ones. One important piece of research in the realm of information security is the IDS, which has the ability to detect intrusions, whether they are ongoing or have already happened (Sisavath & Yu, 2021).

Figure 2. Intrusion detection system

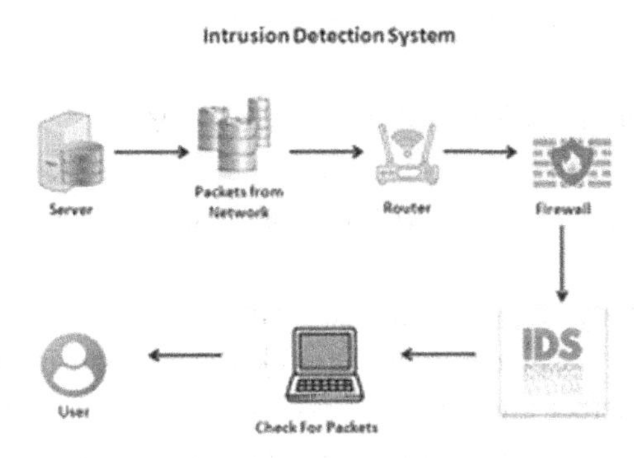

Using supervised and unsupervised machine learning to implement IDS in IoT settings improves security by utilising these algorithms' capacity to learn from past data and identify abnormalities instantly.

3.1.1 Supervised Learning

For the IoT, supervised machine learning-based IDS are essential for improving network security. The purpose of these systems is to keep an eye on network traffic and spot unusual activity that could point to security vulnerabilities. In order to train classification algorithms like Decision Trees, Random Forests, SVM's, and Neural Networks on labelled datasets, they require gathering and preprocessing data, extracting features, and more (He et al., 2023). Once implemented, these models evaluate incoming and outgoing communications to find any breaches.

To identify IoT threats, a supervised machine learning IDS is suggested that has a 99.99% detection accuracy and a 99.97% maximum confidence level (MCC) (Elrawy et al., 2018). In order to prevent information leakage on the test set, an efficient IDS for the IoT network is created using the minimum-maximum normalisation technique for feature scaling.

3.1.2 Unsupervised Learning

Unsupervised machine learning-based IDS for the IoT improves security by identifying irregularities in network traffic without the need for tagged data. Unsupervised learning is especially well-suited for dynamic IoT contexts because it can adjust to changing attack techniques without requiring manual tagging (Ibrahim & Elhafiz, 2023). Unsupervised learning is especially well-suited for dynamic IoT contexts because it can adjust to changing attack techniques without requiring manual tagging. In order to find odd patterns, these systems collect and preprocess network data, extract features, and use methods such as k-means clustering, DB-SCAN, isolation forests, and Autoencoders.

3.2 IoT Applications With IDS Technology

3.2.1 Network-Based Intrusion Detection Systems

Monitoring and analysing network data for indications of questionable behaviour or policy infractions is the function of Network-Based Intrusion Detection Systems (NIDS). In order to detect known attack patterns or abnormalities, it records data packets as they go over the network. Network infrastructure is protected against cyberattacks by NIDS, which detects any threats instantly (Gala et al., 2023). Providing a crucial layer of protection, it is an indispensable part of an all-encompassing cybersecurity plan.

Figure 3. Network-based intrusion detection system

The NIDS has the following advantages and disadvantages:

✓ Benefits:
- NIDS has the ability to monitor a whole network, giving an all-encompassing perspective of all traffic and activity. This results in comprehensive network visibility. It makes it possible to identify hazards from the outside as well as the inside.
- **Real-Time Analysis:** The NIDS scans data packets as they go across the network, identifying suspicious patterns or abnormalities promptly (Khuntia et al., 2021). This allows for prompt reactions to any attacks.
- **Centralised Management:** NIDS is frequently centralised, which facilitates control and management. This centralised method guarantees uniform security rules throughout the network for a Dubai-based company with several branches.

✓ Adverse effects
- **Limited Encrypted Traffic Inspection:** Since NIDS can only read encrypted packets after they have been decrypted, it has difficulty examining encrypted traffic. Encrypted communication channels may have security holes as a result of this restriction.
- **Performance Impact:** Because NIDS must analyse a lot of data in real-time, its implementation in networks with high traffic volumes may cause performance problems (Elrawy et al., 2018). It could perform bet-

ter on networks with strict bandwidth restrictions and a need for strong hardware.

HIDS improves overall security posture by offering comprehensive insights into threats particular to a host. It is a vital part of any complete cybersecurity plan because of its real-time detection and response capabilities to questionable activity. Strong security against external and internal threats is ensured at the host level via HIDS implementation.

3.2.2 Host-Based Intrusion Detection Systems

A security tool that keeps an eye on your devices for questionable activity is called a Host-Based Intrusion Detection Systems (HIDS). It keeps on monitoring for things like unknown processes, unwanted modifications, and possibly malicious software (He et al., 2023). HIDS analyses this data to assist you in locating security concerns and taking appropriate action, as shown in Figure 4.

Figure 4. Host-based intrusion detection system

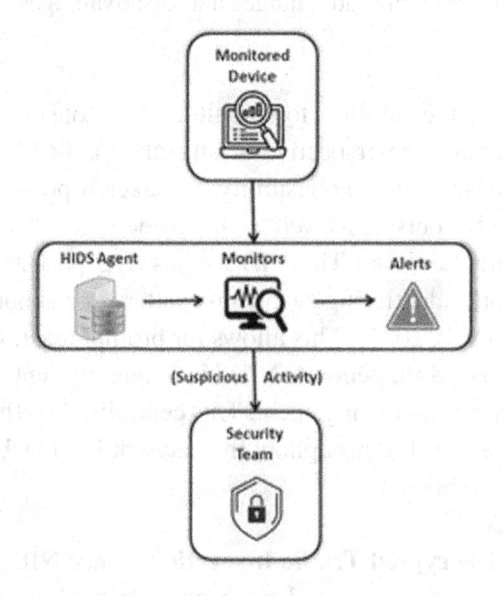

Some advantages and disadvantages of HIDS are listed below:

✓ Benefits:

- **Comprehensive Host Monitoring:** HIDS focuses on monitoring specific hosts, offering thorough insights into the operations and procedures on every system. It is capable of detecting illegal activity on certain computers and insider threats (Nicholls & Nicholls, 2024).
- **Enhanced Log Analysis:** In order to identify potentially intrusive events or actions, HIDS can access and analyse system logs in real-time.

✓ Adverse effects:

- **Minimal Network Overhead:** Because HIDS focuses mostly on individual hosts, it reduces network overhead and is thus appropriate for systems with constrained bandwidth.
- **Inadequate Network Visibility:** HIDS's capacity to identify network-wide threats that do not directly affect the host is restricted since, in contrast to NIDS, it just keeps an eye on activity on the host on which it is placed.
- **High Management Complexity**: It might be difficult and time-consuming to manage several HIDS installations on different hosts. To guarantee any system's efficacy, regular maintenance and upgrades are required.

However, they continue to be an essential part of a multi-layered security strategy, offering accurate and timely identification of known risks. Their total effectiveness can be increased by interaction with other security measures and regular upgrades.

3.2.3 Signature-Based Intrusion Detection Systems

Like a security bouncer, a Signature-Based Intrusion Detection Systems (SIDS) operates. It compares network traffic to a catalogue of known harmful attack patterns, or signatures. It can overlook completely new dangers, but it is excellent at identifying established ones, as shown in Figure 5.

Figure 5. Signature-based intrusion detection system

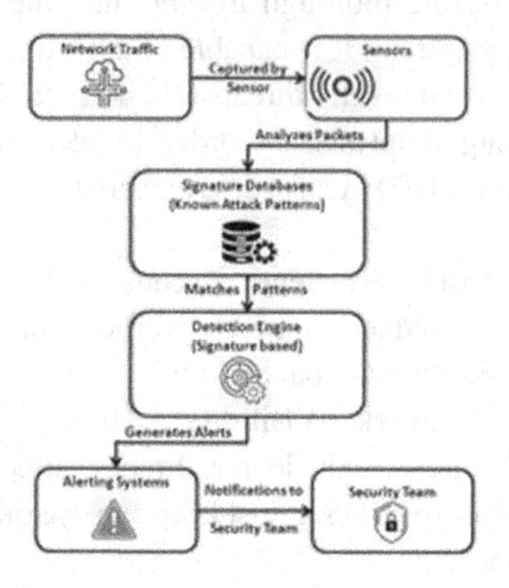

The following are the benefits and adverse effects of the signature-based intrusion detection system:

✓ Benefits:

- **Established Threat Identification:** By using a database of recognised threat signatures, signature-based IDS are able to identify known and verified assaults. It makes it possible to quickly identify typical dangers.
- **Low False-Positive Rate:** Compared to other detection techniques, signature-based IDS often has a lower false-positive rate because of its unique methodology. It lessens the likelihood of erroneous warnings and alarms.

✓ Adverse effects

- **Limited Zero-Day Threat Detection:** IDSs that rely solely on signatures have trouble identifying novel and unidentified threats in the absence of signatures. Networks might become susceptible until new signatures are created and deployed.
- **Unable to Identify Polymorphic Assaults**: Signature-based detection is less successful against complex threats because polymorphic assaults, which alter their appearance continually, might elude detection.
- They need to be updated often to stay effective, although they are quite good at identifying known risks through pattern matching. They provide quick identification of known dangers, making them an essential part of a multi-layered security approach.

4. IDS IN DEEP LEARNING

In order to offer the appropriate level of network security, IDSs watch over network traffic for any unusual activity. An IDS is being developed using machine learning techniques in order to identify and categorise cyberattacks at both the network and host levels in an automated and timely way. It aids in the creation of an adaptable and powerful IDS for the purpose of identifying and categorising unanticipated and unplanned cyberattacks. As seen in Figure 6, the focus is on how DL, or Deep Neural Networks (DNNs), may provide flexible IDS with the capacity to learn and identify known and novel or zero-day network behavioural traits. This will ultimately force the system intruder to leave and lower the chance of penetration.

Figure 6. Machine learning or deep learning techniques

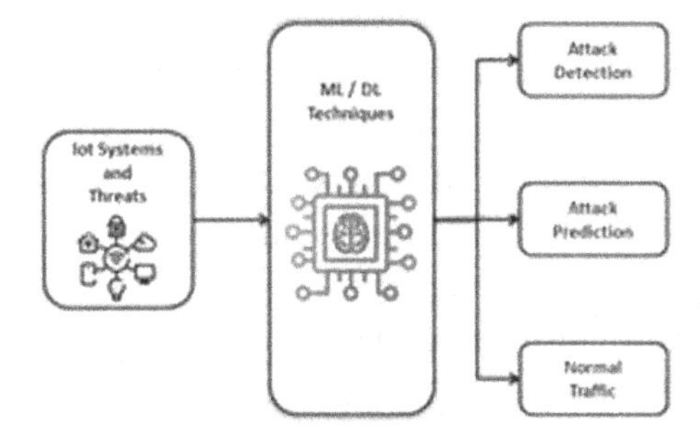

4.1 Deep Learning (DL) Techniques for IDS

Owing to their capacity to automatically recognise and adjust to intricate patterns in data, DL techniques have proven remarkably useful in augmenting information forensic skills.

Some DL methods that are frequently used for IDS include the following:

4.1.1 Convolutional Neural Networks (CNNs)

With their extensive feature extraction and classification capabilities, CNNs provide effective tools for enhancing the accuracy and efficiency of IDS, which are crucial for safeguarding against dynamic cyber threats in more intricate network settings. CNNs are helpful for processing network traffic because they can automatically extract characteristics from raw data and convert them into representations (He et al.,

2023). After features are retrieved, CNNs can distinguish between suspicious and typical network data, spotting intricate patterns and correlations that conventional techniques would overlook. Processing time-series data is a particularly good use for this DL skill.

CNNs have demonstrated potential in network traffic analysis for IDSs and are frequently employed in image recognition.

✓ **Benefits:** Automatic feature learning, effective management of high-dimensional data, variation resistance, and real-time detection are just a few of the advantages that CNNs in IDS provide (Aljanabi et al., 2021). Data collection, preprocessing, appropriate architecture design, training, evaluation, tweaking, and deployment in real-world settings are all part of implementation. Data asymmetry, changing attacker evasion strategies, and computing resource needs are challenges.

✓ Applications:

- **Network Traffic Analysis:** CNNs are able to identify harmful activity by analysing raw network traffic data. CNNs are able to recognise patterns that indicate different kinds of cyberattacks, such as DDoS, (Chuan-long et al., 2017) port scanning, and unauthorised access, by converting network traffic into a format that is appropriate for CNNs, such as pictures or matrices depicting packet flows.
- **Anomaly Detection:** CNNs are used by anomaly-based IDS to find departures from standard patterns of network activity (Henry et al., 2023). CNNs are able to identify anomalous activity that might indicate an intrusion by learning the features of typical traffic. This is especially helpful in locating zero-day attacks that don't correspond with recognised signatures.
- **Malware Detection:** CNNs may be used to analyse executable files or their actions in order to detect malware. Converting binary data into grayscale graphics, for example, (Chuan-long et al., 2017; Susilo & Sari, 2020), can change it into a visual representation that CNNs can use to find patterns that separate harmful software from useful programmes.

In summary, by automatically learning and extracting characteristics from complicated network data, CNNs offer tremendous capabilities for boosting IDS, enabling high accuracy and efficiency in the detection of a wide range of cyber threats.

4.1.2 Recurrent Neural Networks (RNNs)

Utilising their capacity to handle sequential input and preserve temporal context, Recurrent Neural Networks (RNNs) offer a potent tool for improving IDS. They enable efficient identification of irregularities and new incursions in network traffic through their use in IDS.

For sequential data processing, RNNs are appropriate, especially Long Short-Term Memory (LSTM) networks (Ibrahim & Elhafiz, 2023). They are useful in identifying complicated assaults that span several packets or happen over time because they are good at catching temporal dependencies in network data.

✓ **Benefits:** By efficiently managing temporal dependencies in data, RNNs provide a number of advantages for IDS. This makes RNNs perfect for time-series network traffic analysis, where the sequence of events plays a critical role. As a result of their exceptional memory retention—which keeps prior inputs' information in their internal state—RNNs are able to analyse current packets while taking into account the context that they provided (Madhu et al., 2023). When looking for trends and irregularities that appear over time, this skill is quite helpful. Moreover, real-time intrusion detection is made possible by RNNs' ability to analyse data sequentially, which is crucial for prompt reactions to new threats.

✓ **Applications of RNNs in IDS**
- **Sequential Data Analysis:** Naturally, network traffic may be thought of as a series of packets sent over a period of time. To examine such sequences and find abnormalities or patterns suggestive of intrusions, RNNs are a good fit.
- **Anomaly Detection:** RNNs are able to learn the typical patterns of network traffic and recognise patterns that deviate from these patterns, indicating possible intrusions (Fan et al., 2024). This skill is especially useful for identifying new or unidentified assault patterns.
- **Feature Extraction:** RNNs, like CNNs, are capable of automatically extracting pertinent features from unprocessed input, but they prioritise temporal relationships in this process (Wassan et al., 2024). They are therefore perfect for comprehending how network traffic changes over time.

To sum up, RNNs are excellent at retaining information from past inputs in their internal state. This enables them to analyse current packets while taking into account the context supplied by prior packets. This skill is especially useful for seeing trends and irregularities that appear over time (Hnamte & Hussain, 2023).

Real-time intrusion detection is made possible by RNNs' ability to analyse data sequentially, which is crucial for prompt reactions to new threats.

4.1.3 Autoencoders

As Autoencoders can identify abnormalities in network data, they have drawn a lot of attention from the IDS community. Unsupervised DL models called Autoencoders are employed to acquire effective data representations (Kasongo, 2023). They may be used for anomaly detection in the context of IDS by recreating typical network traffic patterns and marking differences as possible intrusions. With an encoder that compresses input data into a lower-dimensional representation and a decoder that reconstructs the original data from the compressed form, these neural networks are used for unsupervised learning. Reconstruction error reduction is the main objective, which aims to provide a result that closely resembles the original input.

✓ **Benefits:** In contrast to labelled data, Autoencoders may be used for unsupervised learning without the requirement for labelled data. By spotting deviations from standard operating procedures, they are able to uncover abnormalities and maybe even unidentified assaults (Arai et al., 2023). Furthermore, in order to adjust to changing network behaviours, Autoencoders may be routinely retrained using fresh normal traffic data.

✓ **Application**

- **Network Traffic Anomaly Detection:** Finding anomalous patterns in network traffic that can point to possible security breaches is known as application network traffic anomaly detection.
- **System Log Analysis:** System log analysis is the process of looking for irregularities in system logs that can indicate malfunctions or malicious activity (Saheed et al., 2022).
- **Behavioural Analysis:** Activity analysis involves keeping an eye out for variations in user or device activity that might point to compromised accounts or gadgets (General International Group, n.d.).

Autoencoders have the potential to improve the resilience and efficiency of IDS with further development and appropriate use.

5. ASSESSING THE DETECTING CAPACITY OF IDS

An extensive assessment procedure is required to determine the detection capability of IDS enabled by DL. It involves primarily examining the detection accuracy, which gauges how well the system can distinguish between typical network activity and possible intrusions (Pinto et al., 2023). In order to maintain efficacy in dynamic network settings, it entails assessing detection accuracy, reducing false positives and negatives, testing resistance against adversarial assaults, and guaranteeing scalability and interpretability. Here's how to evaluate an IDS's threat detection capability:

1. Detection Methods:
    - **Signature-based:** This technique recognises known attack patterns using pre-established rules. It performs poorly against zero-day attacks but well against recognised threats.
    - **Anomaly-based:** This approach looks for departures from the usual, allowing it to identify new threats. Nevertheless, (Siva Shankar et al., 2024) it may produce false positives as a result of odd but genuine behaviour.
2. Methods of Evaluation:
    - **Penetration testing:** Using tools and tactics to mimic real-world assaults, determine if the IDS picks them up.
    - **Benchmark Datasets:** To evaluate the true positive and false positive rates of the IDS, publicly accessible datasets with labelled network traffic (attack and normal) are used (Ibrahim & Elhafiz, 2023).
3. Increasing the Power of Detection:
    - **Frequent Updates:** By keeping IDS signatures current, you can be sure it detects the newest threats.
    - **Fine-tuning Rules:** Modifying IDS rules to preserve accurate detection while lowering false positives (Wassan et al., 2024).
    - **Combining Techniques:** For a more thorough approach, combine anomaly-based and signature-based detection.
    - **Integration with Other Security Tools:** For a layered defence, IDS can cooperate with vulnerability scanners and firewalls.

By being aware of these elements, you may evaluate and enhance your IDS's detection capabilities and strengthen your system's defence against cyberattacks.

6. IDS CHALLENGES

Figure 7. Intrusion detection systems challenges

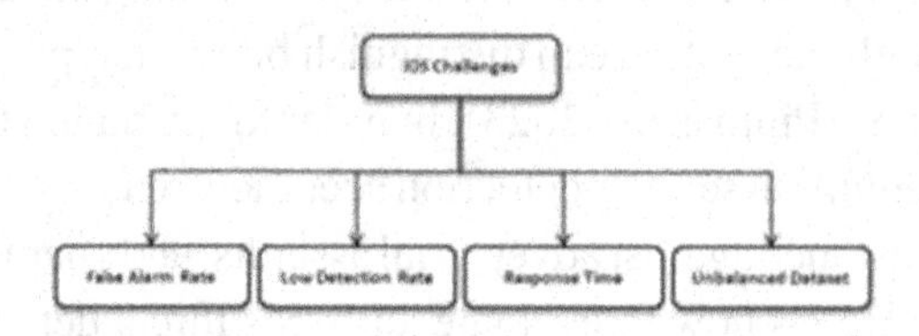

The main issues with IDSs are false positives and false negatives (Fan et al., 2024). The performance of an IDS can be significantly hampered by false positives, which increase noise. On the other hand, false negatives happen when an IDS misses an incursion and treats it as legitimate (Nicholls & Nicholls, 2024).

In order to successfully detect and mitigate attacks on computer networks, IDS must overcome a number of obstacles. Below are a few typical difficulties:

- **Negatives and False Positives:** IDS may produce false positive alerts (false positives) or false negative alarms (false negatives), which might result in ineffective resource use and security lapses.
- **Complexity of Network Traffic:** Particularly on high-speed networks, networks create enormous volumes of data, which makes it difficult for IDS to examine all of it in real time (General International Group, n.d.).

Maintaining IDS's efficacy is a constant struggle because of these obstacles. To deal with them, use these strategies:

- **Invest in qualified staff:** Security experts are able to adjust IDS rules, give alarms top priority, and remain abreast of emerging threats.
- **Adopt threat intelligence:** Information about novel attack techniques is shared, which aids in IDS adaptation and raises detection rates (Nicholls & Nicholls, 2024).
- **Use automation:** By streamlining incident response and accelerating reaction times, security orchestration and automation technologies (SOAR) may be used.
- **Periodic evaluations and audits:** IDS vulnerabilities may be found, and optimal performance can be ensured by regularly testing and monitoring the system (Pathmudi et al., 2023).

To maximise the benefits of IDS and maintain enhanced system security, you must recognise these obstacles and take appropriate action to address them.

7. RESEARCH OPENING

IDS research offers a bright future full of possibilities for growth and innovation. It is wide and complex. Here are a few prospective fields of study and upcoming opportunities in IDS:

1. The research offers a DCNNBiLSTM model (Hnamte & Hussain, 2023) for network intrusion detection that combines the capabilities of LSTM and CNN to identify anomalous attack patterns. Although it performs better than single models, more study is required to improve forecasting and preventive models.
2. A hybrid identification framework that uses a convolutional recurrent neural network (CRNN) to identify hostile network assaults is presented in Wassan et al. (2024). Experiments using publicly available data show promising results in accuracy and data loss. Future improvements include expanding the model to handle zero-day assaults and using actual backbone network traffic to demonstrate its value.
3. In resource constrained IoT contexts, the study provides an intelligent IDS that makes use of feature dimensionality reduction and machine learning (Saheed et al., 2022). In the future, a new IoT dataset and an ensemble model including DL models can be used. The BoT-IoT dataset can be compared with UNSWNB-15 for the purpose of classifying network traffic.
4. The IoTID20 dataset (Abdulla & Jameel, 2023) is shown to be the most successful when discussing the latest research on IoT intrusion detection utilising a variety of methodologies, datasets, and machine learning algorithms. To gather actual IoT intrusion detection datasets, create lightweight, real-time IDSs, and determine the best deployment strategies for IoT security, more study is required.
5. The study (Berahmand et al., 2024) investigates the concepts, uses, and future development prospects of Autoencoders, an essential unsupervised learning technique with applications in computer vision and natural language processing. The future of technology is anticipated to be shaped by Autoencoders as they continue to advance and provide intelligent solutions that benefit society and machine learning innovation.

8. CONCLUSION

IDS is a decades-old technology; in fact, some of the more contemporary solutions of today still bear some of the fundamental characteristics of the original systems. IDS may detect any suspicious activity and notify the system administrator by evaluating network traffic patterns. By offering insights and enhancing network performance, IDS may be a useful complement to any organisation's security setup. IDS, which provides proactive threat detection, contextual awareness, and scalability, is essential to strengthening the security of IoT installations inside the 5G ecosystem. Organisations may strengthen their defences against cyber-attacks and protect vital assets and data by incorporating IDS into IoT 5G settings and resolving related difficulties. This chapter starts with an overview of IDS, followed by a section that describes the various IDS kinds and methods, particularly those that are based on Deep Learning. It also incorporates the benefits and drawbacks of using IDS in conjunction with IoT applications. In order to improve detection accuracy and enable adaptive responses to develop security threats in real-time, the chapter emphasises how the integration of intelligent algorithms works.

REFERENCES

Abdulla & Jameel. (2023). A Review on IoT Intrusion Detection Systems Using Supervised Machine Learning: Techniques, Datasets, and Algorithms. *UHD Journal of Science and Technology, 7*(1), 53–65. 10.21928/uhdjst.v7n1y2023.pp53-65

Al-Shareeda, M. (2022). DDoS Attacks Detection Using Machine Learning and Deep Learning Techniques: Analysis and Comparison. *Bulletin of Electrical Engineering and Informatics, 12*(2). https://ssrn.com/abstract=4515135

Aljanabi, M., Ismail, M. A., & Ali, A. H. (2021, January 1). Intrusion Detection Systems, Issues, Challenges, and Needs. *International Journal of Computational Intelligence Systems.* 10.2991/ijcis.d.210105.001

Alotaibi, A., & Rassam, M. A. (2023). Adversarial Machine Learning Attacks against Intrusion Detection Systems: A Survey on Strategies and Defense. *Future Internet*, 15(2), 62. 10.3390/fi15020062

Amazon Web Services, Inc. (n.d.). *What is IoT? - Internet of Things Explained - AWS.* https://aws.amazon.com/what-is/iot/#:~:text=with%20AWS%20IoT-,What%20is%20the%20Internet%20of%20Things%20(IoT)%3F,as%20between%20the%20devices%20themselves

Azam, Z., Islam, M. M., & Huda, M. N. (2023). Comparative Analysis of Intrusion Detection Systems and Machine Learning-Based Model Analysis Through Decision Tree. *IEEE Access : Practical Innovations, Open Solutions*, 11, 80348–80391. 10.1109/ACCESS.2023.3296444

Berahmand, K., Daneshfar, F., Salehi, E. S., Li, Y., & Xu, Y. (2024). Autoencoders and their applications in machine learning: A survey. *Artificial Intelligence Review*, 57(2), 28. 10.1007/s10462-023-10662-6

Chuan-long, Yue-fei, Jin-long, & Xin-zheng. (2017). A Deep Learning Approach for Intrusion Detection Using Recurrent Neural Networks. *IEEE Access.* IEEE. .10.1109/ACCESS.2017.2762418

Cloudflare. (n.d.). *What is IoT security?* https://www.cloudflare.com/learning/security/glossary/iot-security/#:~:text=Internet%20of%20Things%20(IoT)%20devices%20are%20computerized%20Internet%2Dconnected,introduce%20threats%20into%20a%20network

Cybrosys. (n.d.). *An Overview of Network-based Intrusion Detection & Prevention Systems.* https://www.cybrosys.com/blog/an-overview-of-network-based-intrusion-detection-and-prevention-systems

Debicha, I., Cochez, B., Kenaza, T., Debatty, T., Dricot, J.-M., & Mees, W. (2023). Adv-Bot: Realistic adversarial botnet attacks against network intrusion detection systems. *Computers & Security, 129*.10.1016/j.cose.2023.103176

Desai, P., Sonawane, A., Mane, T., & Jaiswal, R. (2023). Network based intrusion detection system. *International Research Journal of Modernization in Engineering Technology and Science*, 5, 3851–3857. 10.56726/IRJMETS35232

Dixit, S., Jain, R., & Patel, H. B. (2024). Impact of 5G Wireless Technologies on Cloud Computing and Internet of Things (IOT). *Advances in Robotic Technology*. https://ssrn.com/abstract=4700149

Elrawy, M., Awad, A., & Hamed, H. (2018). Intrusion Detection Systems for IoT-based smart environments: A survey. *Journal of Cloud Computing (Heidelberg, Germany)*, 7(1), 21. 10.1186/s13677-018-0123-6

Fan, C., Cui, J., Jin, H., Zhong, H., Bolodurina, I., & He, D. (2024). Auto-Updating Intrusion Detection Systems for Vehicular Network: A Deep Learning Approach Based on Cloud-Edge-Vehicle Collaboration. *IEEE Transactions on Vehicular Technology*. IEEE. 10.1109/TVT.2024.3399219

Flinders, M., & Flinders, M. (2024, March 14). The future of 5G: What to expect from this transformational technology. *IBM Blog*. IBM. https://www.ibm.com/blog/5g-future/

Gala, Y., Vanjari, N., Doshi, D., & Radhanpurwala, I. (2023). *AI based Techniques for Network-based Intrusion Detection System: A Review*. 10th International Conference on Computing for Sustainable Global Development (INDIACom), New Delhi, India.

General International Group. (n.d.). *Advantages & Disadvantages of Intrusion Detection Systems (IDS) Types*.https://generalintlgroup.com/en/blog/advantages-and-disadvantages-of-intrusion-detection-system-ids types#:~:text=Limited%20Zero%2DDay%20Threat%20Detection,and%20deploy%2C%20leaving%20networks%20vulnerable

Goldena, N. J. (2022, January 4). *Essentials of the Internet of Things (IoT)*. Auerbach Publications eBooks. 10.1201/9781003119784-3

Gutierrez-Garcia, J. L., Sanchez-DelaCruz, E., & Pozos-Parra, M. P. (2023). A Review of Intrusion Detection Systems Using Machine Learning: Attacks, Algorithms and Challenges. In Arai, K. (Ed.), *Advances in Information and Communication. FICC 2023. Lecture Notes in Networks and Systems* (Vol. 652). Springer. 10.1007/978-3-031-28073-3_5

He, K., Kim, D. D., & Asghar, M. R. (2023). Adversarial Machine Learning for Network Intrusion Detection Systems: A Comprehensive Survey. *IEEE Communications Surveys & Tutorials*. IEEE. 10.1109/COMST.2022.3233793

Henry, A., Gautam, S., Khanna, S., Rabie, K., Shongwe, T., Bhattacharya, P., Sharma, B., & Chowdhury, S. (2023). Composition of Hybrid Deep Learning Model and Feature Optimization for Intrusion Detection System. *Sensors (Basel)*, 23(2), 890. 10.3390/s2302089036679684

Hnamte, V., & Hussain, J. (2023). DCNNBiLSTM: An Efficient Hybrid Deep Learning-Based Intrusion Detection System. *Telematics and Informatics Reports, 10*. 10.1016/j.teler.2023.100053

Hnamte, V., & Hussain, J. (2023). Dependable Intrusion Detection Systems using deep convolutional neural network: A Novel framework and performance evaluation approach. *Telematics and Informatics Reports, 11*. 10.1016/j.teler.2023.100077

Ibrahim, M., & Elhafiz, R. (2023). Modeling an intrusion detection using recurrent neural networks. *Journal of Engineering Research, 11*(1). 10.1016/j.jer.2023.100013

Kasongo, A. (2023). A Deep Learning technique for Intrusion Detection Systems using a Recurrent Neural Networks based framework. *Computer Communications, 199*, 113-125. ,10.1016/j.comcom.2022.12.010

Khuntia, M., & Singh, D., & Sahoo, S. (2021). *Impact of Internet of Things (IoT) on 5G*. Springer. 10.1007/978-981-15-6202-0_14

Madhu, B., Chari, M. V. G., Vankdothu, R., Silivery, A. K., & Aerranagula, V. (2023, February 1). Intrusion detection models for IOT networks via Deep Learning approaches. *Measurement. Sensors*, 25, 100641. 10.1016/j.measen.2022.100641

Nicholls, M., & Nicholls, M. (2024, March 20). *The key challenges of intrusion detection and how to overcome them*. Redscan. https://www.redscan.com/news/the-key-challenges-of-intrusion-detection-and-how-to-overcome-them/

Pathmudi, V. R., Khatri, N., Kumar, S., Antar, S. H. A.-Q., & Vyas, A. K. (2023). A systematic review of IoT technologies and their constituents for smart and sustainable agriculture applications. *Scientific African, 19*. 10.1016/j.sciaf.2023.e01577

Paya, A., Arroni, S., García-Díaz, V., & Gómez, A. (2024). Apollon: A robust defense system against Adversarial Machine Learning attacks in Intrusion Detection Systems. *Computers & Security, 136*. 10.1016/j.cose.2023.103546

Pinto, A., Herrera, L.-C., Donoso, Y., & Gutierrez, J. A. (2023). Survey on Intrusion Detection Systems Based on Machine Learning Techniques for the Protection of Critical Infrastructure. *Sensors (Basel)*, 23(5), 2415. 10.3390/s2305241536904618

Saheed, Y. K., Abiodun, A. I., Misra, S., Holone, M. K., & Colomo-Palacios, R. (2022, December 1). A machine learning-based intrusion detection for detecting internet of things network attacks. Alexandria Engineering Journal. *Alexandria Engineering Journal*, 61(12), 9395–9409. 10.1016/j.aej.2022.02.063

Santos, V. F., Albuquerque, C., Passos, D., Quincozes, S. E., & Mossé, D. (2023). Assessing Machine Learning Techniques for Intrusion Detection in Cyber-Physical Systems. *Energies*, 16(16), 6058. 10.3390/en16166058

Sisavath, C., & Yu, L. (2021). Design and implementation of security system for smart home based on IOT technology. *Procedia Computer Science, 183*. https://www.sciencedirect.com/science/article/pii/S187705092100487710.1016/j.procs.2021.02.023

Siva Shankar, S., Hung, B. T., Chakrabarti, P., Chakrabarti, T., & Parasa, G. (2024). A novel optimization based Deep Learning with artificial intelligence approach to detect intrusion attack in network system. *Education and Information Technologies*, 29(4), 3859–3883. 10.1007/s10639-023-11885-4

Susilo, B., & Sari, R. F. (2020, May 21). Intrusion Detection in IoT Networks Using Deep Learning Algorithm. *Information (Basel)*, 11(5), 279. 10.3390/info11050279

Verma, A., & Ranga, V. (2019). On evaluation of Network Intrusion Detection Systems: Statistical analysis of CIDDS-001 dataset using Machine Learning Techniques. *TechRxiv*.

Wassan, S., Dongyan, H., Suhail, B., Jhanjhi, N. Z., Xiao, G., Ahmed, S., & Murugesan, R. K. (2024). Deep convolutional neural network and IoT technology for healthcare. *Digital Health*, 10. 10.1177/20552076231220123 38250147

Section 3
Antenna Design and Power Allocation for 5G

Chapter 7
The Performance Analysis of PSO-Based Power Allocation for Alamouti Decode and Forward Relaying Protocol

Shoukath Ali K.
https://orcid.org/0000-0001-9256-373X
Presidency University, India

Arfat Ahmad Khan
Khon Kaen University, Thailand

ABSTRACT

Alamouti decode and forward (DF) relaying protocol using cooperative-maximum combining ratio (C-MRC) for three user cooperative system is considered. The performance of approximate SER for Alamouti DF relaying protocol with M-PSK modulation over Rayleigh fading and Rician fading channel is presented and compared with upper bound SER results. The approximate SER of proposed protocol is asymptotically tight at high SNR compared to the upper bound. The Alamouti DF relaying protocol provides full diversity gain compared with the existing Protocols. The PSO algorithm is used to calculate the optimum PA factor based on the minimum approximate SER. The simulated SER is compared with the theoretical approximate SER. It is shown that the Alamouti DF relaying protocol outperforms the existing Alamouti AF, AF, and DF relaying protocols. Modified cooperative subchannel allocation (CSA) algorithm for Alamouti DF relaying protocol using C-MRC technique is proposed, which maximizes the total throughput of the multiuser orthogonal

DOI: 10.4018/979-8-3693-2786-9.ch007

frequency division multiplexing access (OFDMA) systems.

1. INTRODUCTION

To overcome the limitations caused by fluctuating signal strength in wireless channels, researchers are exploring novel techniques to achieve maximum diversity gain. One such method to increase diversity gain in wireless communication is called MIMO (Multiple Input Multiple Output) systems. In the technique employing many transceiver antennas, high data rates, bandwidth efficiency, diversity gain, and reliability of the wireless system in fading environments are provided. However, conventional MIMO antenna systems are challenging to implement. Alamouti (1998), Anghel and Kaveh (2004), Anghel et al. (2003), and Ashourian et al. (2013) suggested cooperative communication to the generalized MIMO system. This method offers a diversity technique known as cooperative diversity (Atapattu & Rajatheva, 2008; Farhadi, 2007; Levin, 2012). Among the various types of Cooperative diversity schemes, Alamouti coding is utilized in this chapter (Himsoon et al., 2005; Priyanka, 2013). This approach is known for its ability to accommodate high speeds in wireless communication systems, resulting in improved Bit Error Rate (BER) and higher data rates.

Cooperative Communication are classified into two protocols such as: Amplify-and-Forward (AF) and Decode-and-Forward (DF). In AF, relay station simply amplifies signal received from source and forward to destination. While this is a simpler approach, it also amplifies any noise present in the signal. DF, decodes signal from source before re-encoding. This allows the relay to remove noise before sending the information on to the destination. At destination, signals combined by a technique called Maximum Combining Ratio (MRC) to improve overall transmission quality (as shown in Bletsas et al., 2006; Brennan, 2003; Deng & Haimovich, 2005; Dohler & Li, 2010).

This chapter focuses on the approximate Symbol Error Rate (SER) of a system employing Alamouti's Decode-and-Forward (DF) relaying protocol with Cooperative Maximum Ratio Combining (C-MRC) and M-Phase Shift Keying (PSK) modulation. The approximate SER is shown to be a bound for high Signal-to-Noise Ratio (SNR) values. Additionally, we utilize Particle Swarm Optimization (PSO) to find optimal Power Allocation (PA) value that minimizes approximate SER for Alamouti DF relaying protocol with C-MRC.

Additionally, Non-Cooperative Subchannel Allocation (Non-CSA) and Cooperative Subchannel Allocation (CSA) for AF Relaying Protocol with Equal Power Allocation (EPA) do not achieve maximum throughput and optimize power in the wireless system. To overcome this problem, a modified CSA algorithm and PSO based optimum PA for multi-user Orthogonal Frequency Division Multiplexing (OFDM) Alamouti DF Relaying Protocol are proposed. The usage of the modified CSA algorithm in the proposed protocol provides increased throughput in wireless systems compared to existing Relaying Protocols.

2. LITERATURE SURVEY

In wireless communication systems, cooperative diversity protocols combat multipath fading channels. Space-time coded diversity protocols achieve spatial diversity gain and high bandwidth efficiency of the wireless system (Ray Liu & Ahmed, 2009; Sadek et al., 2005; Sendonaris et al., 2003). Outage probabilities and outage events are derived to measure the strength of transmission signals with respect to the fading environment at high SNR conditions. Siriwongpairat et al. (2006) and Stefanov and Erkip (2004) suggested a new simple diversity technique called Alamouti coding scheme, consisting of a pair of antennae, to enhance the diversity gain, data rate, channel capacity, and reduce BER. Ibrahim et al. (2005), Ikki and Ahmed (2007, 2008), Kennedy (2010), Laneman et al. (2004), and Lu and Nikookar (2009) discussed the performance analysis and comparison of SER for AF and DF Relaying Protocols with optimum PA with MRC technique. However, SER upper bound and approximate SER achieve asymptotic performance in relaying protocols.

Cooperative relaying protocols promise considerable capacity and increased multiplexing gain. Several studies have investigated the performance of Alamouti coded Decode-and-Forward (DF) relaying protocol at an optimal relay selection (Nostratinia et al., 2004; Su et al., 2005; Su et al., 2008; Swasdio & Pirak, 2010; Swasdio et al., 2011; Swasdio, Pirak, Jitapunkul, & Ascheid, 2014). These studies compared its outage probability to existing protocols. Other research (Bai, 2010; Bai, 2013; Fei et al., 2008; Sadr, Anpalagan, & Raahemifar, 2009; Sadr, Anpalagan, & Raahemifar, 2009) explored the concept of power allocation for DF relaying protocols.

Mobile communication environments aim to achieve high data rates and minimum total transmit power. OFDM is preferred due to its multicarrier modulation technique (Dutt, 2014; Lin, 2009; Swasdio, Pirak, Jitapunkul, & al Gerd, 2014; Wong & Cheng, 1999). In multiuser OFDM systems, the total data rate is maximized when only one user selects each subcarrier with high channel gain. S. A. K, (2021, 2022), Shoukath Ali (2019), and Shoukath Ali and Sampath (2023) analyze downlink communication in non-cooperative multi-user OFDMA systems. These studies employ dynamic Resource Allocation Algorithms (RAA) and optimization techniques (Ali et al., 2019; Shoukath Ali & Sampath, 2021) to achieve goals. First goal is to maximize throughput under constraint on total transmit power and second goal is to minimize total transmit power while guaranteeing a certain level of throughput. The text also discusses limitations of using Equal Power Allocation (EPA) for subcarrier at AF relaying protocol within multi-user OFDMA systems. Unlike EPA, RAA offers improved throughput and power allocation as shown in Ali et al. (2019), S. A. K, (2021, 2022), Shoukath Ali (2019), Shoukath Ali and Sampath (2021, 2023). While Maximum Ratio Combining (MRC) technique offers benefits for combining signals, it cannot guarantee the highest Signal-to-Noise Ratio (SNR) and improved throughput, as noted by references (Ali et al., 2023; Hamed, 2012).

Effective cooperative communication is pivotal for optimizing the performance of smart grid systems, bolstering their efficiency, reliability, and resilience. However, safeguarding the security of such communication is imperative given the sensitive nature of the involved data and control signals. Key considerations for ensuring the security of cooperative communication in smart grids encompass data confidentiality and integrity, the security of relay nodes, resilience against cyber threats, and authentication mechanisms (Thantharate & Beard, 2022; Zeb et al., 2021). Similarly, in 5G networks, cooperative communication protocols present both opportunities and challenges regarding security. These include addressing privacy concerns, ensuring the trustworthiness of relay nodes, and thwarting potential attacks like Denial of Service (DoS) and Man-in-the-Middle (MitM). While these protocols offer enhanced network performance, it's critical for encryption, authentication, access control, and intrusion detection to mitigate associated risks. Furthermore, in smart grids and 5G networks alike, careful consideration of security vulnerabilities, threats, and mitigation strategies is essential (AlQahtani, 2023; Guo et al., 2022; Li et al., 2021). Taking a comprehensive approach that encompasses secure communication protocols and robust security measures is vital for enhancing the trustworthiness and resilience of cooperative communication in both smart grids and 5G networks.

3. RECEIVED SIGNAL MODEL

This research investigates a system with Alamouti Decode-and-Forward (DF) relaying protocol for wireless communication. In Figure 1, the system involves a source (S), two-way relay stations (RS), and a destination (D).

Figure 1. Source (S), two-way relay stations, (RS), and destination (D)

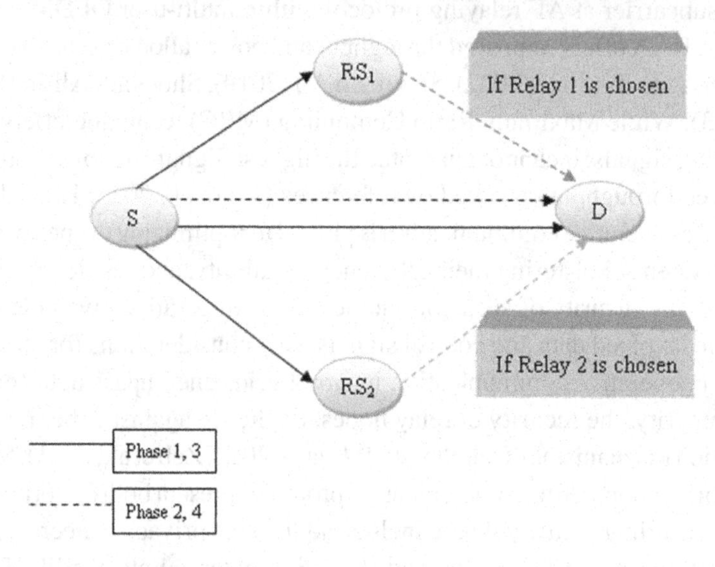

To achieve high data rates and diversity gain in wireless systems, Orthogonal Space Time Block Codes (OSTBC) are employed. This technique utilizes multiple transmit and receive antennas. Specifically, the Alamouti scheme employed improves received signal quality. Encoder, operating both space and time domains, transmits two modulated symbols (X1 and X2) at a time. This encoding process is represented by the matrix in Equation (1),

$$X = \begin{pmatrix} X_1 & X_2 \\ -X_2^* & X_1^* \end{pmatrix} \tag{1}$$

where X* stands for complex conjugate of the variable X.

Figure 2 depicts the Time Division Multiple Access (TDMA) transmission for the Alamouti DF that are divided into four phases. In phase 1, source (S) transmits its information to a chosen relay station using an optimal selection technique. During phase 2, selected relay station decodes information from source and then forwards it using the Alamouti scheme. Phases 3 and 4 repeat the processes of phases 1 and 2, respectively.

In phase 1, S transmits symbol (X_1) to RS and to D. The transmits power of S is defined byP_1. $h_{(s\text{-}d)1}$ is channel coefficient of S-D and $n_{(s\text{-}d)1}$ is an additive noise of S-D in phase1. $Y_{(s\text{-}d)1}$ implies received signal at D as in Equation (2).

$$Y_{(s-d)1} = \sqrt{P_1}\, h_{(s-d)1} X_1 + n_{(s-d)1} \tag{2}$$

$h_{(s-r)1}$ is channel coefficient of S-RS and $n_{(s-r)1}$ is noise component of the S-RS in phase1. $Y_{(s-r)1}$ is received signal at RS as expressed in Equation (3).

$$Y_{(s-r)1} = \sqrt{P_1}\, h_{(s-r)1} X_1 + n_{(s-r)1} \tag{3}$$

Phase 2, demonstrates that S transmits symbol(X_2) to D with P_1 as transmitting power and RS should decode the symbol($\widehat{X}_1$) and forward that to D with its power asP_1.

Figure 2. System model: Alamouti DF relaying protocol in four different phases

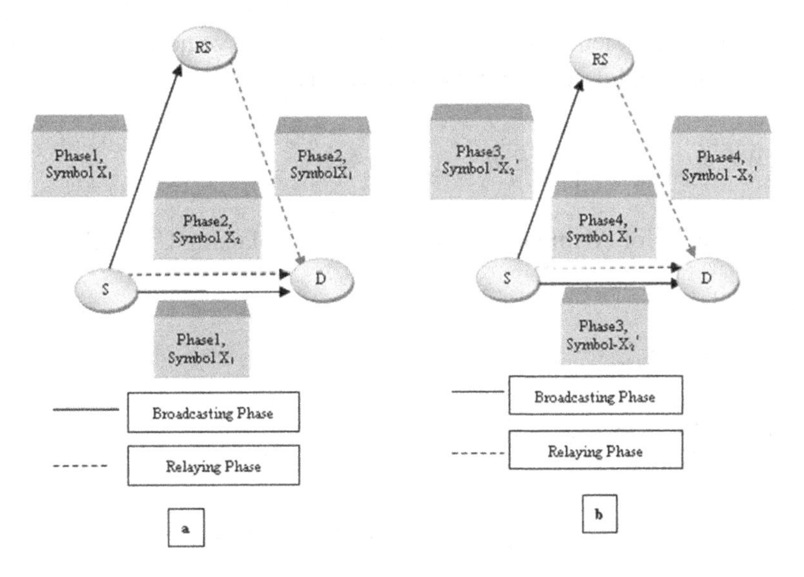

Phase 2 introduces weighting factors for S-D and RS-D links to maximize system's overall Signal-to-Noise Ratio (SNR).The expression of$W = (\Gamma_{eq}, \Gamma_{r-d})$, where$\Gamma_{eq} = \min(\Gamma_{s-r}, \Gamma_{r-d})$and Γ_{s-r} and $\Gamma_{r-d}$are instantaneous SNR between S-RS and RS-D nodes respectively.$h_{(s-d)2}$is channel coefficient of S-D and$h_{(r-d)2}$is channel coefficient of RS-D during phase 2.n_{d2} is an additive noise at D .Y_{d2} is amount of received signal, and is given by

$$Y_{d2} = \sqrt{WP_1}\, h_{(s-d)2} X_2 + \sqrt{WP_2}\, h_{(r-d)2} \widehat{X}_1 + n_{d2} \tag{4}$$

S transmits symbol $(-X_2^*)$ with its transmitting power ofP_1 to RS and D in phase3. $h_{(s-d)3}$ and $h_{(s-r)3}$ are channel coefficient as S-D and

S-RS respectively in phase3. $n_{(s\text{-}d)3}$ and $n_{(s\text{-}r)3}$ are noise component at S-D and S-RS respectively in phase 3. $Y_{(s\text{-}d)3}$ and $Y_{(s\text{-}r)3}$ are signals at D and RS respectively as narrated by

$$Y_{(s-d)3} = \sqrt{P_1}\, h_{(s-d)3}(-X_2^*) + n_{(s-d)3} \tag{5}$$

$$Y_{(s-r)3} = \sqrt{P_1}\, h_{(s-r)3}(-X_2^*) + n_{(s-r)3} \tag{6}$$

Then, S transmits a symbol(X_1^*)to D with transmitting power ofP_1, and RS has to decode symbol $(-\hat{X}_2^*)$and forward to D with its power as in phase 4. $h_{(s\text{-}d)4}$ and $h_{(r\text{-}d)4}$ is channel coefficient of S-D and RS-D.n_{d4} is additive noise element at D.Y_{d4} is received signals and described in Equations (7) as

$$Y_{d4} = \sqrt{WP_1}\, h_{(s-d)4} X_1^* + \sqrt{WP_2}\, h_{(r-d)4}(-\hat{X}_2^*) + n_{d4} \tag{7}$$

Total SNR at D for Alamouti DF Relaying Protocol using C-MRC technique is given by

$$\Gamma_{Total} = \Gamma_{(s-d)} + \Gamma_{Alamouti} \tag{8}$$

Where is total SNR received at D from S and RS.

Received signals at D in Equations (2), (4), (5), and (7) are merged using $_{\text{C-MRC}}$ as

$$y_{\hat{X}_1(CMRC)} = W_{11(CMRC)}(Y_{(s-d)1}) + W_{12(CMRC)}(Y_{d2}) + W_{13(CMRC)}(Y_{d4}^*) \tag{9}$$

$$y_{\hat{X}_2(CMRC)} = W_{21(CMRC)}(-Y_{(s-d)3}^*) + W_{22(CMRC)}(Y_{d2}) + W_{23(CMRC)}(-Y_{d4}^*) \tag{10}$$

Where$y_{\hat{X}_2(CMRC)}$and$y_{\hat{X}_1(CMRC)}$are resultant of C-MRC combiner at D for symboland . Also - are C-MRC weights expressed from Equations (11)-(16).

$$W_{11(CMRC)} = \frac{\sqrt{P_1}}{N_0} h^*_{(s-d)1} \tag{11}$$

$$W_{12(CMRC)} = \frac{\sqrt{WP_2}}{N_0} h^*_{(r-d)2} \tag{12}$$

$$W_{13(CMRC)} = \frac{\sqrt{WP_1}}{N_0} h_{(s-d)4} \tag{13}$$

$$W_{21(CMRC)} = \frac{\sqrt{P_1}}{N_0} h_{(s-d)3} \tag{14}$$

$$W_{22(CMRC)} = \frac{\sqrt{WP_1}}{N_0} h^*_{(s-d)2} \tag{15}$$

$$W_{23(CMRC)} = \frac{\sqrt{WP_2}}{N_0} h_{(r-d)4} \tag{16}$$

Substitute the Equations (11) -(16) in Equation (9) and Equation (10), and final signal received at D is written as,

$$y_{\hat{X}_1(CMRC)} = \frac{P_1 |h_{(s-d)1}|^2 X_1}{N_0} + \frac{WP_1 |h_{(s-d)4}|^2 X_1}{N_0} + \frac{WP_2 |h_{(r-d)2}|^2 X_1}{N_0} + N_{X_1(CMRC)} \tag{17}$$

$$y_{\hat{X}_2(CMRC)} = \frac{P_1 |h_{(s-d)3}|^2 X_2}{N_0} + \frac{WP_1 |h_{(s-d)2}|^2 X_2}{N_0} + \frac{WP_2 |h_{(r-d)4}|^2 X_2}{N_0} + N_{X_2(CMRC)} \tag{18}$$

Where and are the total noises at D for symbol(X_1)and (X_2) and described as follows:

$$N_{X_1(CMRC)} = \frac{\sqrt{\Gamma_1} h^*_{(s-d)1} n_{(s-d)1}}{N_0} + \frac{\sqrt{WP_1} h_{(s-d)4} n_{d4}}{N_0} + \frac{\sqrt{WP_2} h^*_{(r-d)2} n_{d2}}{N_0} \tag{19}$$

$$N_{X_2(CMRC)} = \frac{\sqrt{P_1} h_{(s-d)3} n_{(s-d)3}}{N_0} + \frac{\sqrt{WP_1} h^*_{(s-d)2} n_{d2}}{N_0} + \frac{\sqrt{WP_2} h_{(r-d)4} n_{d4}}{N_0} \tag{20}$$

Finally, denotes total SNR at D for symbol(X_1)and is total SNR at D for symbol (X_2)using C-MRC technique:

$$\Gamma_{X_1} = \frac{(1+W)P_1 |h_{(s-d)}|^2 + WP_2 |h_{(r-d)}|^2}{N_0} \tag{21}$$

$$\Gamma_{X2} = \frac{(1+W)P_1|h_{(s-d)}|^2 + WP_2|h_{(r-d)}|^2}{N_0} \tag{22}$$

Quality of decoded symbol($\hat{X}_1$)and($\hat{X}_2$)measure rely on SNR of S-RS links. If SNR value of S-RS is less than that of RS-D, then there occur more errors on symbol($\hat{X}_1$)and($\hat{X}_2$). In C-MRC technique, if the channel variance of S-D is greater on compared with channel variance of RS-D, then more power allocated to S and less power allocated to RS link. If the channel variance of S-D is less than the channel variance of RS-D, then less power will be allocated to S and more power to RS links, that leads to the maximization of SNR and full diversity gain at D. After completion of C-MRC combining at D node, it detects symbol(X_1)and (X_2)by a Maximum-Likelihood (ML) decoder.

3.1 Alamouti DF Relaying Protocol

This work analyses the approximate SER for a system employing Alamouti DF relaying protocol with M-PSK modulation. The derived approximate SER is then compared to the existing upper bound on SER. We use a PSO algorithm to find the optimal PA factor that minimizes the approximate SER for the proposed protocol. Interestingly, the optimal PA ratio depends on the specific channel conditions of the system, and can either increase or decrease accordingly.

Furthermore, the approximate SER shows significant improvement at high Signal-to-Noise Ratio (SNR) values.

$$M - PSK = \psi_{M-PSK}(\Gamma_X) \tag{23}$$

$$M - PSK = \psi_{M-PSK}(\Gamma_X) = 1/\pi \int_0^{(M-1)\pi/M} \exp\left(-\frac{b_{psk}\Gamma_X}{\sin^2\theta}\right) d\theta \tag{24}$$

$$M - PSK = \psi_{M-PSK}(\Gamma_X) = 1/\pi$$

$$\int_0^{(M-1)\pi/M} \exp\left(-\frac{b_{psk}((1+W)P_1|h_{(s-d)}|^2 + WP_2|h_{(r-d)}|^2)}{N_0 \sin^2\theta}\right) d\theta \tag{25}$$

In phase 2, the transmits symbol of S is decoded correctly at the RS. Then decrypted symbol to D with transmitting power of is considered as. Otherwise, the RS does not send information to the D and D considered as. Possible of decoding

M-PSK symbol correctly at the RS is and possible of decoding symbol incorrectly at the RS is .

Now, calculate conditional SER for M-PSK modulation by taking the two scenarios of and in Equation (25).

$$M - PSK = \psi_{M-PSK}(\Gamma_X)\big|_{\tilde{P}_2=0}\psi_{PSK}\left(\frac{P_1|h_{(s-r)}|^2}{N_0}\right) + \psi_{PSK}(\Gamma_X)\big|_{\tilde{P}_2=P_2}\left[1 - \psi_{PSK}\left(\frac{P_1|h_{(s-r)}|^2}{N_0}\right)\right] \quad (26)$$

$$M - PSK = \psi_{M-PSK}(\Gamma_X) = 1/\pi^2 \int_0^{(M-1)\pi/M} \exp\left(-\frac{b_{psk}P_1|h_{(s-d)}|^2}{N_0\sin^2\theta}\right)d\theta$$

$$\times \int_0^{(M-1)\pi/M} \exp\left(-\frac{b_{psk}P_1|h_{(s-r)}|^2}{N_0\sin^2\theta}\right)d\theta \times \int_0^{(M-1)\pi/M} \exp\left(-\frac{b_{psk}P_1|h_{(s-d)}|^2}{N_0\sin^2\theta}\right)d\theta$$

$$\times \int_0^{(M-1)\pi/M} \exp\left(-\frac{b_{psk}P_1|h_{(s-r)}|^2}{N_0\sin^2\theta}\right)d\theta + 1/\pi \int_0^{(M-1)\pi/M} \exp\left(-\frac{b_{psk}((1+W)P_1|h_{(s-d)}|^2 + WP_2|h_{(r-d)}|^2)}{N_0\sin^2\theta}\right)d\theta \times$$

$$\left[1 - 1/\pi \int_0^{(M-1)\pi/M} \exp\left(-\frac{b_{psk}P_1|h_{(s-r)}|^2}{N_0\sin^2\theta}\right)d\theta\right] \quad (27)$$

$$\int_0^{\infty} \exp\left(-\frac{b_{psk}P_1 z}{N_0\sin^2\theta}\right)P|h|^2(z)\,dz = \frac{1}{1 + \frac{b_{psk}P_1|h|^2}{N_0\sin^2\theta}} \quad (28)$$

Averaging the conditional SER of Alamouti DF Relaying Protocolwith *M*-PSK modulation system is given by,

$$M - PSK = F_1\left(1 + \frac{b_{psk}P_1|h_{(s-d)}|^2}{N_0\sin^2\theta}\right)F_1\left(1 + \frac{b_{psk}P_1|h_{(s-r)}|^2}{N_0\sin^2\theta}\right)$$

$$+F_1\left(\left(1+\frac{b_{psk}(1+W)P_1|h_{(s-d)}|^2}{N_0\sin^2\theta}\right)\left(1+\frac{b_{psk}WP_2|h_{(r-d)}|^2}{N_0\sin^2\theta}\right)\right) \left[1-F_1\left(1+\frac{b_{psk}P_1|h_{(s-r)}|^2}{N_0\sin^2\theta}\right)\right] \tag{29}$$

From Equation (29), closed form expression for SER of Alamouti DF Relaying Protocol is calculated numerically. For simplicity, upper bound SER is derived by neglecting negative term in Equation (29). From the assumption of large power of , the negative term approximately equal to one.

$$M-PSK \le F_1\left(1+\frac{b_{psk}P_1|h_{(s-d)}|^2}{N_0\sin^2\theta}\right)F_1\left(1+\frac{b_{psk}P_1|h_{(s-r)}|^2}{N_0\sin^2\theta}\right)$$

$$+F_1\left(\left(1+\frac{b_{psk}(1+W)P_1|h_{(s-d)}|^2}{N_0\sin^2\theta}\right)\left(1+\frac{b_{psk}WP_2|h_{(r-d)}|^2}{N_0\sin^2\theta}\right)\right) \tag{30}$$

$$1-F_1\left(1+\frac{b_{psk}P_1|h_{(s-r)}|^2}{N_0\sin^2\theta}\right) = 1-\frac{N_0}{\pi b_{psk}P_1|h_{(s-r)}|^2}\int_0^{(M-1)\pi/M}\sin^2\theta d\theta \approx 1 \tag{31}$$

Where, $M-PSK = SER = I_1(P_1/N_0)+I_2(P_1/N_0, P_2/N_0)$

$$I_1(x)= F_1\left(1+\frac{(1+W)xb_{psk}|h_{(s-d)}|^2}{\sin^2\theta}\right)F_1\left(1+\frac{xb_{psk}|h_{(s-r)}|^2}{\sin^2\theta}\right) \tag{32}$$

$$I_2(x,y)= F_1\left(1+\frac{(1+W)xb_{psk}|h_{(s-d)}|^2}{\sin^2\theta}\right)\left(1+\frac{Wyb_{psk}|h_{(r-d)}|^2}{\sin^2\theta}\right) \tag{33}$$

From the assumption of large power ofand, the value 1's is omitted in Equation (32) and Equation (33).

Where, $1+\frac{(1+W)xb_{psk}|h_{(s-d)}|^2}{\sin^2\theta} \approx \frac{(1+W)xb_{psk}|h_{(s-d)}|^2}{\sin^2\theta}$,

$$1 + \frac{Wyb_{psk}\left|h_{(r-d)}\right|^2}{\sin^2\theta} \approx \frac{Wyb_{psk}\left|h_{(r-d)}\right|^2}{\sin^2\theta},$$

$$1 + \frac{Wxb_{psk}\left|h_{(s-r)}\right|^2}{\sin^2\theta} \approx \frac{Wxb_{psk}\left|h_{(s-r)}\right|^2}{\sin^2\theta}$$

$$\lim_{x\to\infty} x^2 I_1(x) = \frac{A^2}{b_{psk}^2(1+W)\left|h_{(s-d)}\right|^2\left|h_{(s-r)}\right|^2} \tag{34}$$

$$\lim_{x,y\to\infty} I_2(x,y) = \frac{B}{b_{psk}^2(1+W)\,W\left|h_{(s-d)}\right|^2\left|h_{(r-d)}\right|^2} \tag{35}$$

$$A = \frac{1}{\pi}\int_0^{(M-1)\pi/M} \sin^2\theta d\theta = \frac{(M-1)}{2M} + \sin\frac{\left(\frac{2\pi}{M}\right)}{4\pi} \tag{36}$$

$$B = \frac{1}{\pi}\int_0^{(M-1)\pi/M} \sin^4\theta d\theta = \frac{3(M-1)}{8M} + \sin\frac{\left(\frac{2\pi}{M}\right)}{4\pi} - \sin\frac{\left(\frac{4\pi}{M}\right)}{32\pi} \tag{37}$$

Therefore, for large value and, asymptotically tight approximations are as follows,

$$I_1(x) \approx \frac{1}{x^2}\frac{A^2}{b_{psk}^2(1+W)\left|h_{(s-d)}\right|^2\left|h_{(s-r)}\right|^2} \tag{38}$$

$$I_2(x,y) \approx \frac{1}{xy}\frac{B}{b_{psk}^2(1+W)\,W\left|h_{(s-d)}\right|^2\left|h_{(r-d)}\right|^2} \tag{39}$$

This approximation assumes that the error terms become negligible as x and y approach positive infinity. We can achieve this by substituting x with γ_s and y with γ_r in Equations (38) and (39), respectively. These represent SNR values of S-RS and RS-D links. Finally, an asymptotically tight approximation for Symbol Error Rate (SER) of the system with M-PSK modulation is obtained.

$$SER = \frac{N_0^2}{b_{psk}^2}\frac{1}{P_1\left|h_{(s-d)}\right|^2(1+W)}\left(\frac{A^2}{P_1\left|h_{(s-r)}\right|^2} + \frac{B}{WP_2\left|h_{(r-d)}\right|^2}\right) \tag{40}$$

Equation (44) shows an approximate SER of Alamouti DF Relaying Protocol.

3.2 Optimum Relay Selection Technique

The increased number of relay achieves diversity gain of cooperative communication system. Source node selects optimal relay depends on channel quality of each RS link.

The optimum relay selection for Alamouti DF Relaying Protocol is given by

$$\min(SER(R_1), SER(R_2)) \tag{41}$$

Wherestands for the RS1, andstands for the RS2. From Equation (40), calculate SER approximation for RS1 and RS2 using PSO based PA method. The selection of optimum relay is done based on the performance of minimum approximate SER for Alamouti DF Relaying Protocol.

3.3 Power Allocation using PSO

This research compares performance of two PA techniques for DF relaying protocol: EPA and PSO based PA.

PSO is an optimization technique that excels at finding the best solution for a given problem. It utilizes an objective function to evaluate potential solutions. Similar to population search methods, PSO iteratively searches for optimal solution. During iteration, all particles in swarm update their positions and velocities to improve their optimization capabilities.

Each particle in PSO keeps track of its current position, fitness value (how well it solves the problem), and velocity. Random numbers are U1 and U2 that are uniformly distributed between 0 and 1. These random numbers influence the updates to the particle's velocity (Equation 42) and its current position (Equation 43), guiding it towards the optimal solution.

$$V_i(t+1) = \omega V_i(t) + C_1 r_1 [y_i(t) - P_i(t)] + C_2 r_2 [\hat{y}(t) - P_i(t)] \tag{42}$$

$$P_i(t+1) = P_i(t) + V_i(t+1) \tag{43}$$

Figure 3. PSO algorithm

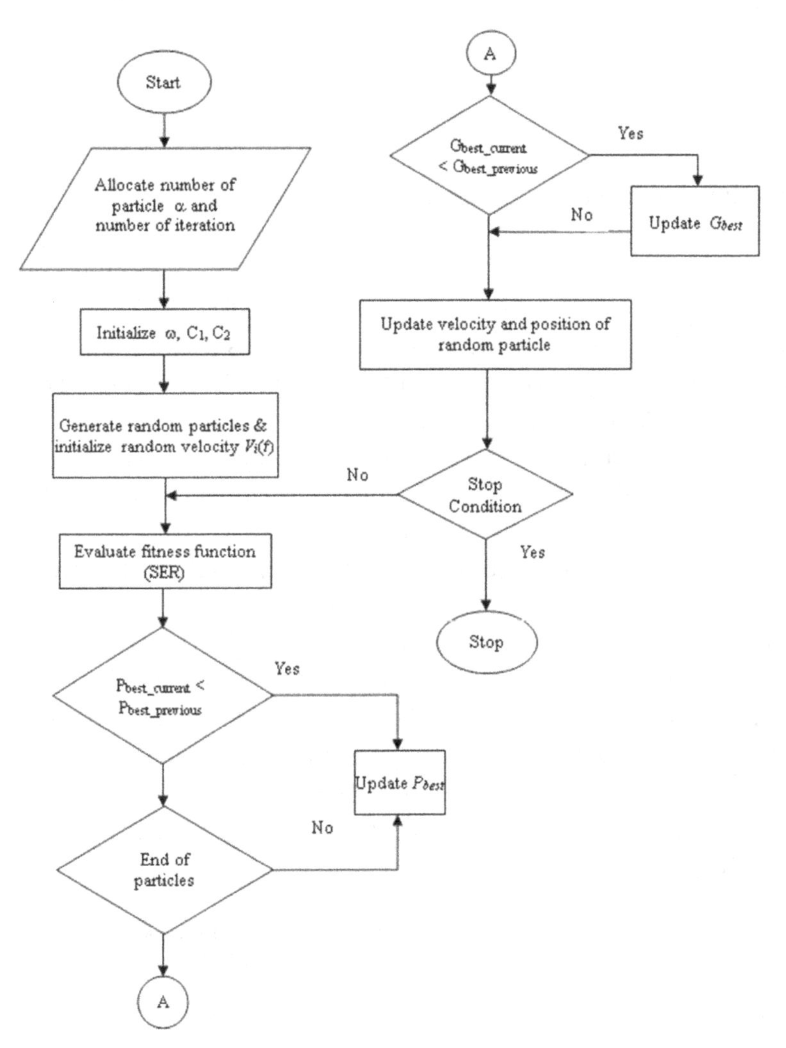

3.3.1. Problem Statement

The problem statement is to minimize the approximate in Equation (40)
Where (α = PA factor), 0 <α< 1

$$P_1 = \alpha P_T \tag{44}$$

$$P_{2i} = (1 - \alpha)P_T \tag{45}$$

$$P_1 + \sum_{i=1}^{N} P_{2i} = P_T \tag{46}$$

For optimum single relay selection case,

$$P_1 + P_2 = P_T \tag{47}$$

Within the PSO algorithm, the approximate SER is used as fitness metric for each particle. The PSO algorithm leverages Equation (40) to evaluate each particle and identify the PA factor that minimizes the approximate SER.

When a particle's current SER (fitness value) is lower (better) than its previously recorded best (pbest), the current SER becomes the particle's new best fitness. This signifies that the particle has discovered a solution with a lower error rate.

The PSO algorithm also tracks the best fitness value (gbest) discovered by any particle within the entire swarm. Essentially, the gbest represents the particle with the minimum SER found so far. If a particle's current SER is lower than the current global best, the current SER is then designated as the new global best value.

For this specific implementation, 40 particles and 300 iterations are chosen for optimal PA. These parameters, number of particles and iterations, influence the search process and convergence speed of the PSO algorithm.

Refer to Figure 3 for a visual representation of the PSO algorithm flowchart, which further details the involved steps.

3.4 Modified CSA for Alamouti DF Relaying Protocol

Modified CSA algorithm enhances throughput of users in all subchannels of the wireless systems.

Where *B* is total channel bandwidth of the system,

N is each subchannel,

B/*N* is total bandwidth is divided by subchannel,

N_0 is noise power spectral density

3.4.1. User Data Rate

The subchannel allocation matrix$A = [d_{k,n}]_{K \times N}$ specifies that the user can be allocated to all subchannels. $d_{k,n} = 1$, if and only if subchannel$n$is allocated to user k, or else it is zero. Users not sharing the same subchannel, which mean $d_{k,n} = 1$and

$d_{l,n} = 0$ for all $l \neq k$. The total transmit power of P_T that are divided between all the N subchannels.

In Alamouti DF Relaying Protocol, data rate for k^{th} user, R_k is given by

$$R_k = \frac{B}{2N}\sum_{n}^{N} d_{k,n}\log_2(1 + r_{k,n}) \tag{48}$$

Where $r_{k,n}$ is SNR of nth sub channel for k^{th} user given by,

$$r_{k,n} = \frac{(1+W)P_1|h_{(s-d)}|^2}{N_0} + \frac{WP_2|h_{(r-d)}|^2}{N_0} \tag{49}$$

Where, R_k is given by

$$R_k = \frac{B}{2N}\sum_{n}^{N} d_{k,n}\log_2\left(1 + \frac{(1+W)P_1|h_{(s-d)}|^2}{N_0} + \frac{WP_2|h_{(r-d)}|^2}{N_0}\right) \tag{50}$$

Equation (50) is the function of indirect path and direct path. Therefore, the B is divided by $2N$.

3.5 Throughput Oriented Modified CSA

Throughput oriented modified CSA algorithm is used in proposed protocol to increase the throughput of OFDMA systems.

Subject to:

$$\left[R_k = \sum_{k=1}^{K}\sum_{n=1}^{N} d_{k,n}\frac{B}{2N}\log_2\left(1 + \frac{(1+W)P_1|h_{(s-d)}|^2}{N_0} + \frac{WP_2|h_{(r-d)}|^2}{N_0}\right)\right] \tag{51}$$

$$D_1 : d_{k,n} \in \{0,1\}, \forall k, n$$

$$D_2 : \sum_{i=1}^{I} d_{k,n} = 1, \forall n \tag{52}$$

The proposed approach ensures that subchannels in multi-user Orthogonal Frequency Division Multiple Access (OFDMA) systems are not entirely shared among all users. This is achieved through a careful subchannel allocation strategy. Equation (54) reveals that the data rate in this system is not solely determined by gain of subchannel between source (S) and destination (D) (represented by γ_sd). It also depends on subchannel gain between relay station (RS) and destination (D) (represented by γ_rd), with a weight factor (ω) applied. In essence, the channel between S-D and RS-D links influence the overall data rate.

During subchannel allocation, the selected channel value for k^{th}user is, $c\,h_{k,n} = (1 + W)P_1 |h_{(s-d)}|^2 + WP_2 |h_{(r-d)}|^2$.

The initial step involves assigning channel values to all subchannels for each user. Additionally, all variables used in the following steps are initialized (as shown in the provided pseudocode). Subsequently, users are assigned subchannels in descending order of channel value. This two-step process continues for all users until the Channel State Awareness (CSA) is achieved.

4. SIMULATION RESULTS AND DISCUSSION

This work investigates the approximate SER of the Alamouti DF Relaying Protocol among M-PSK modulation against Rayleigh and Rician fading channels. The simulation results are compared with those of Alamouti AF and DF.

Monte Carlo simulations were conducted using MATLAB software. Key parameters include:

- Total number of symbols (N1) = 10000
- QPSK modulation scheme (for achieving bandwidth efficiency of 1 bit/s/Hz)
- Total transmitted power (Pt) = 2W
- Noise power spectral density (σ^2)

In the PSO algorithm, 40 particles and 300 iterations were used to find the optimal PA factor that minimizes the approximate SER for the proposed protocol.

The approximate SER serves as the fitness value for each particle in the PSO algorithm and identifies the PA factor that leads to the minimum approximate SER. During each iteration, position and velocity of each particle are updated. Upon completion of all iterations, PSO algorithm provides the optimal PA factor for minimizing the approximate SER.

Figure 4. Optimum PA factor vs. iterations

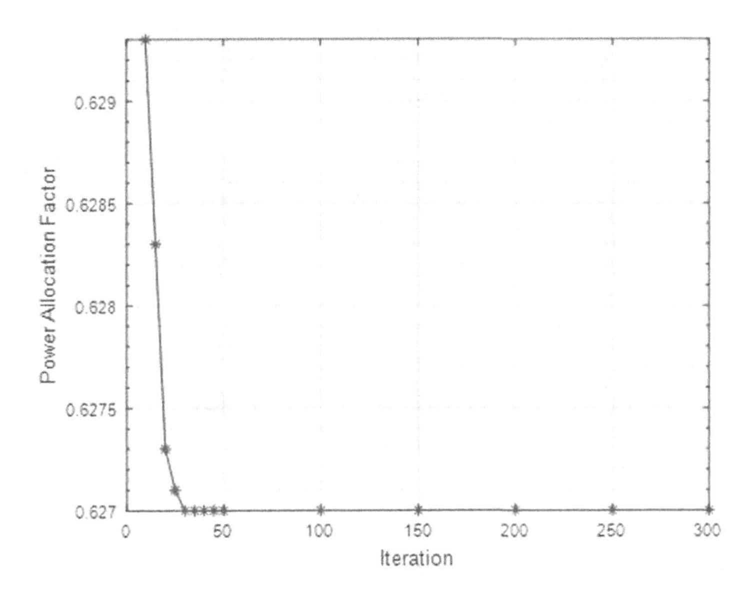

Figure 4 depicts convergence behaviour of PSO algorithm, showing an optimal PA factor versus the number of iterations. The graph indicates that the optimal PA factor is approximately 0.6270. The source (Ps) and relay power (Pr) are then determined using the PSO algorithm.

Figure 5. Upper bound and the approximate SER for the Alamouti DF relaying protocol

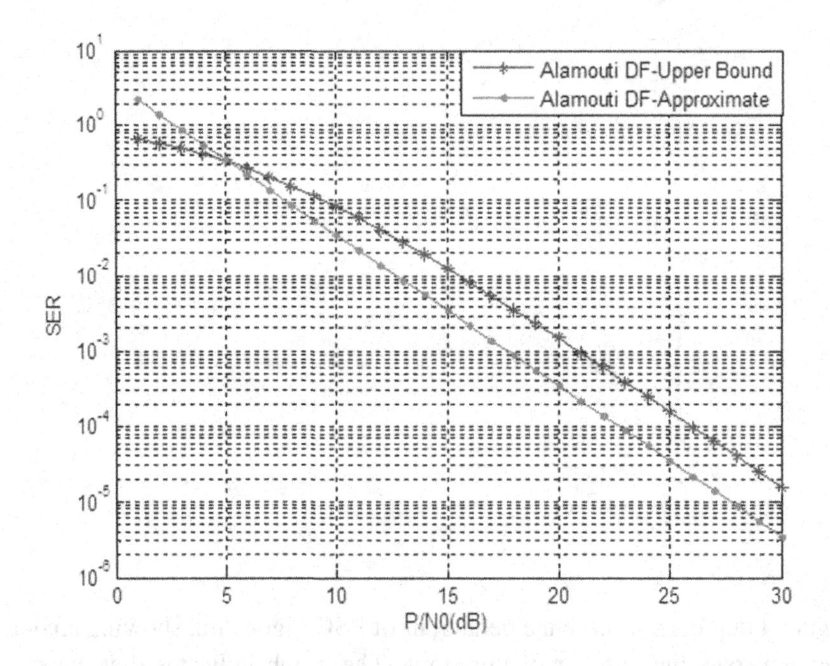

Figure 5 compares asymptotic performance of Alamouti DF Relaying Protocol with QPSK to the SER upper bound. The figure shows approximate SER is lower at minimal SNR and improves at higher values of SNR due to availability of all channel variances and the use of high power. The source-to-destination link contributes a diversity order of one. Similarly, two-channel links also contribute a diversity order of one each. Therefore, Alamouti DF relaying protocol exhibits an overall diversity order of two.

Figure 6 compares simulated SER with theoretical SER for the Alamouti DF Relaying Protocol using QPSK with both EPA and PSO-based PA. The figure demonstrates that the PSO-based PA method achieves a minimal approximate SER compared to EPA. Additionally, performance of PSO-based PA offers an SNR improvement of about 0.3 dB compared to EPA.

Figure 6. SER vs. approximate SER with and without PSO PA

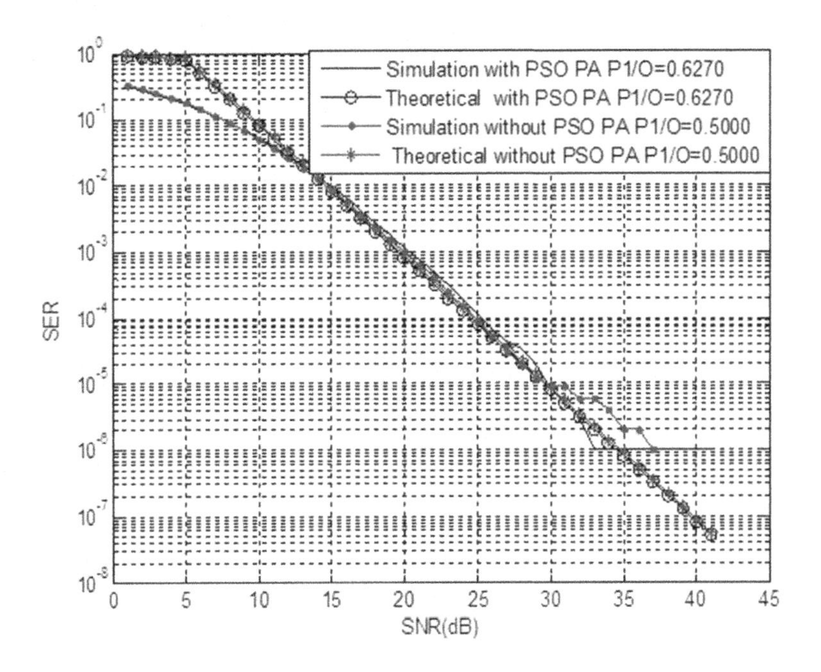

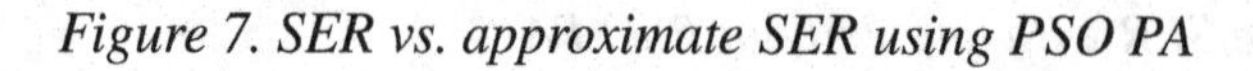

Figure 7. SER vs. approximate SER using PSO PA

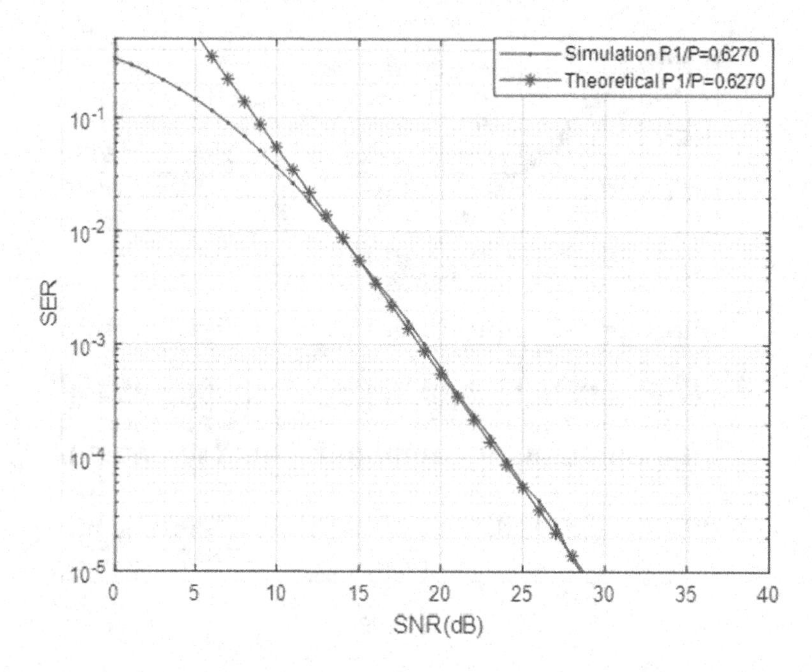

Figure 7 compares the simulation results with theoretical approximate SER for proposed protocol with QPSK modulation using PSO-based PA over a Rician fading channel. It also validates accuracy of the theoretical analysis.

Figure 8 explores the approximate SER of the Alamouti DF Relaying Protocol with various M-PSK schemes. It shows that a higher symbol rate leads to a higher data rate but also increases the approximate SER. Additionally, the QPSK modulation scheme exhibits a lower approximate SER in comparison with M-PSK schemes under Alamouti DF Relaying Protocol.

Figure 8. Alamouti DF relaying protocol using PSO based PA

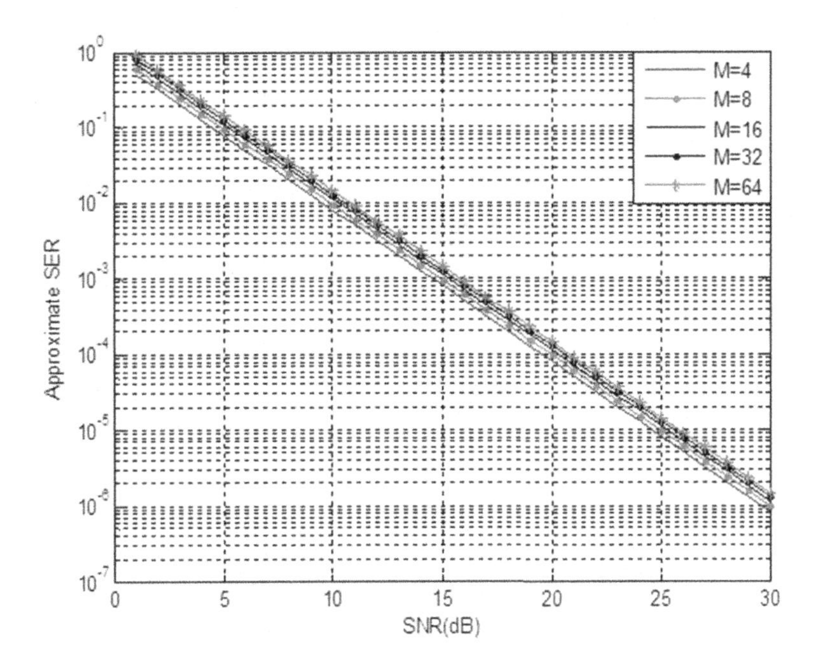

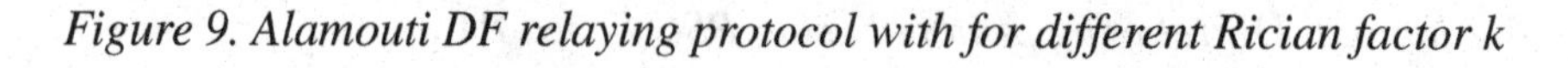

Figure 9. Alamouti DF relaying protocol with for different Rician factor k

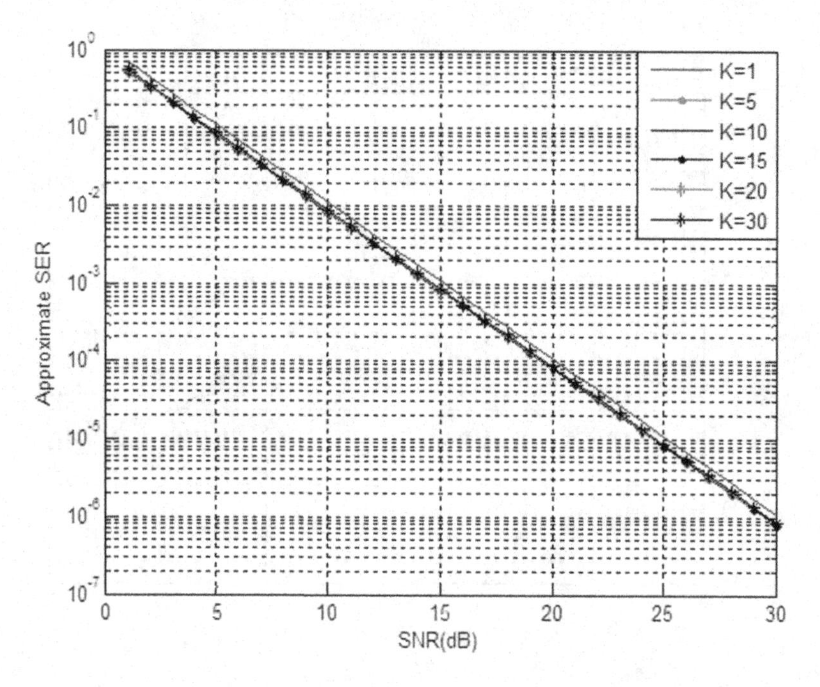

Figure 9 investigates performance of approximate SER for Alamouti AF Relaying Protocol with QPSK across different Rician factor (k) values. The figure reveals that higher Rician factor values suppress the Rayleigh fading channel effect. A Rician factor (k) of 5 offers a lower approximate SER compared to other k values in the proposed protocol, making it a suitable choice.

Figure 10 depicts an approximate SER versus SNR of Alamouti DF Relaying Protocol for various channel qualities. Channel variance and mean value represent the link's quality, with higher channel gains signifying a good link due to lower path loss. Rayleigh and Rician fading channel coefficients as average power gain (P) and channel gain (h). This gain (P) can be manipulated using

SNR = transmit power * channel gain / noise power.

The channel gain itself depends on propagation model, with a typical path loss exponent value ranging from 2 to 4. The free space path loss for the proposed system is 19 dB.

Figure 10. Alamouti DF relaying protocol for different channel quality

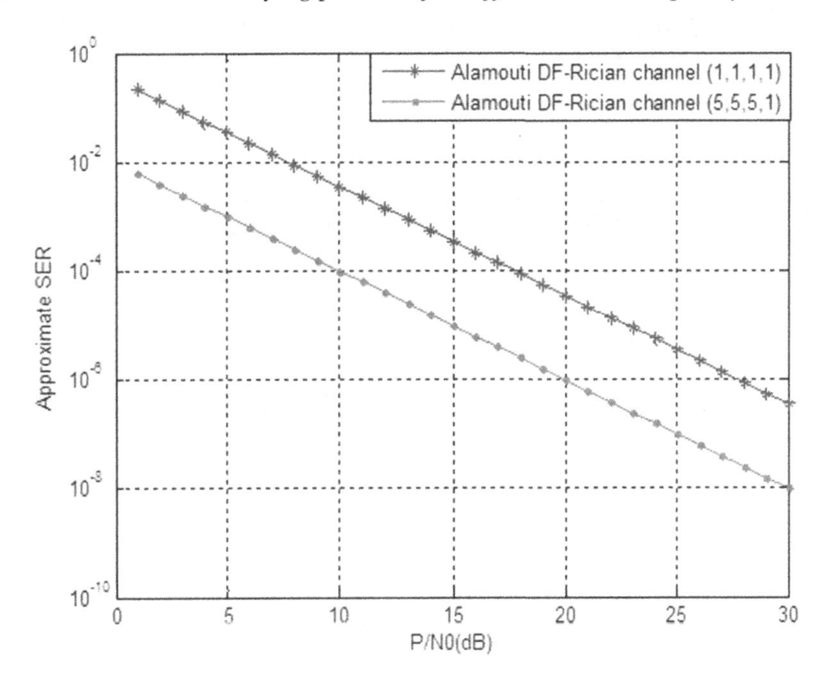

Figure 11 compares performance among relaying protocols (Alamouti DF, Alamouti AF, DF, and AF) using a QPSK and PSO-based power allocation over a Rician fading channel. These results demonstrate that Alamouti DF relaying protocol outs others, achieving SNR improvements of 0.4 dB, 1.8 dB, and 2 dB compared to AF, DF, and Alamouti AF relaying protocols, respectively.

This advantage stems from the Alamouti scheme and decoding method employed at the relay station (RS). These techniques effectively mitigate effects of additive white noise on RS link, leading to a higher overall SNR and a reduction in the approximate SER. Consequently, Alamouti DF relaying protocol achieves full diversity gain, translating to improved performance.

Figure 11. Alamouti DF, Alamouti AF, DF, and AF relaying protocols using PSO PA over Rician fading channel

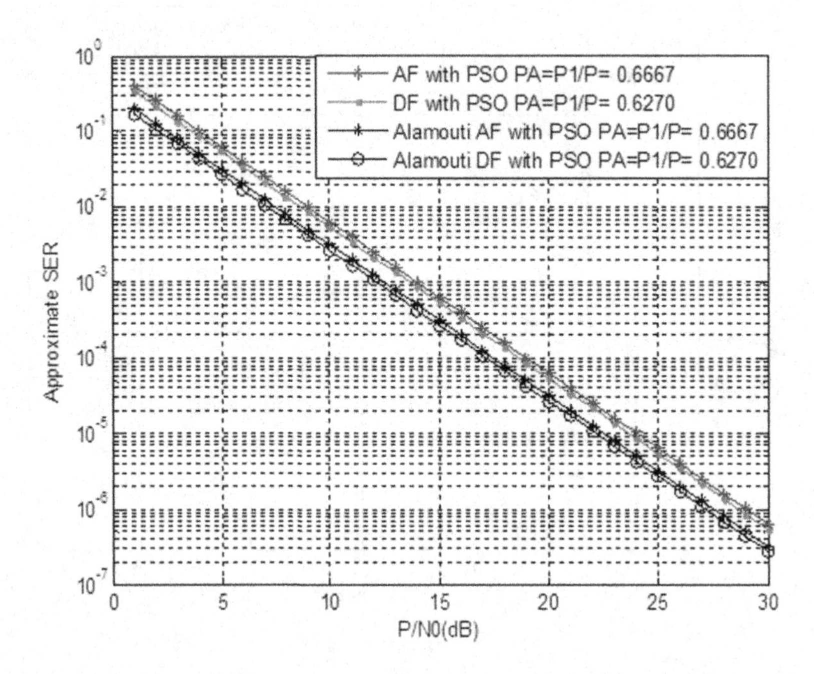

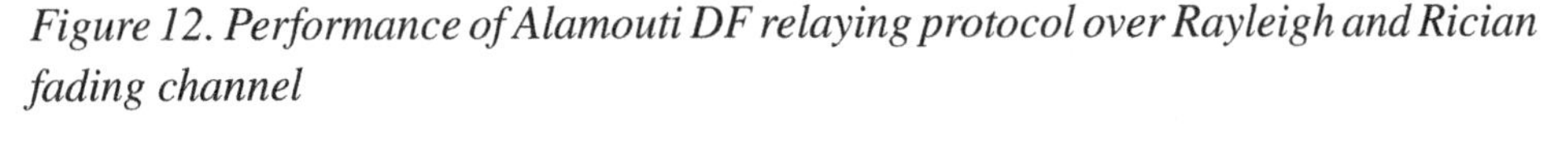

Figure 12. Performance of Alamouti DF relaying protocol over Rayleigh and Rician fading channel

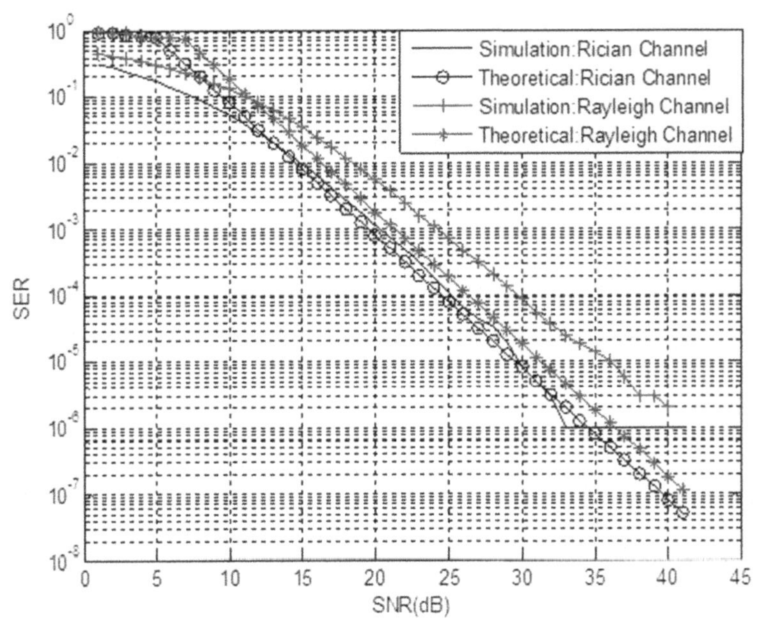

Figure 12 compares the simulated SER with theoretical approximate SER for the Alamouti DF relaying protocol using QPSK modulation over both Rayleigh and Rician fading channels. The results reveal that Rician fading channel offers superior performance among Rayleigh fading channel as Rician channel combines signal strengths from both NLOS and LOS paths, resulting in a lower approximate SER for proposed protocol.

Simulations assumed a total transmitted power (Pt) of 2W, a channel bandwidth (B) of 20 MHz, and 256 subchannels allocated to users with a specific noise variance (σ^2).

Figure 13 showcases the performance of modified Throughput-based CSA algorithm in multi-user OFDMA Alamouti DF relaying protocol. Compared to the DF and AF relaying protocols, this modified CSA algorithm achieves significant throughput improvements of 25% and 7%, respectively.

Figure 13. Total throughputs versus number of users

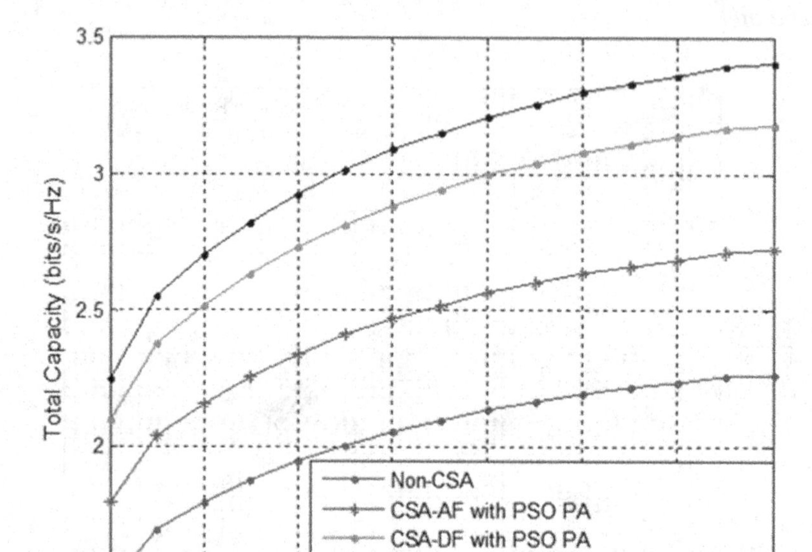

5. CONCLUSION

Cooperative communication plays a vital role in enhancing efficiency, reliability, and resilience of smart grid systems. This work presented the approximate SER of Alamouti DF relaying protocol with M-PSK using C-MRC, comparing it to existing protocols. The approximate SER exhibits asymptotic tightness at high SNR values. Additionally, a PSO-based algorithm was employed to determine the optimal power allocation factor that minimizes the approximate SER for proposed protocol. Theoretical results were validated by comparing them with the simulated SER for the Alamouti DF relaying protocol using both EPA and PSO-based power allocation. Proposed protocol demonstrates an SNR improvement of 0.3 dB compared to the EPA method. Furthermore, derived approximate SER for Alamouti DF relaying protocol closely matches simulated SER curve.

Modified CSA algorithm used in the multi-user OFDMA Alamouti DF relaying protocol offers enhanced throughput compared to traditional relaying protocols. While cooperative communication protocols provide benefits such as improved coverage, reliability, and spectral efficiency, it's crucial to address the security implications introduced by these protocols, which will be discussed in a separate chapter.

REFERENCES

Alamouti, S. M. (1998). A simple transmit diversity technique for wireless communications. *IEEE Journal on Selected Areas in Communications*, 16(8), 1451–1458. 10.1109/49.730453

Ali, K. S., Khan, A. A., Ur, T. P., Rehman, A., & Ouahada, K. (2023). Learned-SBL-GAMP based hybrid precoders/combiners in millimeter wave massive MIMO systems. *PLoS One*, 18(9), e0289868. 10.1371/journal.pone.028986837682816

Ali, K. S., Sampath, P., & Poongodi, C. (2019). Symbol Error Rate Performance of Hybrid DF/AF Relaying Protocol Using Particle Swarm Optimization Based Power Allocation. *2019 International Conference on Advances in Computing and Communication Engineering (ICACCE)*, Sathyamangalam, India. 10.1109/ICACCE46606.2019.9079970

AlQahtani, S. A. (2023). Cooperative-Aware Radio Resource Allocation Scheme for 5G Network Slicing in Cloud Radio Access Networks. *Sensors (Basel)*, 23(11), 5111. 10.3390/s2311511137299838

Anghel, P., & Kaveh, M. (2004). Exact symbol error probability of a cooperative network in a Rayleigh-fading environment fading channels. *IEEE Transactions on Wireless Communications*, 3(5), 1416–1421. 10.1109/TWC.2004.833431

Anghel, P. A., Leus, G., & Kaveh, M. (2003). Multi-user space-time coding in cooperative networks. In *Proceedings of the IEEE International Conference on Acoustics, Speech and Signal Processing (ICASSP)*. IEEE.

Ashourian, M., Salimian, R., & Nasab, H. M. (2013). A Low Complexity Resource Allocation Method for OFDMA System Based on Channel Gain. *Wireless Personal Communications*, 71(1), 519–529. 10.1007/s11277-012-0826-9

Atapattu, S., & Rajatheva, N. (2008). *'Analysis of Alamouti code transmission over TDMA-based Cooperative Protocol', Vehicular Technology. VTC Spring 2008*. IEEE.

Bai, Q. (2010). Analysis of Particle Swarm Optimization Algorithm. *Computer & Information Science*, 3(1), 3. 10.5539/cis.v3n1p180

Bai, Z. (2013). Particle swarm optimization-based power allocation in DF cooperative communications. *Ubiquitous and Future Networks (ICUFN), 2013 Fifth International Conference*. IEEE.

Bletsas, A., Khisti, A., Reed, D. P., & Lippman, A. (2006). simple cooperative diversity method based on network path selection. *IEEE Journal on Selected Areas in Communications*, 24(3), 659–672. 10.1109/JSAC.2005.862417

Brennan, D. G. (2003). Linear Diversity Combining Techniques. *Proceedings of the IEEE*, 91(2), 331–356. 10.1109/JPROC.2002.808163

Deng, X., & Haimovich, A. M. (2005). Cooperative relaying in wireless networks with local channel state information. *Proc.IEEE Veh. Tech. Conf. (VTC)*. IEEE.

Dohler, M., & Li, Y. (2010). *Cooperative Communications: Hardware, Channel and PHY*. Wiley. 10.1002/9780470740071

Dutt, T. A. (2014). PSO based Power Allocation for Single and Multi Relay AF Cooperative Network. *IEEE Sponsored International Conference on Green Computing, Communication and Electrical Engineering (ICGCCEE 2014)*. IEEE.

Farhadi, G. (2007). Selective Decode-and-Forward Relaying Scheme for Multi-Hop Diversity Transmission Systems. *Global Telecommunications Conference, GLOBECOM '07*. IEEE.

Fei, L., Xi-mei, L., Tao, L., & Guang-xin, Y. (2008). Optimal Power Allocation to minimize SER for multimode amplify-and-forward cooperative communication systems. *Journal of China Universities of Posts and Telecommunications*, 15(4), 14–23. 10.1016/S1005-8885(08)60395-7

Levin, G. (2012). *Amplify-and-Forward Versus Decode-and-Forward Relaying: Which is Better*? ETH.

Guo, L., Zhu, Z., Lau, F. C. M., Zhao, Y., & Yu, H. (2022). Joint Security and Energy-Efficient Cooperative Architecture for 5G Underlaying Cellular Networks. *Symmetry*, 14(6), 1160. 10.3390/sym14061160

Hamed, R. (2012). Cooperative Subcarrier and Power Allocation in OFDM Based Relaying Systems. [Unpublished Dissertation, Ryerson University, Canada].

Himsoon, T., Su, W., & Liu, K. J. R. (2005). Differential transmission for amplify-and-forward cooperative communications. *IEEE Signal Processing Letters*, 12(9), 597–600. 10.1109/LSP.2005.853067

Ibrahim, A., Sadek, A., Su, W., & Liu, K. J. R. (2005). Cooperative communications with partial channel state information: when to cooperate? *In Proceedings of the IEEE Global Telecommunications Conference, GLOBECOM*. IEEE. 10.1109/GLOCOM.2005.1578321

Ikki, S., & Ahmed, M. (2008). Performance of multiple-relay cooperative diversity systems with best relay selection over Rayleigh fading channels. *EURASIP Journal on Advances in Signal Processing*, 2008(1), 580368. 10.1155/2008/580368

Ikki, S., & Ahmed, M. (2007). Performance analysis of cooperative diversity wireless networks over Nakagami-m fading channel. *IEEE Communications Letters*, 11(4), 334–336. 10.1109/LCOM.2007.348292

Kennedy, J. (2010). *'Particle Swarm Optimization', Encyclopedia of Machine Learning*. Springer US.

Laneman, J., Tse, D., & Wornell, G. (2004). Cooperative diversity in wireless networks: Efficient protocols and outage behavior. *IEEE Transactions on Information Theory*, 50(12), 3062–3080. 10.1109/TIT.2004.838089

Li, Z., Hu, H., Hu, H., Huang, B., Ge, J., & Chang, V. (2021). Security and energy-aware collaborative task offloading in D2D communication. *Future Generation Computer Systems*, 1, 1–28. 10.1016/j.future.2021.06.009

Lin, W. (2009). SER Performance Analysis and Optimal Relay Location of Cooperative Communications with Distributed Alamouti Code. *43rd Annual Conference on Information Sciences and Systems,* (pp. 646-651). IEEE. 10.1109/CISS.2009.5054798

Lu, H., & Nikookar, H. (2009). *A thresholding strategy for DF-AF hybrid cooperative wireless networks and its performance*. Proc IEEE SCVT '09, UCL, Louvain.

Nostratinia, A., Hunter, T. E., & Hedayat, A. (2004). Cooperative communication in wirelesss networks. *IEEE Communications Magazine*, 42(10), 74–80. 10.1109/MCOM.2004.1341264

Priyanka, M. (2013). BER analysis of Alamouti space time block coded 2×2 MIMO systems using Rayleigh dent mobile radio channel. *Advance Computing Conference (IACC), 2013 IEEE 3rd International*. IEEE.

Ray Liu, K. J., & Ahmed, K. (2009). *Cooperative Communication and Networking*. Cambridge University Press, the Edinburgh Building, Cambridge.

S. A. K. (2021). Approximate Message Passing for mmWave Massive MIMO Architecture using Optimal Hybrid Precoder/Combiner. 2021 Smart Technologies, Communication and Robotics. STCR. 10.1109/STCR51658.2021.9588908

S. A. K. (2022). GM-LAMP with Residual Learning Network for Millimetre Wave MIMO Architectures. 2022 Smart Technologies, Communication and Robotics. STCR. 10.1109/STCR55312.2022.10009163

Sadek, A. K., Su, W., & Liu, K. J. R. (2005). Performance analysis for multimode decode-and-forward relaying in collaborative wireless networks. Proc. IEEE Int. Conf. Acoust., Speech, Signal Process. (ICASSP), Philadelphia, PA.

Sadr, S., Anpalagan, A., & Raahemifar, K. (2009). A novel subcarrier allocation algorithm for multiuser OFDM system with fairness: User's perspective. *Vehicular Technology Conference*, (pp. 1772-1776). Research Gate.

Sadr, S., Anpalagan, A., & Raahemifar, K. (2009). Radio Resource Allocation Algorithms for the Downlink of Multiuser OFDM Communication Systems. *IEEE Communications Surveys and Tutorials*, 11(3), 92–106. 10.1109/SURV.2009.090307

Sendonaris, A., Erkip, E., & Aazhang, B. (2003). User cooperation diversity - Part I: System description. *IEEE Transactions on Communications*, 51(11), 1927–1938. 10.1109/TCOMM.2003.818096

Shoukath Ali, K. (2019). Sampath Palaniswami, Particle swarm optimization-based power allocation for Alamouti amplify and forward relaying protocol. *International Journal of Communication Systems*, 32(10).

Shoukath Ali, K., & Sampath, P. (2021). Time Domain Channel Estimation for Time and Frequency Selective Millimeter Wave MIMO Hybrid Architectures: Sparse Bayesian Learning-Based Kalman Filter. *Wireless Personal Communications*, 117(3), 2453–2473. 10.1007/s11277-020-07986-9

Shoukath Ali, K., & Sampath, P. (2023). Sparse Bayesian Learning Kalman Filter-based Channel Estimation for Hybrid Millimeter Wave MIMO Systems: A Frequency Domain Approach. *Journal of the Institution of Electronics and Telecommunication Engineers*, 69(7), 4243–4253. 10.1080/03772063.2021.1951367

Siriwongpairat, W. P., Himsoon, T., & Su, W. (2006). Optimum threshold-selection relaying for decode-and-forward cooperation protocol. In *Proc. IEEE Wireless Communications and Networking Conference*. Las Vegas, NV.

Stefanov, A., & Erkip, E. (2004). Cooperative coding for wireless networks. *IEEE Transactions on Communications*, 52(9), 14701476. 10.1109/TCOMM.2004.833070

Su, W., Sadek, A. K., & Liu, K. J. R. (2005). SER performance analysis and optimum power allocation for decode-and-forward cooperation protocol in wireless networks. Proc. *IEEE Wireless Commun. Netw. Conf. (WCNC'05)*, New Orleans, LA.

Su, W., Sadek, A. K., & Ray Liu, K. J. (2008). Cooperative Communication Protocols in Wireless Networks: Performance Analysis and Optimum Power Allocation. *Wireless Personal Communications*, 44(2), 181–217. 10.1007/s11277-007-9359-z

Swasdio, W., & Pirak, C. (2010). *A Novel Alamouti-Coded Decode-and-Forward Protocol for Cooperative Communications.* In *TENCON 2010-2010 IEEE Region 10 Conference*, (pp. 2091-2095). IEEE. 10.1109/TENCON.2010.5686616

Swasdio, W., Pirak, C., & Ascheid, G. (2011). Alamouti-Coded Decode-and-Forward Protocol with Optimum Relay Selection for Cooperative Communications. In *Ultra Modern Telecommunications and Control Systems and Workshops (ICUMT), 3rd International Congress*. IEEE.

Swasdio, W., Pirak, C., Jitapunkul, S., & al Gerd, A. (2014). Alamouti-coded decode-and-forward protocol with optimum relay selection and Power Allocation for cooperative communications. *EURASIP Journal on Wireless Communications and Networking*, 1(1), 1–13. 10.1186/1687-1499-2014-112

Swasdio, W., Pirak, C., Jitapunkul, S., & Ascheid, G. (2014). Alamouti-coded decode-and-forward protocol with optimum relay selection and Power Allocation for cooperative communications. *EURASIP Journal on Wireless Communications and Networking*, 2014(1), 112. 10.1186/1687-1499-2014-112

Thantharate, A., & Beard, C. (2022). ADAPTIVE6G: Adaptive Resource Management for Network Slicing Architectures in Current 5G and Future 6G Systems. *Journal of Network and Systems Management*, 31(1), 9. 10.1007/s10922-022-09693-1

Wong, C. Y., & Cheng, R. S. (1999). Multiuser OFDM with adaptive subcarrier, bit and power allocation. *IEEE Journal on Selected Areas in Communications*, 17(10), 1747–1758. 10.1109/49.793310

Zeb, J., Hassan, A., & Nisar, M. D. (2021). Joint power and spectrum allocation for D2D communication overlaying cellular networks. *Computer Networks*, 184, 1–13. 10.1016/j.comnet.2020.107683

Chapter 8
Metamaterial-Inspired Concatenated Dual Ring Antenna Design for 5G NR C-Band Applications:
Design, Analysis, Development

B. Meenambal
https://orcid.org/0009-0009-6664-710X
Thiagarajar College of Engineering, India

Sherene Jacob
https://orcid.org/0009-0003-3655-403X
Thiagarajar College of Engineering, India

K. Vasudevan
Thiagarajar College of Engineering, India

ABSTRACT

This book chapter explores the intricacies of designing a compact triple band antenna specifically tailored for 5G NR C-Band applications. As communication technologies advance towards 5G networks, the demand for compact and efficient antennas in the C-Band spectrum becomes increasingly vital. This chapter delves into the challenges and opportunities, emphasizing the need for antennas with reduced height profiles without compromising performance. The antenna's design structure has Concatenated SRR with partial ground plane and it is defected by Circular Complementary Split Ring Resonator that achieves DGS technique. The objective is to attain high gain, low sidelobe levels, and efficient power radiation. To enhance performance metrics, this chapter employs various optimization tech-

DOI: 10.4018/979-8-3693-2786-9.ch008

niques, including parameter tuning, metamaterial technique and electromagnetic simulations. A reduced dimension of $0.179\lambda 0 \text{ X } 0.318\lambda 0 \text{ X } 0.006\lambda 0 \text{ mm3}$ *is achieved. This reduction in size holds promising implications for applications with limited space constraints, particularly in 5G NR C-Band scenarios.*

1. INTRODUCTION

Innovative antenna technology developed especially to meet the growing needs of 5G networks is shown by the Metamaterial Inspired Concatenated Dual Ring Antenna Design for 5G NR C-Band Applications. Strong, high-performing antennas are becoming more and more necessary as 5G networks spread over the world. This is because these antennas are necessary for maintaining the security and integrity of the networks in addition to facilitating seamless communication. The 5G network infrastructure and devices communicate with each other using antennas. On the other hand, they also offer possible openings for malicious individuals to use in order to jeopardize network security. In this regard, it is imperative to strengthen 5G networks against possible cyber-attacks through the design and installation of antennas with integrated security features. In addition to meeting the performance demands of 5G NR (New Radio) technology in the C-Band band, the Metamaterial Inspired Concatenated Dual Ring Antenna design incorporates security considerations into its construction. This antenna design offers increased functions including higher bandwidth, efficiency, and radiation characteristics by utilizing metamaterial-inspired design concepts. These features are crucial for providing 5G-style high-speed, low-latency communication services.

Furthermore, the 5G network environment is further protected by the antenna design's incorporation of security element.

1. By dynamically altering the operating frequency, frequency hopping makes it harder for enemies to intercept or jam communications. The antenna design may include devices for this feature.
2. The antenna's built-in encryption features might protect data transmissions and ensure that only those who possess the appropriate authorization can see and decode sensitive data.
3. To reduce the possibility of unwanted access to the network, the antenna may use authentication protocols to confirm the identity of linked devices.
4. Using sensors or algorithms integrated into the antenna system to identify and react to unusual activity that could be a sign of an impending cyberattack or an attempt at unauthorized entry.
5. To stop undesired physical access to the antenna components, design considerations for tamper-evident or physically resistant features are included.

2. LITERATURE REVIEW

The 5G New Radio (NR) C-Band, spanning from 3.4 to 5.0 GHz, occupies a pivotal position in the spectrum landscape, offering a harmonious blend of coverage and capacity crucial for a diverse array of 5G applications. Positioned as a mid-band spectrum, the C-Band strikes a balance between the extensive coverage of Sub-6 GHz and the high-capacity characteristics of mm Wave bands. The C-Band is particularly well-suited for applications such as Enhanced Mobile Broadband (eMBB) with its ability to deliver high data rates, Fixed Wireless Access (FWA) for providing last-mile connectivity, and Internet of Things (IoT) scenarios requiring a massive number of connected devices. Its ability to deliver extremely fast data rates, little latency, and widespread device connectivity has been extensively researched. Creating cutting-edge antennas that can fulfil all of the demands of evolving technology is one of the most crucial components of the 5G network, (Wang et al., 2017) (Kumar et al., 2020). Antennas based on metamaterials provide an answer to all the problems associated with the rapid advancement of technology. One type of electromagnetic structure that is not found naturally is a metamaterial. It has unique features that are artificially produced and effectively homogenous. The metamaterial cell has a dimension of only a quarter wavelength. A variety of microwave components are designed using metamaterials based on Split Ring Resonators (SRRs). Switchable mushroom reflectors surround a T-shaped SRR, which is utilized as an omnidirectional radiator to create a variety of steerable beams for 5G NR bands, (Wang & Dong, 2022). A ZOR-based dual polarization omnidirectional antenna coupled with a shared aperture hybrid metamaterial-TL construction is used for 5G indoor applications, (Wang, Zhao, & Dong, 2022) (Wang, Ning, & Dong, 2022).

To increase isolation, CSRR is incorporated into the monopole to boost antenna performance, (Dong, Toyao, & Itoh, 2012) and metamaterial structure serves as a decoupling element between the antennas, (Milias et al., 2022). To increase the gain, a metamaterial lens based on the zero-index metamaterial element with an AMC reflector is suggested, (Li & Chen, 2022). For portable X-band wireless sensor applications, miniaturization is accomplished by the loading of a defective ground structure with a highly capacitive modified Minkowski fractal antenna, (Mishra & Mangaraj, 2019). In order to confirm the metamaterial structure's property, different techniques are employed to calculate the metamaterial element's effective permittivity and permeability from the scattering parameters S11 and S21, (Chen et al., 2004) (Smith et al., 2002) (Veselago, 1968). Utilizing an eight-antenna array, the 5G NR bands n77/n78/n79 are covered, (Liu et al., 2019) (Fadhil et al., 2022) (Sim, Liu, & Huang, 2020). NR C bands for 5G connectivity, however, require larger antennas, array antennas, and lower performance. Several methods, including the incorporation of metamaterial within the antenna, can be used to solve this issue.

This chapter introduces the metamaterial-inspired Concatenated Dual Ring Resonator structure as the radiating element and Another metamaterial element, the CSRR structure, is used to defect the ground plane. Improved impedance matching is achieved by the microstrip feeding. Using the Nicholson-Ross-Weir approach, the permeability properties of a concatenated split ring resonator (SRR) are analysed and simulated. Analysis demonstrates that Single Negative Metamaterial (SNG) is produced by the property of Concatenated SRR. Concatenated SRR and CSRR, on the other hand, results in triple band resonance with improved bandwidth at 2.45GHz, 3.5GHz, and 5.9GHz. The antenna's total performance is achieved through its downsizing, specifically tailored for 5G NR C band applications. A parametric study of the intended antenna's numerous parameters has been conducted.

3. ANTENNA DESIGN AND SIMULATED RESULTS

3.1 Evolution of Concatenated Dual Ring Antenna

The evolution of the proposed low-profile Concatenated Dual Ring Antenna inspired by metamaterials is shown in Figure 1(a). To generate triple band resonance frequencies from the monopole antenna, metamaterial elements known as Concatenated SRR and Circular CSRR are developed, as illustrated in Figure 1(a). The first component of the design is a microstrip feeding monopole antenna. The monopole antenna's square ring construction is added in the second step. A split is added to the square ring construction in the third phase to create an SRR metamaterial-based antenna that produces resonance at 3.5 GHz. To accomplish dual band resonance, a concatenated split ring resonator (SRR) is introduced in the fourth phase.

Concatenated Split Ring Resonator (SRR) dimensions are changed in the fifth phase to generate resonance at 2.45GHz. The sixth phase involves introducing CSRR into the ground plane, which produces resonance at 5.9GHz, in order to deform the ground structure. The antenna's resonance characteristics are altered by the metamaterial elements Concatenated Split Ring Resonator (SRR) and CSRR, which are used to create triple band characteristics and reach the necessary resonance frequencies of 2.45GHz, 3.5GHz, and 5.9GHz with return losses of -25dB, -29dB, and -20dB, respectively. The return loss is improved by using CSRR. The suggested Concatenated Dual Ring metamaterial antenna is constructed and simulated using the complete wave tool, CST Microwave Studio. The simulated S11 characteristics at various antenna stages are displayed in Figure 1(b).

Figure 1a. Evolution stages of metamaterial inspired concatenated dual ring antenna

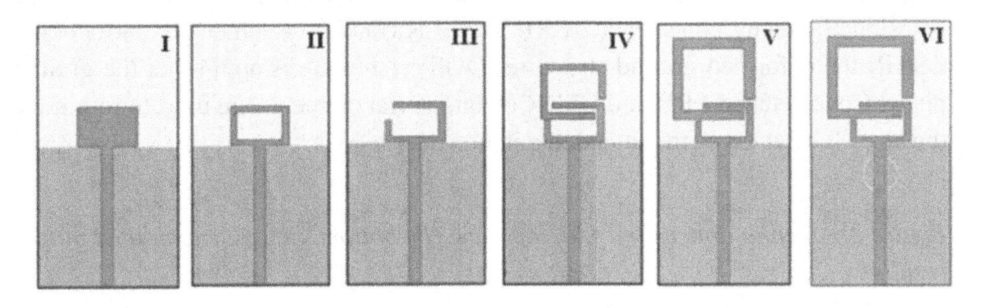

Figure 1b. Simulated $S_{11}(dB)$ at various stages of antenna

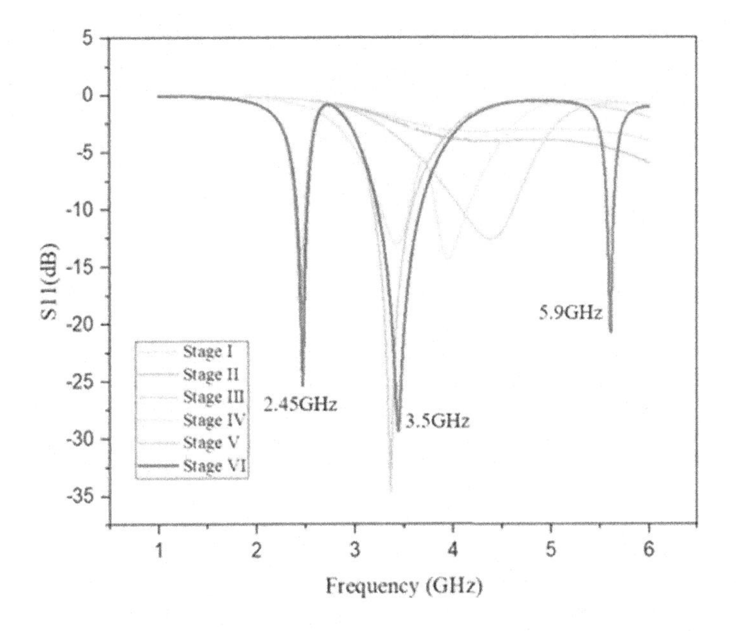

3.2 Antenna Configurations

The stated metamaterial-inspired concatenated dual ring antenna, operating at the resonance frequencies of 2.45GHz, 3.5GHz, and 5.9GHz, is depicted structurally in Figure 2. The intended antenna's dimensions are 0.179λ0 X 0.318λ0 X 0.006λ0 mm^3. The antenna is designed using a Rogers RO4350 substrate, thickness of 0.8

mm, a relative dielectric constant value of $\varepsilon_r = 3.66$ and a loss tangent value of tan $\delta = 0.004$. The 50Ω microstrip feeding technique is used to feed the antenna. In the envisioned antenna, concatenated SRR is used as a radiating element. By introducing CSRR, the deformed ground structure (DGS) technique is applied in the ground plane. Concatenated SRR and CSRR metamaterial elements are used to minimize the size. Table.1 gives the intended antenna dimensions.

Figure 2. Antenna dimension: (A) Top view (B) bottom view (C) simulated output results

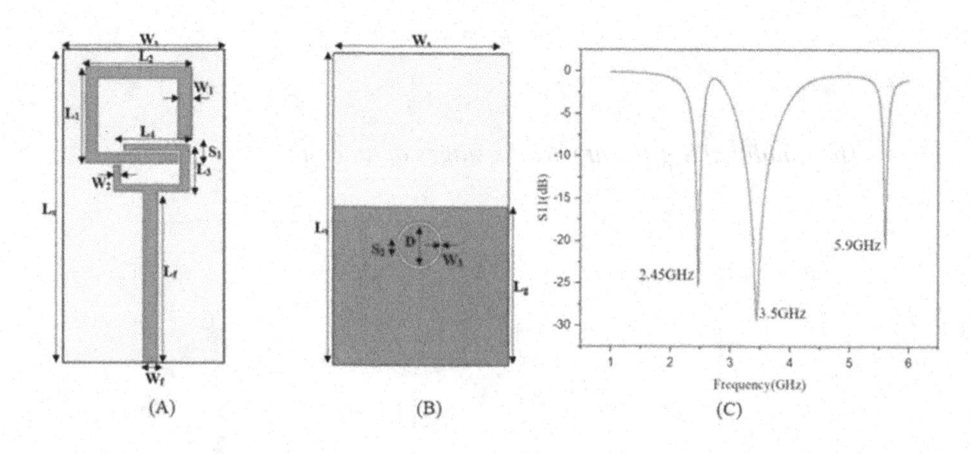

Table 1. Parameters with dimensions of intended antenna

Parameter	*Dimension(mm)*	*Parameter*	*Dimension(mm)*
L_s	39	L_4	9.2
W_s	22	W_1	1.5
L_f	21.2	W_2	0.9
W_f	1.8	W_3	0.3
L_1	12	S_1	1.7
L_2	13	S_2	0.6
L_3	5.8	D	2.5
L_g	21		

3.3 Radiation Mechanism

Figure 3(a), (b) & (c) resp. shows the current density results of the triple-band concatenated dual ring metamaterial inspired antenna at 2.45 GHz, 3.5 GHz, and 5.9 GHz. An understanding of the radiation process of the concatenated dual ring metamaterial inspired antenna can be gained by looking at the image, which shows the parts of the antenna that play a vital role in making the specified antenna resonance at the appropriate frequency. Surface current distribution at three distinct frequencies is the responsibility of each part of the antenna. The upper ring in the concatenated dual ring has a high current density at 2.45GHz. At 3.5GHz, the high current density is visible in the lower ring. At 5.9GHz, the ground plane's CSRR displays a high current density.

Figure 3. Distribution of surface currents at (a) 2.45 GHz (b) 3.5 GHz (c) 5.9 GHz

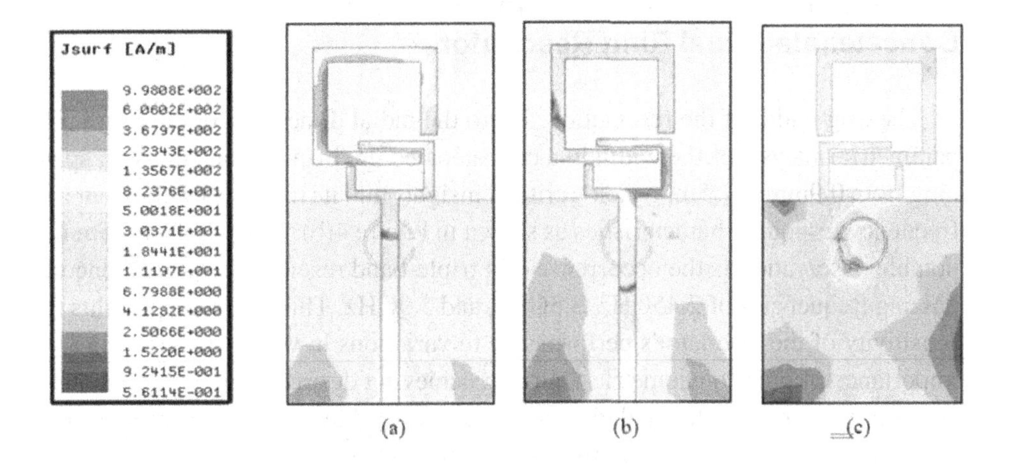

4. PARAMETRIC ANALYSIS OF DESIGNED ANTENNA

With a concatenated dual ring serving as the resonator and a CSRR structure in the antenna's ground section, parametric study is performed on a number of the structure's parameters. Numerous impacts are examined, including the position of the CSRR, split width of concatenated SRR and the ring width of the concatenated SRR and CSRR.

4.1 Parametric Analysis of Split Width of the Concatenated Dual Ring Resonator

A concatenated dual-ring resonator consists of two interconnected ring resonators with a coupling gap between them. The gap or split width refers to the distance between the two rings of the resonator. In this parametric analysis, we investigate the influence of varying the split width of a concatenated dual-ring resonator (S1), ranging from 3mm to 1.7mm as shown in Fig 4(a). By using electromagnetic simulation, we explore the effects of this parameter on the resonator's return loss, transmission properties, and resonant frequency. Through a methodical adjustment of the split width, we are able observe changes in the resonator's performance and determine the split width of 1.7mm as the point at which the resonant frequency matches the required range. Interestingly, for frequencies of 2.45GHz, 3.5GHz, and 5.9GHz, this setup produces a triple-band response.

4.2 Parametric Analysis of Ring Width of the Concatenated Dual Ring Resonator

The ring width of the resonator refers to the radial dimension of the rings. The parametric analysis of the width in a concatenated dual-ring resonator(W1), spanning from 0.9mm to 1.5mm, yields critical insights into its resonance behaviour and frequency response characteristics as shown in Figure 4(b). At a width of 1.5mm, a notable observation is the occurrence of a triple-band resonant frequency, encompassing frequencies of 2.45GHz, 3.5GHz, and 5.9GHz. This finding highlights the sensitivity of the resonator's performance to variations in width, underscoring the importance of precise geometric tuning in achieving desired frequency responses. Through electromagnetic simulation or analytical modelling, the resonator's behaviour across this width range can be thoroughly investigated, providing valuable guidance for optimizing its design for specific applications. The triple-band resonance observed at 1.5mm width offers potential advantages for applications requiring multiband operation.

4.3 Parametric Analysis of Position of the Circular CSRR

In this analysis, we focus on varying the position of the CSRR along the y-axis on the ground plane. The y-position changes from 8.55mm to 14.55mm, allowing us to investigate how shifting the CSRR's location affects its resonance behaviour and frequency response as shown in Fig 4(c). Through electromagnetic simulation, we explore how the resonant frequency of the CSRR changes with variations in its y-position. By calculating the resonance frequencies at different y-positions, we aim to identify the point where the desired triple-band resonance occurs, specifically at y = 14.55mm. A notable finding of this analysis is the observation of a triple-band resonance at frequencies of 2.45ghz, 3.5ghz, and 5.9ghz when the CSRR is positioned at y = 14.55mm. The observed triple-band resonance emphasizes the sensitivity of the CSRR's performance to variations in its position on the ground plane. Changes in the y-position influence the EM coupling between the CSRR and the surrounding environment, affecting the resonant frequencies and bandwidths of the structure.

4.4 Parametric Analysis of Ring Width of the Circular CSRR

Figure 4(d) shows the parametric analysis of the ring width in a circular CSRR, ranging from 0.9mm to 0.3mm. It offers crucial insights into its resonance behaviour and frequency response characteristics. Notably, at a width of 0.3mm, a significant observation is the occurrence of resonant frequencies with high return loss, indicative of efficient energy transfer into the resonator. Additionally, this configuration exhibits a triple-band resonance, featuring frequencies of 2.45GHz, 3.5GHz, and 5.9GHz. This finding emphasizes the sensitivity of the CSRR's performance to variations in ring width and highlights the importance of precise geometric tuning for achieving desired resonance characteristics. Overall, this parametric analysis enhances the understanding of the interplay between geometric parameters and resonance properties in antenna structures, guiding the design and optimization of concatenated dual ring antenna to achieve triple band characteristics.

Figure 4. Parametric analysis of (a) split width of the concatenated dual ring resonator (b) ring width of the concatenated dual ring resonator (c) position of the CSRR (d) ring width of the CSRR

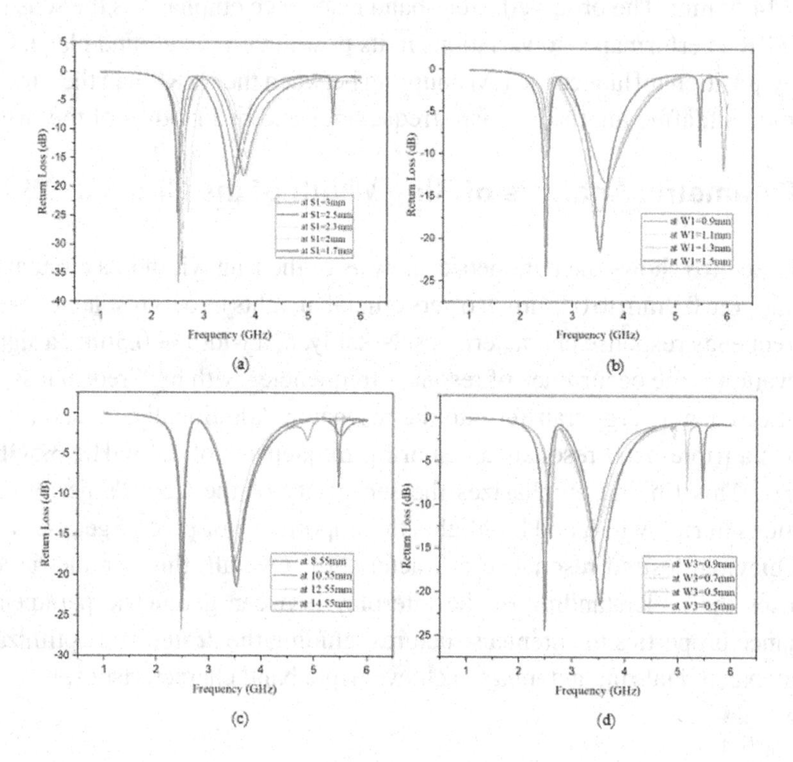

5. EXTRACTION OF METAMATERIAL CHARACTERISTICS

The Concatenated dual ring antenna has the properties of a Mu-negative metamaterial. The reflection and transmission coefficients are used to compute the negative permeability properties. The antenna's mu-negative characteristics are retrieved using the waveguide configuration depicted at Figure 5. Additionally, it demonstrates how the limits of the intended structure are assigned excitations and conditions.

Figure 5. Waveguide setup to obtain metamaterial characteristics

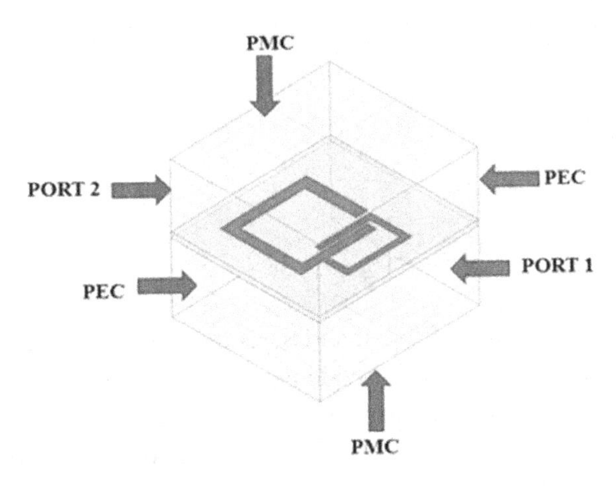

Figure 6. Negative permeability characteristics

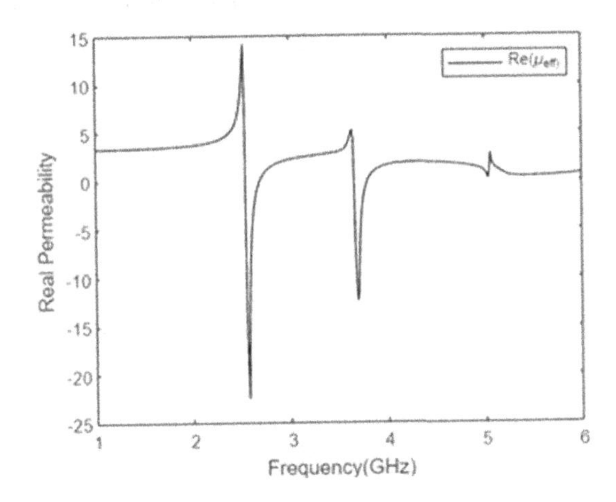

The specified antenna's Mu-negative parameter is computed using the Nicholson-Ross-Weir technique. To obtain the negative permeability, use the following equation:

$$\mu = n \times z \quad (1)$$

where z is impedance, n is refractive index.

The values of impedance and refractive index can be determined by using the parameter S_{11} and S_{21} as,

$$z = \sqrt{\frac{(1 + S_{11})^2 - S_{21}^2}{(1 - S_{11})^2 - S_{21}^2}} \tag{2}$$

$$n = \frac{1}{kd} cos^{-1}\left[\frac{1}{2S_{21}}\left(1 - S_{11}^2 + S_{21}^2\right]\right. \tag{3}$$

where S_{11} is refection coefficient and S_{21} is transmission coefficient.

As seen in Figure 6, MATLAB code is utilized to determine the real values of permeability based on the retrieved S parameter. Permeability values at 2.45 GHz, 3.5 GHz, and 5.9 GHz are shown to be negative. The Single Negative Metamaterial (SNG) feature is attained via the concatenated dual ring antenna configuration. Figure 7 displays the results of S11 and S21 parameter simulations. At the intended resonance frequencies, the refection coefficient S11 is attained with a substantial return loss. In a similar vein, the transmission coefficient approaches 0 dB. Thus, it is demonstrated that pass band features, as seen in Figure 7, lead to the achievement of a new resonance frequency.

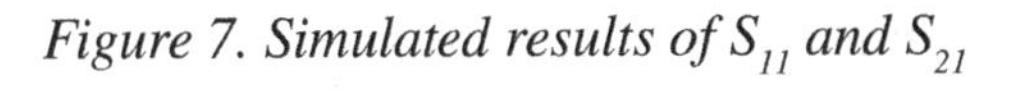

Figure 7. Simulated results of S_{11} and S_{21}

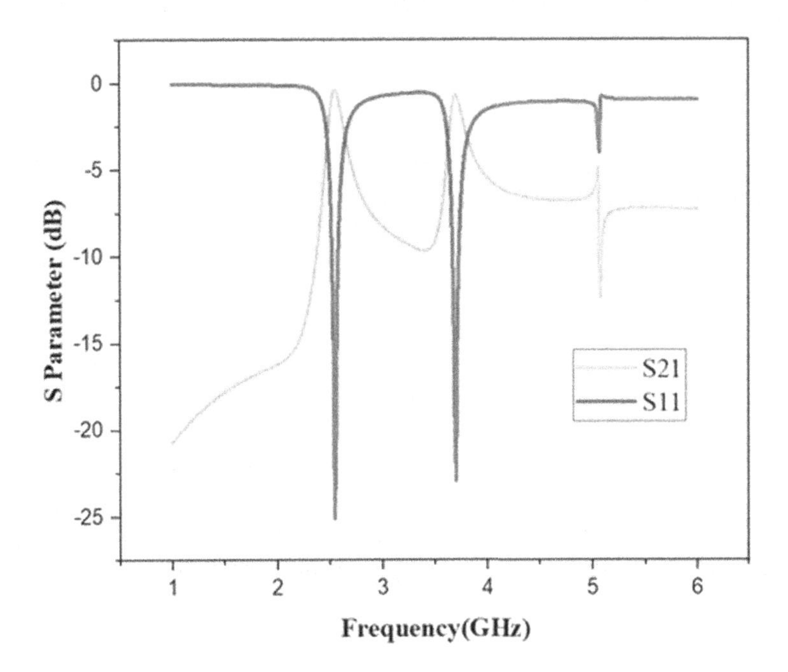

6. RESULTS AND DISCUSSION

The simulated output results of the constructed antenna's far-field radiation patterns in the E- and H-planes at resonance frequencies of 2.45GHz, 3.5GHz, and 5.9GHz are displayed in Figure 8.

Figure 8. Simulated E-plane and H-plane results of far field radiation pattern at (a) 2.45GHz, (b) 3.5GHz and (c) 5.9GHz

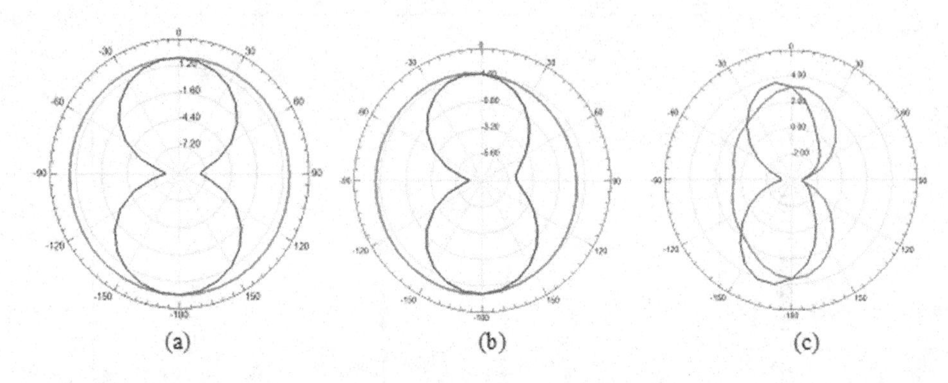

Figure 9. Simulated 3D gain plot at resonance frequency (a) 2.45GHz (b) 3.5GHz (c) simulated 3D gain plot at resonance frequency at 5.9GHz

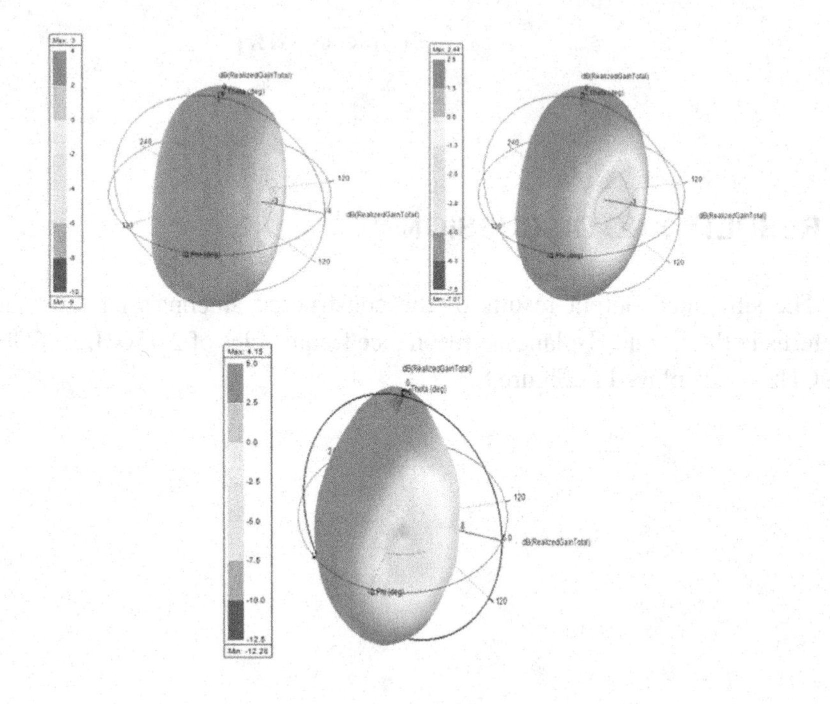

Figure 10 illustrates the proposed antenna's gain across frequency. The simulated figure that the resonating frequencies of 2.45GHz, 3.5GHz, and 5.9GHz yield substantial gain. Additionally, 99.7% efficiency at 2.45GHz, 99.8% efficiency at

3.5GHz and 99.15% efficiency at 5.9GHz are reached. The measured axial ratios at 2.45GHz, 3.5GHz and 5.9GHz for the intended frequency are 31.33dB, 22.55dB and 24.36dB, respectively is shown in Figure 11. It demonstrates that the polarization is linear. The gain values and antenna's radiation efficiency demonstrate that the suggested antenna performs better at the targeted frequencies.

The graphic makes it evident that H-Plane radiation pattern is observable in all directions, whereas the E-Plane radiation pattern is formed like an eight-shaped dipole like pattern. In Figure 9, the simulated 3D plot of gain at resonance frequencies 2.45 GHz, 3.5 GHz, and 5.9 GHz is shown in (a), (b), and (c), and it is sidelobe-free. The findings show that the antenna's radiating gain is 3dB, 2.44dB and 4.15dB at the resonance frequencies 2.45 GHz, 3.5 GHz, and 5.9 GHz, respectively.

Figure 10. Gain over frequency of the antenna

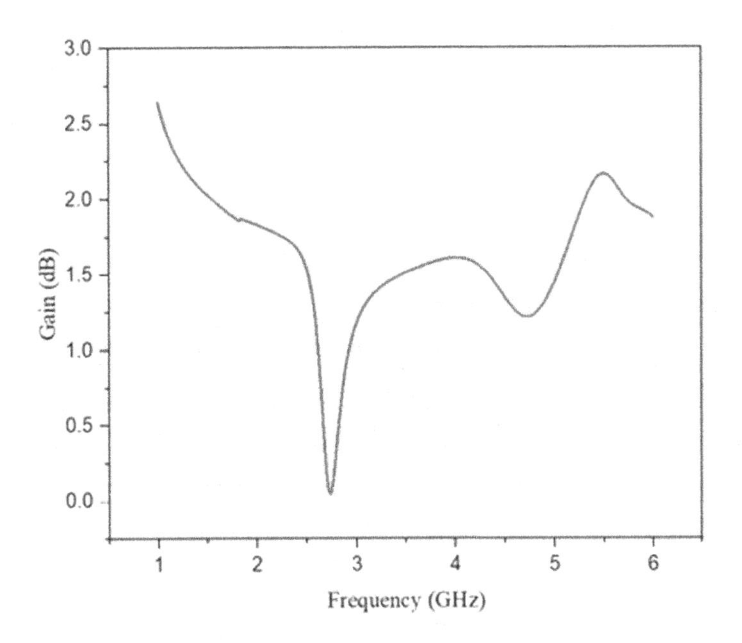

Figure 11. Axial ratio over frequency of the antenna

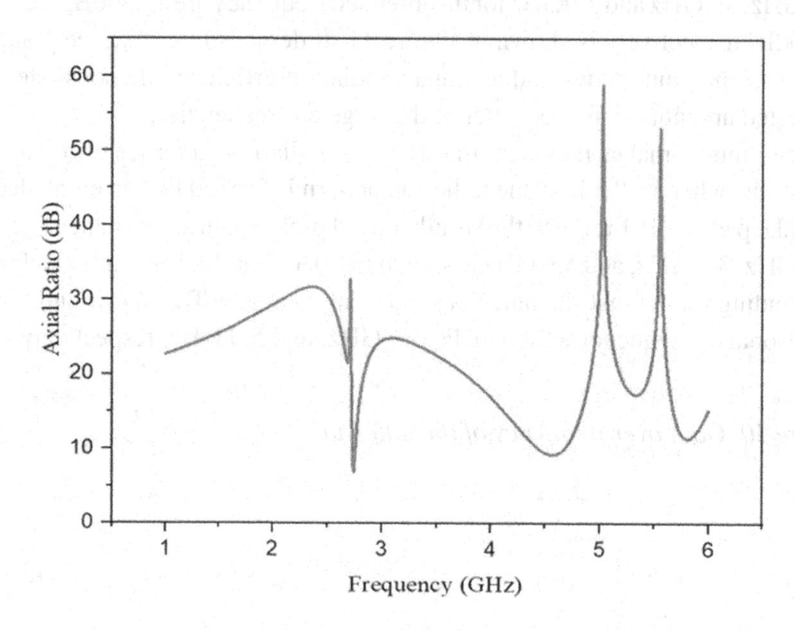

Table 2. Comparison of state of art

Ref	Antenna Dimensions (mm²)	Substrate Used	Resonating frequency (GHz)	Bandwidth (MHz)	Resonance Band	Radiation Efficiency	Verification of Metamaterial property
16	25 x 21	FR4	5.75	167	Single	-	No
17	65 x 50	FR4	3.5	300	Single	60%	Verified
18	40 x 10	FR4	2.4	300	Single	79%	No
19	30 x 30	FR4	3.6	400	Single	60%	No
20	52 x 40	FR4	3.5	200	Single	78%	Verified
21	54 x 38	FR4	3.3	200	Single	-	No
22	30 x 20	RO5880	5.6	250	Single	50%	No
Proposed Antenna	**39 x 22**	**Rogers RO4350**	**2.45,3.5 and 5.9**	**150,300 & 100**	**Triple**	**99.7%, 99.8% & 99.15%**	**Verified**

The comparison between the intended and reported antennas is displayed in Table 2. When compared to published antennas, it is evident that the intended antenna exhibits triple band resonance and good efficiency. The metamaterial property also contributes to the size reduction, and its verification is carried out.

Figure 12. Schematic picture of anechoic chamber setup for farfield measurements of antenna

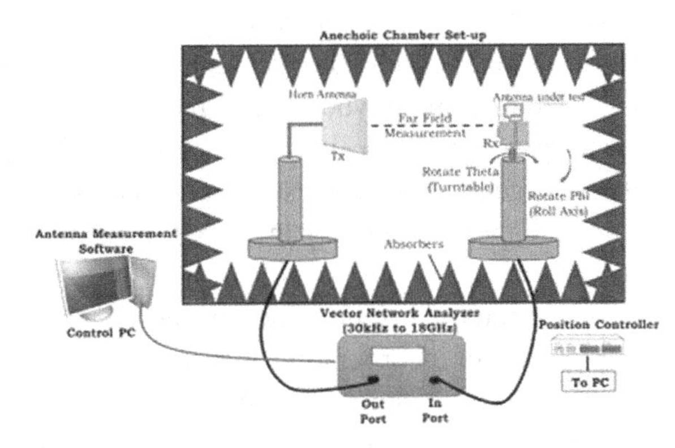

Figure 12 shows the measurement of antenna setup within the anechoic chamber typically involves connecting the antenna to measurement equipment such as a vector network analyser (VNA) or spectrum analyser. Calibration procedures are then performed to account for system losses and inaccuracies. Using reference antennas or calibration kits, among other renowned standards, calibration entails making sure that the measurement setup effectively captures the antenna's performance. With the setup calibrated, measurements of various antenna parameters can be taken. These parameters may include radiation patterns, gain, voltage standing wave ratio (VSWR), efficiency, and polarization. The collected data is then analysed to assess the antenna's performance and compared against design specifications or industry standards.

7. CONCLUSION

In conclusion, the metamaterial-inspired concatenated dual-ring antenna design presents a promising solution for 5G NR C-band applications. By leveraging metamaterial structures, this antenna design offers enhanced performance characteristics, including improved bandwidth, reduced size, and enhanced radiation efficiency. The intended antenna achieves the triple band resonance with a miniaturized antenna dimension of $0.179\lambda_0$ X $0.318\lambda_0$ X $0.006\lambda_0$ mm^3. The concatenated dual-ring configuration enables versatile tuning capabilities, allowing for optimization to meet the stringent requirements of 5G networks operating in the C-band spectrum. By directly integrating security features like intrusion detection, authentication procedures, frequency hopping, signal encryption, and physical security upgrades into the antenna architecture, 5G networks are made more resilient to future cyber-attacks. In addition to satisfying 5G NR technology's performance requirements, the Metamaterial Inspired Concatenated Dual Ring Antenna Design reinforces 5G networks overall security posture, facilitating stakeholders to deploy and operate these networks with assurance in spite of transforming cybersecurity threats. Through rigorous analysis and testing, this innovative antenna design demonstrates its ability to address the challenges of subsequent generation wireless communication systems, paving the way for efficient and reliable connectivity in 5G NR deployments moreover in regard to security aspects.

The future scope of the proposed Metamaterial Inspired Concatenated Dual Ring Antenna design includes further exploration of incorporating advanced encryption techniques directly into the antenna structure to ensure end-to-end data protection. Additionally, the antenna's unique characteristics can be leveraged for developing secure beamforming and MIMO configurations, enhancing privacy and preventing unauthorized access. Future research may also focus on integrating artificial intelligence and machine learning algorithms into the antenna system for real-time threat detection and adaptive security responses.

REFERENCES

Ayomikun, E. (2020). Mokayef Mastaneh. Miniature microstrip antenna for IoT application. *Materials Today: Proceedings*, 29(43–47).

Aznabet, M., Mrabet, O. E., Floc'H, J. M., Falcone, F., & Drissi, M. (2014). A coplanar waveguide-fed printed antenna with complementary split ring resonator for wireless communication systems. *Waves in Random and Complex Media*, 25(1), 43–51. 10.1080/17455030.2014.956846

Chen, X., Grzegorczyk, T. M., Wu, B., Pacheco, J., & Kong, J. A. (2004). Robust method to retrieve the constitutive effective parameters of metamaterials. *Physical Review. E*, 70(1), 016608. 10.1103/PhysRevE.70.01660815324190

Dong, Y., Toyao, H., & Itoh, T. (2012). Design and characterization of miniaturized patch antennas loaded with complementary Split-Ring resonators. *IEEE Transactions on Antennas and Propagation*, 60(2), 772–785. 10.1109/TAP.2011.2173120

Fadhil, T. Z., Murad, N. A., Rahim, M. K. A., Hamid, M. R., & Nur, L. O. (2022). A Beam-Split metasurface antenna for 5G applications. *IEEE Access : Practical Innovations, Open Solutions*, 10, 1162–1174. 10.1109/ACCESS.2021.3137324

Islam, M. S., Ullah, M. A., Beng, G. K., Amin, N., & Misran, N. (2019). A modified Meander Line microstrip patch antenna with enhanced bandwidth for 2.4 GHz ISM-Band internet of things (IoT) applications. *IEEE Access : Practical Innovations, Open Solutions*, 7, 127850–127861. 10.1109/ACCESS.2019.2940049

Kumar, S., Dixit, A. S., Malekar, R. R., Raut, H. D., & Shevada, L. K. (2020). Fifth Generation Antennas: A comprehensive review of design and performance enhancement techniques. *IEEE Access : Practical Innovations, Open Solutions*, 8, 163568–163593. 10.1109/ACCESS.2020.3020952

Li, Y., & Chen, J. (2022). Design of miniaturized high gain Bow-Tie antenna. *IEEE Transactions on Antennas and Propagation*, 70(1), 738–743. 10.1109/TAP.2021.3098595

Liu, L., Liu, T., Zheng, Y., Jin, Z., & Sun, Z. (2022). Archimedean spiral antenna based on metamaterial structure with wideband circular polarization. *AEÜ. International Journal of Electronics and Communications*, 152, 154257. 10.1016/j.aeue.2022.154257

Liu, Y., Ren, A., Liu, H., Wang, H., & Sim, C. (2019). Eight-Port MIMO array using characteristic mode theory for 5G smartphone applications. *IEEE Access : Practical Innovations, Open Solutions*, 7, 45679–45692. 10.1109/ACCESS.2019.2909070

Liu, Y., Ren, A., Liu, H., Wang, H., & Sim, C. (2019). Eight-Port MIMO array using characteristic mode theory for 5G smartphone applications. *IEEE Access : Practical Innovations, Open Solutions*, 7, 45679–45692. 10.1109/ACCESS.2019.2909070

Milias, C., Andersen, R. B., Lazaridis, P. I., Zaharis, Z. D., Muhammad, B., Kristensen, J. T. B., Mihovska, A., & Hermansen, D. D. S. (2022). Miniaturized Multiband Metamaterial Antennas With Dual-Band Isolation Enhancement. *IEEE Access : Practical Innovations, Open Solutions*, 10, 64952–64964. 10.1109/ACCESS.2022.3183800

Mishra, G. P., & Mangaraj, B. B. (2019). Miniaturised microstrip patch design based on highly capacitive defected ground structure with fractal boundary for X-band microwave communications. *IET Microwaves, Antennas & Propagation*, 13(10), 1593–1601. 10.1049/iet-map.2018.5778

Raviteja, V., Kumar, S. A., & Shanmuganantham, T. (2019). *CPW-fed inverted six shaped antenna design for internet of things (IoT) applications*. In: *International Conference on Microwave Integrated Circuits, Photonics and Wireless Networks (IMICPW)*, Tiruchirappalli, India. 10.1109/IMICPW.2019.8933248

Sim, C., Liu, H., & Huang, C. (2020). Wideband MIMO Antenna Array Design for Future Mobile Devices Operating in the 5G NR Frequency Bands n77/n78/n79 and LTE Band 46. *IEEE Antennas and Wireless Propagation Letters*, 19(1), 74–78. 10.1109/LAWP.2019.2953334

Smith, D. R., Schultz, S., Markos, P., & Soukoulis, C. M. (2002). Determination of effective permittivity and permeability of metamaterials from reflection and transmission coefficients. *Physical Review*, 65(19).

Veselago, V. G. (1968). The electrodynamics of substances with simultaneously negative values of e and μ. *Soviet Physics - Uspekhi*, 10(4), 509–514. 10.1070/PU1968v010n04ABEH003699

Wang, H., Jiang, Z. H., Yu, C., Zhou, J., Chen, P., Yu, Z., Zhang, H., Yang, B., Pang, X., Jiang, M., Cheng, Y. J., Al-Nuaimi, M. K. T., Zhang, Y., Chen, J., & He, S. (2017). Multibeam antenna technologies for 5G wireless communications. *IEEE Transactions on Antennas and Propagation*, 65(12), 6231–6249. 10.1109/TAP.2017.2712819

Wang, Z., & Dong, Y. (2022, November). Metamaterial-Based, Vertically Polarized, Miniaturized Beam-Steering Antenna for Reconfigurable Sub-6 GHz Applications. *IEEE Antennas and Wireless Propagation Letters*, 21(11), 2239–2243. 10.1109/LAWP.2022.3188548

Wang, Z., Ning, Y., & Dong, Y. (2022). Hybrid Metamaterial-TL-Based, Low-Profile, Dual-Polarized omnidirectional antenna for 5G indoor application. *IEEE Transactions on Antennas and Propagation*, 70(4), 2561–2570. 10.1109/TAP.2021.3137242

Wang, Z., Zhao, S., & Dong, Y. (2022). Pattern reconfigurable, Low-Profile, vertically polarized, ZOR-Metasurface antenna for 5G application. *IEEE Transactions on Antennas and Propagation*, 70(8), 6581–6591. 10.1109/TAP.2022.3162332

Chapter 9
Artificial Magnetic Conductor-Backed Microstrip-Fed Dual-Band Antenna for 5G Wearable Sensor Nodes and C Band Application in WBAN Network

Sherene Jacob
https://orcid.org/0009-0003-3655-403X
Thiagarajar College of Engineering, India

Meenambal Bose
https://orcid.org/0009-0009-6664-710X
Thiagarajar College of Engineering, India

Vasudevan Karuppiah
Thiagarajar College of Engineering, India

ABSTRACT

This chapter is the development of a microstrip fed dual band antenna with a low-profile, small, and flexible structure, utilizing an artificial magnetic conductor (AMC) and microstrip feed. The antenna is biocompatible and operates at resonant

DOI: 10.4018/979-8-3693-2786-9.ch009

at 3.5 GHz for 5G applications and 5.7 GHz catering to C band applications like Wi-Fi. In order to render the suggested antenna biocompatible, a 3x3 array of AMC is integrated. This array functions to augment the FBR, thereby improving overall parameters when in contact with living human tissues. From the simulation of the microstrip fed dual band antenna with AMC, the result indicates the improved parameters of the antenna gain is 3.91dB and 4.87dB, bandwidth of 340MHz and 480MHz, Front to Back ratio is 12.6dB and 161.8dB with a radiation efficiency of 84% and 85% in 3.5GHz and 5.7GHz resp. The antenna having flexibility, improved gain and improved FBR makes it good for wearable antenna applications.

1. INTRODUCTION

Wireless Body Area Networks (WBAN) were established as an outcome of the growing importance of a number of technologies in sectors including sports, entertainment, health monitoring, rescue and emergency operation networks, etc. WBAN has attracted a great deal of developers as a result of numerous challenges confronted in its development and production process, (Yalduz et al., 2020). The individual using them is enabled to wear any of these devices on-body as a result of WBAN sensors. This can be counted as an advantage in telemedicine and ensures the patient/wearer to be constantly monitored for indicators of health by professionals in real time. Furthermore, it can be utilized for transmitting data to remote or close devices, facilitating communication between devices. Communication is divided into two categories depending on the mode of communication, (Shahzad et al., 2021) (Yeboah-Akowuah et al., 2021) (Zu et al., 2021). A) Data shared between wearable devices on the body is referred to as "on-body communication," and B) communication via off-body connection enables data to be transmitted between body-worn sensors and the sensor hub base unit. Considering the antenna is in charge of transmitting and receiving data from one device to another, it serves an essential function for enhancing the functioning of each specific WBAN sensor, (Yeboah-Akowuah et al., 2021). The development of a wearable antenna encompasses a range of attributes of design that distinguish it apart from traditional antenna designs, which includes flexibility, low profile, safety, biocompatibility, its capacity to withstand bending or crushing while in use, and good radiation characteristics when positioned over human tissues. It should be simple to incorporate the antenna with human tissues. It can be difficult to develop and manufacture an antenna that fulfills every one of these requirements at the same time.

The designers of the WBAN antenna face several obstacles that have been thoroughly understood and analyzed since these nodes are in close range to the human body. The wearer's need for movement might deform the antenna, which might impair certain capabilities. The choice of the substrate is crucial for characterizing the functionality of the antenna to be designed in bent condition. Numerous substrates are suitable for antenna design, and several more successful studies have been completed with them. The comprehensive examination of the literature provides insight regarding multiple substrate options for the flexible antenna design. The most utilized wearable antenna substrate is the textile (Yalduz et al., 2020) (Smida et al., 2020) (Kim et al., 2020) photo-graphic paper-based, (Kim et al., 2012), polyester films, (Abirami & Sundarsingh, 2017), polyimides (Zahran, Abdalla, & Gaafar, 2019) (Genovesi et al., 2016) (El Atrash et al., 2019) (Wang et al., 2018), latex rubbers, (Agarwal, Guo, & Salam, n.d.). High-conducting textile substrates are used for the antenna and the ground plane such as Zelt, (Ali et al., 2019), Taffeta, (Moro et al., 2018), Flectron, (Soh et al., 2013), and Shieldit, (Zhang, Soh, & Yan, 2020). Polyimide (PI) was one of the different substrates utilized in this work. Among the many advantages they offer the capacity to reduce latency and provide thermal stability, which guarantees that even when a person has a high body temperature the performance of the antenna remains unaffected.

One problem with wearables is the Specific Absorption Rate (SAR). The FCC and ICNIRP have determined safe usage guidelines of about 1.6 W/Kg and 2 W/Kg, respectively, for single gram and ten grams of tissue. One of the key things into account in the wearable antenna's design is SAR reduction. It is necessary to lower the back radiation for the purpose to regulate the SAR value. A number of techniques have been employed to enhance isolation between the human body & antenna, minimize back radiation, enhance the FBR value, and preserve the SAR value within the range considered ideal. A responsive method utilized extensively in the recent is the Metamaterial (MMT) structures. Due to the enhanced FBR value, these artificial structures assist in reducing the SAR, (Zhang, Soh, & Yan, 2020). A different acronym for these structures is AMC. With the reflection BW ranges from 90 to -90 degrees, and performs the best as a reflector at the necessary frequencies. Incorporating AMC for the ground plane improves other antenna parameters, isolates the human body & antenna, and lowers the SAR (Abdelghany et al., 2022).

This chapter presents AMC-based microstrip fed dual-band antenna that resonates at 3.5GHz and 5.7GHz. The antenna has a flexible geometrical structure, enhances gain, operates better despite body effects, lowers back radiations, improves FBR, and overcomes issues like frequency detuning. Using 3D CST software, AMC based dual-band antenna were simulated. Both the on-body and off-body analysis were carried out.

2. CHARACTERIZATION OF ANTENNA AND AMC

The Antenna and AMC design is explained in this section. The 3.5 GHz and 5.7 GHz are the resonant frequencies for the antenna and the Artificial Magnetic Conductor. With the help of the traditional approach of design, the dual band microstrip fed dual band antenna and the AMC was developed for the same frequency.

2.1 Antenna Design

Table 1. Dimension of the structure

Parameters	Length (mm)	Parameters	Length (mm)
f_w	4.5	L_1	8.9
f_L	14	L_2	2.8
A_L	5.7	L_3	1.2
A_w	2.8	P_L	26
F_1	4.0	P_w	25
F_2	6.0	G_w	15
F_3	1.7		

Figure 1. Microstrip-fed dual-band antenna with ground

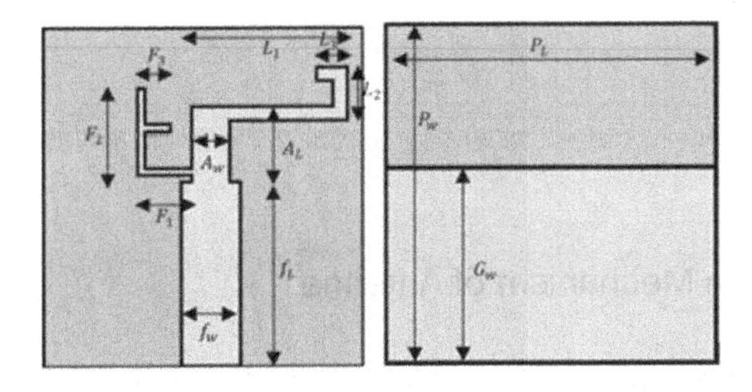

The Figure 1 is the dual-band microstrip-fed-antenna, and its dimensions are tabulated in Table 1. The substrate of 0.8 mm Polyimide with a tan δ of 0.008 and Ɛr of 3.5 is used. The Polyimide is chosen as the substrate due to the characteristics like flexibility, strength, thermal stability, and chemical resistance. The volume is

$0.303\ \lambda_0 \times 0.291\ \lambda_0 \times 0.0093\ \lambda_0\ mm^3$, where λ_0 denotes the wavelength at 3.5 GHz in free space.

With help of the the current distribution technique, the structure of the dual-band antenna is designed with two radiating arms—a F-shaped arm and a L-shaped arm of different lengths. The resonance is produced at 3.5 GHz and 5.7 GHz and the evolution is explained through a step-by-step procedure, and the corresponding return loss (dB) graph is illustrated in Figure 2(a) & (b) resp. With the presence of only the long arm L-shaped resonator peaks return loss S_{11} level which is almost equal to – 20 dB obtained at 3.5 GHz with bandwidth of 320MHz from 3.43 to 3.75 GHz and no sharp resonance was produced in the upper frequency. With the addition of another F-Shaped arm near the other end of the structure, the antenna has a dual resonance at frequencies where the lower frequency at 3.56 GHz with a return loss of – 22.6 dB with BW of 320 MHz and the upper band 5.7 GHz with a S_{11} of about -34 dB with BW of 480 MHz from 5.50 GHz to 6.02 GHz. The size of the L section and the F section in the antenna's structure are optimally chosen after analyzing parametrically to attain better performance in the desired bands.

Figure 2. (a)The stages of evolution of the dual-band-antenna (b) return loss (dB) vs frequency plot

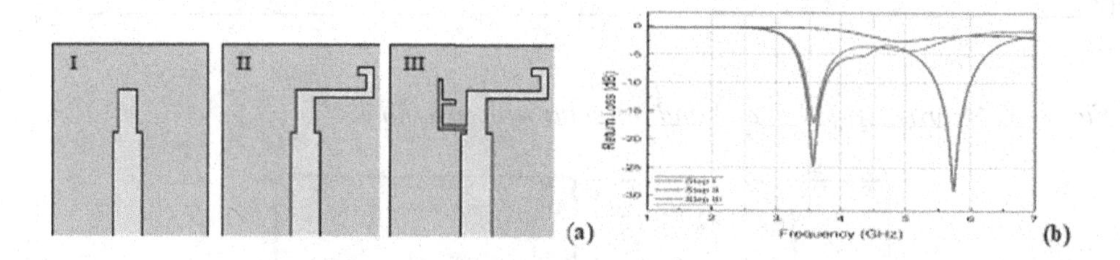

2.1.1 Radiation Mechanism of Antenna

The depicted current density for the proposed antenna for 3.5 GHz and 5.7 GHz in Figure 3 (a) and (b), respectively. The resonating lengths of the antenna that influences the resonance at the desired frequencies. It provides details into the radiation mechanisms of both the L-shaped and F-shaped arms.

(a) (b)

Figure 3. (a) Current density – 3.5 GHz (b) current density - 5.7 GHz

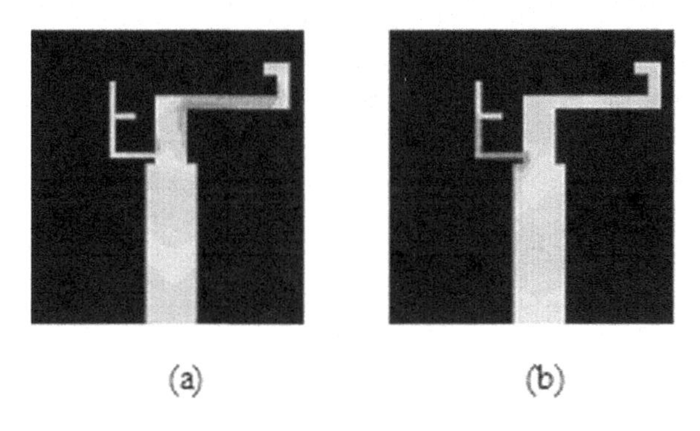

At 3.5 GHz, the upper arm of the antenna shows greater current density. This indicates it is the resonant length at that frequency. In Figure 3(b), at 5.7 GHz, the F-shaped arm displays elevated current density, having a significant role in the resonance at that particular frequency. This shows that different parts of the antenna resonate at different frequencies.

2.1.2 Parametric Analysis of the Anti-Symmetric Dual L-Shaped Antenna

The structural dimensions of the designed antenna were optimally chosen based on the parametric analysis to get optimally better performance. The parameters involved in the analysis were the dimension of the ground conductor, the position of the F shaped and the L-shaped arms.

A. Parametrical Study of the Dimension of Ground

The ground plane leads to a low $S_{11}(dB)$ value at the specified frequency. The dimension 'w' of the ground plane is systematically va. The optimal width that produces desired performance is chosen. Figure 4 illustrates the impact of varying ground width on the S_{11} vs. frequency graph. To enhance performance at the required frequencies, a width of 15 mm is determined as the optimal choice.

Figure 4. Parametric analysis of the ground plane

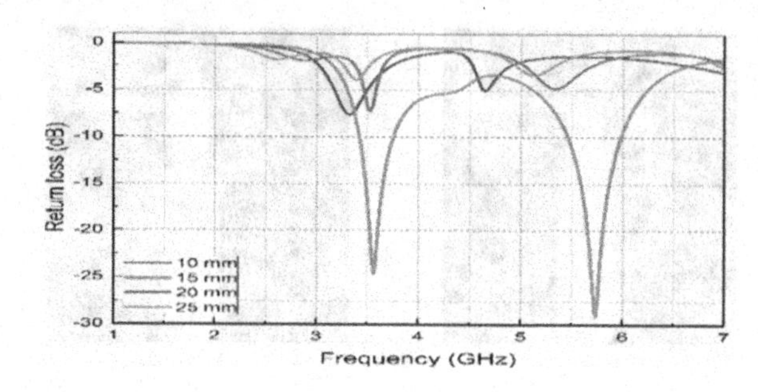

B. Parametrical Study of the Position of the L Arm of the Antenna

The position where the long L resonator should be placed has an impact on the resonance. The position of the L-structure is chosen in such a way that it produces a resonance at 3.5 GHz. The distance from the feed 'd' is chosen as 5.63 mm from the freed as it resonates at 3.5 GHz. Figure 5 (a) represents the frequency shift concerning the position of the L-shaped arm from the feed.

C. Parametrical Study of the Position of the F Arm of the Antenna

The position where the long F resonator should be placed has an impact on the resonant frequency of the designed anti-symmetric dual L antenna. The position of the F-shaped structure is chosen in such a way that it produces a resonance at 5.7 GHz. The distance from the feed 'd' is chosen such that it resonates at 5.7 GHz. Figure 5 (b) represents the frequency shift concerning the position of the F-shaped arm from the feed.

(a) (b)

Figure 5. Parametrical study of the (a) position of the L shape resonator with respect to the feed (b) position of the F shape resonator with respect to the feed

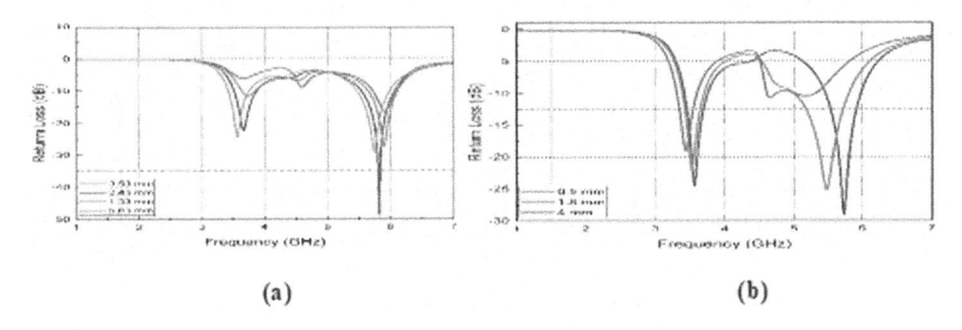

2.2 Design of Artificial Magnetic Conductor (AMC)

An antenna's backward lobe radiation needs to be considered in mind before it gets selected for a wearable application. Optimizing the antenna characteristics and minimizing the back radiation are both essential. A 3x3 AMC array has been installed to be positioned behind the antenna in order to verify EM wave propagation, enhance the gain of the antenna, and shield back radiation whilst providing an insulation between the wearer's tissues and the radiation. The zero-phase reflection characterization is employed in this chapter to develop the AMC structure. The proposed antenna is then combined with the envisioned AMC, and the antenna's performance is evaluated. The AMC is designed on a 0.8 mm Polyimide substrate. The polyimide is chosen as the substrate because of its flexibility. The single AMC unit cell has a volume of 0.70 λ_0x 0.70 λ_0 x 0.0093 $\lambda_0 mm^3$. The AMC structure is in Figure 6 (a) and (b) and the optimal dimensions are in Table 2.

Table 2. Dimensions of the AMC

Parameters	Length (mm)	Parameters	Length (mm)
l	20	l_2	13
b	20	l_3	8
l_1	17	l_4	5

Figure 6. (a) Top view (b) ground plane of the AMC unit cell

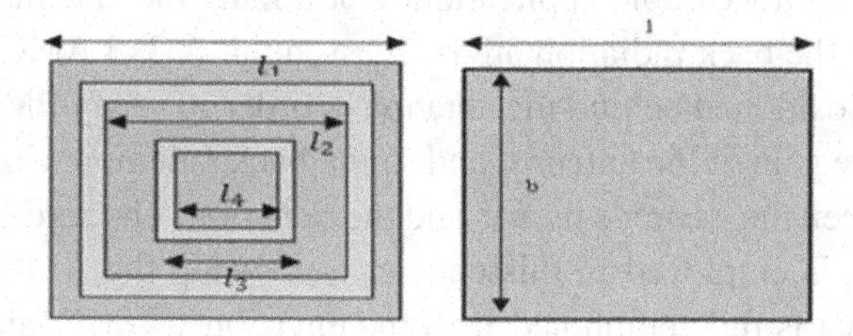

The two consecutive square rings constitutes the design of the AMC. The 0 degree reflection appears at 3.5 GHz and 5.7 GHz. In order for the electromagnetic waves to be reflected by AMC to positively interfere with the stimulating signal, in-phase reflection is necessary. The outer ring has a circumference of 0.6 λ_0, with each ring having a different wavelength of λ_0 frequencies. The excitation of the AMC structure is shown in Figure 7.

Figure 7. Excitation - AMC unit cell

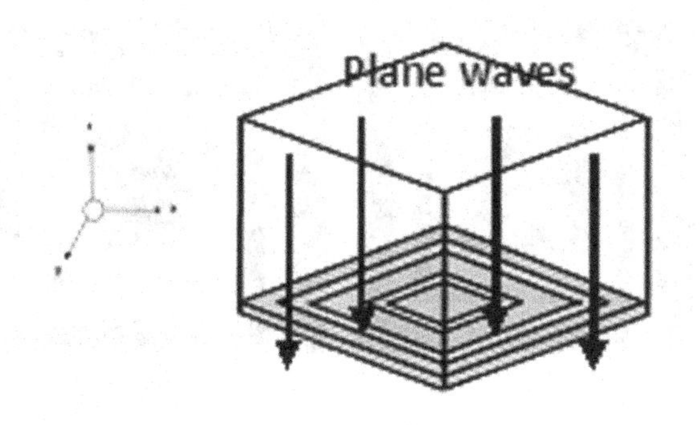

Figure 8. Reflection phase vs. frequency of the AMC.

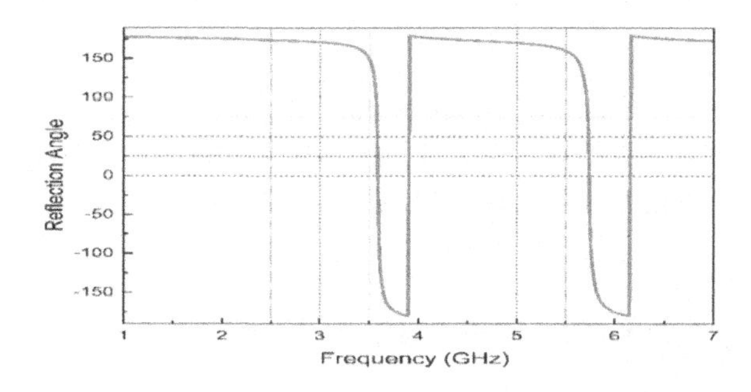

Using TE 10 mode plane waves, every single unit cell layout of the intended AMC is simulated from upwards to down, as illustrated in Figure 7. The reflection phase obtained by the frequency-dependent designed AMC structure is in Figure 8. At the required frequencies, the AMC reflects with a zero-degree reflection phase. The unit cell behaves as a reflector for the dual band antenna in the range. Figure 8 depicts that it is a Perfect Magnetic Conductor (PMC) at these desired bands.

Figure 9. Distribution of the J (current density) of the AMC (a) upper (b) lower band

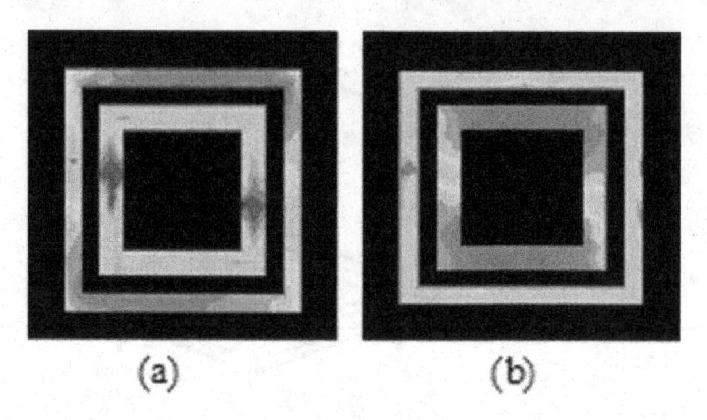

The reflection phase bandwidth BW_θ with the given mathematical expression.

$$BW_\theta = \left(\frac{f(-90) - f(+90)}{f_c}\right) \times 100 \quad (1)$$

The $f(-90)$ is the frequency straight to the -90° phase and $f(+90)$ is the frequency straight to the +90° phase. The phase bandwidth of the AMC structure at 3.5 GHz is 1.25% and that of at 5.7 GHz is 2.201%.

3. RESULTS AND DISCUSSIONS

3.1 Off- Body Analysis

The effectiveness of the proposed antenna in both on and off-body situations has been contrasted in this work with and without the presence of this AMC structure. A 3x3 array had been incorporated into the previously described planned L-shaped and F shaped resonator. The suggested antenna construction is positioned over the AMC with a layer of Styrofoam in between, measuring is 0.70 λ_0x 0.70 λ_0 x 0.0093 $\lambda_0 mm^3$. The AMC and the antenna are isolated from each other by the Styrofoam. There is a 2.5 mm isolation gap. The AMC integrated dual band antenna's top view is displayed in Figure 10.

Figure 10. Top view of the AMC integrated

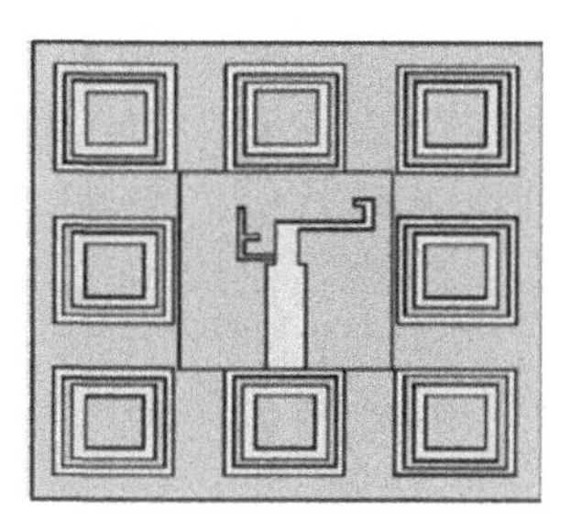

Figure 11. The return loss characteristics proposed anti-symmetric dual L-shaped antenna for the AMC and Antenna.

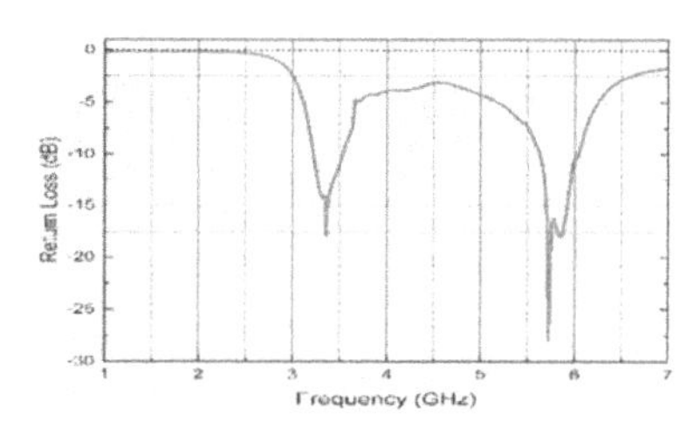

Figure 11 shows the S_{11} curve of the antenna on an AMC. It is observed that s_{11} at 3.3 GHz is about -19dB with a 260 MHz bandwidth (3.24 – 3.50 GHz). Similarly at the upper band from 5.56GHz to 6.1GHz a with a BW of 540MHz centered at 5.72 GHz with $s_{11}$of -24.7 dB.

Figure 12. 2D radiation plot – 3.5 GHz

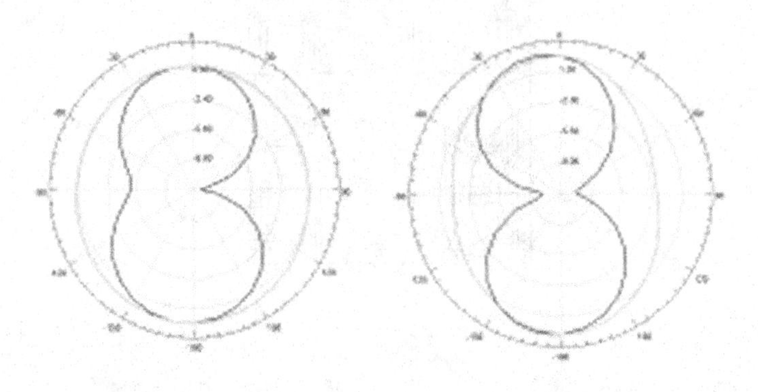

Figure 13. 2D radiation plot – 3.5 GHz and 5.7 GHz without AMC and 5.7 GHz with AMC.

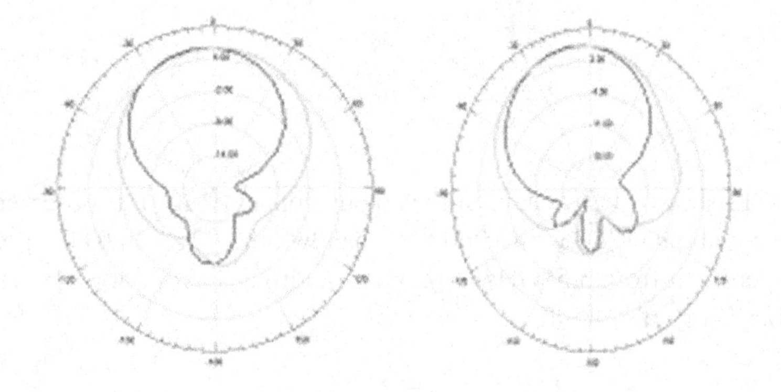

The characteristics of radiation for the suggested dual band antenna were investigated in the E and H-plane. The radiation plot at resonance with and without AMC is displayed in Figures 12 and 13, respectively. The antenna's 3D radiation pattern is shown in Figures 14 and 15, respectively, with and without AMC. The gain at different frequencies is in Figures 16(a) and (b), resp, without and with AMC. The radiation plot of the suggested antenna is shaped like an eight at both frequencies. Without an AMC, the gain is 2.4 dB and 3.6 dB at 3.5 GHz and 5.7 GHz. After the addition of the AMC the antenna's emission pattern changes to reduce the rear lobes more significantly, increasing the antenna gain to 4.7 dB and 5.7 dB at 3.5GHz and 5.7GHz, respectively

Figure 14. 3D gain plot – without AMC

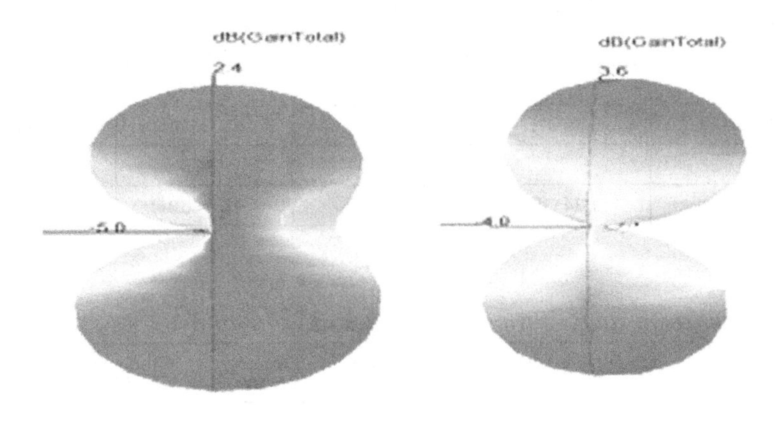

Figure 15. 3D gain plot - with AMC.

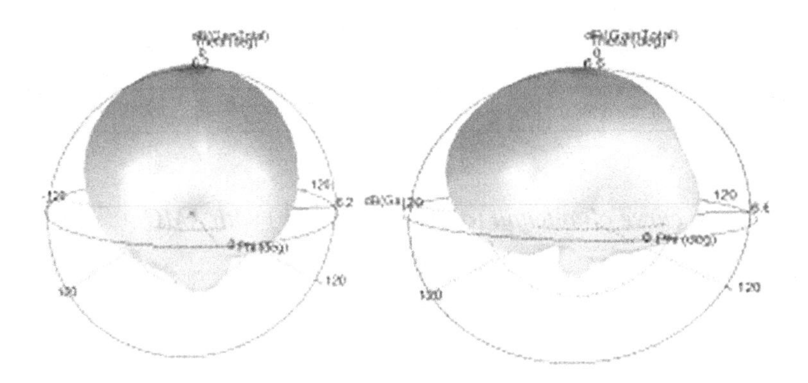

Table 3 tabulates the antenna performance comparison between AMC and non-AMC. Table 4 presents a comparison between the most recent developments and the developed dual band antenna. It is conceivable that our approach outperforms the previously discussed antennas. In the dual band, factors like gain and bandwidth are both improved.

Table 3. Antenna parameters without and with AMC

Parameter	Antenna		Antenna with AMC	
	3.5GHz	5.7GHz	3.5GHz	5.7GHz
Gain	2.1 dB	3.6 dB	4.4 dB	5.7 dB
VSWR	1.195	1.1481	1.1417	1.018
Bandwidth	320 MHz	480 MHz	320 MHz	480 GHz
% efficiency	84%	84.8%	91%	93.7%
FBR	1.345 dB	1.287 dB	14.04 dB	188.9 dB

Table 4. Comparison study: Proposed antenna vs. state-of-art

Ref	Overall Dimension	Substrate	Frequency (GHZ)	Bandwidth (MHz)	Gain
(Aprilliyani1 et al., 2020)	50x35x2 mm3	Felt	2.4	80	5.2 dBi
(Haiyan et al., 20220	$(0.64 \times 0.64 \times 0.0125)\lambda g$	Denim	2.45,5.8, 3.3, 3.85	90,190,230,570	-0.81,-2.81, -1.16,2.83 dBi
(Li & Li, 2018)	102x85x3.6 mm3	Pellon fabric	5.8	56	6.02 dB
(Mersani et al., 2018)	50x75x6 mm3	Felt	2.4, 5.2	120,125	7, 9.4 dB
This work	40x40x3 mm3	Felt	4, 8	520, 520	4.76,1.76 dBi
This work	$(0.70 \times 0.70 \times 0.0093)\lambda_0$	Polyimide	3.5, 5.7	260, 540	6.2,6.6 dB

Figure 16. Gain curve of antenna (a) without and (b) with AMC

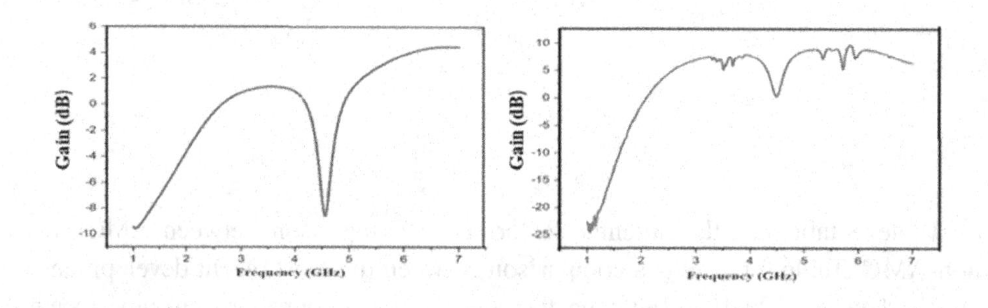

3.2 On-Body Analysis

It is researched how the antenna functions when it is positioned along the tissues. The dual band antenna is installed with a 3 mm gap (caused by clothing) from a three-layered human model. The antenna is configured in Figure 16(a) and (b), respectively, with and with no AMC. The electrical characteristics of muscles, fat, and skin are covered in Table 5. When the antenna is kept near the 3-layere tissue model, the dual band antenna has a s_{11} -18 dB in 3.43GHz, and s_{11} of about -11 dB at 4.49 GHz, and 5.6GHz. The antenna alone with the human tissue has another new resonance that is undesirable at 4.49GHz as evident from Figure 18 (a). Backing with AMC, the antenna is analyzed near the tissue model, the resonant frequencies have not shifted in Figure 18 (b). The S11 is -24.9 dB at 3.4 GHz and another resonance with return loss of -19dB at 3.6 GHz having an overall BW of 330MHz while at 5.76GHz the return loss of about -21.9 dB is obtained with a BW of 580MHz.

Figure 17. On-body analysis of the antenna (a) with AMC (b) without AMC

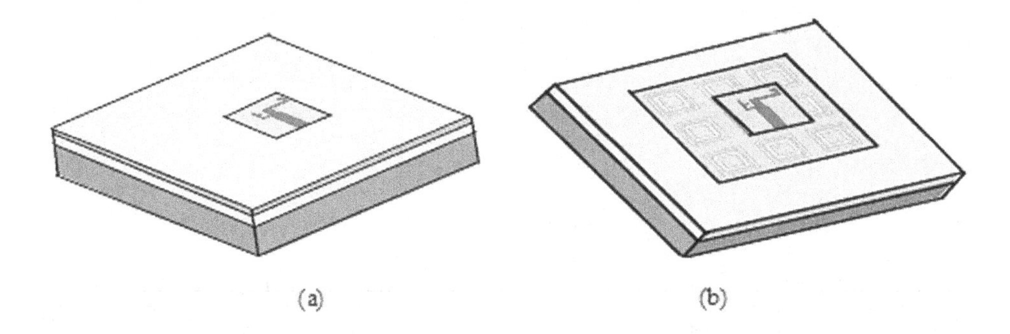

Figure 19 plots the gain of the dual band antenna while kept over the human tissue. The gain of the antenna and AMC backing is 4.2dB at 3.6GHz and 3.4dB at 5.7dB GHz. The antenna parameters without and with AMC when in contact with the human tissues are tabulated in Table 6.

Table 5. Dielectric properties of the body tissues (Yalduz et al., 2020)

	ε_r	Conductivity (S/m)	ε_r	Conductivity (S/m)
	At 3.5 GHz		At 5.7 GHz	
Skin	38.5	1.46	35.13	3.71
Fat	5.4	0.12	4.96	0.27
Muscle	52.7	1.77	48.48	4.95

Figure 18. Return loss of antenna on 3-layer human phantom (a) no AMC (b) with AMC

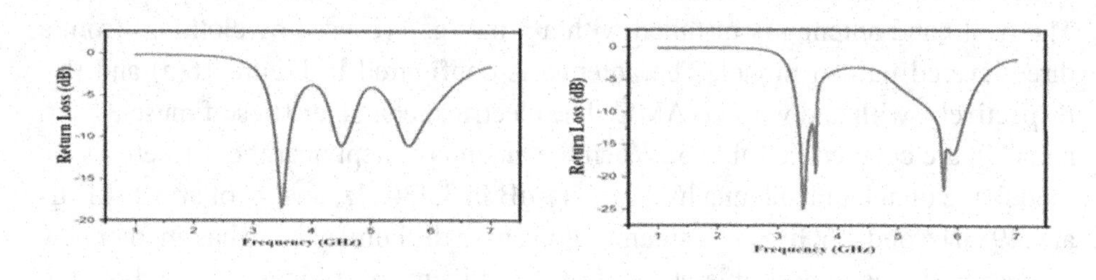

Figure 19. Gain characteristics of the AMC dual band antenna structure in on-body analysis

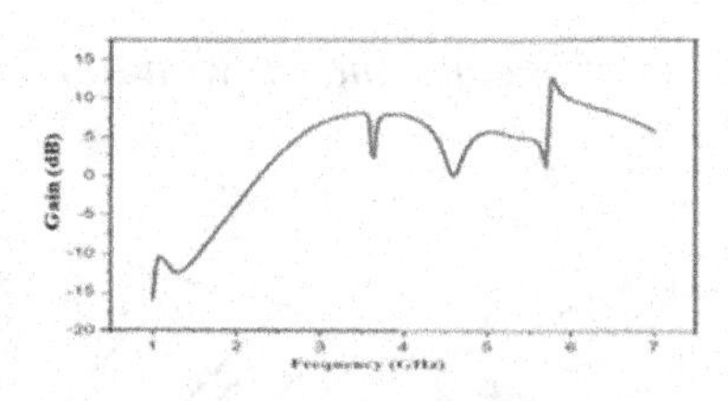

Table 6. Antenna parameters analysis on body without and with AMC backing

Parameter	Proposed Antenna		Proposed Antenna with AMC	
	3.5 GHz	5.7 GHz	3.5 GHz	5.7 GHz
Gain	0.6 dB	3.3 dB	4.2 dB	3.4 dB
VSWR	1.21	1.2	1.26	1.13
Bandwidth	360 MHz	700 MHz	260 MHz	580 MHz
% efficiency	7%	43.2%	84%	65%
FBR	27	29.4	129	115

4. CONCLUSION

This work discusses the design of an AMC-incorporated WBAN antenna that resonates at 3.5 GHz and 5.7 GHz. The antenna's volume of (0.303x0.291x0.0093) λ_0 mm^3 and that of the antenna with AMC surface is (0.70x0.70x0.0093) $\lambda_0 mm^3$

. To limit the back-lobes and improve the performance, and stability while worn by humans a 3x3 AMC array is introduced. The novelty of the dual-band antenna with AMC is the fact that it has improved parameters as gain is 4.2dB and 5.7dB, bandwidth of 320 MHz and 480 MHz, Front to Back ratio is 14.04 dB and 155.9 dB with a radiation efficiency of 91% and 93.7% in 3.5 GHz and 5.7 GHz resp with a reduced size. Having flexibility, improved gain and improved FBR makes it good for WBAN application. The working of the antenna both off and on the body is analyzed. The gain (4.2 dB and 3.4 dB) and the bandwidth (260 MHz and 580 MHz) are unaffected even at the vicinity of the human phantom model with only a marginal frequency shift. The proposed dual band antenna design is considered suitable as wearable as inferred from the simulated results.

REFERENCES

Abdelghany, M. A., Fathy Abo Sree, M., Desai, A., & Ibrahim, A. A. (2022, July). Gain improvement of a dual-band CPW monopole antenna for sub-6 GHz 5G applications using AMC structures. *Electronics (Basel)*, 11(14), 2211. 10.3390/electronics11142211

Abirami, B. S., & Sundarsingh, E. F. (2017, July). EBG-backed flexible printed Yagi– Uda antenna for on-body communication. *IEEE Transactions on Antennas and Propagation*, 65(7), 3762–3765. 10.1109/TAP.2017.2705224

. Agarwal, K. (2020). Wearable AMC backed near-endfire antenna for on-body communications on latexsubstrate. *IEEE Trans. Compon., Packag., Manuf. Technol.*, (*vol. 6*, no. 3, pp. 346–358). IEEE.

Ali, U., Ullah, S., Shafi, M., Shah, S. A. A., Shah, I. A., & Flint, J. A. (2019, November). Design and comparative analysis of conventionaland metamaterial-based textile antennas for wearable applications. *International Journal of Numerical Modelling*, 32(6), 2567. 10.1002/jnm.2567

El Atrash, M., Abdalla, M. A., & Elhennawy, H. M. (2019, October). A wearable dual-band low profile high gain low SAR antenna AMC-backed for WBAN applications. *IEEE Transactions on Antennas and Propagation*, 67(10), 6378–6388. 10.1109/TAP.2019.2923058

Genovesi, S., Costa, F., Fanciulli, F., & Monorchio, A. (2016). Wearable inkjet-printed wideband antenna by using miniaturized AMC for subGHz applications. *IEEE Antennas and Wireless Propagation Letters*, 15, 1927–1930. 10.1109/LAWP.2015.2513962

Kim, S., Ren, Y., Lee, H., Rida, A., Nikolaou, S., & Tentzeris, M. M. (2012). Monopole antenna with inkjet-printed EBG array on paper substrate for wearable applications. *IEEE Antennas and Wireless Propagation Letters*, 11, 663–666. 10.1109/LAWP.2012.2203291

Kim, Y.-S., Basir, A., Herbert, R., Kim, J., Yoo, H., & Yeo, W.-H. (2020, January). Soft materials, stretchable mechanics, and optimized designs for bodywearable compliant antennas. *ACS Applied Materials & Interfaces*, 12(2), 3059–3067. 10.1021/acsami.9b2023331842536

Li, H. (2022). Low-Profile All-Textile Multiband Microstrip Circular Patch Antenna for WBAN Applications. *IEEE Antennas And Wireless Propagation Letters*, (*vol. 21*, no. 4).

Li, S.-H., & Li, J. (2018). Smart patch wearable antenna on Jeans textile for body wireless communication. *12th International Symposium on Antennas, Propagation and EM Theory (ISAPE)*. IEEE. 10.1109/ISAPE.2018.8634084

Mersani, A. (2018). Improved Radiation Performance of Textile Antenna Using AMC Surface. *15th International Multi-Conference on Systems, Signals & Devices (SSD)*. AMC.

Moro, R., Agneessens, S., Rogier, H., & Bozzi, M. (2018). Circularly-polarised cavity-backed wearable antenna in SIW technology. *IET Microwaves, Antennas & Propagation*, 12(1), 127–131. 10.1049/iet-map.2017.0271

Sabban, A. (2019). Small New Wearable Antennas for IOT, Medical and Sport Applications. 13th European Conference on Antennas and Propagation (EuCAP). Research Gate.

Shahzad, A., Paracha, K. N., Naseer, S., Ahmad, S., Malik, M., Farhan, M., Ghaffar, A., Hussien, M., & Sharif, A. B. (2021, November). An artificial magnetic conductor-backed compact wearable antenna for smart watch IoT applications. *Electronics (Basel)*, 10(23), 2908. 10.3390/electronics10232908

Smida, A., Iqbal, A., Alazemi, A. J., Waly, M. I., Ghayoula, R., & Kim, S. (2020). Wideband wearable antenna for biomedical telemetry applications. *IEEE Access : Practical Innovations, Open Solutions*, 8, 15687–15694. 10.1109/ACCESS.2020.2967413

Soh, P. J., Vandenbosch, G. A. E., Wee, F. H., Zoinol, M., Abdul, A., & Campus, P. (2013). Bending investigation of broadband wearable all-textile antennas. *Australian Journal of Basic and Applied Sciences*, 7(5), 91–94.

Wang, M., Yang, Z., Wu, J., Bao, J., Liu, J., Cai, L., Dang, T., Zheng, H., & Li, E. (2018, June). Investigation of SAR reduction usingcflexible antenna with metamaterial structure in wireless body area network. *IEEE Transactions on Antennas and Propagation*, 66(6), 3076–3086. 10.1109/TAP.2018.2820733

Yalduz, H., Tabaru, T. E., Kilic, V. T., & Turkmen, M. (2020). Design and analysis of low profile and low SAR full-textile UWB wearable antenna with metamaterial for WBAN applications. *AEÜ. International Journal of Electronics and Communications*, 126(Nov), 153465. 10.1016/j.aeue.2020.153465

Yeboah-Akowuah, B., Tchao, E. T., Ur-Rehman, M., Khan, M. M., & Ahmad, S. (2021, September). Study of a printed split-ring monopole for dual-spectrum communications. *Heliyon*, 7(9), e07928. 10.1016/j.heliyon.2021.e0792834589621

Zahran, R., Abdalla, M. A., & Gaafar, A. (2019, July). New thin wide-band braceletlike antenna with low SAR for on-arm WBAN applications. *IET Microwaves, Antennas & Propagation*, 13(8), 1219–1225. 10.1049/iet-map.2018.5801

Zhang, K., Soh, P. J., & Yan, S. (2020, December). Meta-wearable antennas—A review of metamaterial based antennas in wireless body area networks. *Materials (Basel)*, 14(1), 149. 10.3390/ma1401014933396333

Zu, B., Wu, B., Yang, P., Li, W., & Liu, J. (2021, August). Wideband and high-gain wearable antenna array with specific absorption rate suppression. *Electronics (Basel)*, 10(17), 2056. 10.3390/electronics10172056

Chapter 10
Design of Patch Antenna and Its Implementation in Spatial Multiplexing for 5G NR Applications

S. Krithiga
Department of Electronics and Communication Engineering, SRM Institute of Science and Technology, Chennai, India

Dhinakaran Vijayalakshmi
Department of Electronics and Communication Engineering, SRM Institute of Science and Technology, Chennai, India

R. Dayana
Department of Electronics and Communication Engineering, SRM Institute of Science and Technology, Chennai, India

K. Vadivukkarasi
https://orcid.org/0000-0002-4618-7208
Department of Electronics and Communication Engineering, SRM Institute of Science and Technology, Chennai, India

ABSTRACT

This chapter proposes a microstrip patch antenna suitable for 5G NR applications operating at 66 GHz. This frequency range lies in mm-Wave spectrum, also called as high band spectrum that has the advantage of very high speed over short distance. The antenna has been designed with a reflector grid which provides increased efficiency at less cost. This antenna design has been developed to the next stage as an antenna array with 64 elements, which yields improved directivity. The proposed

DOI: 10.4018/979-8-3693-2786-9.ch010

design employs row and tapered column tapering providing a better input impedance matching for wider frequency range. Directivity of 20dB has been obtained through this proposed antenna design with Kaiser windowing technique. The spatial multiplexing increases data rate within a limited bandwidth. The simulation results have shown that the designed antenna will be suitable for 5G systems offering a higher data rate in the range of gigabits per second

1. INTRODUCTION

1.1 Overview

In the realm of telecommunications, Fifth Generation (5G) technology has ushered in a new era of connectivity, with multiplexing playing a pivotal role in its architecture. Multiplexing in 5G enables the simultaneous transmission of multiple signals over a single communication channel, optimizing bandwidth usage and enhancing data transfer efficiency. Through techniques such as time-division, frequency-division, and spatial-division multiplexing, 5G networks can accommodate a vast array of devices and applications with varying bandwidth requirements (Muhammad Iqbal Rochman, et al., 2023; Shikha Sargam et al., 2023; Abdul Ahad, et al., 2023; Nessrine Trabelsi 2024).

The Multiple Input and Multiple Output (MIMO) technology (Saad Hassan Kiani, et al., 2024; S. Leones Sherwin (2023); Mohamed Nasr Eddine Temmar, et al., 2021) enhances the quality of radio links. These techniques are employed in 5G NR networks to enhance the data rates. Spatial diversity technique in MIMO is the physical separation between multiple antennas at transmitter and receiver. It increases the transmission or received signal quality by overcoming the channel link issues due to signal fading.

The other main advantage of MIMO is spatial multiplexing in which more than one antenna at different locations will be used at transmitter or receiver. These spatially separated antennas can handle multiple data streams using same frequency and time resources but work as individual communication channel to transfer data between the transmitter and receiver. The parallel data streams from a MIMO transmitter can be aimed to a single user or multiple user devices. The multiple data streams, transmitted to a user device increases the data rate. The increase in antenna separation within a given array length does not increase the number of degrees of freedom in the channel but reduces the beam width.

By harnessing the power of multiplexing, 5G networks (Peng Liu, 2023; Parveez Shariff B. et al., 2023) can deliver unprecedented speeds, low latency, and seamless connectivity, thereby revolutionizing the way we communicate and interact in the digital age.

1.2 Literature survey

Pandey (2021) presented an architecture for resource allocation dynamically suitable for various traffic loads. This paper demonstrates the performance of RoF network employed with quadrature amplitude modulation technique. Shanbhag (2023) implemented Layered Division Multiplexing in 5G. BER constellation diagram of channel with additive white gaussian noise and Rayleigh models has been obtained in MATLAB simulink. An antenna which shares radiator with four ports tuned to different frequency ranges and acting as multiplexing filter has been proposed by Mao (2021). The deigned antenna provides good isolation of more than 25dB between the channels.

The research work by Chen (2023) has proved that side lobes can be reduced by removing central part of the cross-shaped antenna. The authors have also explored that cross-shaped pattern increases the isolation between two ports. The concept of generating magnetic current array by blocking microstrip has been proposed by Wang (2023). This magnetic current excites the sides-shorted microstrip patch antennas through coupling slots. Simulation results show that a wide bandwidth of 26% and a 1 dB gain bandwidth of .3% has been obtained.

The idea of transmitting main signal in different directions through single-pole dual-throw switch has been proposed by Trzebiatowski (2021). To reduce the cost, fabrication of the proposed antenna has been done in a Cu Clad 217 substrate. The main beam is switched between -45° and +45° in horizontal plane with beam width of 80° and gain of around 3 dBi in both directions.

This paper by Serghiou (2020) presents a double band 8 × 8 antenna array operating in sub-6 GHz spectrum to be suitable for fifth generation smartphones. The orthogonal pairs of antennas are closely and symmetrically placed on long edges and corners of mobile phones. Every antenna is operated in 3.1 - 3.85 GHz (low band) and 4.8 - 6 GHz (high band). Improved results has been obtained for various parameters such as channel capacity. envelope correlation coefficient total efficiency and mean effective gain.

Shen (2021) proposed a miniature microstrip antenna with 2 elements which can be used in 5G communication. An electromagnetic band gap structure is used as ground for two closely patch antennas, operated in 5G spectrum. The proposed EBG ground, can provide close proximity to elements provides 10dB high mutual coupling than normal ground. A resonant cavity antenna designed by Wen (2021) provides increased gain, gain-bandwidth product when operated around 28 GHz and 38 GHz band of frequencies. Gain has been still improved with circle shaped patch on surface which is partially reflected constructed with high permittivity substrate.

The research work by Kim (2022) presents an ultra-thin and wider band 2×2 array antenna in millimetre wave devices. A well tight coupled frequency selective surface, feeding is capacitively coupled and H-plane walls are used to obtain wide bandwidth. A frequency selective surface integrated with dual-band reflection antenna designed by Zicheng (2020) provides good reflection and transmission at Ka-band and X-band respectively. Results has proved that proposed antenna exhibits good performance characteristics and can be implemented as broadband antennas.

Yamamoto (2021) designed a reflector antenna fed by TM01 circular waveguide to address the issue of polarization loss. This design adds a metal grid on reflecting surface which provides improved gain in comparison with conventional antennas.

Multiplexing techniques such as GFDM, CPOFDM, WOLA, FOFDM and FBMC suitable for 5G to produce orthogonal and non-orthogonal waveforms has been explained by Mojtaba Vaezi (2018). Moreover this book gives an insight about multiple access techniques such as CSMA, CSMA/CD and LoRA and use cases of 5G networks.

This research work by Chitra (2020) focuses on Generalized Frequency Division Multiplexing for 5G networks. A precoder with differential encoding and correlative coding providing imbalance compensation and improved spectral efficiency has been proposed to overcome the disadvantages due to RF impairments. The obtained results proved that presented model has reduced BER, PAPR and radiations.

Microstrip patch antenna array for 5G network applications has been designed by Rovin Tiwari (2023). The proposed antenna array model is a dumbbell shaped 2X2 and 4X4 which yields improved simulation results with high gain and increased bandwidth.

Leeladhar Malviya (202) has designed a MIMO array with 2 transmit and 8 receive antennas with substrate as Rogers Duroid 5880 and dimension as 23.61 x 55.18. The resonant frequency range is at 28 GHz. This design has provided good results suitable for both indoor and outdoor applications.

Mouaaz Nahas (2022) developed a rectangular patch antenna having additional slots in L and I shapes to provide high gain. The obtained results outperform the conventional types yielding increased gain and directivity. High voltage standing wave ratio and efficiency are also obtained through this model.

The model proposed by Hassan Sani Abubakar (2024) has 8 MIMO symmetrical antennas with 50Ω feeding input line. The main advantage of this design is that more than 18 dB isolation has been obtained. The output of this model has efficiency greater than 60% with correlation coefficient less than 0.04, improved gain greater than 5 dBi and channel capacity around 38.5 bps/Hz that is superior by triple times compared to 2 × 2 antenna configuration.

A 3 stage coplanar waveguide with microstrip antenna array operating above 100 GHz has been presented by Uri Nissanov (2023) Simulated results show that peak gain of 12.1 dB and directivity of 12.98 dBi had been obtained This proposed model finds suitable application in integrated vehicular autonomous and sixth generation communication system.

The planar configuration of 8 loop antennas placed at various corners of cellular phone mainboard having FR- 4 substrate and dimension as 75 × 150 × 0.8 mm^3 has been introduced by Naser Ojaroudi Parchin (2020). Additionally, specifically designed arrow-shaped strips are placed to have improved bandwidth and isolation. This design provides full coverage and directivity with easy integration in smart phones.

A new patch array antenna operating at 25GHz, 28GHz and 38GHz suitable for cellular systems has been designed by Sanae Dellaoui (2018). The proposed design consists of 2 patch antennas. Wilkinson power divider model is applied to feed the antennas. Electromagnetic Band Gap dielectric is used as superstrate to improve the gain greater than 12 dBi.

Kapil Jain (2021) designed a antenna for operating frequency of 6 GHz in dual band. For fast data transmission, The signal with one frequency value changes to dual and multi frequencies. Improved results have been obtained than art of state works. Two patches antenna in rectangle shape are combined by corporate feed and these square-shaped quarter-wavelength antennas has been designed by Sourav Ghosh (2023). This proposed work yielded a high gain and efficiency of 11 dBi and 96% respectively at 28 GHz . This model also provides improved diversity including envelop correlation coefficient, diversity gain and channel capacity loss.

Ce Lakpo Bamy (2021) propose Dolly-shaped antenna operating at dual band such as 23.52 GHz and 28.39 GHz for fifth generation applications. The antenna dimensions are 7 × 7 × 1.28 mm^3 which is. Rogers RO3010 is used as the substrate Two parasitic elements with F-shape and a rectangular slot are used to obtain the improved performance. The design has also been validated through High-Frequency Structure Simulator. Additionally prototype was also developed to validate the simulation results obtained.

1.3 Motivation of this paper

5G technology, being currently used for communication. Many research works are being carried out to get the maximum benefits overcoming the limitations in it. Among these researches, antenna design is on main focus. Nevertheless spatial multiplexing with microstrip patch antennas with reflector grid has not been introduced so far. This motivates us to work on the design and implementation of 8x8 antenna array for 5G applications.

Spatial multiplexing offers several advantages in the realm of telecommunications and wireless communication systems such as increased throughput, support for multiple users, enhanced reliability and diversity, improved spectral efficiency, improved coverage and range, compatibility with existing standards.

1.4 Organization of this paper

Section 2 provides the design of proposed microstrip patch antenna operating in mm wave range suitable for 5G communication. Section 3 presents the spatial multiplexing of antenna array which improvise the data rate. Section 4 discusses the simulation results obtained for proposed model. The last section 5 provides the conclusions and future scope.

2. DESIGN OF MICROSTRIP PATCH ANTENNA

This paper involves the design of rectangular microstrip patch antenna for mm-Wave spectrum which supports very high speed over short distance. Spectrum in the range of 26GHz, 40 GHz, 50 GHz, and 66 GHz are supported by the service providers. The extended V band (66-71 GHz) is not attenuated by oxygen absorption, and suitable for medium and long backhaul applications. Design parameter of proposed rectangular micro patch antenna are specified in Table 1.

Table 1. Simulation parameters of microstrip patch antenna

Parameter	values
Dielectric permittivity of substrate (air)	1
Feed location	[0.0005, 0, 0.0013]
Impedance	50 ohms
Operating frequency	66 GHz

Basic dimensions of proposed patch antenna are calculated based on the following equations.

$$l = \frac{c_0}{2f_p\sqrt{\epsilon_r}} \quad (1)$$

$$w = \frac{c_0}{2f_p\sqrt{\frac{2}{(\epsilon_r + 1)}}} \quad (2)$$

where l and w denotes length and width respectively. f_p is the frequency of proposed antenna and $c_0$denoted the light speed. The dimensions of ground plane are calculated from the equations mentioned below

$$l_g = 6h + l \quad (3)$$

$$w_g = 6h + w \quad (4)$$

in which l_g and w_g denotes length and width of the ground plane, h is thickness of substrate. Additionally, antenna is supported with rectangular reflector grid structure. The design parameters of grid structure are mentioned in Table 2.

Table 2. Specifications of proposed reflector grid structure

Parameter	**values**
Grid type	HV
Grid width	9.084e-04
Grid spacing	6.041e-04
Number of grids	[4,4]
spacing	0.013m
Length of Ground Plane	0.065 m
Width of Ground Plane	0.065 m
Tilt axis	[1,0,0]

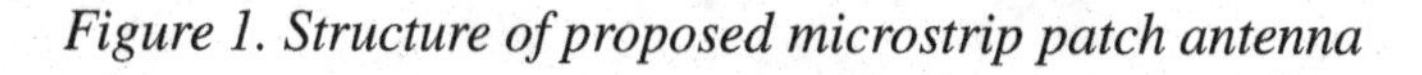

Figure 1. Structure of proposed microstrip patch antenna

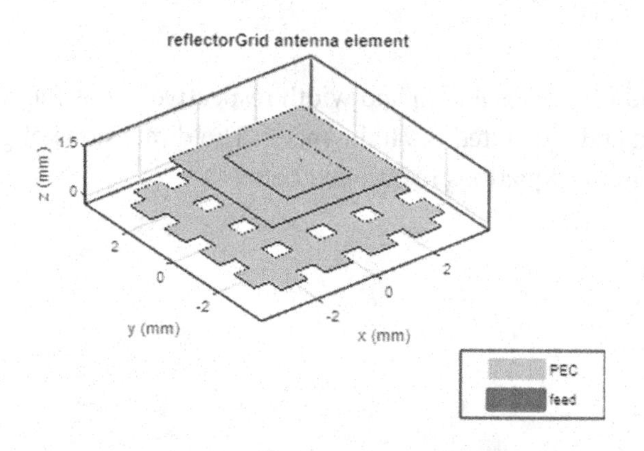

The structure of proposed patch antenna is shown in Figure 1. The reflector grid is implemented with a grid of parallel wires in one direction. The main advantage of grid reflector is to provide high gain. The linearly polarized waves that are radiated from antenna elements is reflected with a parallel electric field. The grid antennas provide the benefits of less cost, small size, suitable for batch production and easy integration.

3. DESIGN OF 8X8 ANTENNA ARRAY

In this paper, design of an 8x8 antenna array with 64 antenna elements to achieve desired radiation characteristics, beam forming capabilities, and coverage has been presented. The design procedure is as following:

1. **Consideration of design parameters**: The design parameters such as operating frequency, antenna height, width must be decided.
2. **Choice of antenna elements**: The type of antenna elements suitable for the application and frequency range are very important. This design involves microstrip patch antennas as discussed in section 2.
3. **Configuration of antenna array**: The array structure with 64 elements has to be decided. There are many possible configurations such as linear, planar, circular conformal arrays. This research work implements rectangular lattice structure.

4. **Spacing and Geometry**: Determine the distance between antenna elements and the geometry of the array. Spacing between elements affects the radiation pattern, beam steering capabilities, and side lobe levels. Proper spacing is crucial to achieve desired performance characteristics. These design parameter values are mentioned in Table 3.
5. **Feed Network Design**: Design the feed network to connect each antenna element to the transmitter or receiver. The feed network controls the phase and amplitude of the signals to enable beam forming and radiation pattern shaping.

The simulation parameters taken for consideration has been mentioned in Table 3. This design procedure is adopted with careful consideration of various parameters and design aspects to achieve optimal performance and functionality for the intended application in 5G networks. The advantage of employing Kaiser window is to achieve the flexibility in the selection and continuous variation inside lobe level. The 2-D model of 8 X 8 antenna array structure developed is presented in Figure 2.

This proposed rectangular microstrip patch antenna array has centreline feed and excitation due to other modes are reduced. They are also light in weight with minimal fabrication cost. This design involves row and column tapering through which different gains are assigned to each element present in the array. It is worthy to note that the center antenna elements are allocated with high gain values and outer elements have lesser gains .

Table 3. Simulation parameters of proposed antenna array

Parameter	values
size	[8,8]
spacing	[0.5 0.5]λ
Lattice	Rectangular
Row and column taper	Kaiser
Beta	1
Azimuth angles	[-180:180]
Elevation angles	[-90:90]
Propagation speed(m/s)	3e8

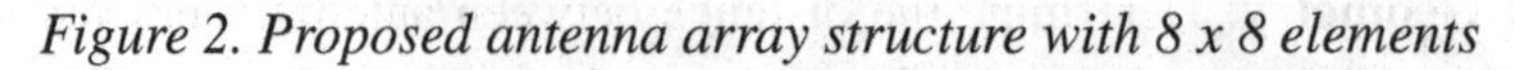

Figure 2. Proposed antenna array structure with 8 x 8 elements

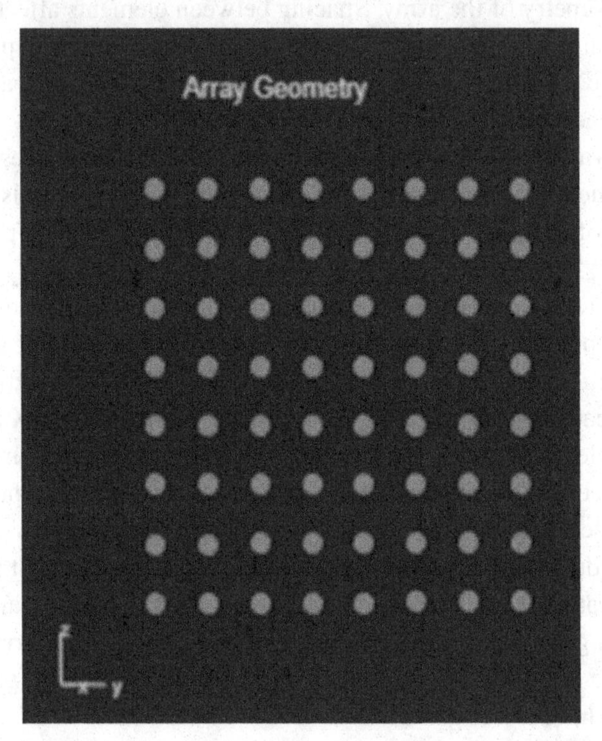

4. SIMULATION RESULTS

The graph between resistance in the order of ohms and frequency spectrum in the range of GHz of proposed antenna design is presented as Figure 3. This result shows that peak resonance occurs at 66 GHz which is the cut off frequency for the designed patch antenna. The reactance curve in this Figure 3 clearly indicates that at 66.5 GHz, the resistive inductance and resistive capacitance is very less, and the antenna becomes purely resistive at desired frequency.

Figure 3. Impedance vs. frequency response of proposed antenna array structure with 8 x 8 elements

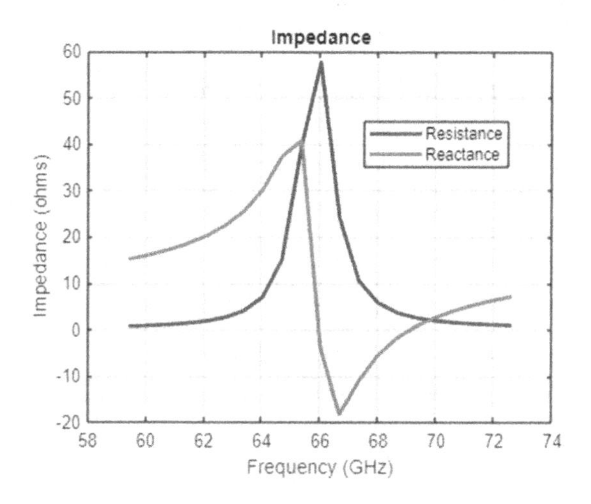

Figure 4. Directivity of proposed antenna in 3D plot

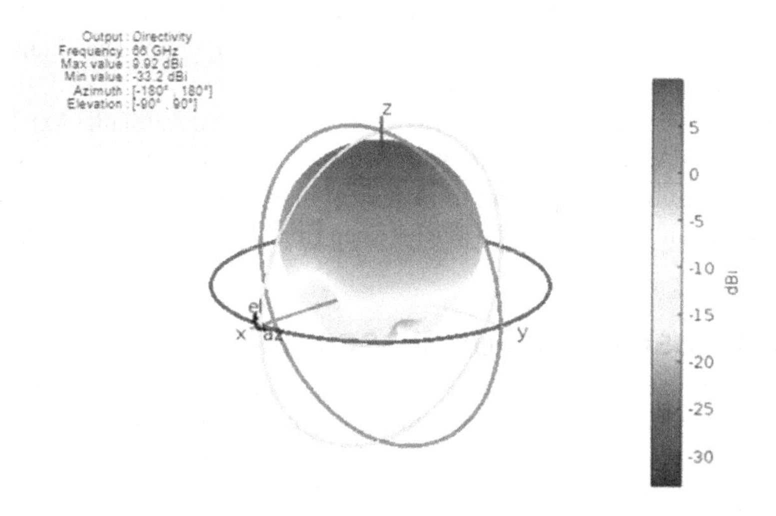

Directivity is an important antenna design parameter which specifies the ratio of radiation intensity in a particular direction from a single antenna or antenna array to the radiation intensity averaged over all directions. The 3D plot of directivity for proposed antenna is shown in Figure 4. The maximum Directivity of this single patch antenna is 9.92dBi. A high value of directivity indicates that maximum radiated power is concentrated in a particular direction, whereas low directivity implies that antenna radiation is equal in all directions. Directivity depends on factors such as antenna shape, size, and construction.

Figure 5. Directivity of proposed antenna array in 3D plot

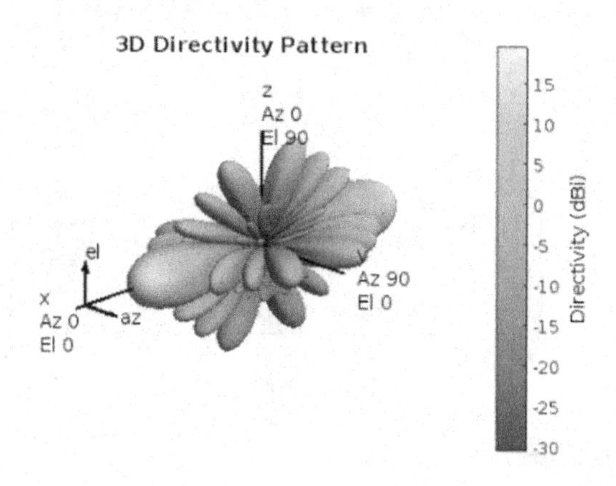

Increase in the number of antennas has improved the directivity. The 3D plot for directivity of proposed antenna array is shown in Figure 5. The obtained result of proposed antenna array (64 elements) with array span x = 0, y = 15.91 mm, z = 15.93 mm has yielded an array Directivity= 19.76 dBi at 0 Azimuth (Az); Elevation (El) at 13.34° Az/14.00° El Half Power Beam Width, which is the magnitude at angular separation of 50 percent from peak of main beam and 30.83° Az/30.00° El First Null Beam Width, angular span between first lobe nulls adjacent to main lobe.

5. CONCLUSION AND FUTURE SCOPE

This work presents the design of a microstrip patch antenna operating at 66 GHz, a mm wave frequency range suitable for 5G applications, The 66 GHz frequency range is an optimal value for real time applications in short distance and fast speed transmission such as biomedical applications and wireless sensor networks. The design involves rectangular reflector grid on Fr4 substrate. Simulation result obtained show that desired peak frequency value is obtained with impedance of 50 ohms.

After designing the single antenna element, 8 X 8 antenna array with 64 elements has been developed to support spatial multiplexing in the unlicensed 57 ~ 66 GHz frequency band. The proposed rectangular antenna array is tapered by row and column using Kaiser window to exhibit a directivity of 19.76 dBi. To best of author's knowledge, this is a novel antenna array design which provides spatial multiplexing for 5G networks published ever.

This work can further extend with beam forming and precoding techniques aiming increased directivity and better signal transmission.

REFERENCES

Abubakar, H. S., Zhao, Z., Kiani, S. H., Rafique, U., Alabdulkreem, E., & Elmannai, H. (2024). Eight element dual-band MIMO array antenna for modern fifth generation mobile phones. *International Journal of Electronics and Communications*, 175, 155083. 10.1016/j.aeue.2023.155083

Bamy, C. L., Mbango, F. M., Dominic, B. O. K., & Mpele, P. M. (2021). A compact dual-band Dolly-shaped antenna with parasitic elements for automotive radar and 5G applications. *Heliyon*, 7(4), e06793. 10.1016/j.heliyon.2021.e0679333948514

Chen, C. (2023). A Single-Layer Single-Patch Dual-Polarized High-Gain Cross-Shaped Microstrip Patch Antenna. *IEEE Antennas and Wireless Propagation Letters*, 22(10), 2417–2421. 10.1109/LAWP.2023.3289861

Chitra, S., Ramesh, S., & Beulah Jackson, S. (2020). Performance enhancement of generalized frequency division multiplexing with RF impairments compensation for efficient 5G wireless access. *AEÜ. International Journal of Electronics and Communications*, 127, 153467. 10.1016/j.aeue.2020.153467

Dellaoui, S., Kaabal, A., El Halaoui, M., & Asselman, A. (2018). Patch array antenna with high gain using EBG superstrate for future 5G cellular networks. *Procedia Manufacturing*, 22, 463–467. 10.1016/j.promfg.2018.03.071

Ghosh, S. (2023). Design of a highly-isolated, high-gain, compact 4-port MIMO antenna loaded with CSRR and DGS for millimeter wave 5G communications. *AEU - International Journal of Electronics and Communications, 169*, 154721.

Jain, K., & Kushwah, V. S. (2021). Design and development of dual band antenna for sub-6 frequency band application. *Materials Today: Proceedings*, 47, 6795–6798. 10.1016/j.matpr.2021.05.133

Kiani, S. H., Ibrahim, I. M., Savcı, H. Ş., Rafique, U., Alsunaydih, F. N., Alsaleem, F., Alhassoon, K., & Mostafa, H. (2024). Side-edge dual-band MIMO antenna system for 5G cellular devices. *AEÜ. International Journal of Electronics and Communications*, 173, 154992. 10.1016/j.aeue.2023.154992

Kim, S., & Nam, S. (2022). Wideband and Ultrathin 2×2 Dipole Array Antenna for 5G mm Wave Applications. *IEEE Antennas and Wireless Propagation Letters*, 21(12), 2517–2521. 10.1109/LAWP.2022.3199695

Leones Sherwin Vimalraj, S. (2023). Power allocation algorithm for capacity maximization in 5G MIMO systems. *Measurement: Sensors*, *30*, 100919.

Liu, P., Shen, C., Liu, C., Cintrón, F. J., Zhang, L., Cao, L., Rouil, R., & Roy, S. (2023). Towards 5G new radio sidelink communications: A versatile link-level simulator and performance evaluation. *Computer Communications*, 208, 231–243. 10.1016/j.comcom.2023.06.005

Malviya, L. (2020). *Highly isolated inset-feed 28 GHz MIMO-antenna array for 5G wireless application*. Research Gate.

Malviya, L., Parmar, A., Solanki, D., Gupta, P., & Malviya, P. (2020). Highly isolated inset-feed 28 GHz MIMO-antenna array for 5G wireless application. *Procedia Computer Science*, 171, 1286–1292. 10.1016/j.procs.2020.04.137

Mao, C. X., Zhang, L., Khalily, M., Gao, Y., & Xiao, P. (2021). A Multiplexing Filtering Antenna. *IEEE Transactions on Antennas and Propagation*, 69(8), 5066–5071. 10.1109/TAP.2020.3048589

Nahas, M. (2022). Design of a high-gain dual-band LI-slotted microstrip patch antenna for 5G mobile communication systems. *Journal of Radiation Research and Applied Sciences*, 15(4), 100483. 10.1016/j.jrras.2022.100483

Nissanov, U., & Singh, G. (2023). Grounded coplanar waveguide microstrip array antenna for 6G wireless networks. *Sensors International*, 4, 100228. 10.1016/j.sintl.2023.100228

Parchin, N. O., Basherlou, H. J., Al-Yasir, Y. I. A., & Abd-Alhameed, R. A. (2020). A broadband multiple-input multiple-output loop antenna array for 5G cellular communications. *AEÜ. International Journal of Electronics and Communications*, 127, 153476. 10.1016/j.aeue.2020.153476

Parveez Shariff, B. (2023). Planar MIMO antenna for mmWave applications: Evolution, present status & future scope. *Heliyon, 9*(2), e13362.

Rochman, M. I., Sathya, V., Fernandez, D., Nunez, N., Ibrahim, A. S., Payne, W., & Ghosh, M. (2023). A comprehensive analysis of the coverage and performance of 4G and 5G deployments. *Computer Networks*, 237, 10060. 10.1016/j.comnet.2023.110060

Sargam, S., Gupta, R., Sharma, R., & Jain, K. (2023). A Comprehensive review on 5G-based Smart Healthcare Network Security: Taxonomy, Issues, Solutions and Future research directions. *Array (New York, N.Y.)*, 18, 100290. 10.1016/j.array.2023.100290

Serghiou, D., Khalily, M., Singh, V., Araghi, A., & Tafazolli, R. (2020). Sub-6 GHz Dual-Band 8 × 8 MIMO Antenna for 5G Smartphones. *IEEE Antennas and Wireless Propagation Letters*, 19(9), 1546–1550. 10.1109/LAWP.2020.3008962

Shanbhag, J. (2023). Layered Division Multiplexing in 5G NR. *International Conference for Advancement in Technology (ICONAT)*, Goa, India.

Shen, X., Liu, Y., Zhao, L., Huang, G.-L., Shi, X., & Huang, Q. (2019). A Miniaturized Microstrip Antenna Array at 5G Millimeter-Wave Band. *IEEE Antennas and Wireless Propagation Letters*, 18(8), 1671–1675. 10.1109/LAWP.2019.2927460

Tiwari, R. Sharma, R., & Dubey, R. (2023). 2X2 & 4X4 dumbbell shape microstrip patch antenna array design for 5G Wi-Fi communication application. *Materials Today*.

Trabelsi, N., Fourati, L. C., & Chen, C. S. (2024). Interference management in 5G and beyond networks: A comprehensive survey. *Computer Networks*, 239, 110159. 10.1016/j.comnet.2023.110159

Trzebiatowski, K., Rzymowski, M., Kulas, L., & Nyka, K. (2021). Simple 60 GHz Switched Beam Antenna for 5G Millimeter-Wave Applications. *IEEE Antennas and Wireless Propagation Letters*, 20(1), 38–42. 10.1109/LAWP.2020.3038260

Vaezi, M., Ding, Z., & Poor, H. V. (2018). *Multiple Access Techniques for 5G Wireless Networks and Beyond*. Springer Cham publications.

Wang, B., Zhao, Z., Sun, K., Du, C., Yang, X., & Yang, D. (2023). Wideband Series-Fed Microstrip Patch Antenna Array With Flat Gain Based on Magnetic Current Feeding Technology. *IEEE Antennas and Wireless Propagation Letters*, 22(4), 834–838. 10.1109/LAWP.2022.3226461

Wen, L., Yu, Z., Zhu, L., & Zhou, J. (2021). High-Gain Dual-Band Resonant Cavity Antenna for 5G Millimeter-Wave Communications. *IEEE Antennas and Wireless Propagation Letters*, 20(10), 1878–1882. 10.1109/LAWP.2021.3098390

Yamamoto, S., Nuimura, S., & Takikawa, M. (2021). A Design Concept of Grid-loaded Step Reflector Antenna with Coaxial-Mode Excitation. *2020 International Symposium on Antennas and Propagation (ISAP)*, Osaka, Japan. 10.23919/ISAP47053.2021.9391172

Zicheng, Z., Keyan, T., & Yang, N. (2020). An FSS integrated with reflector antenna. *2020 9th Asia-Pacific Conference on Antennas and Propagation (APCAP)*, Xiamen, China. 10.1109/APCAP50217.2020.9246126

Section 4
Applications in Smart Grids and Cyber Defense

Chapter 11
Leveraging Fiber Bragg Grating Sensors for Enhanced Security in Smart Grids

Serif Ali Sadik
https://orcid.org/0000-0003-2883-1431
Kutahya Dumlupinar University, Turkey

ABSTRACT

Fiber Bragg grating (FBG) sensors have emerged as a promising technology for enhancing the monitoring and maintenance of smart grid systems. In this study, a comprehensive simulation-based investigation into the application of FBG sensors for monitoring critical components of the smart grid infrastructure, including high-voltage transmission towers, insulators, and overhead lines was presented. A wavelength division multiplexed FBG sensor network was designed and deployed to capture key parameters such as strain, temperature, and sag, enabling real-time monitoring and early detection of potential issues. The simulation results demonstrate the effectiveness of FBG sensors in accurately detecting changes in these parameters, with high sensitivity and reliability. The findings of this study underscore the importance of FBG sensors in enhancing the reliability, efficiency, and resilience of smart grid systems, paving the way for their widespread adoption in modern power networks.

DOI: 10.4018/979-8-3693-2786-9.ch011

1. INTRODUCTION

In the dynamic landscape of modern energy distribution, smart grids represent transformative infrastructures that use advanced technologies to optimize how electricity gets generated, distributed and consumed. Unlike traditional grids, smart grids operate on both on upstream and downstream flow of information, facilitating real-time communication among the smart grid components. This bidirectional communication is fundamental for achieving optimal efficiency and responsiveness. However, the complexity and interconnectivity inherent in smart grids also introduce new challenges, necessitating a comprehensive monitoring approach (Tuballa & Abundo, 2016).

The capacity to swiftly adjust to fluctuations in real time, generation and unforeseen events will determine the effectiveness of a smart grid. Continuous monitoring of grid components, such as power lines, transformers, and substations, becomes paramount in this context. Monitoring not only ensures the efficient utilization of resources but also identifies potential issues before they escalate, thereby minimizing downtime and enhancing overall operational reliability (Kabalci & Kabalci, 2017).

Undoubtedly, the advantages of 5G technology have created opportunities for great progress in communication systems of the monitoring and control of smart grids. One of the principal advantages of 5G technology is its capacity to transmit data at exceedingly high speeds, frequently reaching gigabit-per-second rates. This high-speed data transmission is crucial for smart grid systems, as it enables the rapid exchange of information between various grid components, including sensors, control systems, and the mainframe. In addition to the capacity for high-speed data transmission, 5G technology offers a reduction in the latency of wireless communications compared to previous generations. This refers to the time taken for data to travel between its source and destination. With 5G, this latency can be reduced to milliseconds, which enables a near-instantaneous communication between grid devices (Meng et al., 2019).

Another significant attribute of 5G technology is its capacity to facilitate extensive connectivity, enabling a considerable number of devices to be connected simultaneously. This objective is achieved via techniques such as network slicing, which enables the creation of virtual networks targeting specific applications or user groups. For example, a utility could create dedicated network slices for grid monitoring, distribution automation, and demand response, ensuring that each application receives the needs and quality of service (Dragičević et al., 2019; Leligou et al., 2018; Meng et al., 2019).

Figure 1. Representation of smart grid sensor network

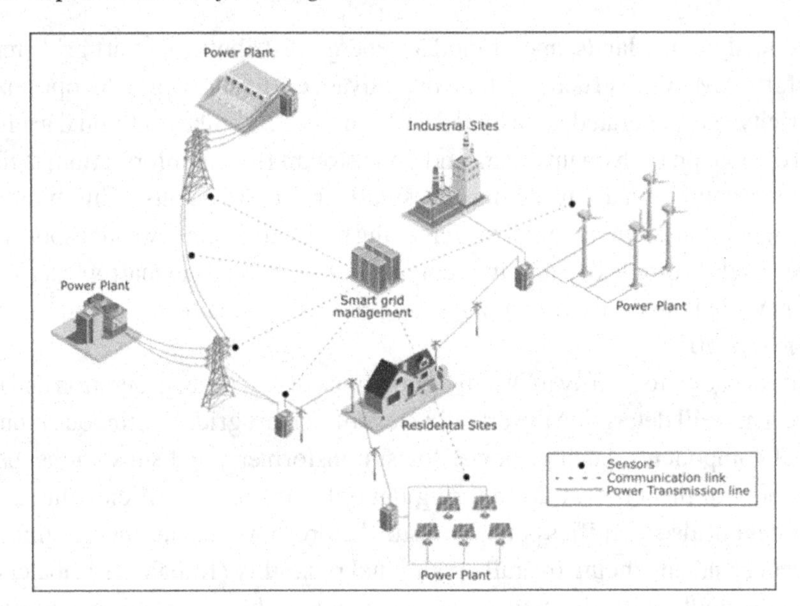

Smart grids demand a paradigm shift from reactive to proactive maintenance strategies. Monitoring technologies, such as Fiber Bragg Grating (FBG) sensors, play an important role in this transition. The provision of on-line data regarding the condition of critical infrastructure, these sensors enable utilities to schedule maintenance activities proactively, addressing potential faults and mitigating the risk of system failures. In an era where cybersecurity threats pose a significant risk to critical infrastructures, monitoring becomes a linchpin for grid resilience. The ability to detect and respond to anomalous activities or potential cyber intrusions is essential for safeguarding the integrity of smart grids (Figure 1). As we navigate the intricacies of monitoring technologies in this chapter, it is essential to recognize the critical role they play in ensuring the robustness and adaptability of smart grids in the face of evolving challenges (Huang et al., 2010).

In the evolving landscape of smart grids, FBG sensors are emerging as indispensable tools for advancing the realms of monitoring and security. Operating on a simple yet powerful principle, these sensors harness the capabilities of fiber optics to precisely capture data related to alterations in temperature, strain, or vibrations within the grid. The straightforward nature of FBG sensor technology conceals a wealth of potential applications that extend beyond mere simplicity (R. Zhang, 2013). The ability of FBG sensors to continuously monitor changes in their environment plays a crucial role in smart grid operations. Whether it be variations in temperature along power lines, strains on critical infrastructure components, or subtle vibrations indicative of potential faults, FBG sensors excel in providing real-time, high-precision data. This level of accuracy is fundamental for optimizing the performance of smart grids, ensuring timely responses to dynamic conditions (Elsayed & Gabbar, 2022).

FBG sensors are not confined to a singular application within smart grids; instead, they find versatile use across various components of the infrastructure. Power lines, essential conduits for energy transmission, benefit from ability of FBG sensors to detect temperature fluctuations and strains, allowing for proactive maintenance and minimizing downtime. Similarly, in substations, FBG sensors contribute to the identification of vibrations or strains on critical equipment, enhancing overall operational reliability (Mahin et al., 2022). Beyond their immediate applications, FBG sensors are instrumental in fortifying the resilience of smart grid infrastructures. By continuously monitoring and promptly detecting changes in the physical environment, these sensors play a crucial role in preventing potential failures or disruptions. This proactive approach contributes significantly to the overall reliability and security of smart grids, mitigating risks associated with both natural elements and potential cyber threats. Within the intricate web of smart grid elements, the monitoring of temperature, vibrations, and strain emerges as a critical necessity. Each of these parameters provides essential insights into the health and performance of key components, contributing to the overall efficiency and reliability of the smart grid (Chai et al., 2019; Juraszek, 2020; Kuhn et al., 2020).

Temperature variations along power lines can offer valuable information about their operational status. Elevated temperatures may indicate excessive load or the presence of faults, necessitating prompt intervention to prevent potential failures. By monitoring temperature, utilities can optimize energy transmission, identify areas prone to overheating, and proactively address issues to ensure uninterrupted power supply (El-Gammal et al., 2020; Floridia et al., 2013; Gangopadhyay et al., 2009; Kalizhanova et al., 2024).

Vibrations serve as early indicators of potential issues within smart grid equipment, especially in substations where various components operate in tandem. Anomalous vibrations may signal mechanical stress, misalignments, or impending failures. Timely detection allows for proactive maintenance, preventing equipment breakdowns and ensuring the continuous and safe functioning of critical elements within the grid (Min et al., 2018; Yan et al., 2022; Zou & Lu, 2021).

Monitoring strain on infrastructure components, such as power line towers and transformers, is essential for assessing their structural integrity. Strain variations may result from external forces, weather conditions, or potential tampering. Early detection of abnormal strain levels enables utilities to address structural concerns promptly, reducing the risk of catastrophic failures and enhancing the overall resilience of the smart grid (Juraszek, 2020; Luo et al., 2013; L. Zhang et al., 2023).

Collectively, the monitoring of temperature, vibrations, and strain supports a proactive maintenance paradigm in smart grids. By identifying potential issues before they escalate, utilities can implement targeted interventions, optimizing the lifespan of equipment and minimizing the risk of unplanned outages. Moreover, proactive maintenance enhances the overall resilience of the smart grid, ensuring its adaptability to changing conditions and minimizing vulnerabilities. As we delve deeper into the applications of monitoring technologies, particularly FBGs, the significance of tracking temperature, vibrations, and strain becomes increasingly evident. FBGs stand out as versatile and advantageous instruments, particularly when measuring temperature, vibrations, and strain in smart grid elements (Ramnarine et al., 2023).

FBG sensors exhibit a remarkable level of sensitivity and precision, making them well-suited for applications where accurate measurements are paramount. In temperature monitoring, FBG sensors can detect subtle variations with exceptional accuracy, providing real-time data crucial for optimizing the performance of power lines, transformers, and substations. Their high precision ensures that even minor changes in temperature are captured, allowing utilities to proactively address potential issues (Aimasso et al., 2023; F. Liu et al., 2023; Mishra et al., 2016).

When it comes to vibrations, FBG sensors offer the advantage of real-time monitoring with high resolution. The dynamic nature of smart grid equipment requires continuous and instant feedback on vibrations. FBG sensors, with their ability to capture rapid changes, enable utilities to identify and respond to anomalies promptly. This real-time monitoring capability is essential for preventing equipment failures and ensuring the longevity of critical components (Guo et al., 2023).

FBG sensors excel in measuring strain in structural elements, providing a reliable gauge of their integrity. The ability to accurately assess strain levels in power line towers, transformers, and other infrastructure components is crucial for preemptive maintenance. FBG sensors can detect even small increments in strain, allowing utilities to address structural concerns before they escalate, thereby contributing to the resilience of the smart grid (L. Zhang et al., 2023; Zhao et al., 2023).

One of the standout features of FBGs is their immunity to electromagnetic interference (EMI). In smart grid environments where electronic equipment and power lines generate electromagnetic fields, the interference can impact the accuracy of measurements. FBG sensors, being immune to EMI, provide reliable and interference-free data, ensuring the fidelity of the information gathered from the smart grid elements (Fang et al., 2019; Zhu et al., 2023).

FBG sensors enable long-distance distributed sensing, allowing for comprehensive coverage of smart grid components. This distributed sensing capability is particularly advantageous for large-scale infrastructures, such as extensive power transmission networks. By deploying FBG sensors strategically, utilities can monitor vast areas with a single sensor network, enhancing the overall cost-effectiveness of smart grid monitoring (Barrias et al., 2016; Gui et al., 2023).

2. LITERATURE REVIEW

Monitoring partial discharges (PD) resulting from insulation failure in a power system is a top priority in predictive maintenance and outage management. Researchers have explored various methods for detecting and measuring partial discharges in high-voltage (HV) equipment. The results of a pertinent study indicate the potential of an optical PD sensor based on FBG for the detection of PDs in a dielectric test cell, which may be regarded as a model of a transformer. The sensitivity of the FBG sensor was found to be 0.015 pm/pC. This indicates that the developed system is well-suited for PD detection, and the proposed method can be readily expanded for online PD detection in practical applications (Sarkar et al., 2015).

In 2017, Shi et al. proposed a novel approach for the detection of PDs utilizing FBG based on ultrasound detection in power transformers. The study involved a comprehensive analysis of the theoretical basis of FBG-based ultrasound detection, with experimental investigations conducted in transformer oil to analyze the performance of the novel methodology. Additionally, to showcase its sensing capabilities, PD sensing experiments were carried out using a distributed FBG array with varying center wavelengths. The study's findings indicate that the FBG ultrasonic detection system outperforms traditional piezoelectric transducers (PZT) sensors, and it demonstrates successful sensing coherence with different center wavelengths (Shi et al., 2017).

In a separate investigation focusing on PD detection, an experiment was conducted to compare the sensitivity performance of a FBG sensor with a length of 10 mm in PD ultrasonic detection. The findings revealed that within the range of 20 kHz to 500 kHz, the FBG sensor exhibited -79.503 dB sensitivity at 180 kHz. In comparison, the PZT sensor demonstrated as -83.5 dB sensitivity at 200 kHz. It can be inferred that the sensitivity of the 10-mm-length FBG sensor within the 20 kHz to 500 kHz frequency range was slightly superior for the compared to the conventional PZT sensor to identical ultrasonic signals (Zheng et al., 2015).

Another study focusing on PD detection, presented a sensor system utilizing phase-shifted FBGs (PS-FBGs). It was observed that the PS-FBG exhibited a sensitivity 8.46 dB higher than the traditional PZT sensor through frequency response measurement. Moreover, a comparative experiment was conducted with the objective of assessing the sensitivity of the PS-FBG for real-time monitoring. The findings revealed that when mounted on the external surface of the oil tank, the PS-FBG demonstrated sensitivity 4.5 times greater than that of the PZT sensor (G.-M. Ma et al., 2018).

In a study conducted in 2017, three FBG sensors were coated with polyether ether ketone (PEEK) and positioned on a transformer winding for axial force measurement. The sensor demonstrated a sensitivity of 0.133 pm/kPa, with a repeatability error of 2.7% FS (Y. Liu et al., 2017).

Kuhn et al. employed a FBG sensor to observe mechanical defects in a 3 kVA transformer. The application of nonlinear loads to the transformer induced harmonic distortions and elevated the mechanical vibration of the windings, effectively detected by the FBG sensor. While this laboratory-scale investigation implies the feasibility of winding monitoring, it also indicates that the bare installation of fiber may not be robust enough for prolonged, real-world installations (Kuhn et al., 2018).

In 2009, successful implementation of online monitoring for temperature and sag in a 400KV power transmission line was achieved. The sensor system, designed as a comprehensive device, was constructed using an FBG sensor with an aluminum mount. This system was connected via a fiber-optic cable and installed on the ACSR power conductor, enabling continuous measurements for two years (Gangopadhyay et al., 2009).

In another study, researchers tried to create a system to measure temperatures on high-voltage transmission lines and bus bars from a distance with FBG sensors along with free-space optics (FSO). The FSO system worked well within certain distances, similar to the length of insulator chains on power lines up to 230 kV. The system effectively monitored bus bar temperatures in a substation, even during interruptions. Although the system faced timeouts about 45% of the time, including a significant interruption lasting nearly 15 hours, most outages were shorter, lasting less than 18 minutes (Floridia et al., 2013).

The study conducted in 2020 presented a distributed online temperature monitoring system for overhead transmission lines (OTL). Continuous temperature monitoring was deemed crucial due to the harsh meteorological and geographical conditions these transmission lines endured, as high temperatures could lead to detrimental effects like flashovers, line breakages, and tower collapses. The system was based on a novel hybrid FBG technology, leveraging the advantages of FBG sensing over traditional electronic sensors. The hybrid FBGs demonstrated optimal performance in terms of peak reflectivity, full width at half maximum, side lobes analysis, and ripple factor. The proposed method demonstrated resilience to operational fluctuations in response to elevated temperatures, even when the parameters governing its control, such as the grating length and the amplitude of refractive index modulation, were subject to variation. The hybrid FBGs that were the focus of this study exhibited a performance profile that outstripped that of traditional uniform FBGs.

As a consequence, they have the potential to serve as more effective temperature sensors in OTL applications. (El-Gammal et al., 2020).

In a recent study, the application of an inclined FBG grid for monitoring wire extension in overhead lines was presented. The study illustrated that, through the careful selection of mechanical parameters for the extension transformer, consideration of optical parameters, and the use of a filter, the proposed system could effectively manage the sagging. In simulated measurements emulating the operation of a power line wire, the temperature was intentionally varied between 10-60°C. This temperature change resulted in the elongation of the test wire from 38.987 m to 49.275 m, with an error margin within 4% (Kalizhanova et al., 2024).

In a study focused on measuring sag on OTLs with an approach relied on chirped FBG sensors with a novel structure, ensuring insensitivity to the temperature, electromagnetic fields, and other external factors. The sensing head's structure, which included clamps and a specially designed steel plate to which the Chirped FBG was attached, achieved this insensitivity. The use of a FBG with a linearly variable period allowed for a more significant variation in the optical spectrum compared to standard FBGs, resulting in increased sensitivity and accuracy. The Full Width at Half Maximum (FWHM) width exhibited a sensitivity of 0.00603 nm/μm to elongation. Moreover, the FWHM was found to remain constant with respect to temperature fluctuations. These findings indicated that wire elongation could be determined independently of temperature variations, thus facilitating the study. (Wydra et al., 2018).

In 2020, the interrogation part of the aforementioned sag estimation system was presented. The study considered various interrogation methods utilizing adjusted filters. The experimental phase involved the examination of three types of FBG pairs, including those with a small shift in spectra and gratings with exact matching. The findings of this study indicate that, through the meticulous selection of mechanical parameters for the elongation transformer, sensor parameters and a filter, it is possible to tune the optical system to observe the sag of overhead line wires within the desired range. The extent of sag is influenced by factors such as the distance between poles, wire type and the actual length of the wire in the span, which effectively determines the observed sag (Skorupski et al., 2020).

Another external factor that affects OTLs and needs to be observed is dynamic stresses, including Aeolian vibrations and galloping. Bjerkan showcased the efficacy of fiber-optic sensors in measuring stress caused by galloping and vibrations on a high-power OTL in 2000. An experimental setup using three FBGs, one of them is the temperature reference, was deployed across a 60-kV line. Multiple recordings on a conductor were collected under different stress conditions caused by wind. Notably, the study frequently observed aeolian vibrations, and several measurements of this phenomenon were obtained. The outcomes demonstrated a strong correlation with straightforward theoretical predictions and visual observations (Bjerkan, 2000).

In a work focusing on OTL galloping, an optimization method was applied to the development of a high-frequency FBG acceleration sensor. Through vibration experiments, the study focused on sensing capabilities of the FBG acceleration sensor. The high-frequency FBG acceleration sensor exhibited high sensitivity and provided high accuracy according to experimental results (Zou & Lu, 2021).

In a more recent study, a distributed optical fiber sensing setup was presented and experimentally realized by combining the Φ-OTDR with a FBG array. With the proposed setup, wind effected galloping of OTLs successfully monitored in real-time. The enhanced Φ-OTDR sensing system demonstrated its utility in measuring galloping behavior, aiding in monitoring ice-induced galloping frequency response characteristics and assessing cable motion tendencies. The feasibility and reliability of the online monitoring system were verified through experiments, showcasing its effectiveness in predicting destructive vibration behavior, preventing accidents like cable breaking and tower toppling, and contributing to the safe operation of smart grids (Yan et al., 2022).

Overhead insulators prevent unwanted electrical leakage and ensure the efficient transmission of electricity. Further, insulator failures can lead to flashovers, where electricity arcs across the surface of the insulator. Monitoring their health ensures they are functioning properly in a smart grid structure. In a study conducted in 2011, the authors proposed a novel optical monitoring system based on FBG sensors. Their innovative design combined a shear beam FBG tension sensor for ice-induced strain detection with an additional element measuring line tilt angle, providing a more comprehensive picture of icing severity. This immune-to-electromagnetic-interference system was enabling early warning of potential line failures. Experimental results showcase the sensor's sensitivity, with successful detection of icing events on a test line. Overall, this FBG-based solution presents a promising alternative to traditional methods, potentially enhancing grid reliability and safeguarding against power outages. Notably, the paper reports a tension sensitivity of 0.0413 pm/N and a resolution of 24.21 N, while the sensitivity of the FBG sensor was reported as 16.17 pm/° for the tilt angle (G. Ma et al., 2011).

In 2013, researchers propose a novel online monitoring system utilizing the combined power of Brillouin scattering and FBG sensors. This hybrid system, a fusion of distributed Brillouin Optical Time Domain Reflectometry (BOTDR) for temperature mapping and point specific FBG strain sensing, provided a real-time picture of line health. The BOTDR, deployed within the line itself, showed temperature variations along its entire length, aiding in dynamic ampacity adjustments and proactive ice forecasting. The temperature measurement accuracy was found to be ±2°C. Meanwhile, FBG sensors strategically placed near insulators measured strain changes, providing early warning of potential line deformation due to excessive ice accumulation or strong winds. The measurements showed that the central wavelength would shift by 1 nm when a voltage of 107.3 kN was applied. This powerful duo operated with immunity to electromagnetic interference, delivering accurate and reliable data even in harsh environments. Experimental trials validated the system's effectiveness, showcasing its ability to detect icing events and monitor line tension with impressive precision. By harnessing the synergy of distributed and point-specific sensing, this novel approach paved the way for enhanced grid reliability and resilience, safeguarding against the perils of icy threats and ensuring uninterrupted power transmission (Luo et al., 2013).

Another study addressed the issue of decreased electrical performance in transmission line insulators during ice-covered conditions. The researchers employed interface FBG sensor setup with three sensors for icing loads on insulators. A more complex system with 13 FBGs placed on six optical fibers was proposed and implemented. Tests, including minimum load tests, FBG axial/environmental sensing tests, and weight simulation tests demonstrated accurate positioning of glaze and correct detection of icing in the laboratory condition (Hao et al., 2022).

In a more recent study, six optical fibers with FBGs were implanted into a 110 kV composite insulator, and three methods for measuring icing loads were proposed and tested. The results demonstrated accurate detection of serious icing on conductors and suggested the necessity of six optical fibers for precise measurement of both umbrella skirt and conductor icing (Hao et al., 2023).

The primary role of high-voltage transmission towers is to serve as support structures for transmission lines, ensuring their stability and functionality. These towers are designed to withstand various additional loads, including the weight of ice, the forces exerted by hurricanes, and the impact of seasonal winds. This structural resilience is crucial for maintaining the integrity of the power transmission infrastructure and preventing disruptions caused by external environmental factors. Thus, monitoring the health of high-voltage transmission towers is essential for ensuring the sustainability and resilience of power grid systems. By continuously assessing the condition of these support structures, potential issues or vulnerabilities can be detected early, allowing for timely interventions and maintenance. In this study, the vulnerability of high-voltage transmission towers in coastal areas to destructive hurricanes and seasonal winds was explored. The focus was on real-time condition monitoring systems to detect severe and critical damage at an early stage. The researchers investigated the potential for real-time monitoring of wind-induced vibration using an all-optical fiber sensing system. The optical sensing probe, created by splicing a thin-core fiber with tilt fiber gratings to a lead-in single mode fiber, demonstrated a linear response exceeding 97% in a range of 0.1-6.5 m/s^2 (Nan et al., 2020).

Another study introduced an online monitoring system for tower strain using FBG strain monitoring. It proposed both medium-short term and ultra-short term strain prediction models based on ARIMA and a combination of ARIMA-LSTM, providing a foundation for early warning of tower failure. The designed fixture facilitated sensor installation on a 500 kV transmission tower, and measured strain data of weak components were obtained. Comparison with the measured data demonstrated the accuracy of the proposed prediction models, with the ARIMA-LSTM combined model showing superior predictive performance. The research outcomes hold significant importance for the condition maintenance and disaster prevention of transmission lines (L. Zhang et al., 2023).

Table 1. Summary of recent studies on FBG utilization on monitoring smart grids

Element	Measurand	Reference
Transformer	Temperature	(Santamargarita et al., 2023; Shen et al., 2024)
	Partial Discharge	(G.-M. Ma et al., 2018; Sarkar et al., 2015; Shi et al., 2017)
	Mechanical Deformation	(Kuhn et al., 2018, 2020; Y. Liu et al., 2017)
Overhead Transmission Lines	Temperature	(El-Gammal et al., 2020; Floridia et al., 2013; Gangopadhyay et al., 2009; Kalizhanova et al., 2024)
	Sag	(Gangopadhyay et al., 2009; Skorupski et al., 2020; Wydra et al., 2018)
	Galloping or Vibrations	(Bjerkan, 2000; Yan et al., 2022; Zou & Lu, 2021)
Overhead Insulators	Icing	(Hao et al., 2022, 2023; G. Ma et al., 2011)
	Strain	(Luo et al., 2013)
Towers	Vibration	(Nan et al., 2020)
	Strain	(L. Zhang et al., 2023)

The brief literature summary also provided in Table 1, demonstrates that FBG sensors prove to be a highly effective and practical technology for monitoring the components constituting smart grids. Despite the exploration of alternative technologies, whether optical fiber-based or electronic device-based, for measuring parameters such as temperature, strain, and vibration on these elements, FBG sensors stand out for their versatility. Consequently, FBG sensors offer the capability for comprehensive multi-parameter condition monitoring of assets.

This study specifically aims to classify whether container ships are empty or full, a classification that, to the author's knowledge, has not been explored in the literature. The ability to identify and distinguish between empty and full container ships holds significant importance, influencing decision-making processes in shipping, logistics, and port management. Traditional classification methods often struggle with the complexities and scale of satellite imagery. Our approach leverages deep cognitive modeling, integrating the power of Convolutional Neural Networks (CNNs).

3. MATERIAL AND METHODS

3.1 FBG Sensing Fundamentals

At their core, FBG sensors operate on a straightforward yet powerful principle. These sensors consist of periodic variations in the refractive index along the core of an optical fiber, forming a grating structure. In the case where the refractive index of the core region of optical fibers is n_1, and the refractive index of the cladding region is n_2, there exists a relationship where $n_1 > n_2$ between the refractive indices. When a sufficiently powerful ultraviolet laser beam is divided into two and directed onto the germanium (Ge)-doped fiber using mirrors, creating an interference pattern, a periodic change occurs in the core refractive index. These periodic changes in the core refractive index are introduced at specific intervals (<1μm) using the masking technique, resulting in the formation of the FBG structure. This interval (Λ) is referred to as the grating period. The general structure of a FBG is illustrated in Figure 2. Under certain conditions known as the Bragg condition, a portion of the light incident on the optical fiber reflects from each fringe forming the grating and adds up to each other. The power of the reflected light reaches its maximum value at a specific wavelength depending on the grating period. This wavelength is called the Bragg wavelength (λ_B) and can be calculated as in Eq. 1.

Figure 2. Grating structure lies in FBG

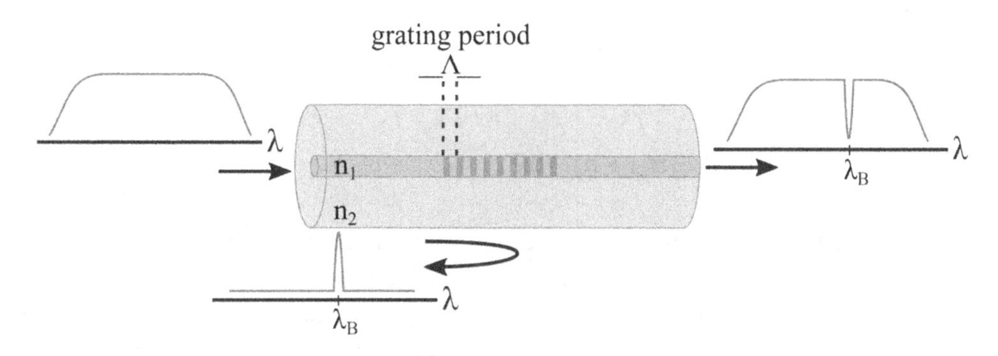

$$\lambda_B = 2n_{eff}\Lambda \tag{1}$$

In the equation, λ_B represents the Bragg grating center wavelength, n_{eff} represents the effective refractive index of the core region at the central wavelength, and Λ represents the grating period. Due to the response of the silica, which constitutes the structure of the fiber, to mechanical stresses and temperature, it is evident that the grating period will increase or decrease under these effects. Proportional to the change in the grating period, the center wavelength of the FBG will also undergo changes. This feature makes FBGs natural temperature or strain sensors (Jin et al., 2006).

The key to their functionality lies in the Bragg wavelength, which is determined by the spacing of these periodic variations. Figure 3 shows a basic experimental FBG sensor setup. When subjected to changes in temperature, strain, or vibrations, the Bragg wavelength of the FBG sensors shifts accordingly. This shift in wavelength serves as a precise indicator of the alterations occurring in the monitored elements.

Figure 3. Basic sensing principle of FBG sensors

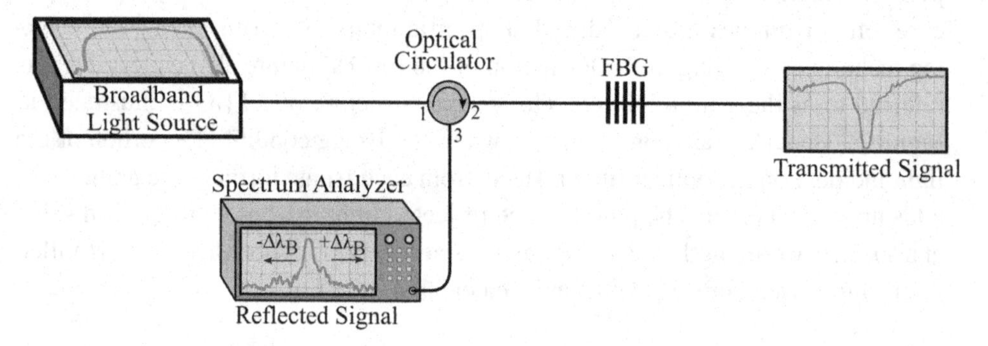

The equation below (Eq.2) illustrates how the alteration in ambient temperature or strain on the optical fiber leads to changes in the grating period and, consequently, the Bragg wavelength.

$$\Delta\lambda_B = \lambda_B(1 - p_e)\varepsilon + \lambda_B(\alpha_{TE} + \alpha_{TO})\Delta T \quad (2)$$

In the equation, $\Delta\lambda_B$ indicates the Bragg wavelength shift, p_e is the photo-elastic coefficient, ε is the longitudinal strain, α_{TE} and α_{TO} are thermal expansion coefficient and thermo-optic coefficient, respectively. Lastly, ΔT is the temperature shift (Yucel et al., 2016). By analyzing these wavelength shifts, FBG sensors provide accurate and real-time data on parameters such as temperature, strain, and vibrations. The versatility of FBG sensors allows for multi-parameter monitoring, enabling a comprehensive understanding of the condition of assets within the smart grid infrastructure.

3.2 FBG Sensor Network Setup for Comprehensive Smart Grid Monitoring

In the pursuit of analyzing the smart grid monitoring capabilities of FBGs, a wavelength division multiplexed FBG sensor network has been designed to capture and analyze critical parameters influencing the health and performance of key components within the smart grid infrastructure. To monitor the strain and temperature of the fundamental elements of the high-voltage transmission system, several FBG sensors were connected in a serial fashion and placed on the elements to be observed. Figure 4 shows the proposed FBG sensor system for monitoring smart grid infrastructure. Two FBG sensors were affixed to the high-voltage tower structure to measure strains induced by structural deformations. The FBGs were chosen to be insensitive to temperature to observe only the strain effects. Positioned on insulators were two FBGs employed for precise temperature measurements. The temperature-sensitive FBGs provide continuous monitoring of ambient conditions around the insulators, facilitating early identification of overheating or temperature-related stress. Two FBGs are deployed along the high-voltage line cable to monitor sag, a critical parameter influenced by load and environmental factors. Real-time sag data assist in assessing the cable's mechanical strain and predicting potential issues related to sag-induced stress. All six FBGs were interconnected and interrogated using the Wavelength Division Multiplexing (WDM) method. WDM allows for simultaneous data collection from multiple sensors, ensuring efficient and comprehensive monitoring of the smart grid's condition. The Bragg wavelength of the FBG sensors was chosen to be 1520, 1523, 1526, 1529, 1532, and 1535 nm. Furthermore, the reflectivity of the FBG sensors was chosen as 95%, ensuring a reflected signal with a high Optical Signal-to-Noise Ratio (OSNR). A broadband optical laser source was utilized for interrogation, and the reflected signals were analyzed with a spectrum analyzer.

Figure 4. The proposed FBG sensor array setup for smart grid infrastructure monitoring simulations

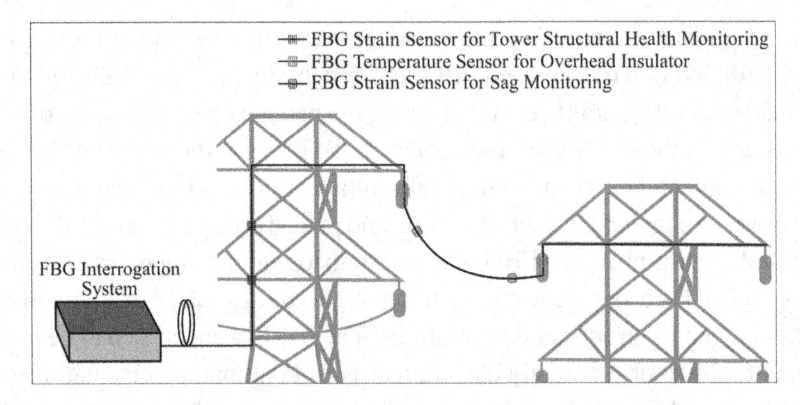

FBG sensor array interrogation simulations were run using OptiSystem software. OptiSystem is a powerful and comprehensive simulation software designed for optical communication system design and analysis. Developed by Optiwave, it provides a versatile platform for modeling and simulating various optical components, devices, and systems. With a user-friendly interface and advanced features, OptiSystem enables researchers and engineers to explore the behavior of optical networks, components, and sensors in a virtual environment.

In the context of our study, OptiSystem serves as the simulation environment for running FBG simulations. Leveraging the software's capabilities, we model and analyze the performance of FBG sensors embedded within the smart grid infrastructure. This simulation setup allows us to assess the impact of different parameters such as temperature, strain, and structural deformations on the FBGs, providing valuable insights into their suitability for smart grid monitoring applications.

4. SIMULATION RESULTS

The simulation results obtained from the proposed FBG sensor network setup in OptiSystem software provide valuable insights into the monitoring capabilities of smart grid components. In this section, we present the data acquired through simulations conducted using the established setup. The simulations were designed to emulate real-world scenarios and capture critical parameters influencing the health and performance of key elements within the smart grid infrastructure.

Figure 5. The reflection spectrum of the 6 FBG sensors for the initial state

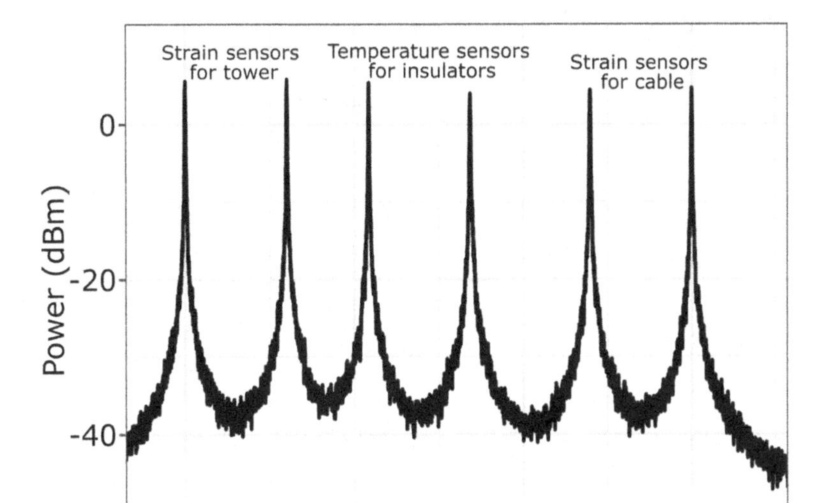

Figure 5 displays the reflection spectrum of the FBG sensors in their initial state. In iteration no. 1, no strain was applied to the tower and cable, and the temperature around the insulators was set to -20 °C as the starting point.

Figure 6. The reflection spectra of the FBG sensor placed on the power tower for increasing strain applied to the tower

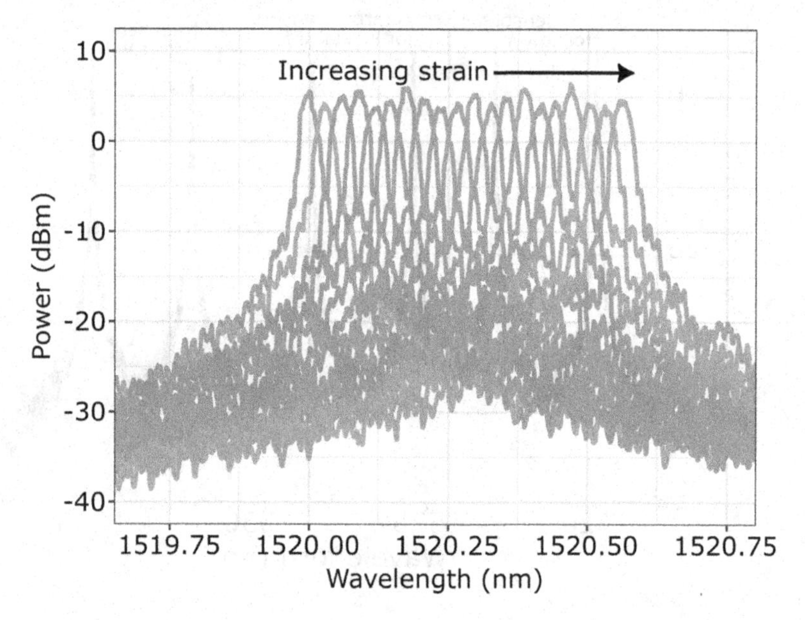

The reflection spectra of the first FBG sensor placed on the high-voltage tower are depicted in Figure 6. Strain was incrementally increased from 0 με to 475 με over 20 iterations. It is evident from the figure that as the strain increased the peak wavelength of the sensor shifted towards longer wavelengths.

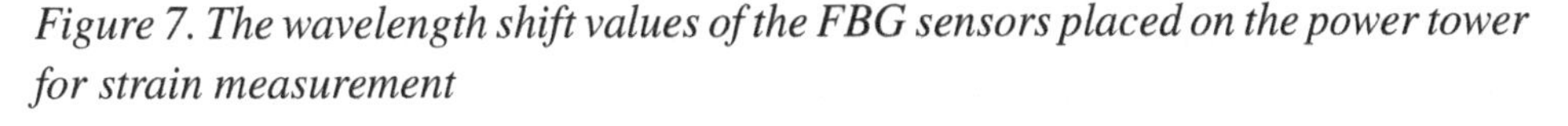

Figure 7. The wavelength shift values of the FBG sensors placed on the power tower for strain measurement

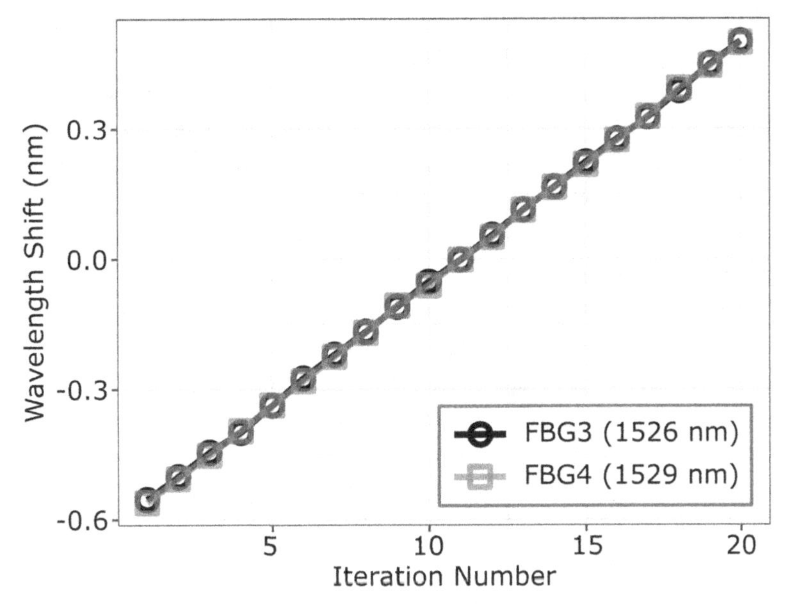

The Bragg wavelength shifts of the FBG sensors placed on the high-voltage power tower under increasing strain values are illustrated in Figure 7. One can see from the figure that, the wavelength shift increased linearly with the increasing strain. The Bragg wavelength shifts were measured as 0.556 nm and 0.561 nm for FBG1 and FBG2 under 475 με strain, respectively.

In the simulation study, the strain applied to the tower was incrementally increased from 0 με to 475 με over 20 iterations. From Eq. (2) the measured strain was calculated with Bragg wavelength shift measurements. Figure 8 displays the measured strain values obtained from the Bragg wavelength shift measurements for FBG1. It can be seen that the measured strain values exhibit a linear relationship with the actual strain values applied. The coefficient of determination (R^2) was found to be 0.99, indicating a strong linear correlation, and the root mean square error (RMSE) was calculated to be 3.92.

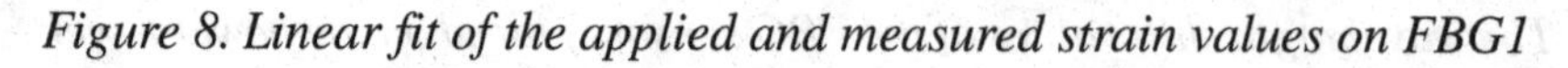

Figure 8. Linear fit of the applied and measured strain values on FBG1

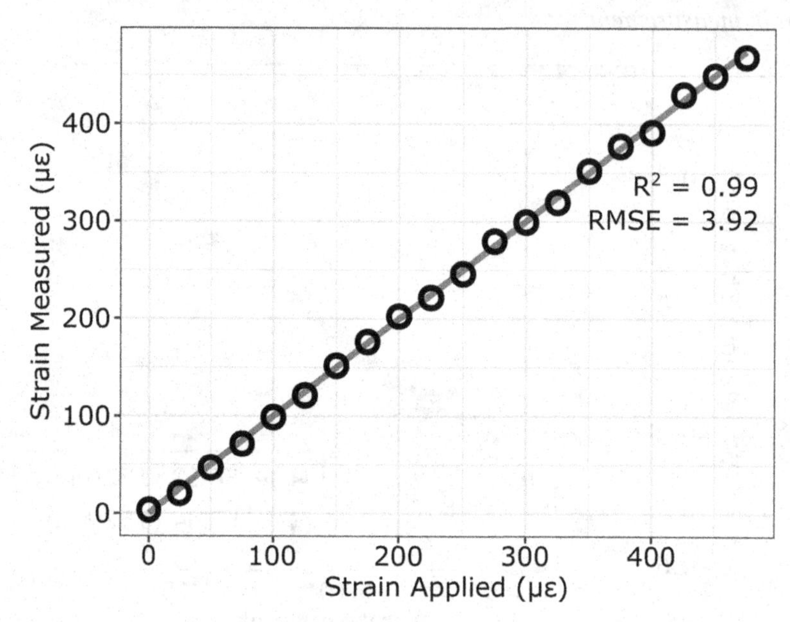

FBG sensors with Bragg wavelengths of 1526 nm and 1529 nm were installed on the overhead insulators to measure temperature. Figure 9 illustrates the reflection spectra of the FBG sensor with a Bragg wavelength of 1526 nm, showcasing temperature effects ranging from -20 °C to 56 °C across 20 iterations. The reference temperature was set to 20 °C, representing room temperature. As depicted in the figure, with increasing temperature, the Bragg wavelength of the sensor shifted towards longer wavelengths.

Figure 9. The reflection spectra of the FBG sensor placed on the overhead insulators for increasing temperature

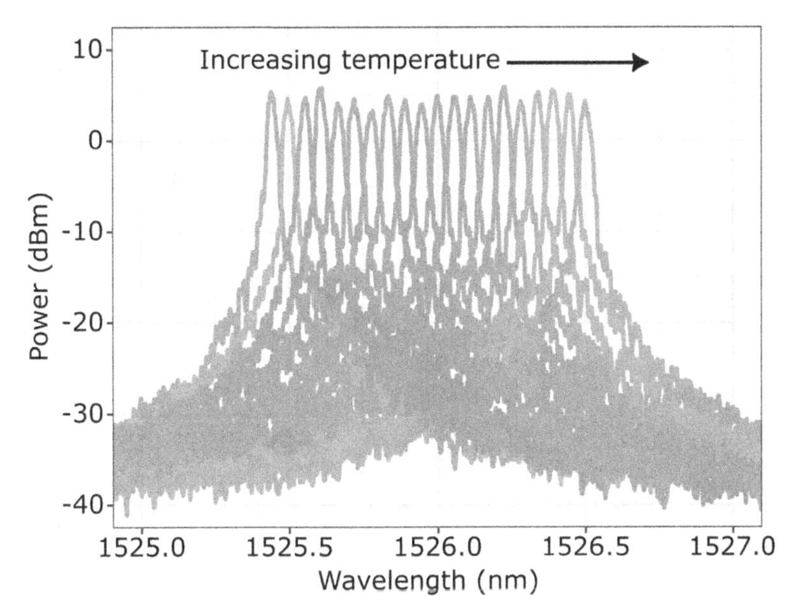

The Bragg wavelength shifts of the FBG sensors placed on the insulators under increasing temperature values are illustrated in Figure 10. One can see from the figure that the wavelength shift increased linearly with the increasing temperature. For the -20 °C temperature, the Bragg wavelength shifts were measured as -0.554 nm and -0.561 nm for FBG3 and FBG4, respectively. Further, for the 56 °C temperature, the Bragg wavelength shifts were measured as +0.501 nm and +0.498 nm for FBG3 and FBG4, respectively.

Figure 10. The wavelength shift values of the FBG sensors placed on the overhead insulators for temperature measurement

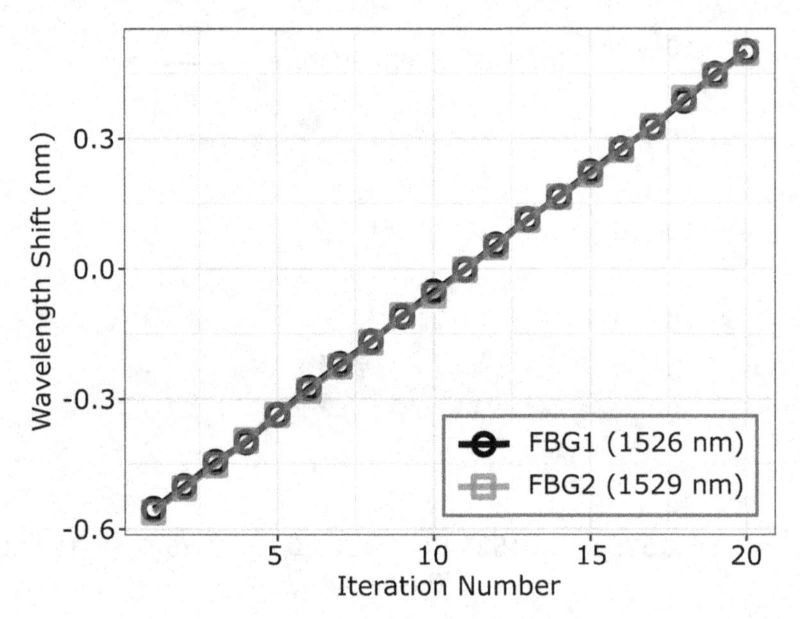

The temperature applied to the insulators was incrementally increased from -20 °C to +56 °C over 20 iterations. Using Eq. (2), the measured strain was calculated based on the Bragg wavelength shift measurements, while maintaining zero strain change to simulate a strain-insensitive FBG sensor. Figure 11 displays the measured temperature values obtained from the Bragg wavelength shift measurements for FBG3. It is evident that the measured strain values exhibit a linear relationship with the actual applied strain. The high R^2 value of 0.99 indicates a strong linear correlation, with an RMSE of 0.25.

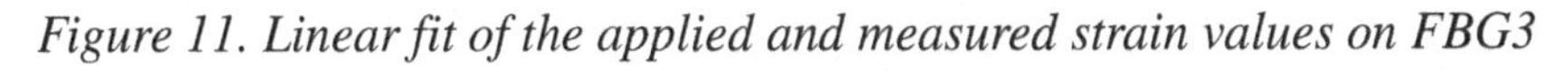

Figure 11. Linear fit of the applied and measured strain values on FBG3

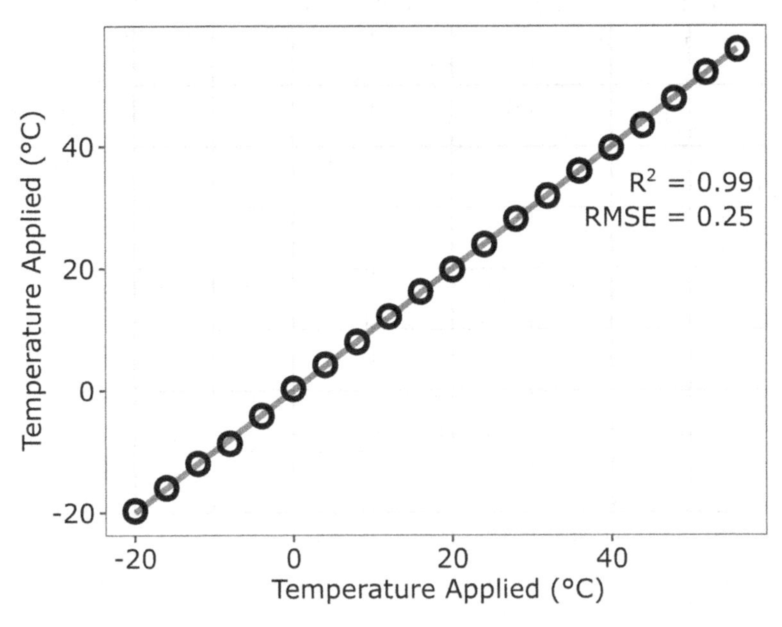

Figure 12. The reflection spectra of the FBG sensor placed on the overhead transmission lines under the effect of strain

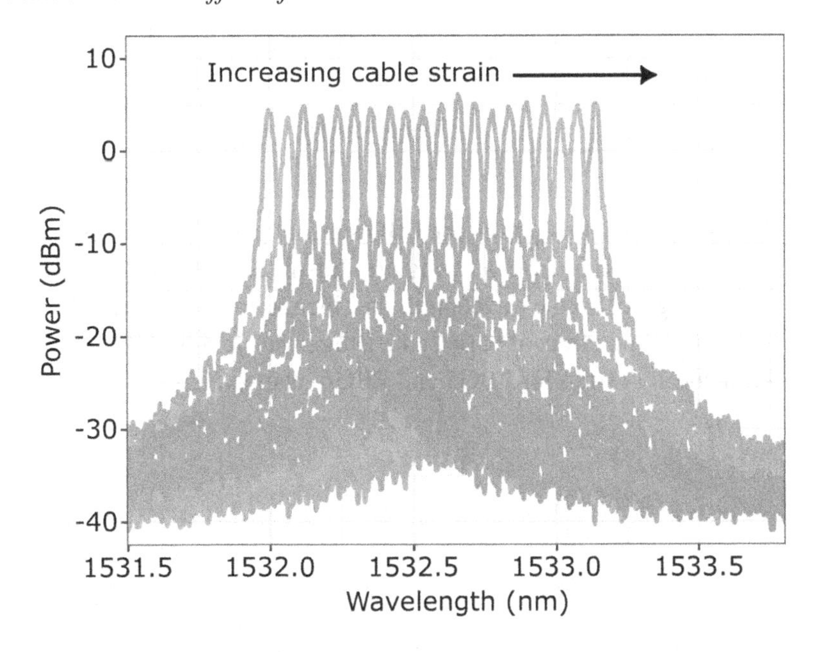

Lastly, FBG sensors with Bragg wavelengths of 1532 nm and 1535 nm were installed on the overhead transmission line to detect sagging. In Figure 12, the reflection spectra of the FBG sensor with a Bragg wavelength of 1532 nm was given. The shift of the Bragg wavelength indicates strain increase resulting from sagging. The strain variation was set from 0 με to 950 με over 20 iterations. As depicted in the figure, with increasing strain, the Bragg wavelength of the sensor shifted towards longer wavelengths.

Figure 13. The wavelength shift values of the FBG sensors placed on the overhead transmission lines for monitoring cable sagging

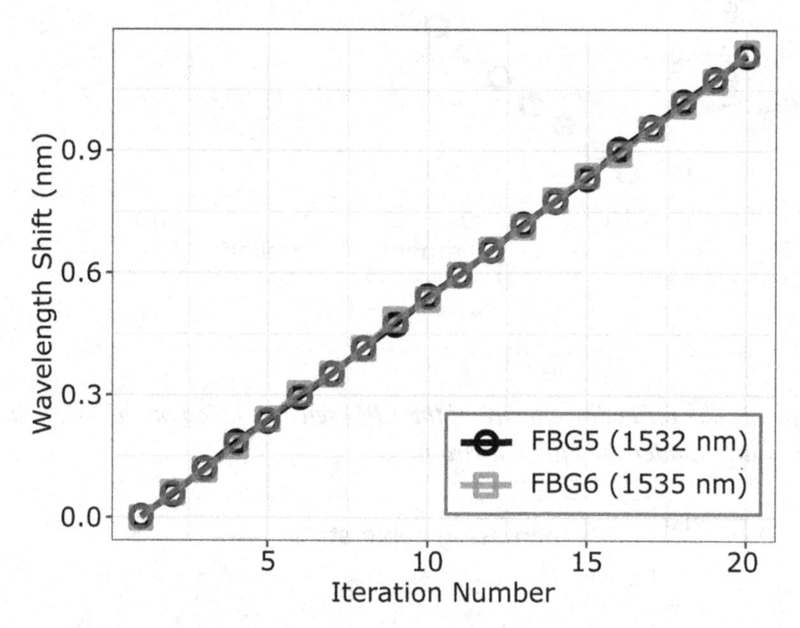

The Bragg wavelength shifts of the FBG sensors placed on the insulators under increasing strain resulting from sagging are illustrated in Figure 13. It can be observed from the figure that the wavelength shift increased linearly with the increasing strain. Under the strain effect of 950 με, the Bragg wavelength shifts were measured as 1.134 nm and 1.141 nm for FBG5 and FBG6, respectively.

Figure 14. Linear fit of the applied and measured strain values on FBG5

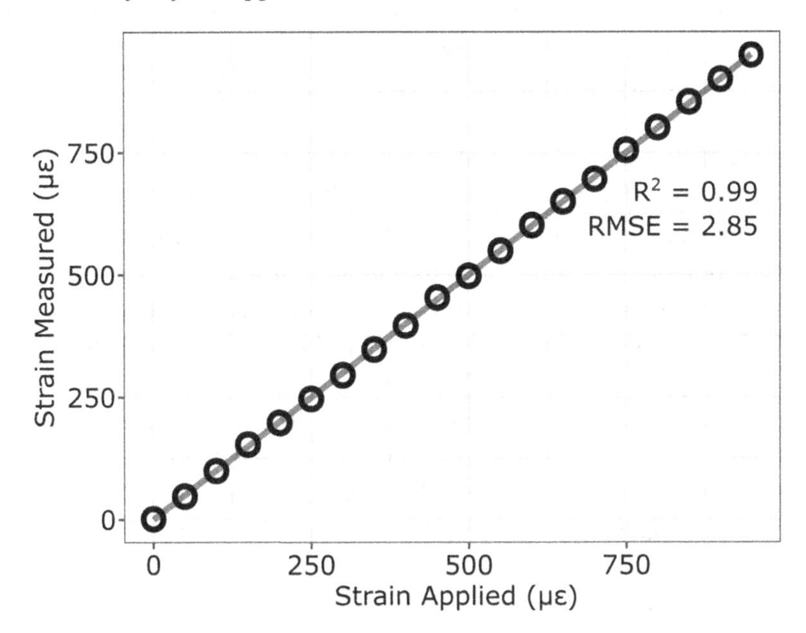

The strain applied to the overhead transmission lines was incrementally increased from 0 με to 950 με in 50 με steps. Figure 14 displays the measured strain values obtained from the Bragg wavelength shift measurements for FBG5. It is evident that these measured strain values exhibit a linear relationship with the actual applied strain. The high R^2 value of 0.99 indicates a strong linear correlation, with an RMSE of 2.85.

5. DISCUSSION

FBG sensors play a pivotal role in enhancing the monitoring and maintenance of critical components within smart grid systems. The real-time data provided by FBG sensors enables utilities to efficiently manage the operation of transmission lines, insulators, and other infrastructure, thereby ensuring grid reliability and resilience. By continuously monitoring parameters such as strain, temperature, and sag, FBG sensors facilitate early detection of potential issues, allowing utilities to implement proactive maintenance strategies and minimize downtime.

In this simulation study, we designed a comprehensive FBG sensor network to monitor key components of the smart grid infrastructure. The deployment of FBG sensors on high-voltage transmission towers, insulators, and overhead lines allowed us to capture critical data related to structural deformations, temperature variations, and mechanical strain. The simulation results demonstrated the effectiveness of FBG sensors in accurately detecting changes in these parameters simultaneously, with high sensitivity and reliability with the advantage of wavelength division multiplexing method.

Compared to traditional monitoring methods, FBG sensors offer several advantages in terms of accuracy, versatility, multiplexing capabilities and resilience to environmental conditions. Unlike conventional sensors, FBG sensors can simultaneously monitor multiple parameters and provide real-time data with high precision. Additionally, FBG sensors are immune to electromagnetic interference and offer long-term stability, making them ideal for deployment in harsh operating environments typical of smart grid systems.

In conclusion, FBG sensors represent a valuable tool for smart grid monitoring, offering unparalleled capabilities in capturing critical data for infrastructure health assessment. Our simulation study underscores the importance of FBG sensors in enhancing the reliability, efficiency, and resilience of smart grid systems. Continued research and innovation in FBG sensor technology will further advance the capabilities of smart grid monitoring, ultimately contributing to the sustainable and efficient operation of modern power networks.

6. LIMITATIONS OF THE PRESENT STUDY

While our simulation study yielded promising results, several limitations should be acknowledged. The placement of FBG sensors and calibration procedures may impact the accuracy of data collected, requiring careful consideration in practical implementations. Furthermore, future research should focus on enhancing the robustness and scalability of FBG sensor networks, as well as integrating advanced analytics techniques to extract actionable insights from the collected data.

REFERENCES

Aimasso, A., Vedova, M. D. L. D., & Maggiore, P. (2023). Sensitivity analysis of FBG sensors for detection of fast temperature changes. *Journal of Physics: Conference Series*, 2590(1), 012006. 10.1088/1742-6596/2590/1/012006

Barrias, A., Casas, J. R., & Villalba, S. (2016). A Review of Distributed Optical Fiber Sensors for Civil Engineering Applications. *Sensors (Basel)*, 16(5), 748. 10.3390/s1605074827223289

Bjerkan, L. (2000). Application of fiber-optic Bragg grating sensors in monitoring environmental loads of overhead power transmission lines. *Applied Optics*, 39(4), 554–560. 10.1364/AO.39.00055418337925

Chai, Q., Luo, Y., Ren, J., Zhang, J., Yang, J., Yuan, L., & Peng, G. (2019). Review on fiber-optic sensing in health monitoring of power grids. *Optical Engineering (Redondo Beach, Calif.)*, 58(7), 072007. 10.1117/1.OE.58.7.072007

Dragičević, T., Siano, P., & Prabaharan, S. R. (2019). Future generation 5G wireless networks for smart grid: A comprehensive review. *Energies*, 12(11), 2140. 10.3390/en12112140

El-Gammal, H. M., El-Badawy, E.-S. A., Rizk, M. R. M., & Aly, M. H. (2020). A new hybrid FBG with a π-shift for temperature sensing in overhead high voltage transmission lines. *Optical and Quantum Electronics*, 52(1), 53. 10.1007/s11082-019-2171-7

Elsayed, Y., & Gabbar, H. A. (2022). FBG Sensing Technology for an Enhanced Microgrid Performance. *Energies*, 15(24), 9273. Advance online publication. 10.3390/en15249273

Fang, X., Xiong, F., & Chen, J. (2019). An Experimental Study on Fiber Bragg Grating-Point Heat Source Integration System for Seepage Monitoring. *IEEE Sensors Journal*, 19(24), 12346–12352. 10.1109/JSEN.2019.2937155

Floridia, C., Rosolem, J. B., Leonardi, A. A., Hortencio, C. A., Fonseca, R. F., Moreira, R. O. C., Souza, G. C. L., Melo, A. L., & Nascimento, C. A. M. (2013). Temperature sensing in high voltage transmission lines using fiber Bragg grating and free-space-optics. *Proceedings of the Society for Photo-Instrumentation Engineers*, 8722, 87220N. 10.1117/12.2017934

Gangopadhyay, T. K., Paul, M. C., & Bjerkan, L. (2009). Fiber-optic sensor for real-time monitoring of temperature on high voltage (400KV) power transmission lines. *Proceedings of the Society for Photo-Instrumentation Engineers*, 7503, 75034M. 10.1117/12.835447

Gui, X., Li, Z., Fu, X., Guo, H., Wang, Y., Wang, C., Wang, J., & Jiang, D. (2023). Distributed Optical Fiber Sensing and Applications Based on Large-Scale Fiber Bragg Grating Array [Review]. *Journal of Lightwave Technology*, 41(13), 4187–4200. 10.1109/JLT.2022.3233707

Guo, Y., Mao, X., Tong, X., & Wu, H. (2023). A real time digital vibration acceleration fiber sensing system based on a multi-carrier modulation/demodulation technique. *Optics & Laser Technology*, 167, 109724. 10.1016/j.optlastec.2023.109724

Hao, Y., Huang, L., Wei, J., Liang, W., Pan, R., & Yang, L. (2023). The Detecting System and Method of Quasi-Distributed Fiber Bragg Grating for Overhead Transmission Line Conductor Ice and Composite Insulator Icing Load. *IEEE Transactions on Power Delivery*, 38(3), 1799–1809. 10.1109/TPWRD.2022.3222774

Hao, Y., Huang, L., Wei, J., Pan, R., Zhang, W., & Yang, L. (2022). Interface quasi-distributed fibre Bragg grating positioning detection of glaze icing load on composite insulators. *IET Science, Measurement & Technology*, 16(5), 316–325. 10.1049/smt2.12106

Huang, Q., Zhang, C., Liu, Q., Ning, Y., & Cao, Y. (2010). New type of fiber optic sensor network for smart grid interface of transmission system. *IEEE PES General Meeting*, (pp. 1–5). IEEE. 10.1109/PES.2010.5589596

Jin, L., Zhang, W., Zhang, H., Liu, B., Zhao, J., Tu, Q., Kai, G., & Dong, X. (2006). An embedded FBG sensor for simultaneous measurement of stress and temperature. *IEEE Photonics Technology Letters*, 18(1), 154–156. 10.1109/LPT.2005.860046

Juraszek, J. (2020). Fiber Bragg Sensors on Strain Analysis of Power Transmission Lines. *Materials (Basel)*, 13(7), 1559. 10.3390/ma1307155932230998

Kabalci, Y., & Kabalci, E. (2017). Modeling and analysis of a smart grid monitoring system for renewable energy sources. *Solar Energy*, 153, 262–275. https://doi.org/https://doi.org/10.1016/j.solener.2017.05.063. 10.1016/j.solener.2017.05.063

Kalizhanova, A., Kunelbayev, M., Wojcik, W., & Kozbakova, A. (2024). Experimental study of a temperature measurement system for an overhead power line using sensors based on TFBG. *International Journal of Innovative Research and Scientific Studies*, 7(1), 180–188. 10.53894/ijirss.v7i1.2596

Kuhn, G. G., de Morais Sousa, K., & da Silva, J. C. C. (2018). Dynamic Strain Analysis of Transformer Iron Core with Fiber Bragg Gratings. *Advanced Photonics 2018 (BGPP, IPR, NP, NOMA, Sensors, Networks, SPPCom, SOF)*, JTu2A.74. https://opg.optica.org/abstract.cfm?URI=Sensors-2018-JTu2A.74

Kuhn, G. G., Sousa, K. M., Martelli, C., Bavastri, C. A., & da Silva, J. C. C. (2020). Embedded FBG Sensors in Carbon Fiber for Vibration and Temperature Measurement in Power Transformer Iron Core. *IEEE Sensors Journal*, 20(22), 13403–13410. 10.1109/JSEN.2020.3005884

Leligou, H. C., Zahariadis, T., Sarakis, L., Tsampasis, E., Voulkidis, A., & Velivassaki, T. E. (2018). Smart grid: a demanding use case for 5G technologies. *2018 IEEE International Conference on Pervasive Computing and Communications Workshops (Percom Workshops)*, (pp. 215–220). IEEE. 10.1109/PERCOMW.2018.8480296

Liu, F., Wei, S., Li, B., Tan, Y., Guo, X., & Fu, X. (2023). A novel fast response and high precision water temperature sensor based on Fiber Bragg Grating. *Optik (Stuttgart)*, 289, 171257. 10.1016/j.ijleo.2023.171257

Liu, Y., Li, L., Zhao, L., Wang, J., & Liu, T. (2017). Research on a new fiber-optic axial pressure sensor of transformer winding based on fiber Bragg grating. *Photonic Sensors*, 7(4), 365–371. 10.1007/s13320-017-0427-z

Luo, J., Hao, Y., Ye, Q., Hao, Y., & Li, L. (2013). Development of Optical Fiber Sensors Based on Brillouin Scattering and FBG for On-Line Monitoring in Overhead Transmission Lines. *Journal of Lightwave Technology*, 31(10), 1559–1565. 10.1109/JLT.2013.2252882

Ma, G., Li, C., Quan, J., Jiang, J., & Cheng, Y. (2011). A Fiber Bragg Grating Tension and Tilt Sensor Applied to Icing Monitoring on Overhead Transmission Lines. *IEEE Transactions on Power Delivery*, 26(4), 2163–2170. 10.1109/TPWRD.2011.2157947

Ma, G.-M., Zhou, H.-Y., Shi, C., Li, Y.-B., Zhang, Q., Li, C.-R., & Zheng, Q. (2018). Distributed Partial Discharge Detection in a Power Transformer Based on Phase-Shifted FBG. *IEEE Sensors Journal*, 18(7), 2788–2795. 10.1109/JSEN.2018.2803056

Mahin, A. U., Islam, S. N., Ahmed, F., & Hossain, Md. F. (2022). Measurement and monitoring of overhead transmission line sag in smart grid: A review. *IET Generation, Transmission & Distribution, 16*(1), 1–18.

Meng, S., Wang, Z., Tang, M., Wu, S., & Li, X. (2019). Integration application of 5g and smart grid. *2019 11th International Conference on Wireless Communications and Signal Processing (WCSP)*, 1–7.

Min, L., Li, S., Zhang, X., Zhang, F., Sun, Z., Wang, M., Zhao, Q., Yang, Y., & Ma, L. (2018). The Research of Vibration Monitoring System for Transformer Based on Optical Fiber Sensing. *2018 IEEE 3rd Optoelectronics Global Conference (OGC)*, (pp. 126–129). IEEE. 10.1109/OGC.2018.8529978

Mishra, V., Lohar, M., & Amphawan, A. (2016). Improvement in temperature sensitivity of FBG by coating of different materials. *Optik (Stuttgart)*, 127(2), 825–828. 10.1016/j.ijleo.2015.10.014

Nan, Y., Xie, W., Min, L., Cai, S., Ni, J., Yi, J., Luo, X., Wang, K., Nie, M., Wang, C., Peng, G.-D., & Guo, T. (2020). Real-Time Monitoring of Wind-Induced Vibration of High-Voltage Transmission Tower Using an Optical Fiber Sensing System. *IEEE Transactions on Instrumentation and Measurement*, 69(1), 268–274. 10.1109/TIM.2019.2893034

Ramnarine, V., Peesapati, V., & Djurović, S. (2023). Fibre Bragg Grating Sensors for Condition Monitoring of High-Voltage Assets: A Review. *Energies*, 16(18), 6709. Advance online publication. 10.3390/en16186709

Santamargarita, D., Molinero, D., Bueno, E., Marrón, M., & Vasić, M. (2023). On-Line Monitoring of Maximum Temperature and Loss Distribution of a Medium Frequency Transformer Using Artificial Neural Networks. *IEEE Transactions on Power Electronics*, 38(12), 15818–15828. 10.1109/TPEL.2023.3308613

Sarkar, B., Koley, C., Roy, N. K., & Kumbhakar, P. (2015). Condition monitoring of high voltage transformers using Fiber Bragg Grating Sensor. *Measurement*, 74, 255–267. 10.1016/j.measurement.2015.07.014

Shen, T., Yuan, Y., Chen, Q., Xu, S., Zheng, W., Xiao, H., Liu, D., Liu, D., & Zhao, S. (2024). Fiber optic sensor for transformer temperature detection. *Microwave and Optical Technology Letters*, 66(1), e33813. 10.1002/mop.33813

Shi, C., Ma, G., Mao, N., Zhang, Q., Zheng, Q., Li, C., & Zhao, S. (2017). Ultrasonic detection coherence of fiber Bragg grating for partial discharge in transformers. *2017 IEEE 19th International Conference on Dielectric Liquids (ICDL)*, (pp. 1–4). IEEE. 10.1109/ICDL.2017.8124639

Skorupski, K., Harasim, D., Panas, P., Cięszczyk, S., Kisała, P., Kacejko, P., Mroczka, J., & Wydra, M. (2020). Overhead Transmission Line Sag Estimation Using the Simple Opto-Mechanical System with Fiber Bragg Gratings—Part 2: Interrogation System. *Sensors (Basel)*, 20(9), 2652. Advance online publication. 10.3390/s2009265232384715

Tuballa, M. L., & Abundo, M. L. (2016). A review of the development of Smart Grid technologies. *Renewable & Sustainable Energy Reviews*, 59, 710–725. 10.1016/j.rser.2016.01.011

Wydra, M., Kisala, P., Harasim, D., & Kacejko, P. (2018). Overhead Transmission Line Sag Estimation Using a Simple Optomechanical System with Chirped Fiber Bragg Gratings. Part 1: Preliminary Measurements. *Sensors (Basel)*, 18(1), 309. 10.3390/s1801030929361714

Yan, Q., Zhou, C., Feng, X., Deng, C., Hu, W., & Xu, Y. (2022). Galloping Vibration Monitoring of Overhead Transmission Lines by Chirped FBG Array. *Photonic Sensors*, 12(3), 220310. 10.1007/s13320-021-0651-4

Yucel, M., Ozturk, N. F., & Gemci, C. (2016). Design of a Fiber Bragg Grating multiple temperature sensor. *2016 Sixth International Conference on Digital Information and Communication Technology and Its Applications (DICTAP)*, (pp. 6–11). IEEE. 10.1109/DICTAP.2016.7543992

Zhang, L., Ruan, J., Du, Z., Huang, D., & Deng, Y. (2023). Transmission line tower failure warning based on FBG strain monitoring and prediction model. *Electric Power Systems Research*, 214, 108827. 10.1016/j.epsr.2022.108827

Zhang, R. (2013). Application of optical fiber sensors in Smart Grid. *Proceedings of the Society for Photo-Instrumentation Engineers*, 9044, 90440J. 10.1117/12.2037572

Zhao, J., Dong, W., Hinds, T., Li, Y., Splain, Z., Zhong, S., Wang, Q., Bajaj, N., To, A., Ahmed, M., Petrie, C. M., & Chen, K. P. (2023). Embedded Fiber Bragg Grating (FBG) Sensors Fabricated by Ultrasonic Additive Manufacturing for High-Frequency Dynamic Strain Measurements. *IEEE Sensors Journal*, 1, 1. 10.1109/JSEN.2023.3343604

Zheng, Q., Ma, G., Jiang, J., Li, C., & Zhan, H. (2015). A comparative study on partial discharge ultrasonic detection using fiber Bragg grating sensor and piezoelectric transducer. *2015 IEEE Conference on Electrical Insulation and Dielectric Phenomena (CEIDP)*, (pp. 282–285). IEEE. 10.1109/CEIDP.2015.7352071

Zhu, M., Zhang, Y., Zhao, W., Hu, Y., Wan, H., Li, K., & Zhou, A. (2023). Temperature Measurement of Pulsed Inductive Coil Continuous Discharge Based on FBG. *IEEE Sensors Journal*, 23(19), 22524–22532. 10.1109/JSEN.2023.3305092

Zou, H., & Lu, M. (2021). Developing High-Frequency Fiber Bragg Grating Acceleration Sensors to Monitor Transmission Line Galloping. *IEEE Access : Practical Innovations, Open Solutions*, 9, 30893–30897. 10.1109/ACCESS.2021.3055820

Chapter 12
Privacy Preserving Data Aggregation Algorithm for IoT-Enabled Advanced Metering Infrastructure Network in Smart Grid

Subaselvi Sundarraj
https://orcid.org/0000-0002-2084-1232
M. K. College of Engineering, India

ABSTRACT

The integration of emerging IoT (internet of things) technologies into utility control centers have data exchange between smart appliances, smart meters (SM), data collector (DC) and control center server (CCS). The DC controls the receiving and processing of advanced metering infrastructure (AMI) applications data from multiple SM. To address the issues associated with DC, SM are proposed to act as a relay device. The SM face communication challenges during peak hours when a significant amount of data with varying traffic rates and latency is exchanged within the utility control center. The challenges related to AMI in the context of smart grids focus the role of DC and hybrid data aggregation strategy. A hybrid data aggregation strategy is implemented on a cluster head aggregator (CA) within a clustering topology and the aggregated data are sliced to ensure privacy is proposed. In the proposed algorithm, CA reduces the workload of cluster-head (CH), targets interval meter reading (IMR) application data for aggregation, and efficiently utilizes the constrained resources of AMI devices are evaluated using the network simulator.

DOI: 10.4018/979-8-3693-2786-9.ch012

1. INTRODUCTION

Smart Grid (SG) has emerged as a major innovation in the field of energy management. A significant evolution in SG from traditional electrical grids is integrating modern information and communication technologies to create a more efficient, reliable, and sustainable electricity system. The SG is an electricity network that employs digital technology to monitor, control, and analyze the dynamics of power supply and demand. Unlike conventional grids, SGs are equipped with smart meters, and IoT (Internet of Things) that provide real-time information and control over the electricity flow. The technological integration enables the grid to respond quickly when changes in an electricity demand and supply the optimizing energy distribution and consumption. One of the primary benefits of SG is the ability to enhance energy efficiency by providing detailed information about energy usage patterns.

The SGs are pivotal in the integration of renewable sources based on energy and facilitate a smoother transition to green energy. The Smart Meters (SM) allow consumers to monitor their energy consumption in real-time by encouraging energy-saving behaviors leads to cost savings for the consumer but also contributes to the overall efficiency of the grid. Despite their advantages, SG face challenges such as cyber security risks, high implementation costs, and the need for regulatory frameworks. Addressing these challenges is crucial for the widespread adoption of SG. Thus, the SG represents a transformative approach for energy management aligning with the needs of a more digitalized, sustainable, and efficient future. The ability of SG to integrate advanced technologies, empower consumers, and support sustainable energy sources positions in a more resilient and environmentally friendly energy ecosystem.

A mixed Radio Frequency (RF) in a network with 3 hops in a Free space Optics (FSO) channel with RF assisted Reconfigurable Intelligent Surface (RIS) is formulated. The SM transfer the data through RIS assisted RF from the systems to the Data Aggregator Unit (DAU). The Gamma-Gamma distribution and Saleh-Valenzuela (S-V) are used to model the FSO and RF links. The coverage and performance of the system are improved using decode-and-forward and RIS relays (A. K. Padhan, et al., 2023, pp. 48 – 59). A case study as a practical application in bangladesh is performed in SM based IoT. The SM maintain the power quality and peak-clipping by serving as a bidirectional data transmission, DSM, local and online monitoring at the consumer side (M. T. Ahammed and I. Khan, 2022).

Machine learning in IoT is proposed to develop the smart city and the cities allow to build efficient system, urban network, economic power and unique solutions. The IoT is the core for wireless networks, sensor network, telecommunication network, mesh network, broadcast network and mesh network (T. M. Ghazal, et al., 2023, pp. 1953–1968). The Quality of Service (QoS) aware in smart grids for machine learning based AMI applications was proposed. The clustering approach based on machine learning and scheduling approach based on priority based hierarchical architecture is formed for AMI applications. A three tier architecture for cloud infrastructure and IoT communication has SM over the area with control center to network devices. The devices are controlled digitally and monitored over REST APIs (A. Khan, et al., 2021, pp. 1 – 22). In 5G network, the SISO and MIMO channel implemented that offer QoS and power efficiency by designing an optimization based power allocation (S. Yadav and S. Nanivadekar, 2023, pp. 1- 16).

A grid model with limited DAP as a relay deployment scheme for increasing the communication quality is proposed (T.-W. Sung, et al., 2022, pp. 3189–3201). To reduce the latency and congestion level in the AMI network, an aggregation policy at the forwarding nodes are implemented to reduce network delays in order to get necessary information (U. Das and V. Namboodiri, 2019, pp. 245–256). The fog computing technique based on an area and local service components of multi agent AMI is proposed. The problems can be solved by creating scalable solutions by fog computing meets the requirements of AMI in smart grid. Agent based design executes the AMI operations in distributed architecture that improve the performance of AMI solution (I. Popović, et al., 2022, pp. 1–22). In the smart metering cellular network for increasing transmission and smart meter cell throughput, secondary spectrum paired channels for metering through a wireless medium is assigned to optimize the channel assignment in dynamic that explore spectrum reuse in a mobile operator (E. Inga, et al., 2023, pp. 1–14). The IoT with SM and Edge fog cloud computing architecture is proposed to extract data and develop applications. To reduce delay, time, load and utilization interaction between the layers occurs in which edge layer monitor and control IoT applications are performed at fog based layer (S.-V. Oprea and A. Bˇara, 2023, pp. 818-845).

To achieve high accuracy in clustering process a local privacy is enabled and differential privacy is adopted (Mengmeng Yang, et al., 2022, pp. 2524–2537). A comprehensive survey about social network and data anonymization is described (Abdul Majeed and Sungchang Lee, 2020, pp. 8512-8545). An efficient block chain for VANETs is introduced to ensure privacy and security to a real world setting data in the network (Xiaotong Zhou, et al., 2023, pp. 81-92). Dynamic privacy in advanced networks requires innovative approaches to manage and protect information as it continuously moves and changes (Tianqing Zhu, et al., 2020, pp. 2962–2974).

The contribution of proposed algorithm data aggregation for IoT enabled AMI in smart grid is explained as follows

1. Cluster head Aggregator (CA) aggregate the data from the SM within a clustering topology.

2. The data from the SM is sliced into pieces and send to the neighbor nodes. All received data are combined together with own data to form new data to ensure privacy.
3. Each Cluster-Head (CH) has a restricted set of functions including traffic classification, queuing, and relaying AMI smart grid application data's between the Data Collector (DC) and Control Center Server (CCS).
4. An optimization problem is formulated through mathematical modeling using an objective function to optimize the limited resources of CH includes CPU processing, memory, and bandwidth.
5. The proposed algorithm is compared with the existing algorithm and the efficiency of the proposed strategy is evaluated using the optimization of constraint resources, minimization of CH workload, and the maintenance of Quality of Service (QoS) for AMI applications.

2. PROPOSED NETWORK MODEL

Figure 1. AMI network in smart grid

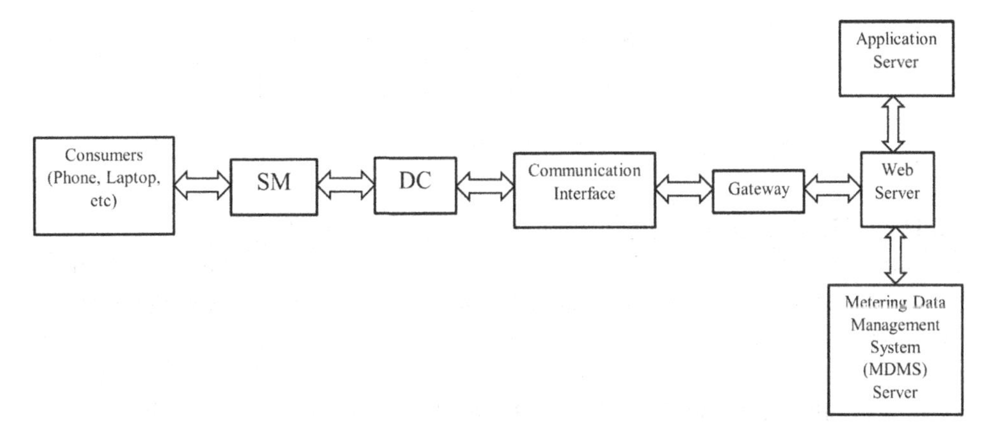

In SG, the AMI facilitates bidirectional data exchange between consumers and utility providers. The communication enables an array of innovative solicitations that improve the functionality and efficiency of the grid compared to traditional systems. The AMI network in smart grid is shown in Figure 1. The IMR allow installed SM in households to transmit electricity consumption data at regular intervals for every 15 to 60 minutes throughout the day including peak hours. The collected data is then sent over the backhaul network to the Control Center Server (CCS) for processing and in the Metering Data Management System (MDMS) server the data is stored.

Figure 2. Proposed network model

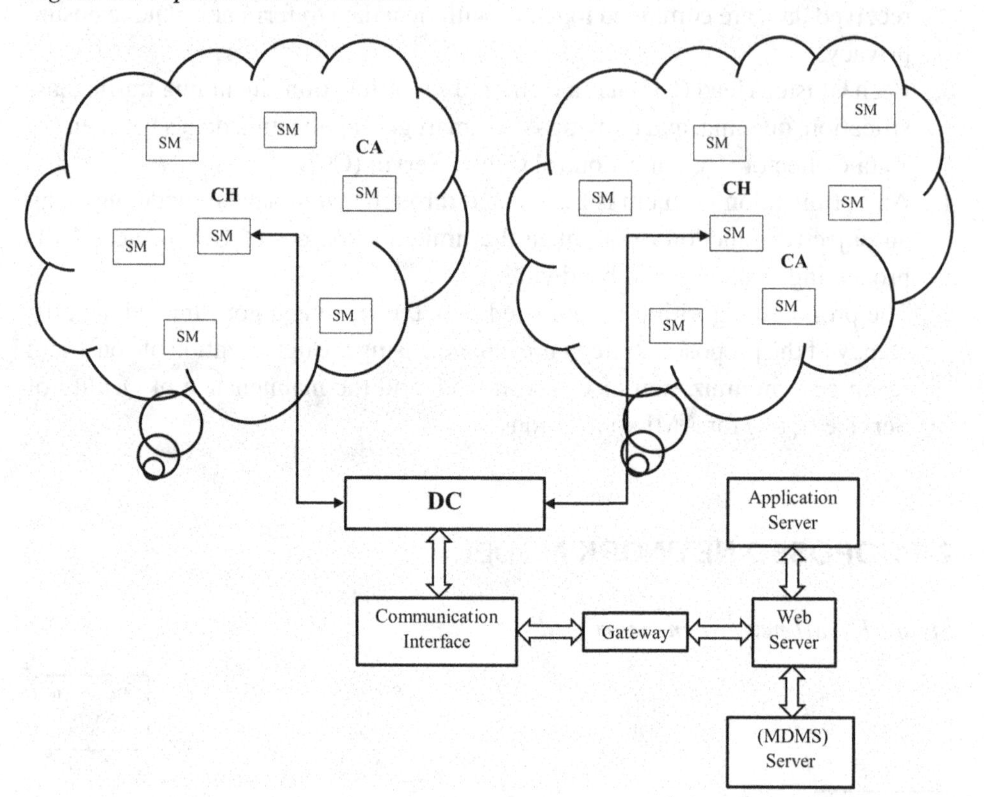

The MDMS utilizes an application to manage a master database containing the data consumption of all smart meters that categorized by the geographical location, residential area, and regional distribution within a city. The SM are equipped with multiple communication technologies and the choice of communication technology depends on the specific SM and the requests of the domestic region in the SG. The data collected through IMR provides CCS with valuable insights into power supply, usage patterns, and demand trends. The information enables more efficient management of the grid and àllows utilities to respond proactively to changes in demand or supply conditions. The AMI and IMR receives real-time and accurate data about their electricity usage and pricing. The information can be delivered through various channels such as text messages, emails, etc. for empowering consumers to make informed decisions about their energy consumption. The AMI applications like IMR are instrumental in transforming the traditional electric grid into a more dynamic, responsive, and efficient Smart Grid.

The proposed AMI deployment network with data aggregation and communication is shown in Fig. 2. The SM are grouped into clusters based on distance using a modified K-Means algorithm. Each cluster has a Cluster Head (CH) which act as a centroid for traffic arrangement and forwards data between the cluster members. Further, a Cluster Aggregator Head (CH) is elected in each cluster on periodic IMR applications to perform data aggregation. The data are sliced and combined at the neighbor nodes with new data to ensure privacy in the system. The DC and communication interface act as a backhaul network aggregate and manage data from multiple clusters. The DC is connected to Application, web and MDMS server through the gateway. The cloud computing is implemented at the CCS is integrated with the IoT services of the AMI network. The IoT-enabled SM measure electricity consumption periodically by generating load profiles that useful for demand-response, load forecasting, and billing applications. The request response mechanisms facilitate the exchange of electricity load patterns between residential consumers SM and the utility provider whenever required and enable the retrieval of specific load profile data such as minimum, maximum, and average consumption.

3. PROBLEM FORMULATION

In order to overcome the challenges in cluster-based AMI networks in resource-constrained CH, an optimization problem is formulated. The objective of the problem is to minimize resource usage at the CH while ensuring efficient distribution of workload, increased throughput, optimized utilization of constrained resources, and guaranteed QoS for various AMI applications.

minimize

$$\sum_{k=1}^{N_C} \sum_{i=1}^{N_{SM}} \left(C_{CPU}\, C_{RAM} + C_{BW} \right)_i^k \tag{1}$$

Subject to

$$\bigcup_{i=1}^{N_C} C_i = N$$

(1a)

$$\sum_{k=1}^{N_C}\sum_{i=1}^{N_{SM}} SM_{i,j}^{k} = 1$$

(1b)

$$\sum_{k=1}^{N_C}\sum_{i=1}^{N_{SM}} SM_{i,j}^{k}\lambda_i \leq \mu_j$$

(1c)

(1d)

(1e)

(1f)

where is the number of clusters, is the number of smart meters, is the CPU utilized by i^th^ SM in k^th^ CH, is the RAM utilized by i^th^ SM in k^th^ CH, is the bandwidth (BW) utilized by i^th^ SM in k^th^ CH, is the ith cluster in the Nth network, is the traffic arrival rate and is the service rate.

The objective function eq. (1) optimize the constrained resources of the CH by minimizing the sum of utilized CPU, RAM, and BW. The constraint (1a) is the AMI network (N) is partitioned into clusters. The eq. (1b) is each cluster member is connected to one CH / CA in cluster and eq. (1c) is average traffic arrival rate from cluster members is offered less than service rate of CH. The eq. (1d) and (1e) resource requirements of all AMI traffic should be less than the CH resource capacity CPU and RAM. The eq. (1f) is the data aggregation traffic from CH towards CA should be lower than the traffic arrival rate. The aggregated packets () from CH is

(2)

where P is the packet payload size and H is the header size.

4. PROPOSED DATA AGGREGATION FOR AMI IN SMART GRID

The objective of the proposed algorithm is to optimize the functionality of resource-constrained CH within the AMI network. Specifically, the aim is to restrict the role of CHs in AMI applications traffic, without performing data aggregation. The objective can be achieved while ensuring the QoS metrics of all AMI applications. Each cluster member generates traffic and time-critical traffic is transmitted directly to the corresponding CH. The periodic traffic is routed towards the assigned CA for data aggregation. The Utility CCS generates on-demand traffic, which is exchanged via the CH and DC to the CA. The CA forwards the traffic to the requested member in cluster, and the reaction is sent back to the Utility CCS.

The proposed algorithm involves coupling existing methods for aggregating traffic from members in a cluster, with each CA automated to perform these tasks. The strategy includes two main methods combining and manipulating method. In combining method, CA interconnect IMR traffic into one aggregated data with a mutual header. The packet containing energy consumption data from all members is sent to the corresponding CH for relay to the CCS. The combining method effectively reduces traffic size and ensures that all relevant consumption data is included in the aggregated packets. In the manipulating method, CA performs manipulation on the aggregated packet, possibly combining, filtering, or processing the data further and ensures that the aggregated data is optimized for transmission and analysis at the Utility CCS.

To ensure data privacy in SM, the data are sliced into random pieces (R). The R-1 pieces are send to random neighbor and remaining one is kept for itself. The DA node collect all m pieces from the SM and combined with its own primitive data to generate the new data. The primitive data is disclosed only if R-1 out degree links in which R is randomly generated and m in degree links are compromised by attackers. Attackers cannot broke all links until they destroy the maximum number. The probability of disclosing the privacy of SM data is (3)

where = 1 / (R_{max} – 2) probability of number of pieces (2,..., R_{max}), is the probability of leaking every pieces, is the probability of leaking every in degree links and data.

The optimal solution for the optimization problem in section 3 is given by the process of the proposed algorithm with data aggregation is shown in Fig.3. Therefore, initially in the proposed algorithm network is constructed according to the specified constraints ensuring each cluster member has connectivity to a single DC via CH and CA. For managing application traffic in the network AMI traffic from each member in a cluster are taken. Then, CH resource utilization are verified and establish exclusive TCP/IP connections for Time-critical traffic. The time grave traffic from the members in a cluster to CH are transferred and hybrid data aggregation strategy for both periodic and on-demand traffic are executed.

Figure 3. Flowchart of the proposed algorithm

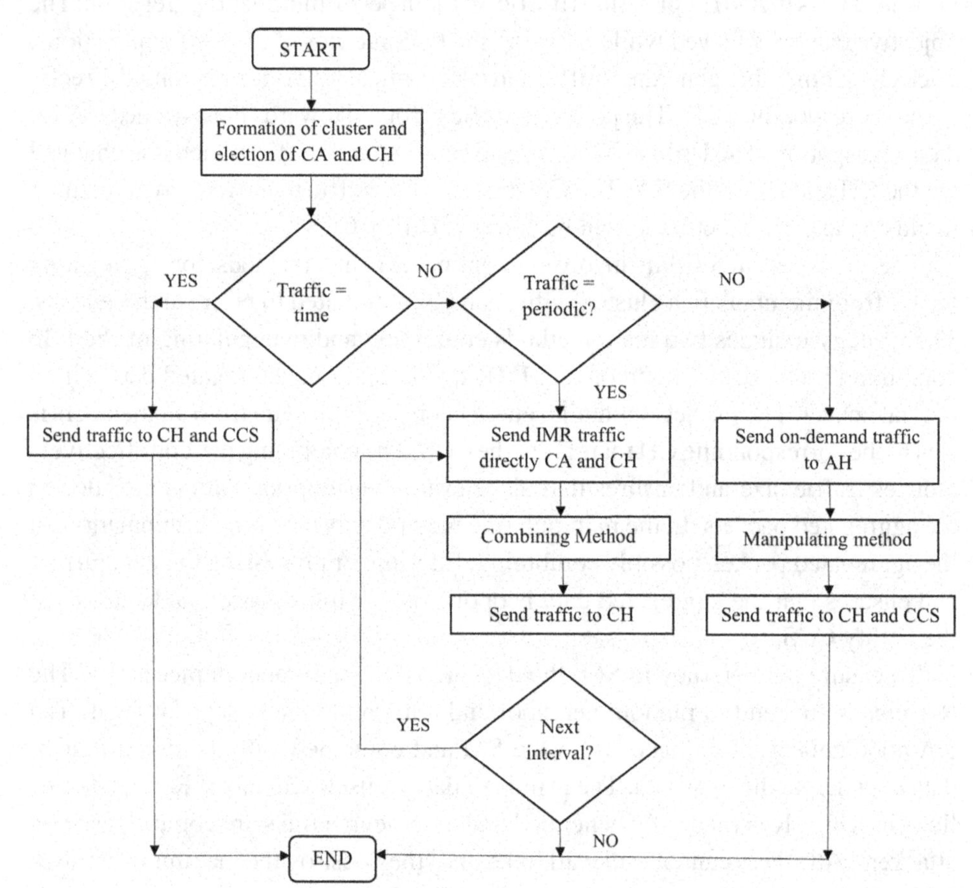

For handling IMR application traffic, TCP/IP connections between cluster members and CA are established. Then transmit electricity consumption at periodic intervals are calculated as

(4)

where a is the household smart appliances and h is the hour in a day. For handling on demand traffic, establish connections between CH and CA then manipulate electricity consumption data at CA in transmit response to requests towards CCS.

The delay in the network is calculated as

(5)

where is the queuing delay, is the processing delay, is the data aggregation delay, is the transmission delay, is the propagation delay occurs at one hop ().

The queuing delay is determined as

(6)

(7)

where N is the average packets queued in the network and is the server utilization of queuing system.

(8)

where and is the packet arrival rate and service rate at the CH.

5. NUMERICAL RESULTS

The performance of the proposed algorithm is compared with the existing algorithm (A. Khan, et al., U. Das and V. Namboodiri, S.-V. Oprea and A. Bˇara) using the Network Simulator. The proposed algorithm is investigated by assuming the clustering topology consists of 200 Smart Meters (SMs) or cluster members. A one root DC connected to a central router over the Internet with a capacity of 2MB. The CCS is connected to the DC via the central router. The IMR application data (electricity consumption) is collected from residential SMs at fixed time intervals (h = 15 minutes). This data is extracted and collected via APIs services. The CA receives and stores the collected IMR application data for a short time to perform data aggregation.

Table 1. Total traffic at CH

Algorithms	Incoming traffic			Outgoing traffic	Total
	Periodic	On-demand	Time critical		
(Khan et al., 2021)	800	252	85	1137	2274
(Das & Namboodiri, 2019)	800	290	110	416	1616
(Oprea & Bâra, 2023)	5567	510	380	2281	8738
Proposed	25	300	69	385	779

The total traffic at CH is shown in Table 1. The traffic is less at the proposed algorithm because switching the functio of data aggregation for IMR application data towards the CA is a key aspect of the proposed strategy. The data aggregation task from the CH to the CA is offloaded, the CHs receive less AMI traffic compared to other scenarios. This reduction in traffic at the CH leads to better performance in terms of reducing the total traffic load in the IoT-enabled AMI network.

Figure 4. Traffic at CH

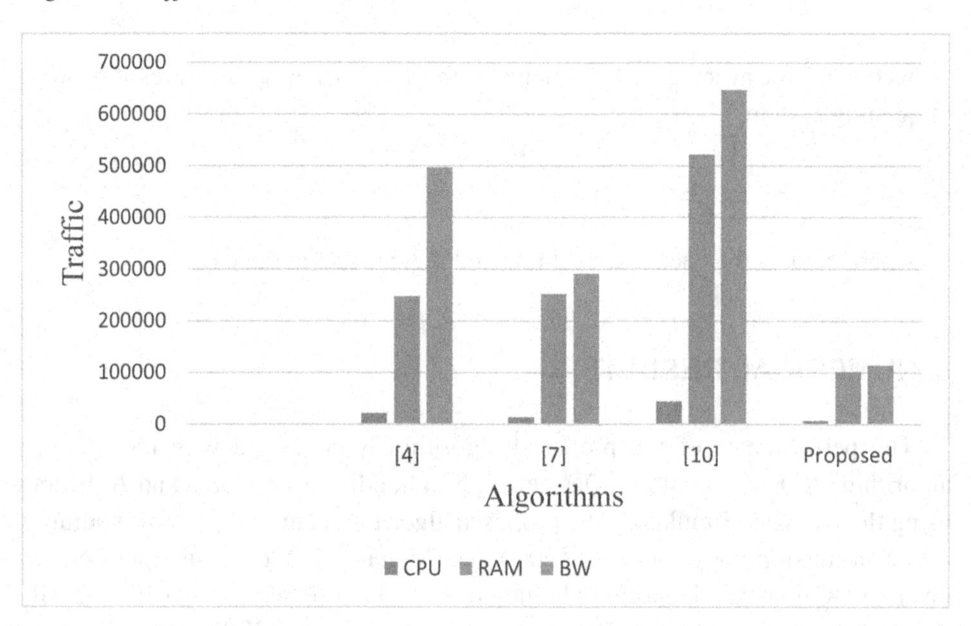

The AMI traffic at the CH in the existing algorithms is 2274 (A. Khan, et al), 1616 (U. Das and V. Namboodiri), 8738 (S.-V. Oprea and A. B˘ara) but for the proposed algorithm 779 which is very less compared to existing algorithms is shown in Fig. 4. Since, by allocating data aggregation to the AH, the CHs are relieved of the burden of processing and aggregating IMR application data. This results in a reduction in the total traffic load experienced by the CHs. For less traffic to handle, the CHs can allocate their resources more efficiently to handle other tasks and maintain QoS for various AMI applications. The less congestion and optimizing resource utilization at the CHs, the proposed strategy ensures better QoS provisioning for AMI applications. The timely delivery of data, minimizing packet loss, and meeting latency requirements, especially for time-critical applications are ensured. The offloading data aggregation to the CA can improve the scalability of the system. As the network grows and the number of SMs increases, the CA can handle the increasing volume of data more effectively than if all processing were done at the CHs. The distribution of data aggregation task and reducing traffic load at the CHs, the proposed strategy contributes to better performance and QoS for AMI applications in the IoT-enabled AMI network.

Table 2. Resource analysis

Algorithms	Resources			Total resources
	CPU	RAM	BW	
(Khan et al., 2021)	21480	248784	497570	767834
(Das & Namboodiri, 2019)	14260	252800	291520	558580
(Oprea & Bâra, 2023)	44538	522360	646478	1213376
Proposed	7560	101030	113400	221990

Figure 5. Utilization of resource

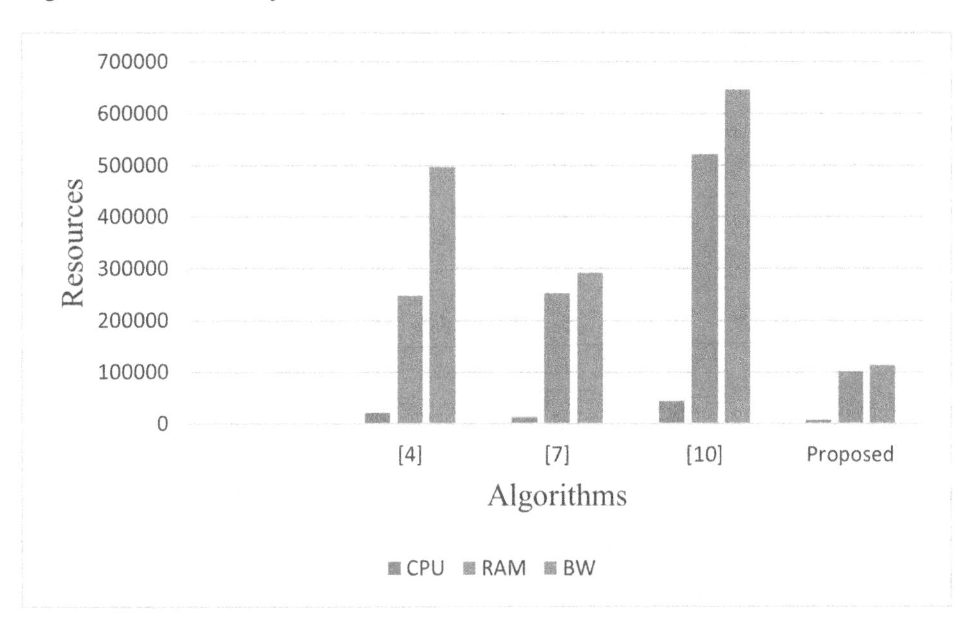

The resource analysis of the proposed algorithm is shown in Table 2. In the proposed algorithm, IMR application data in the AMI network achieves minimization of limited resources usage at the CH. The responsibility of data aggregation for IMR application data is shifted from the CHs to the CA, the CHs are relieved of intensive data processing tasks which reduces the burden on CHs' constrained resources such as CPU and memory. The data aggregation task handled by the CA, the CHs can allocate their resources more efficiently to other critical tasks within the AMI network that includes tasks such as handling real-time data processing, managing network routing, and ensuring QoS for various applications. The offloading data aggregation to the CA helps in reducing network congestion around the CHs and ensures the smoother data transmission and reduces the likelihood of resource contention, leading to improved overall network performance. The IMR application data is aggregated at the CA before transmission to the CCS, the proposed technique helps in optimizing bandwidth usage and ensures that limited network bandwidth resources are utilized more effectively, reducing the strain on CHs' resources. The proposed algorithm enhances the scalability of the AMI network by distributing processing tasks effectively. As the network expands and the number of SMs increases, the CA can scale more efficiently to handle the growing volume of IMR application data, thereby minimizing the impact on CHs constrained resources. The redistributing data processing tasks and optimizing resource usage, the proposed technique minimizes constrained resources usage at the CHs and efficient operation of the AMI network, especially for IMR application data.

Figure 6. Privacy preservation in data aggregation for SM

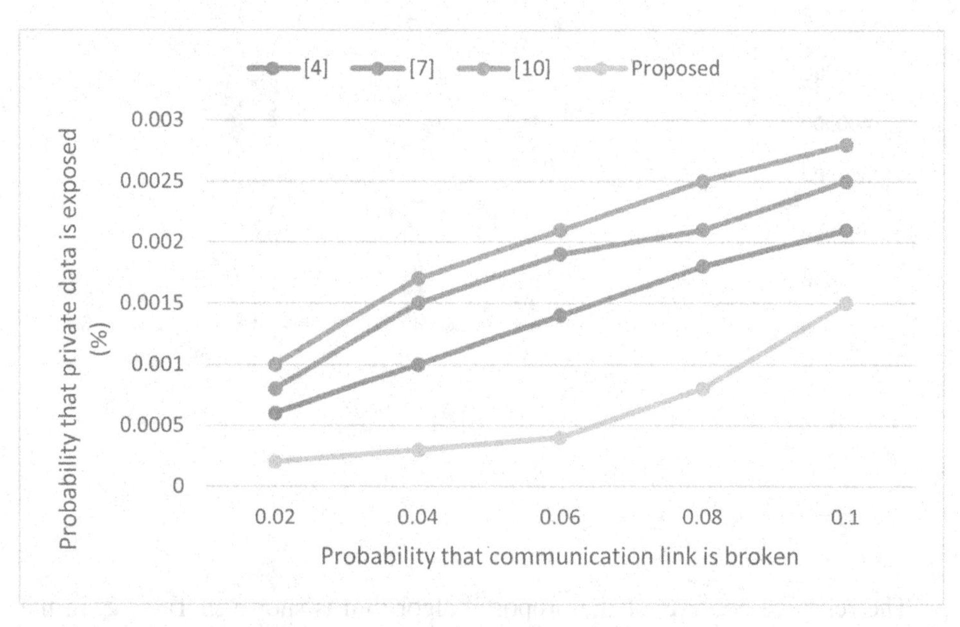

The efficient resource optimization and utilization achieved in the proposed algorithm compared to the existing algorithms is shown in Figure 5 are primarily attributed to the transmission of IMR application data to the CA data aggregation algorithm. The IMR application data is transmitted directly to the CA for aggregation, the traffic load at the CH is significantly reduced. The reduction in traffic alleviates congestion and optimizes resource utilization at the CHs. The hybrid data aggregation algorithm is used at the CA allow for efficient consolidation and processing of IMR application data. The algorithm likely employs techniques such as temporal and spatial aggregation to minimize redundant transmissions and optimize the use of network resources. The proposed algorithm ensures stringent QoS requirements for all time-critical traffic in the AMI network and includes meeting latency, bandwidth, and reliability requirements to ensure timely and reliable delivery of critical data. The IMR application data at the CA are aggregated before transmission that helps to optimize bandwidth utilization in the network and ensures that limited network resources are efficiently utilized to meet the diverse communication needs of AMI applications. The traffic load at the CHs is reduced and ensuring QoS for time-critical traffic, the proposed technique enhances overall network performance and reliability, and crucial for maintaining the integrity and efficiency of the AMI network, especially during peak usage periods or in the presence of network congestion. The CA for data aggregation and implementing stringent QoS measures, the proposed algorithm effectively optimizes resource utilization and ensures efficient operation of the AMI network, meeting the diverse communication requirements of AMI applications while minimizing the strain on CHs' resources.

The privacy preservation in data aggregation for smart grid is shown in Fig.6 in which the proposed algorithm provide better privacy compared to existing algorithm. The number of communication link broken is least in proposed algorithm means the probability of data obtained by eavesdropper is decreased. The data is sliced and send as random pieces to receiver from the transmitter which difficult to find all pieces and malicious attack from eavesdropper. Therefore, the privacy preservation of data from SM are higher in proposed technique compared to the existing techniques.

6. CONCLUSION

A novel privacy preserving hybrid data aggregation algorithm for handling IMR application data in a cluster-based topology within the IoT enabled AMI network for smart grid is proposed. The proposed algorithm aims to address the challenge of high traffic volume during peak hours, which imposes a significant burden on resource limited devices such as SM and DC that used as relay devices in the AMI architecture. A solution to minimize this traffic while ensuring QoS and optimizing resource utilization in the network introduces a hybrid data aggregation for IMR application data in a cluster-based topology. The algorithm aims to reduce the bulk amount of traffic while guaranteeing QoS and optimizing resource utilization. The optimization problem is formulated with an objective function designed to minimize constrained resources at the CHs in the clustering topology. To solve the optimization problem, the trade-off between reducing AMI traffic volume with QoS guarantees and optimizing resource utilization in the AMI network is considered. Experimental and simulation results of reduction of high traffic volume with diverse QoS requirements and resources utilization are provided to evaluate the effectiveness of the proposed strategy. The Simulation results shows that the proposed strategy outperforms existing approaches in the constrained IoT-enabled AMI network.

REFERENCES

Ahammed, M. T., & Khan, I. (2022). Ensuring power quality and demand-side management through IoT-based smart meters in a developing country. *Energy*, 250(Jul), 123747. 10.1016/j.energy.2022.123747

Das, U., & Namboodiri, V. (2019, February). A quality-aware multi-level data aggregation approach to manage smart grid AMI traffic. *IEEE Transactions on Parallel and Distributed Systems*, 30(2), 245–256. 10.1109/TPDS.2018.2865937

Ghazal, T. M., Hasan, M. K., Alzoubi, H. M., Alshurideh, M., Ahmad, M., & Akbar, S. S. (2023). Internet of Things connected wireless sensor networks for smart cities. In Alzoubi, H. M., & Salloum, S. (Eds.), *The Effect of Information Technology on Business and Marketing Intelligence Systems* (pp. 1953–1968). Springer. 10.1007/978-3-031-12382-5_107

Inga, E., Inga, J., & Hincapié, R. (2023). Maximizing resource efficiency in wireless networks through virtualization and opportunistic channel allocation. *Sensors (Basel)*, 23(8), 3949. 10.3390/s2308394937112290

Khan, A., Umar, A., Munir, A., Shirazi, S., Khan, M., & Adnan, M. (2021, December). A QoS-aware machine learning-based framework for AMI applications in smart grids. *Energies*, 14(23), 1–22. 10.3390/en14238171

Majeed, A., & Lee, S. (2020). Anonymization Techniques for Privacy Preserving Data Publishing: A Comprehensive Survey. *IEEE Access : Practical Innovations, Open Solutions*, 9, 8512–8595. 10.1109/ACCESS.2020.3045700

Oprea, S.-V., & Bâra, A. (2023). An Edge-Fog-Cloud computing architecture for IoT and smart metering data. *Peer-to-Peer Networking and Applications*, 16(no. 2), 818–845. 10.1007/s12083-022-01436-y

Padhan, A. K., Kumar Sahu, H., Sahu, P. R., & Samantaray, S. R. (2023). Performance analysis of smart grid wide area network with RIS assisted three hop system. *IEEE Transactions on Signal and Information Processing Over Networks*, 9, 48–59. 10.1109/TSIPN.2023.3239652

Popović, I., Rakić, A., & Petruševski, I. D. (2022, January). Multi-agent real-time advanced metering infrastructure based on fog computing. *Energies*, 15(1), 373. 10.3390/en15010373

Sung, T.-W., Xu, Y., Hu, X., Lee, C.-Y., & Fang, Q. (2022, March). Optimizing data aggregation point location with grid-based model for smart grids. *Journal of Intelligent & Fuzzy Systems*, 42(4), 3189–3201. 10.3233/JIFS-210881

Yadav, S., & Nanivadekar, S. (2023). Hybrid optimization assisted green power allocation model for QoS-driven energy-efficiency in 5G networks. *Cybernetics and Systems*, (Feb), 1–16. 10.1080/01969722.2023.2175147

Yang, M., Tjuawinata, I., & Lam, K.-Y. (2022). K-Means Clustering With Local d -Privacy for Privacy-Preserving Data Analysis. *IEEE Transactions on Information Forensics and Security*, 17(1), 2524–2537. 10.1109/TIFS.2022.3189532

Zhou, X., He, D., Khan, M. K., Wu, W., & Choo, K.-K. R. (2023). An Efficient Blockchain-Based Conditional Privacy-Preserving Authentication Protocol for VANETs. *IEEE Transactions on Vehicular Technology*, 72(1), 81–92. 10.1109/TVT.2022.3204582

Zhu, T., Li, J., Hu, X., Xiong, P., & Zhou, W. (2020). The Dynamic Privacy-Preserving Mechanisms for Online Dynamic Social Networks. *IEEE Transactions on Knowledge and Data Engineering*, 34(6), 2962–2974. 10.1109/TKDE.2020.3015835

Chapter 13
Prospects of Network Function Virtualization (NFV) and Software-Defined Networking (SDN) Techniques for Smart Grid Cyber Defense

Isai Vani Mariyappan
http://orcid.org/0000-0002-1918-2145
Vaigai College of Engineering, Madurai, India

V. Kavitha
Sethu Institute of Technology, Kariapatti, India

R. Aravindaraj
Vaigai College of Engineering, Madurai, India

Sonia C. Chockalingam
Vaigai College of Engineering, Madurai, India

Josep M. Guerrero
Center for Research on Microgrids, Technical University of Catalonia, Spain

DOI: 10.4018/979-8-3693-2786-9.ch013

ABSTRACT

In the enormous growth of smart grid technology, the integration of 5G architecture network technology offers many advantages such as real-time data processing, improved communication facilities, safe data transfer, improved privacy, and diversified services. Yet, it leads to new cybersecurity problems that need to be addressed to ensure the smooth operation of the smart grid. This chapter outlines an extensive study of state-of-the-art network function virtualization (NFV) and software-defined networking (SDN) cybersecurity techniques in smart grid applications. NFV and SDN are key perceptions in the development of 5G networks, promoting flexibility, scalability, programmability, and centralized control over network resources. NFV, along with SDN, promotes the transformation of telecommunication networks, enabling them more adaptable, scalable, and capable of facilitating the diverse services envisioned in the 5G era, protecting critical energy infrastructure from cyber threats.

1. INTRODUCTION

In the prospect of energy infrastructure, the emergence of smart grids implies a paradigm shift, promising unprecedented advancements in sustainability, efficiency, and reliability. However, this transformative development is not without its difficulties. As smart grids seamlessly incorporate state-of-the-art technologies and modern digital communication systems, they unknowingly expose themselves to a complex web of cyber risks. These risks, if appear, have the ability to undermine not just the utility of smart grids but the whole energy supply chain, expanding the need for a strong cyber defense approach. The imperative to ensure the cyber resilience of smart grids has grown to the lead as a critical priority. This necessity is driven by the recognition that possible cyber threats, if successful, could direct intense societal, economic, and environmental consequences.

In the smart grid cyber defense realm, research stipulates crucial perspectives into the various challenges faced by these advanced energy systems. Diverse studies have precisely explored the vulnerabilities enclosed in smart grid components, the potential fallout of cyber threats, and unique defense strategies that promise resilience. For example, Wang et al. (2019), examined the instabilities innate in communication procedures involved in smart grids. The research not only revealed these weaknesses but also examined improved security measures, acting as a marker for strengthening the digital defenses of smart grids. Likewise, the research done by Liu et al. (2020) probed into the area of anomaly detection strategies, offering real-time foresight into strong cyber attacks and imparting the advancement of proactive defense techniques. These academic studies highlight the multidisciplinary nature of smart grid cyber defense, stressing the crucial demand for collaboration among industry stakeholders, researchers, and policymakers. Collectively, they form a generic front, fortifying the flexibility of the energy infrastructure against the continuously evolving and complex landscape of cyber attacks.

The introduction of 5G technology into the era of smart grids implies a monumental stride toward strengthening the cybersecurity endurance of modern energy infrastructures. As smart grids sustain metamorphosis to satisfy the increasing demand for effective and strong energy systems, the crucial part of 5G technology becomes progressively marked in securing cyber defenses. The enhanced data transmission speeds, along with low latency and increased network capacity inherent to 5G, not only improve the functional efficiency of smart grids but also develop a unique canvas for carrying out cutting-edge cybersecurity schemes. In this complex stance between technology and resilience, research takes center part, expanding insights into the complexity of 5G and smart grid cyber defense.

Exploring the domain of academic surveys shows the precise investigations carried out by Li et al. (2021), solving the potential weaknesses and security challenges entangled with the incorporation of 5G into smart grids. This detailed study navigates the network of communication protocols and, the necessity of strong encryption and authentication techniques to set up formidable barriers against cyber attacks. Likely, Zhang et al. (2022) analyze the demand for machine learning algorithms, influencing the capabilities of 5G technology for anomaly identification. This strategic amalgamation not only strengthens the smart grid's capability to swiftly detect but also aptly respond to cyber-attacks. These research findings not only emphasize the transformative potential innate in 5G technology but also serve as the architects facilitating a more stable and dependable energy infrastructure. Despite evolving cyber attacks, this combination of academia and technology approaches makes a flexible future for smart grids, where innovation coordinates with the essentials of cyber defense.

In the dynamic platform of smart grid cyber defense, 5G techniques evolve as a versatile and multifaceted arsenal, strengthening critical infrastructure. The primary cluster of technologies, a harmony of communication efficiency, and reliability develops with the ability of Ultra-Reliable Low Latency Communication (URLLC). Concurrently, Network Slicing gets focus and capably adjusts virtual networks with surgical precision to satisfy the demands of particular smart grid applications. As the performance rises, Edge Computing turns into the spotlight, gracefully decreasing latency by processing data close to the source. This not only accelerates the speed but also strengthens the pattern of both speed and security, addressing the demands for swift, secure, and effective communication in the smart grid ecosystem.

Among the complex design of 5G technologies, two important performers, Network Function Virtualization (NFV) and Software-Defined Networking (SDN), become the focal point, their roles crucial in devising the success of 5G networks. In this technological harmony, NFV plays a key role, implementing an accent of resource efficiency, flexibility, and scalability that reverberates through the digital domain. A growth of adaptability rises, providing the infrastructure with the dynamic capacities to scale, optimize, and evolve. Accessing SDN, gracefully turning onto the stage, establishing a symmetrical merge of programmability and enhanced network management. Ultimately, NFV and SDN make an efficient, vibrant, and adaptive infrastructure resonating in harmonious unity with the various needs of 5G services.

In the prospective technological evolution, the combination of smart grids with 5G services sets the stage for a narrative that resonates with critical significance. This section extends the backdrop of this symbiotic union, highlighting the huge significance of strengthening modern energy infrastructures against the looming specter of evolving cyber-attacks. The focus moves to the realization that as smart grids earnestly adopt cutting-edge technologies, they move dangerously close to the edge of cyber attacks, demanding not merely a reactive but a proactive and actively adaptive defense technique.

The combination of Network Function Virtualization and Software-Defined Networking has a major influence in strengthening smart grid cyber defense. In view of smart grids, where the convergence of information strategy and operational strategy is widespread, NFV permits the extraction of network functions from dedicated hardware to software-oriented instances. This abstraction allows the dynamic layout of security functions, like firewalls, intrusion identification systems, and encryption services, as virtualized operations that can be flexibly scaled to satisfy the emerging security demands of the smart grid infrastructure. By decoupling network functions from physical hardware, NFV improves resource exploitation, accelerates deployment, and promotes effective management of security measures. Furthermore, the combined effort of NFV and SDN in smart grid cyber defense stretches beyond prompt adaptability to include effective resource usage and optimization. NFV's capability to form virtual instances of security operations ensures that computational resources are utilized more efficiently, as these tasks can be dynamically allotted or de-allotted as per emerging security requirements. This flexibility allows the smart grid to sustain optimal performance levels while effectively managing computational resources, which is especially important in energy systems where operational efficiency is prominent.

SDN further improves the potential of smart grid cyber defense by offering a centralized and programmable network control plane. With SDN, the network's actions can be dynamically configured as per security norms and threat intelligence. The programmability enables fast response to cyber-attacks by making the formation of adaptive and context-aware security regulations. SDN's centralized management also improves visibility into network actions, allowing better supervising and analysis of powerful security events. The consolidation of NFV and SDN in the smart grid cyber defense service forms an active and responsive ecosystem that alters to evolving threats, reduces vulnerabilities, and secures the integrity and availability of crucial energy systems.

In practical terms, the combination of NFV and SDN in smart grid cyber defense allows the execution of a defense-in-depth technique. Virtualized security tasks can be distributed deliberately across the network, offering layered protection against diverse attack vectors. This technique improves the overall resilience of the smart grid by reducing the effect of potential breaches and securing the uninterrupted operation of crucial energy services. As the threat landscape expands, the NFV-SDN combination allows for fast adaptation, making it a cornerstone for building a strong and adaptive cyber defense posture in the domain of smart grids.

2. LITERATURE SURVEY

The inclusion of 5G architecture network services into smart grid technology has garnered vital focus due to the various advantages it brings, such as enhanced communication facilities, real-time data processing, improved privacy, safe data transfer, and support for various services. Nevertheless, this combination presents new challenges in terms of cybersecurity, demanding modern techniques to ensure the smooth and secure function of smart grids. This literature survey targets to examine modern cybersecurity strategies as per Network Function Virtualization (NFV) and Software-Defined Networking (SDN) on the subject of smart grid applications with 5G inclusion.

Papers by authors such as Chen et al. (2020) and Zhang et al. (2021) spotlight the benefits and problems associated with the insertion of 5G services into smart grids, stressing the need for strong cybersecurity measures. NFV takes a leading stance in the evolution of 5G services by providing programmability, scalability, flexibility, and centralized control over network resources. Reviews by Zhao et al. (2019) and Liu et al. (2022) explore deeply the implementation of NFV, focusing on its ability to dissociate network operations from specific hardware, dynamically line up network tasks, and aid the creation of service chains for complicated services in 5G networks. Studies by Wang et al. (2020) and Li et al. (2021) highlight the influence of SDN on optimizing network performance, enhancing resource exploitation, and assisting dynamic resource allocation based on real-time network contexts.

Authors Kim et al. (2020) and Sharma et al. (2021) examine the cybersecurity challenges presented by the inclusion of 5G services into smart grids. They spotlight the vulnerabilities linked with improved connectivity and the demand for advanced security measures to defend critical energy infrastructure from cyber attacks. Gadallah et al. (2024) explore an efficient identification method opposing DDoS threat in control and data planes of SDN. The method finds DDoS threats using a Deep Learning (DL) technique through recent features based on traffic statistics for the control plane. A DL technique termed as AE-BGRU for DDoS identification utilises Autoencoder (AE) as well as Bidirectional Gated Recurrent Unit (BGRU) . Studies by Liang et al. (2018) and Song et al. (2019) specifically examine the application of NFV and SDN in resolving cybersecurity issues in smart grids. They explain how NFV enables dynamic security techniques by decoupling security operations from hardware, while SDN offers centralized control for monitoring and responding to security attacks. Recent papers by Wu et al. (2023) and Chen et al. (2024) emphasize the inclusion of NFV and SDN to advance the overall protection of 5G-integrated smart grids. They review the collaborative technique of NFV and SDN in creating scalable and adaptive security solutions, guarding against evolving cyber-attacks. The dynamic allocation of resources is a key aspect of NFV and SDN in 5G-integrated smart grids. Research by Yang et al. (2022) and Liu and Chen (2023) focus on the dynamic resource allocation capabilities of NFV, emphasizing its role in adapting to varying network needs and securing efficient exploitation of resources. SDN, with its view on the control plane, further supports dynamic resource allocation, optimizing network performance as per real-time circumstances.

The creation of service chains, a crucial feature enabled by NFV, is examined broadly in the literature. Works by Zhang and Wang (2021) and Guo et al. (2022) focus on the importance of service chains in making end-to-end network services in 5G environments. This capability is important for delivering complicated services tailored to the various demands of smart grid applications. SDN's capability to make a unified network view and centralized control is focused by Huang et al. (2019) and Xu et al. (2020). This feature not only eases network management but also supports effective cybersecurity measures by enabling administrators to monitor, analyze, and respond to security attacks in a centralized manner. Current advancements in real-time cyber attack monitoring and response techniques are discussed by Zhao and Liu (2023) and Jiang et al. (2024). These studies highlight the significance of leveraging NFV and SDN to make adaptive security measures suitable for detecting and mitigating cyber attacks in real-time, protecting the integrity and availability of smart grid systems.

Examining the future directions of research, papers by Li and Zhang (2023) and Wang et al. (2024) discuss evolving strategies and trends in the combination of 5G, NFV, and SDN for cybersecurity in smart grids. The research highlights the demand for persistent innovation to deal with emerging cyber attacks and improve the resilience of smart grid infrastructures. Studies by the European Telecommunications Standards Institute (ETSI) and International Electrotechnical Commission (IEC) make perceptions into the ongoing normalization efforts, stressing the demand for a unified technique for cybersecurity to eliminate risks and problems (ETSI, 2021; IEC, 2022). Recent research by Zhang and Chen (2023) and Wang et al. (2023) examines machine learning (ML) as well as artificial intelligence (AI) techniques in cybersecurity measures for smart grids. The findings spotlight the potential of ML and AI algorithms in spotting and responding to complicated cyber attacks, offering an additional layer of defense against emerging attack vectors. Privacy issues and data protection are key outlooks of 5G-integrated smart grids. Papers by Liu and Wu (2022) and Li et al. (2024) explain the issues connected with guarding sensitive information in terms of improved data transfer and real-time processing. Techniques for securing privacy and compliance with data protection requirements are examined to upkeep the trust of smart grid users.

Cooperative efforts and industry partnerships have been importantly crucial in addressing cybersecurity vulnerabilities. Research by Li et al. (2022) and Song and Wang (2024) underlines the prominence of collaboration between researchers, industry stakeholders, and government agencies to expand comprehensive security solutions that are scalable, adaptive, and effective in protecting significant energy infrastructure. In view of the pivotal nature of energy infrastructure, recent research by Wu and Zhang (2023) and Chen et al. (2024) examine energy-aware security measures. These measures aim to balance the demand for strong cybersecurity with energy efficiency, ensuring that security protocols do not jeopardize the comprehensive operational efficiency of smart grid systems. The significance of resilience and redundancy in view of cyber attacks is emphasized by Zhao and Wang (2022) and Xu et al. (2023). The research explores techniques for building resilient smart grid architectures and including redundancy in critical components to endure cyber threats and guarantee uninterrupted operation. Securing interoperability between various sectors and strategies is critical for effective cybersecurity. Work by Kim and Lee (2023) and Guo et al. (2024) explains the advantages of cross-sector collaboration, stressing the demand for interoperable solutions that can continuously integrate with different smart grid components and other major infrastructure systems. Mengxiang Liu et al. (2024) tailored an integrated threat modeling technique for the hierarchical DER-based smart grid with a particular focus on vulnerability detection and impact analysis. Then, the defense-in-depth techniques comprising prevention, identification, mitigation, and recovery are surveyed, well-segregated, and outlined in detail.

In conclusion, the emerging domain of cybersecurity in 5G-integrated smart grids needs a multidimensional method, concerning standardization, privacy considerations, evolving strategies, combined efforts, and resilience techniques. The surveyed literature provides useful information on these prospects, contributing to the ongoing discourse on ensuring smart grid infrastructures in the domain of 5G technology integration.

3. NETWORK FUNCTION VIRTUALIZATION (NFV) APPROACH

3.1 Transforming the Landscape of Network Architecture

In the constant evolution of telecommunications and networking, the conventional models controlling network design and management are experiencing a drastic transformation. Network Function Virtualization (NFV) evolves as a groundbreaking technique, revolutionizing the method of building, deploying, and operating networks. Departing from the traditional hardware-centric paradigms, NFV imparts a paradigm shift towards a more scalable, supple, and cost-effective software-driven technique. This comprehensive examination explains the intricacies of NFV, providing an in-depth analysis of its basic principles, the multitude of advantages, the sophisticated difficulties it presents, and its implementations for the future of network architecture.

3.1.1 Unveiling the Essence of Network Function Virtualization (NFV)

A contemporary idea of NFV entails the virtualization of network functions, conventionally confined to committed hardware appliances. This embraces a wide array of functions, including intrusion detection, load balancing, routing, and diverse other network-related operations. The core doctrine of NFV lies in leveraging virtualization strategies to replace particularized network hardware with software instances operating on systematic servers, switches, and storage components. The basic premise of NFV centers on the decoupling of network tasks from specific hardware, and so enabling it to operate as software instances on a shared hardware infrastructure. The abstraction not only promotes greater flexibility and scalability but also supports for best resource exploitation.

3.2 Essential Components of NFV

3.2.1 Virtualized Network Functions (VNFs)

At the core of NFV are considered Virtualized Network Functions (VNFs), depicting the software incarnations of traditional network tasks. These VNFs can be symbolized on regular servers, whether in the cloud or on-premises, and are organized to collaboratively offer the required network services.

3.2.2 NFV Infrastructure (NFVI)

The NFV Infrastructure creates the bedrock upon which VNFs function. Encompassing a mix of hardware and software devices, NFVI comprises computational resources, networking elements, and storage. This framework finds its deployment across various environments, ranging from traditional data centers to evolving edge computing nodes and cloud-based frameworks.

3.2.3 NFV Management and Orchestration (NFV MANO)

The NFV Management and Orchestration (NFV MANO) layer plays a crucial part in the lifespan governance of VNFs. This versatile component encompasses three major functions:

Virtualized Infrastructure Manager (VIM): Directs the allocation and exploitation of NFVI properties, securing best performance and scalability.

Virtual Network Function Manager (VNFM): Manages the extensive lifespan governance of particular VNF samples, containing instantiation, scaling, and cessation.

NFV Orchestrator (NFVO): Coordinating the consistent instantiation, scaling, and termination of VNFs, ordering them to meet the overarching demands of a particular network service.

3.3 Architectural Landscape of NFV

3.3.1 High-Level Overview

Refer to the descriptive figure below, providing a high-level outlook on the NFV architecture:

Figure 1. Overview of NFV architecture

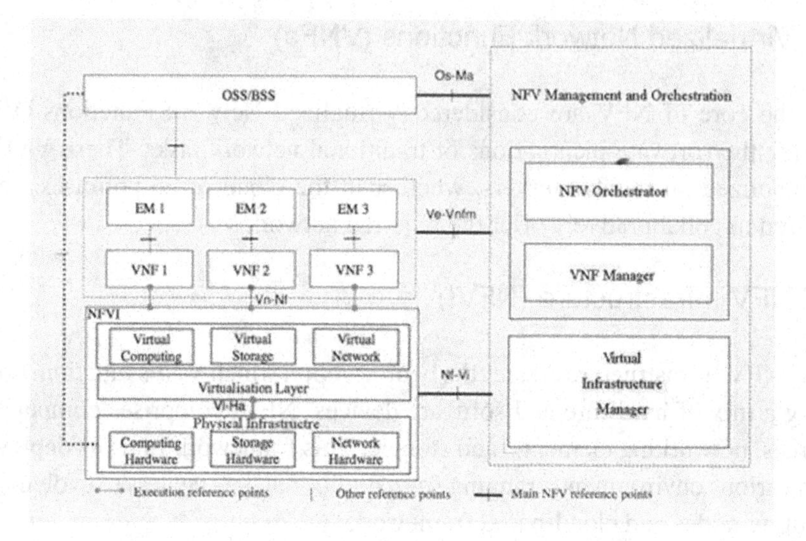

The visual representation explains the NFV architecture in detail with the NFVI layer building the foundation and the NFV MANO layer governing the organization and management of VNFs.

3.3.2 NFV Infrastructure (NFVI)

The NFVI layer aids as the crucible of virtualization, comprising the following units:

Computational Resources: Establishing servers wherein VNFs are extended, these resources could be either virtual machines or physical servers conditional on the deployment context.

Storage Resources: Making the required storage infrastructure for VNFs, cooperating with both persistent and non-persistent storage demands.

Network Resources: Comprising the networking framework essential for interconnecting VNFs and supporting continuous communication between them.

3.3.3 NFV Management and Orchestration (NFV MANO)

The NFV MANO layer is an essential orchestration unit that comprises:

Virtualized Infrastructure Manager (VIM): Manages the dynamic allocation and governance of NFVI properties, engaging responsibilities such as scaling, and resource monitoring operations.

Virtual Network Function Manager (VNFM): Governs the whole lifespan of particular VNF samples, from instantiation to scaling and eventual completion.

NFV Orchestrator (NFVO): Serves as the overarching conductor, coordinating the deployment and interconnection of many VNFs to satisfy the specific demands of a particular network service.

3.4 NFV in Cyber Security

NFV acts as a transformative approach by decoupling network operations from dedicated hardware components and enabling their arrangement as software on general-purpose servers or virtual machines. This section examines the diverse role of NFV in improving the cybersecurity posture of smart grids, highlighting its contributions to scalability, adaptability, and dynamic defense techniques.

3.4.1 Dynamic Deployment for Adaptive Security Measures

The main aspect of NFV in smart grid cyber defense is its own capability to support the dynamic deployment of security measures. Conventional security measures often depend on static setups, making it challenging to adapt to the emerging threat domain. NFV permits the instantiation of security operations as software entities, making the dynamic activation and deactivation of particular security measures as per real-time threat intelligence. This adaptability makes sure that the smart grid's defense strategies remain strong and responsive to evolving cyber challenges.

3.4.2 Resource Allocation and Efficiency

NFV in smart grid cyber defense substantially imparts to effective resource allocation. By decoupling security operations from specific hardware appliances, NFV enables the exploitation of shared computing resources for diverse security tasks. This flexibility secures the best use of resources, enabling the smart grid to allocate computing power as per the particular needs of various security applications. The dynamic resource allocation not only improves the overall efficiency of cyber defense but also assists in cost-effectiveness by eliminating the necessity for dedicated hardware for each security operation.

3.4.3 Creation of Secure Service Chains

NFV's ability to make service chains is specifically relevant in the realm of smart grid cybersecurity. Service chaining comprises linking numerous Virtual Network Functions in a particular arrangement to form end-to-end network services. In the domain of cyber defense, NFV supports the construction of secure service chains, where many security operations are interlinked to create a comprehensive defense technique. This method allows for the organization of security measures modified to the smart grid's specific demands, securing a holistic and layered defense against cyber attacks.

3.4.4 Scalability to Accommodate Rising Smart Grid Networks

The scalability of NFV is critical for dealing with the rising scale of smart grid networks. As these designs continue to rise in complexity and size, conventional cybersecurity techniques may struggle to maintain pace. NFV's scalability permits the continuous inclusion of additional security functions to satisfy the emerging needs of the smart grid. Whether it's accommodating a growing number of linked devices or adapting to modifications in network topology, NFV secures that the cyber defense framework remains scalable and able to manage the requirements of a dynamic smart grid ecosystem.

3.4.5 Integration With Threat Intelligence for Proactive Defense

NFV's key role in smart grid cyber defense expands to proactive measures through continuous inclusion with threat intelligence. By exploiting NFV's dynamic deployment facilities, smart grids can include threat intelligence feeds into their security service chains. This combination permits for real-time investigation of emerging vulnerabilities and the automatic deployment of prompt security measures. NFV's capability to adapt swiftly to new threat detail secures that the smart grid can stay ahead of powerful cyber issues, improving its resilience against emerging attack vectors.

3.4.6 Facilitating Network Slicing for Security Isolation

Network slicing is an alternating prospect of NFV that confirms invaluable in smart grid cybersecurity. NFV allows the formation of isolated network slices, each modified to dedicated operations or user groups within the smart grid ecosystem. This section improves security by powerful breaches and limiting the impact of cyber-attacks. NFV's part in organizing such network slices secures that security measures are implemented selectively, offering a nuanced defense technique that aligns with the various needs of different smart grid components.

3.4.7 Supporting Compliance and Regulatory Needs

In terms of smart grid cybersecurity, loyalty to compliance standards and regulatory demands is predominant. NFV supports attaining and maintaining compliance by providing a flexible and programmable framework. Security rules and regulations can be dynamically altered through NFV, securing that the smart grid aligns with emerging cybersecurity standards. This adaptability not only simplifies compliance efforts but also positions the smart grid to quickly respond to regulatory changes and evolving cybersecurity guidelines.

3.4.8 Enabling Rapid Incident Response and Recovery

NFV's dynamic and programmable nature significantly imparts to the smart grid's incident response and recovery capabilities. In the case of a cybersecurity incident, NFV permits for fast reconfiguration of security services and the development of countermeasures. This ability ensures that the smart grid can quickly contain and reduce the effect of cyber incidents, minimizing downtime and improving overall system flexibility. NFV's part in automating incident response methods imparts a more powerful and effective cyber defense technique for smart grids.

3.5 Case Studies and Real-World Applications

Investigating real-world applications and case studies offers valuable perceptions of the practical applications and victories of NFV.

Sherif Abdelwahab et al. (2016) surveyed service abstraction, structure of NFV, and network virtualization through the network overlay design. As NFV enabling strategies, they explained how to apply SDN and OpenFlow to virtualize and interconnect VNFs. Also, 5G virtualizable radio functions, Coordinated MultiPoint inter-cell device-to-device, and ultra-densified network implementation using NFV are analyzed in detail. In addition, some open research issues particular to NFV in 5G Radio Access Network are explored.

Lingyi Xu et al. (2023) proposed Network function virtualization (NFV-RA), a revolutionary network structure that enhances the intricacy of network function virtualization resource allocation (NFV-RA) while also significantly improving the flexibility of service deployment. Numerous academics have researched the aforementioned issues and proposed solutions. These plans are hard to compare because they are frequently created individually. Research progresses more quickly when a common and effective simulation framework for the NFV-RA issue is available for inclusion, comparison, and evaluation of these methods. To address this, the universal simulation framework SFCSim for NFV-RA is presented by the author. The discrete-time event scheduling engine used by SFCSim allows for the simulation of many situations, including service migration, mobility management, and static and dynamic arrangement of the service function chain.

Organizations across diverse industries, from telecommunications to healthcare and, finances have supported NFV to simplify operations, improve serviceability, and attain cost savings. An in-depth exploration of these case studies sheds light on the troubles faced, lessons learned, and the overall effect of NFV on various business environments.

3.6 Future Directions and Innovations

Network Function Virtualization acts as a catalyst for a huge transformation in network architecture. Its journey comprises a combination of evolving technologies, technological innovation, collective standardization, and an unceasing drive toward adaptive and responsive networks. As NFV keeps evolving, organizations must address the challenges, capitalize on the advantages, and remain active in the face of emerging technologies and regulatory domains. The legacy of NFV is not only in its virtualization of functions but in its capacity to redefine how networks are formed, constructed, and functioned in a field where ability, scalability, and efficiency are major concerns.

Moving forward, ongoing research and innovations in NFV for smart grid cyber defense are expected. Researchers are examining the combination of artificial intelligence and machine learning methodologies within NFV frameworks to improve anomaly identification and predictive threat examination. In addition, advancements in NFV arrangement and governance tools target to ease the deployment and function of security services within smart grids. These future trends highlight NFV's seamless evolution as a cornerstone strategy in strengthening the cybersecurity domain of smart grids.

The diverse role of NFV in smart grid cyber defense is essential to the ongoing efforts to ensure critical energy infrastructure. From dynamic arrangement and resource efficiency to proactive threat intelligence combination and compliance support, NFV's offerings are predominant. As smart grids emerge and face increasingly complex cyber attacks, the never-ending innovation and technical implementation of NFV strategies will take a central role in securing the adaptability, robustness, and flexibility of smart grid cybersecurity technologies.

4. SOFTWARE-DEFINED NETWORKING (SDN) METHODOLOGY

In the modern domain of networking, where requirements for improved flexibility, scalability, and resource optimization dominate, a revolutionary model called Software-Defined Networking (SDN) has turned out to be the focus. Representing a departure from traditional networking methods, SDN presents a revolutionary transformation by decoupling the control plane from the data plane. This section explains an exhaustive examination of the intricacies surrounding SDN, comprising key concepts, architectural nuances, benefits, issues, and future directions.

4.1 Unraveling SDN

4.1.1 A Definitive Context

Software-Defined Networking acts as an avant-garde technique for network management, offering network administrators the capability to programmatically and dynamically govern and manage network resources. At its core, SDN organizes the partition of the control plane, accountable for decision-making respecting traffic routing, from the data plane, the entity employed in the actual forwarding of data packets.

4.1.2 Deconstructing Components

4.1.2.1. The SDN Controller

As the SDN controller becomes the core of SDN, a centralized software entity works as the cognitive center of the network. This controller sets up communication with network components, rendering decisions as per an extensive network view and policies imposed by administrators.

4.1.2.2 Southbound APIs

Enabling communication between the controller and network components like, routers, and switches are southbound Application Programming Interfaces (APIs). These interfaces act as conduits for the controller to provide instructions on traffic handling and update the forwarding tables of components.

4.1.2.3 Northbound APIs

For making contact with applications and business logic, SDN controllers offer northbound APIs. These links authorize developers to craft applications improving the centralized control and programmability accorded by SDN.

4.2 Architectural Echelons of SDN

Figure 2. Overview of SDN architecture

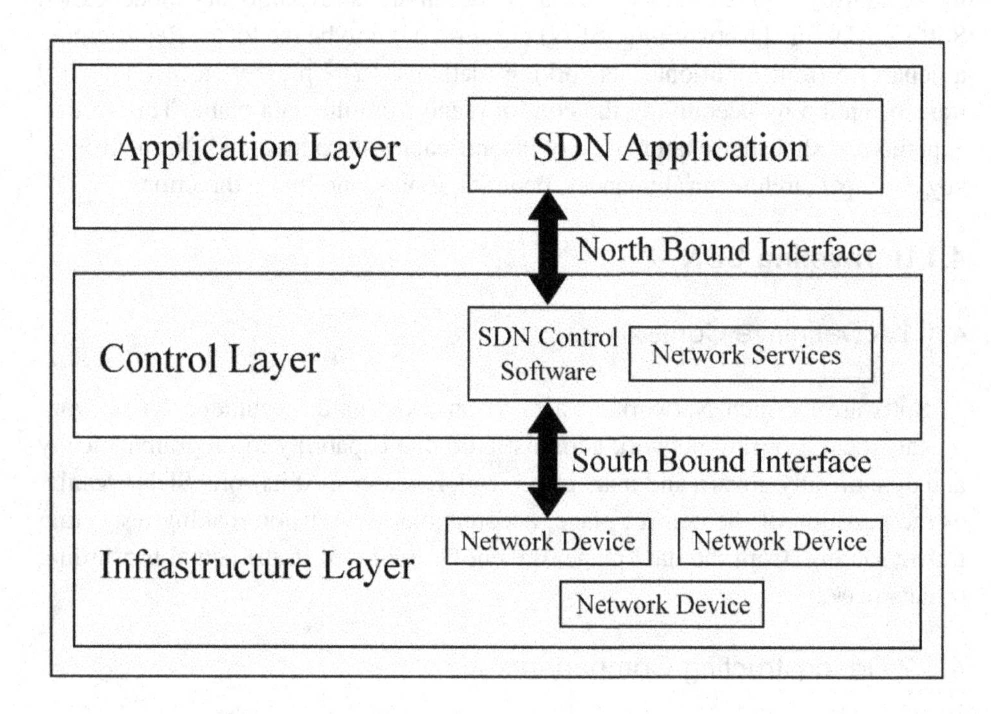

4.2.1 The Control Plane

Conventionally, the control plane and data plane remain within network components, with SDN completely transforming the control plane's locus to the centralized SDN controller. Such separation creates a more active and responsive network infrastructure.

4.2.2 The Data Plane

In SDN, the data plane stays distributed across network components, but these entities now operate as fundamental forwarding elements, yieldingly adhering to instructions distributed from the centralized controller.

4.2.3 Protocols Underpinning Communication

SDN links standard communication conventions for communication between the controller and network components. OpenFlow, a broadly accepted protocol, supports communication, allowing the continuous exchange of instructions between the switches and controller.

4.2.4 Stratified SDN Architecture

Theoretically, SDN architecture composes layers:

4.2.4.1 Application Layer

The peak of the architecture occupies applications and services influencing SDN through northbound APIs. Examples comprise network virtualization, load balancing, and security applications.

4.2.4.2 Control Layer

Placed under the application layer, the control layer comprises the SDN controller, the organizer of global decisions and policy implementation.

4.2.4.3 Infrastructure Layer

At the basis is the physical network infrastructure, comprising routers, switches, and various network devices.

4.3 SDN in Cyber Security

Software-Defined Networking evolves as an essential technology in strengthening the cyber defense techniques within smart grids. As the combination of the latest technologies like 5G converts the domain of energy distribution systems, SDN provides a paradigm shift in network governance. This section examines the key role of SDN in improving the cybersecurity posture of smart grids, highlighting its contributions to centralized control, dynamic resource allocation, and the formation of a secure and responsive network infrastructure.

4.3.1 Centralized Control for Improved Security Oversight

One of the basic contributions of SDN to smart grid cyber defense stands in its formation of centralized control. SDN framework detaches the control plane and data plane, allowing network administrators to control network behavior by utilizing software applications. This centralized control offers a comprehensive prospect of the overall network, allowing improved security oversight. Administrators can supervise, analyze, and answer to security attacks from a centralized point, supporting faster and more effective decision-making regardless of cyber incidents.

4.3.2 Dynamic Resource Allocation Based on Real-Time Conditions

SDN's dynamic resource allotment facilities play a key role in smart grid cyber defense. In answer to real-time network conditions, SDN controls the allotment of latency, bandwidth, and other resources as per the particular demands of various applications. This dynamic resource allotment secures that critical smart grid services receive the mandatory resources to function securely and effectively. SDN's capability to tailor resource allotment in response to varying cybersecurity demands contributes to a resilient defense technique against emerging cyber-attacks.

4.3.3 Creation of Secure Service Chains for Extensive Defense

SDN promotes the formation of secure service chains, linking up with NFV's capabilities, to create an extensive defense mechanism for smart grids. Service chaining includes linking numerous security functions in a particular order to form end-to-end network services. SDN's programmability allows the arrangement of these service chains, enabling the dynamic deployment and cooperation of security measures. This technique makes sure that diverse security operations collaborate continuously, offering a layered and adaptive defense against diverse cyber-attacks aiming at smart grid infrastructures.

4.3.4 Optimizing Network Performance and Resource Usage

SDN's impact on optimizing network performance and resource exploitation contributes greatly to the cyber defense capacities of smart grids. By segregating the control and data planes, SDN enables effective traffic management, minimizing latency and enhancing total network responsiveness. This optimization not only improves the user experience but also offers a more resilient network architectural design. SDN's part in dynamically changing to network conditions secures that cybersecurity measures do not undermine the performance of crucial smart grid services.

4.3.5 Real-Time Governance of Bandwidth, Latency, and Resources

In the area of smart grid cyber defense, SDN's real-time control of bandwidth, latency, and resources is of great significance. This capability secures that crucial applications, such as real-time supervising and control systems, receive the mandatory network resources to function securely and without disturbances. SDN's control over network devices enables traffic prioritization, stopping potential cyber-attacks from influencing necessary smart grid operations.

4.3.6 Integration of SDN for Intrusion Detection and Response

SDN's programmable nature opens paths for innovative techniques for intrusion identification and response within smart grid cyber defense. By exploiting SDN's capability to dynamically adjust network flows, smart grids can carry out real-time intrusion detection systems. When a powerful threat is detected, SDN promotes an automatic and fast response by rerouting traffic or isolating affected portions. This combination improves the speed and accuracy of threat identification, minimizing the potential influence of cyber-attacks on smart grid functions.

4.3.7 SDN's Role in Network Segmentation for Enhanced Security Isolation

Network separation is a crucial technology for improving security, and SDN plays a key role in its implementation within smart grids. By forming logically isolated network portions through SDN, smart grid operators can classify various aspects of the infrastructure. This partition reduces the lateral movement of cyber attacks, confining their scope and impact. SDN's capability to dynamically alter separation policies offers a flexible and adaptive technique to upkeep a secure network architecture.

4.3.8 Ensuring Policy Compliance and Governance

SDN leads to smart grid cyber defense by offering a platform for securing policy compliance and governance. By centralizing governance over network policies, SDN allows uniform enforcement of security measures across the overall smart grid infrastructure. This centralized control secures that security policies are linked up with regulatory demands and industry standards. SDN's programmability permits rapid updates to policies, securing that the smart grid promptly changes to new compliance standards and cybersecurity policies.

4.3.9 Improved Visibility and Monitoring Capabilities

SDN's centralized control plane promotes improved visibility and supervising capabilities within smart grid cyber defense. By summarising network management operations, SDN allows extensive monitoring of anomalies, traffic, and potential security events. Network administrators can procure real-time perceptions of network behavior, promoting proactive detection of doubtful activities. SDN's part in providing a unified prospect of the overall network improves situational awareness, facilitating sooner and more informed responses to powerful cyber threats.

4.4 Industry Perspectives and Vendor Landscape

4.4.1 Key Industry Players

Many industry giants have adopted and supported the SDN cause. Companies such as Juniper Networks, VMware, Cisco, and Arista Networks provide extensive SDN solutions. Each player makes its special features, technique, and ecosystem, imparting to the various prospects of SDN strategies.

4.4.2 Evolving Trends in SDN Offerings

The SDN domain is dynamic, regarding persistent innovation. Evolving trends involve the combination of Artificial Intelligence for prophetic network analytics, intent-oriented networking for streamlined functions, and improved security structures to solve emerging cyber-attacks. Vendors are strenuously emerging their SDN contributions to precede in this competitive domain. Mubashir Husain Rehmani et al. (2019) insisted that network control of the present SG setup is intricate, time-consuming, and manually operated. In addition, the SG communication structure is built upon various vendor-specific components and guidelines. So, the present SG structure is protocol dependent, therefore resulting in interoperability problems. For this application, SDN is suggested to supervise and control the communication networks globally. By segregating the control plane from the data plane, SDN supports the network providers in controlling network adaptability. As the SG system strongly depends on communication network structure, SDN has set the stage for SG. Tianyang Cai et al. (2023) presented the Adaptive DDoS Attack Mitigation(ADAM) method in an SDN cyber physical scheme. By integrating information entropy and unsupervised anomaly identification technique ADAM finds the present state spontaneously and adaptively detects skeptical features which are meant to alleviate DDoS threats in a prompt method. Nikolaos et al. (2017) implemented a proof-of-concept reactive security design for an industrial-grade wind park system which combines SDN and

Supervisory Control and Data Acquisition (SCADA) honeypots, portable to the wind park, permitting uninterrupted supervision of the industrial network system and extensive examination of possible threats, thus separating invaders and allowing to estimate of the level of complexity. Future power systems will be extremely complicated, with many intelligent modules exchanging and processing massive data content as well as vital information in real-time. Accordingly, to enable autonomous supervision, governance, and control and ensure safe power system functioning, dependable, real-time capable, and guaranteed communication systems are needed. Nils Dorsch (2014) presents and analyzes an SDN-based dynamic and adaptable control strategy for satisfying the unique communication needs of the transmission and distribution power grids.

4.5 International Collaborations and SDN Research

Global collaborations and exploration projects within the SDN landscape take the lead in progressing the state-of-the-art research institutions, academia, and organizations, which are involved in collective efforts to address SDN issues, assist in standardization activities, and examine innovative use cases.

4.6 Future Directions and Innovations in SDN for Smart Grid Cyber Defense

The integration of SDN for smart grid cyber defense proceeds with persistent research and innovations. Researchers are examining modern machine learning techniques combined with SDN to allow intelligent and adaptive solutions to cyber challenges. In addition, SDN adaptation tools are being refined to ease the deployment and management of security strategies within intricate smart grid environments. These novelties imply SDN's dynamic part in building the future of smart grid cybersecurity.

SDN evolves as a basic mechanism for strengthening the cybersecurity defenses of smart grids. Its offerings in secure service chains, centralized control, dynamic resource allotment, and improved visibility jointly elevate the adaptability and flexibility of smart grid cyber defense. As smart grids direct the intricacies of the digital age, SDN's part in offering a secure, programmable, and centrally controlled network framework becomes more prominent. The continuing inclusion of SDN innovations assurances a future where smart grids can actively and effectively obstruct evolving cyber threats, securing the ongoing reliability and safety of energy distribution systems.

4.7 Comparison Between NFV and SDN Technologies

From the comparison point of view, NFV follows the function abstraction approach and multiple control protocol whereas, SDN involves the networking abstraction method and openflow protocol. NFV mainly guarantees flexibility and cost minimization while SDN assures to bring interconnected programmable control and open interfaces.

Rashid Mijumbi et al. (2016) stated that SDN and NFV are novel paradigms that aim to handle various prospects of a software-driven networking solution. NFV plans to disentangle NFs from specific hardware components, whereas SDN concentrates on isolating packet and connection management from overall network management. The Open Network Foundation notes that "the NFV concept differs from the virtualization concept as used in the SDN architecture" in their definition of the SDN framework. Virtualization in SDN design refers to assigning abstract properties to specific clients or applications; in Network Function Virtualization (NFV), the objective is to extract network functions (NFs) from particular hardware so that they can be hosted on cloud data center server platforms, for example. It is evident that while telecom providers are spearheading comparable endeavors for NFV, the data center and cloud computing sectors are making the greatest efforts to promote and standardize SDN. Lastly, a crucial contrast to be considered is that SDN necessitates a new network architecture where the data and control planes are isolated, but NFV may operate on current networks since it is hosted on servers and engages with particular traffic that is supplied to them.

Diverse proprietary network equipment leads to issues with network ossification while raising service providers' capital and operating costs. To solve these problems, NFV is suggested, which involves putting network operations as pure software on commodity and hardware in general. Virtual network functions can be deployed, provisioned, and managed centrally with flexibility thanks to NFV. The software-defined NFV design also provides cooperative optimization of network tasks and properties, as well as agile traffic steering when integrated with SDN. This architecture is starting to take over as the most popular type of NFV and benefits from numerous applications like service chaining. An extensive analysis of NFV development under the software-defined NFV architecture is made by Yong Li et al. (2015) with a focus on service chaining as its application.

5. CONCLUSION

In conclusion, the consolidation of Network Function Virtualization and Software-Defined Networking in the domain of smart grid cyber defense provides a potential and transformative solution for ensuring crucial infrastructure. This combination not only impacts the architecture of electricity networks but also presents a dynamic and adaptive technique for cybersecurity. The mutual effect of NFV and SDN permits extended scalability, effective resource exploitation, and simple network management, offering a strong foundation for modern cyber defense techniques.

By decoupling conventional network operations from specific hardware virtualizing them with NFV, and centralizing network control with SDN, the smart grid becomes more resilient against emerging cyber-attacks. The capability to dynamically reply to real-time network scenarios and threats improves the system's ability to resist and restore threats, securing continuous electricity services.

Furthermore, the efficiency gains attained using resource optimization and scalability impact the economic sustainability of smart grid functionalities. The economic rationality of incorporating security operations on a virtualized infrastructure, combined with the on-demand resource allotment enabled by SDN, emphasizes the financial viability of carrying out and maintaining strong cyber defense measures. Economic efficiency is vital for the ongoing growth and advancement of smart grid technologies.

Moreover, the integration of NFV and SDN simplifies the complexity of managing and arranging security measures within the smart grid. It not only enhances the ability and responsiveness of cyber defense techniques but also simplifies the management overheads of electricity operators.

In essence, the combination of NFV and SDN in smart grid cyber defense sets a substantial development towards a highly secure and adaptable crucial infrastructure. As smart grids continue to take the lead in modernizing energy systems, the collective influence of NFV and SDN lays the groundwork for a resilient and potentially productive cyber defense framework, impacting the overall reliability and security of crucial energy delivery systems.

REFERENCES

Abdelwahab, S., Hamdaoui, B., Guizani, M., & Znati, T. (2016, April). Network function virtualization in 5G. *IEEE Communications Magazine*, 54(4), 84–91. 10.1109/MCOM.2016.7452271

Cai, T., Jia, T., Adepu, S., Li, Y., & Yang, Z. (2023). ADAM: An Adaptive DDoS Attack Mitigation Scheme in Software-Defined Cyber-Physical System. *IEEE Transactions on Industrial Informatics*, 19(6), 7802–7813. 10.1109/TII.2023.3240586

Chen, J. (2020). Smart Grid Technology and 5G Integration: Advantages and Challenges. *Journal of Energy Technology*.

Chen, Y. (2024). Collaborative Approach of NFV and SDN for Cybersecurity in Smart Grids. *IEEE Transactions on Smart Grid*.

Dorsch, N., Kurtz, F., Georg, H., Hägerling, C., & Wietfeld, C. (2014). Software-defined networking for Smart Grid communications: Applications, challenges, and advantages. *2014 IEEE International Conference on Smart Grid Communications (SmartGridComm)*, Venice, Italy. 10.1109/SmartGridComm.2014.7007683

European Telecommunications Standards Institute (ETSI). (2021). *Standardization in 5G-Enabled Smart Grids*. ETSI.

Gadallah, W., Ibrahim, H., & Omar, N. (2024). A deep learning technique to detect distributed denial of service attacks in software-defined networks. *Computers & Security*, 137, 103588. 10.1016/j.cose.2023.103588

Guo, H., et al. (2024). Interoperability in 5G-Enabled Smart Grids: Challenges and Solutions. *IEEE Transactions on Industrial Informatics*. IEEE.

Huang, Q. (2019). Real-time Cyber Threat Monitoring with NFV and SDN. *IEEE Transactions on Dependable and Secure Computing*.

International Electrotechnical Commission (IEC). (2022). *IEC Standards for Cybersecurity in Smart Grids*. IEC.

Kim, D., & Lee, J. (2023). Cross-Sector Collaboration for Interoperable Cybersecurity Solutions. *International Journal of Critical Infrastructure Protection*.

Kim, S. (2020). Cybersecurity Challenges in 5G-Enabled Smart Grids. *Journal of Information Security and Applications*.

Li, H. (2021). Securing the Fusion: Vulnerabilities and Security Challenges in the Integration of 5G into Smart Grids. *Journal of Energy Cybersecurity*.

Li, H. (2021). Centralized Control for Cybersecurity in 5G-Enabled Smart Grids: The Role of SDN. *Journal of Cybersecurity and Energy Systems*.

Li, M., & Zhang, S. (2024). Emerging Technologies in 5G-Integrated Smart Grids: A Comprehensive Review. *IEEE Transactions on Power Systems*. IEEE.

Li, W. (2024). Ensuring Data Privacy in Smart Grids: A Comprehensive Approach. *Journal of Information Security and Privacy*.

Li, X. (2022). Collaborative Security Solutions for 5G-Enabled Smart Grids. *International Journal of Critical Infrastructure Protection*.

Li, Y., & Chen, M. (2015). Software-Defined Network Function Virtualization: A Survey. *IEEE Access : Practical Innovations, Open Solutions*, 3, 2542–2553. 10.1109/ACCESS.2015.2499271

Liang, Y. (2018). NFV and SDN for Cybersecurity in Smart Grids: Dynamic Security Measures." . *International Journal of Critical Infrastructure Protection*.

Liu, J., and Wu, H. (2022). Privacy Concerns and Data Protection in 5G-Enabled Smart Grids. *International Journal of Information Management*.

Liu, M. (2024). Enhancing Cyber-Resiliency of DER-Based Smart Grid: A Survey. *IEEE Transactions on Smart Grid*. IEEE. 10.1109/TSG.2024.3373008

Liu, S. (2020). Anomaly Detection Techniques for Cyber Defense in Smart Grids." . *IEEE Transactions on Smart Grid*.

Liu, X. (2022). Dynamic Deployment of Network Functions in 5G: A Focus on NFV. *Journal of Communication Networks*.

Liu, Y., & Chen, Z. (2023). Service Chains in 5G Networks: Enabling End-to-End Network Services. *Wireless Communications and Mobile Computing*.

Mijumbi, R., Serrat, J., Gorricho, J.-L., Bouten, N., De Turck, F., & Boutaba, R. (2016). Network Function Virtualization: State-of-the-Art and Research Challenges. *IEEE Communications Surveys and Tutorials*, 18(1), 236–262. 10.1109/COMST.2015.2477041

Petroulakis, N. E., Fysarakis, K., Askoxylakis, I. G., & Spanoudakis, G. (2017). Reactive Security for SDN/NFV-enabled Industrial Networks leveraging Service Function Chaining. *Transactions on Emerging Telecommunications Technologies*.

Rehmani, M. H., Davy, A., Jennings, B., & Assi, C. (2019). Software Defined Networks-Based Smart Grid Communication: A Comprehensive Survey. *IEEE Communications Surveys and Tutorials*, 21(3), 2637–2670. 10.1109/COMST.2019.2908266

Sharma, R. (2021). Securing Critical Energy Infrastructure: A Study on 5G-Enabled Smart Grids. *IEEE Transactions on Industrial Informatics*. IEEE.

Song, H., & Wang, Y. (2024). Industry Partnerships for Cybersecurity in Smart Grids: A Collaborative Approach. *Journal of Cybersecurity and Energy Systems*.

Song, Q. (2019). Application of NFV and SDN in Enhancing Cybersecurity in Smart Grids. *Smart Grids: Fundamentals and Technologies*.

Wang, J. (2019). Vulnerabilities in Communication Protocols of Smart Grids: A Comprehensive Analysis. *Journal of Cybersecurity and Energy Systems*.

Wang, L. (2020). SDN in 5G Networks: Optimizing Performance and Resource Utilization. *IEEE Transactions on Network and Service Management*. IEEE.

Wang, Q. (2023). AI-Based Strategies for Detecting and Responding to Cyber Threats in 5G-Enabled Smart Grids. *IEEE Transactions on Smart Grid*. IEEE.

Wu, S. & Zhang, J. (2023). Energy-Aware Security Measures in 5G-Enabled Smart Grids. *IEEE Transactions on Sustainable Energy*. IEEE.

Wu, Z. (2023). Integration of NFV and SDN for Enhanced Security in 5G-Enabled Smart Grids. *Journal of Network and Computer Applications*.

Xu, L. (2023). Resilience and Redundancy in Smart Grid Cybersecurity. *Journal of Energy Security and Resilience*.

Xu, L., Hu, H., & Liu, Y. (2023). SFCSim: A network function virtualization resource allocation simulation platform. *Cluster Computing*, 26(1), 423–436. 10.1007/s10586-022-03670-8

Yang, J. (2022). Dynamic Resource Allocation in NFV: Adapting to Varying Network Demands." . *Journal of Network and Systems Management*.

Zhang, H., & Wang, G. (2021). Unified Network View and Centralized Control: SDN's Contribution to Network Management . *Computer Networks*.

Zhang, L., & Chen, W. (2023). Machine Learning and Artificial Intelligence in Cybersecurity for Smart Grids. *Journal of Cybersecurity Research*.

Zhang, Q. (2021). Integration of 5G into Smart Grids: Cybersecurity Measures. *International Journal of Smart Grids and Communications*.

Zhang, Q. (2022). Machine Learning Algorithms for Cyber Defense in 5G-Integrated Smart Grids. *International Journal of Smart Grid and Clean Energy*.

Zhao, X., & Liu, W. (2023). Future Trends in 5G, NFV, and SDN Integration for Cybersecurity in Smart Grids. *Journal of Energy Engineering*.

Zhao, Y. (2019). NFV in 5G Networks: Flexibility, Scalability, and Programmability. *IEEE Transactions on Network and Service Management*. IEEE.

Chapter 14
Securing the Future:
Regulatory Compliance and Standards for 5G and Fiber Optics in Smart Grid Cyber Defense

Nedumal Pugazhenthi
https://orcid.org/0000-0003-4496-5253
Dr. N.G.P. Institute of Technology, India

Vijayakumar Kaliappan
Dr. N.G.P. Institute of Technology, India

Selvaperumal Sundarmoorthy
https://orcid.org/0000-0002-7398-6953
Mohamed Sathak Engineering College, India

Prabhakar Gunasekaran
https://orcid.org/0000-0003-4760-7646
Thiagarajar College of Engineering, India

ABSTRACT

Integrating smart grid technology with 5G and fiber optic systems offers significant improvements in efficiency and resilience, but also presents substantial cybersecurity challenges. This study examines the necessary standards and regulatory compliance for secure implementation, highlighting the importance of adhering to guidelines from organizations like 3GPP, ITU, IEEE, and IEC to ensure robust security and interoperability. Fiber optic infrastructure, essential for fast and secure data transmission, requires stringent physical security and advanced encryption techniques to protect against manipulation and attacks. Additionally, compliance with privacy laws such as GDPR is crucial for handling sensitive data. Organizations must con-

DOI: 10.4018/979-8-3693-2786-9.ch014

tinuously monitor and update their cybersecurity measures to address the evolving threat landscape, ensuring the integrity and reliability of smart grid infrastructures.

1. INTRODUCTION

The relevance of smart grid security is multidimensional and demands top priority. Smart grids, first and foremost, improve the resilience of electricity distribution systems by facilitating faster detection and reaction to disruptions or outages, hence reducing downtime and financial losses. Furthermore, robust cybersecurity measures must be in place as smart grids become increasingly digitally integrated and linked in order to protect sensitive data and essential infrastructure from online threats including malware attacks, hacking, and data breaches. The significance of securing smart grids cannot be overstated. As highlighted by Farhangi (2010), the increasing digitization and interconnectivity of energy infrastructure expose it to a myriad of cyber threats. The potential impact of cyber-attacks on smart grids goes beyond mere service disruptions; it extends to the integrity of critical infrastructure, data privacy, and the overall stability of the energy supply (Mohammed & George, 2022). Additionally, energy distribution and consumption are optimized by protected smart grids, which improve efficiency, decreases energy waste, and lower operating costs for both utilities and customers. Additionally, because smart networks facilitate the efficient management of intermittent renewable energy sources like solar and wind power and promote the adoption of sustainable energy practices, they facilitate the integration of these sources more easily. One of the most important ways to improve the capabilities of smart grids is to combine fiber optics and 5G technologies. 5G provides high-speed, low-latency connection, making it possible for smart grid devices to quickly and effectively transmit real-time data. This speeds up decision-making, improves control and monitoring of the grid, and permits prompt reaction to abnormalities or failures in the grid. With its low-latency communication and high data transfer rates, 5G facilitates real-time monitoring, control, and optimization of the grid, (Ourahou et al., 2020). However, the very connectivity that enhances efficiency also introduces vulnerabilities that can be exploited by malicious actors, (Nagasri & Marimuthu, 2023). Additionally, 5G enables an enormous number of linked devices to operate concurrently, opening the door for the utility networks to widely use smart grid components. However, fiber optics offer dependable, high-bandwidth connectivity that is necessary for sending the massive amounts of data produced by smart grid management systems, meters, and sensors. By strengthening grid resilience, enhancing energy management, and facilitating the smooth integration of renewable energy sources, the combination of 5G and fiber optics produces a strong communication strengthening grid resilience, enhancing energy management, and facilitating the smooth integration of renewable energy sources, the combination of 5G and fiber optics produces a strong communication infrastructure that ultimately contributes to a more effective, dependable, and sustainable energy ecosystem.

Moreover, the reliance on fiber optics for high-speed and reliable communication within smart grids introduces additional complexities. While fiber optics offer numerous advantages, they also present potential entry points for cyber threats that must be diligently addressed, (Letaief et al., 2019). In light of these challenges, regulatory compliance and adherence to cybersecurity standards become imperative components of the overarching strategy to safeguard smart grids. Strong cyber defense for smart grids depends critically on regulatory compliance and adherence to cybersecurity standards. Industry-specific standards, the NIST Cybersecurity Framework, and ISO/IEC 27001 compliance frameworks, among others, offer instructions for putting in place efficient security measures, evaluating risks, and promoting a cybersecurity-aware culture. Following these guidelines encourages industry best practices, information exchange, and interoperability in addition to assisting enterprises in identifying and mitigating vulnerabilities. This chapter will delve into the evolution of smart grids, the specific challenges posed by 5G and fiber optics, and the critical need for robust regulatory frameworks and standards to ensure the cyber defense of smart grids.

2. EVOLUTION OF SMART GRIDS

Over several decades, the idea of a smart grid has developed in response to the demand for an energy infrastructure that is more sustainable, dependable, and efficient. Large-scale power plants and one-way electricity flow to customers were the main emphasis of centralized networks that were developed in the 20th century as a result of improvements in power generation and distribution. Digital technologies such as Supervisory Control and Data Acquisition (SCADA) systems for grid monitoring and control were introduced in the 1970s and 1980s.

Early in the twenty-first century, the phrase "smart grid" became popular, signifying a move toward grid modernization and the incorporation of cutting-edge technologies. An important turning point was reached in the United States with the Energy Independence and Security Act of 2007, which emphasized the development of smart grids as a means of improving energy efficiency.

As mentioned in McDaniel and McLaughlin (2009) the smart grid evolution has five important themes which is given in figure 1. Prioritizing distributed energy supplies from renewable sources and integrating storage is crucial. Grid control and asset optimization come in second. Workforce effectiveness, the third component, is deemed crucial while Smart metering, is the fourth component. Effective management of energy supply curve is also a vital theme of smart grid. These solutions aim to fulfill the energy management demands of the customer by providing more individualized solutions and an environmentally friendly energy supply, all while modernizing the SCE system to increase safety, cost efficiency, and dependability.

Figure 1. Important themes of smart grid

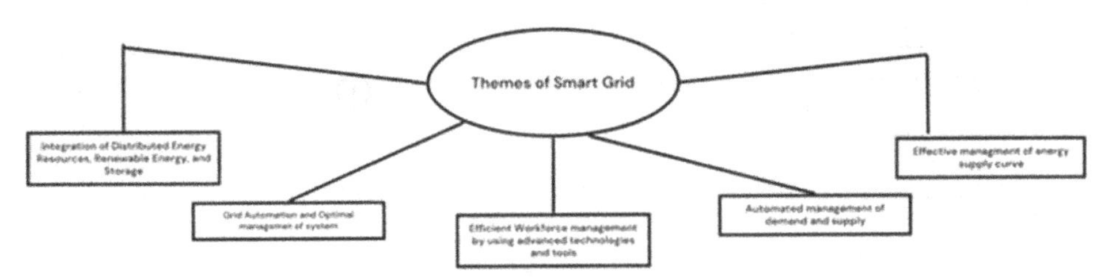

Comprehending the evolution of smart grids, the above themes could also be further explored as the key functionalities and features as shown in figure 2. Also it is crucial to delineate the key components and functionalities that distinguish them from traditional power distribution systems. The Advanced Metering Infrastructure (AMI) is a crucial element of smart grids since it facilitates bidirectional communication between utilities and consumers via smart meters, hence improving grid management and demand response. Grid Automation employs sensors and intelligent devices for real-time monitoring, improving reliability and energy distribution. Distributed Energy Resources (DERs) like solar panels integrate seamlessly, promoting sustainability. Robust Communication Networks, often utilizing 5G and fiber optics, facilitate secure data exchange. Grid Analytics and Data Management platforms process data for predictive maintenance and optimization, ensuring efficient grid performance and resilience.

Figure 2. Key features and functionalities

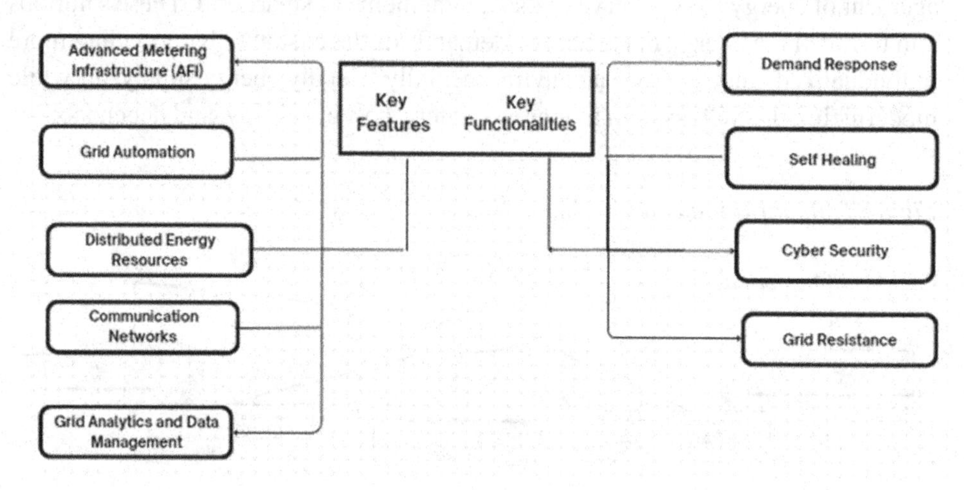

Similarly Smart grids encompass key functionalities essential for modern energy management. Demand Response programs, enabled by dynamic pricing and real-time grid conditions, empower consumers to adjust energy usage, enhancing grid stability and efficiency. Self-Healing capabilities allow smart grids to autonomously diagnose and respond to faults, minimizing downtime and improving reliability through automated power rerouting. Cyber security measures like encryption and intrusion detection systems safeguard against cyber threats in the interconnected grid environment. Grid Resilience, crucial for adapting to disruptions like extreme weather or cyber-attacks, is ensured through redundant systems and rapid response mechanisms, ensuring continuous grid stability and uninterrupted energy supply. The integration of 5G and fiber optics represents a significant advancement in communication and data transfer technologies, promising unprecedented speed, reliability, and efficiency. Fiber optics, known for their high bandwidth and low latency, serve as the backbone for 5G networks, enabling them to reach their full potential. This synergy enhances various applications, from mobile connectivity to smart cities, and supports the growing demand for data-intensive services like augmented reality (AR) and the Internet of Things (IoT).

3. CYBER SECURITY CHALLENGES IN SMART GRIDS

Securing smart grids is imperative as they become more interconnected and reliant on advanced communication technologies. Identifying cyber security threats, understanding vulnerabilities in traditional communication systems, and recognizing risks associated with increased connectivity are crucial aspects of ensuring a resilient smart grid infrastructure.

3.1 Identification of Cyber Security Threats in Smart Grids

Threats originating from within an organization, such as employees or contractors with malicious intent, pose a significant risk to smart grid cybersecurity. Smart grids are susceptible to malware attacks, including ransomware, which can disrupt operations, compromise data integrity, and demand financial extortion Deliberate attempts to overwhelm smart grid systems with excessive traffic, leading to service disruption and degradation, are persistent threats Cyber attackers may employ phishing techniques to trick individuals into divulging sensitive information, posing a direct threat to the integrity of smart grid operations

3.2 Vulnerabilities Associated With Traditional Communication Systems

Traditional communication systems in smart grids may still rely on legacy protocols that lack robust security features, making them susceptible to exploitation. Inadequate encryption mechanisms in traditional communication channels may expose sensitive data to interception and unauthorized access. Weak or absent authentication mechanisms in traditional systems may allow unauthorized entities to gain access to critical infrastructure components

4. REGULATORY LANDSCAPE

As the deployment of smart grids becomes widespread, the need for robust regulatory frameworks governing cybersecurity, 5G, and fiber optics is paramount. This section provides an overview of global and regional regulations, examines frameworks for 5G and fiber optics deployment, and outlines compliance requirements for utility companies and technology providers.

4.1 Global and Regional Regulations Governing Smart Grid Cybersecurity

Internationally, the ISO/IEC 27001 standard offers a framework for information security management that is applicable to smart grids. Power systems in the US are required to implement comprehensive cybersecurity safeguards in accordance with the Critical Infrastructure Protection (CIP) guidelines set out by the North American

Electric Reliability Corporation (NERC). In Europe, the Network and Information Systems (NIS) Directive requires member states to implement strong cybersecurity protocols for essential services, including energy. In Asia, Japan's Ministry of Economy, Trade, and Industry (METI) enforces strict cybersecurity guidelines within its smart grid roadmap. Mandates and recommendations for cybersecurity in the energy industry are issued by CERT-In and the Central Electricity Authority (CEA) of India. Together, these laws strengthen smart grids' defenses against cyberattacks and guarantee safe and effective energy delivery.

4.2 Europe: EU Directive 2009/72/EC

With the goals of promoting investment in renewable energy, ensuring consumer protection, and enhancing competition, the EU Directive 2009/72/EC provides uniform regulations for the internal electricity market. In order to avoid monopolistic tactics and provide equitable access to the grid, it requires the unbundling of energy supply and generation from transmission networks. The directive mandates that autonomous regulatory bodies be established in every member state to supervise market activities and ensure adherence to regulations. It also supports the EU's wider energy and climate goals by highlighting the development of smart networks to increase energy efficiency and integrate renewable energy sources.

4.3 Asia: Smart Grid Roadmap in Japan

The integration of renewable energy, strengthening grid resilience, and increasing energy efficiency within a strong regulatory framework are the main objectives of Japan's smart grid plan. The "Smart Community" initiative, which is spearheaded by the Ministry of Economy, Trade, and Industry (METI), promotes smart meters, demand response technology, and energy storage options. Initiatives like the Energy Conservation Act, which establishes energy efficiency requirements, and the Electricity Business Act, which guarantees market liberalization and system upgrading, provide regulatory assistance. METI's standards also mandate cybersecurity precautions to protect the grid. The goal of smart grid technology demonstration projects is to provide a sustainable, dependable, and efficient energy infrastructure that will aid Japan in its shift to a low-carbon economy. These projects are being carried out in a number of different locations.

4.4 Examination of Regulatory Frameworks for 5G and Fiber Optics Deployment

Different technology landscapes and goals are reflected in the legal frameworks across different countries for the deployment of fiber optics and 5G. The Federal Communications Commission (FCC) is in charge of 5G implementation in the United States. Through its 5G FAST Plan, the FCC prioritizes regulatory modernization, infrastructure siting, and spectrum allocation. The European Electronic Communications Code (EECC) and the "5G for Europe Action Plan," which prioritize coordinated spectrum allocation, cross-border harmonization, and investment promotion, serve as the regulatory framework for the European Union. The Ministry of Industry and Information Technology (MIIT) in China oversees the rollout of 5G technology, placing a high priority on infrastructure development and broad government backing. Global frameworks for fiber optics prioritize infrastructure sharing and high-speed internet connectivity. While the EU's Broadband Cost Reduction Directive seeks to reduce deployment costs and improve broadband coverage, the FCC's Open Internet Order in the United States encourages net neutrality and infrastructure investment. The National Digital Communications Policy, 2018 of India promotes the use of fiber optics by streamlining regulations and forming public-private partnerships. All together, these frameworks seek to improve connectivity, maintain competitive markets, and spur technical advancement in the telecom industry.

4.5 Compliance Requirements for Utility Companies and Technology Providers

In order to guarantee the security and resilience of sophisticated power infrastructure, smart grid cyber defense requires strict adherence to national, international, and local standards. Globally, information security management standards like ISO/IEC 27001, which outline procedures for safeguarding data and controlling cyber threats, must be complied with by smart grid technologies. Adherence to the General Data Protection Regulation (GDPR) is essential for protecting personal data in smart grid networks in the European Union. The Critical Infrastructure Protection (CIP) standards of the North American Electric Reliability Corporation (NERC) are essential in the United States because they provide strict guidelines for protecting crucial electric infrastructure against cyberattacks. policies to safeguard the grid are enforced in India by the Central Electricity Authority (CEA) and the Ministry of Power. These policies include cybersecurity best practices specific to the energy industry. In order to ensure prompt and efficient action against cyber threats, the Indian Computer Emergency Response Team (CERT-In) publishes guidelines for cyber incident response and management. Furthermore, the Information Technology

Act of 2000 regulates data protection and cybersecurity, requiring strong security protocols to safeguard vital information infrastructure. Another important player in protecting assets vital to economic stability and national security is the National Critical Information Infrastructure Protection Centre (NCIIPC). Adherence to these laws guarantees the secure operation of smart grid systems, safeguarding against cyberattacks that may cause disruptions in the electricity supply and jeopardize data integrity. Utility firms and technology suppliers may guarantee the robustness and dependability of smart grid infrastructure by adhering to these strict norms and standards. This will promote innovation and protect against the constantly changing cyber threat scenario.

5. CONCLUSION

In summary, there are many chances to increase efficiency and resilience by combining smart grid technologies with 5G and fiber optic networks, but there are also many cybersecurity risks to consider. In order to guarantee strong security protocols, this study emphasizes the significance of abiding by strict national and international standards, such as those established by 3GPP, ITU, IEEE, and IEC. To safeguard sensitive data, enterprises also need to take care of physical security, use cutting-edge encryption methods, and abide by privacy laws like GDPR. Maintaining the integrity of smart grid infrastructures requires constant monitoring and updating of cybersecurity measures.

REFERENCES

Farhangi, H. (2010, January-February). The path of the smart grid. *IEEE Power & Energy Magazine*, 8(1), 18–28. 10.1109/MPE.2009.934876

Goran, S. (2008). Demand side management: Benefits and challenges. *Energy Policy, 36*(12). 10.1016/j.enpol.2008.09.030

Lakshmi Satya Nagasri, D., & Marimuthu, R. (2023). Review on advanced control techniques for microgrids. *Energy Reports, 10.*10.1016/j.egyr.2023.09.162

Letaief, K. B., Chen, W., Shi, Y., Zhang, J., & Zhang, Y. J. A. (2019). The roadmap to 6G: AI empowered wireless networks. *IEEE Communications Magazine*, 57(8), 84–90. 10.1109/MCOM.2019.1900271

Li, K., Li, H., Chen, X., & Li, Y. (2020). A survey of data-driven methods for fault diagnosis in smart grids. *IEEE Access : Practical Innovations, Open Solutions*, 8, 159437–159453.

Liu, X., Chen, X., & Chen, Y. (2021). A survey of integration of big data analytics with smart grid. *IEEE Access : Practical Innovations, Open Solutions*, 9, 14310–14326.

McDaniel, P., & McLaughlin, S. (2009, May-June). Security and Privacy Challenges in the Smart Grid. *IEEE Security and Privacy*, 7(3), 75–77. 10.1109/MSP.2009.76

Mohammed. (2022). Vulnerabilities and Strategies of Cybersecurity in Smart Grid - Evaluation and Review. *2022 3rd International Conference on Smart Grid and Renewable Energy (SGRE)*, Doha, Qatar. 10.1109/SGRE53517.2022.9774038

Ourahou, M., Ayrir, W., Hassouni, B. E., & Haddi, A. (2020). Review on smart grid control and reliability in presence of renewable energies: Challenges and prospects. *Mathematics and Computers in Simulation*, 167, 19–31. 10.1016/j.matcom.2018.11.009

Priyadarshini, I., Kumar, R., Sharma, R., Singh, P. K., & Satapathy, S. C. (2021). Pradeep Kumar Singh, Suresh Chandra Satapathy, Identifying cyber insecurities in trustworthy space and energy sector for smart grids. *Computers & Electrical Engineering*, 93, 107204. 10.1016/j.compeleceng.2021.107204

Zhang, Y., Huang, T., & Bompard, E. F. (2018). Big data analytics in smart grids: A review. *Energy Informatics*, 1(1), 8. 10.1186/s42162-018-0007-5

Section 5
Advanced Techniques, Future Directions, and Case Studies

Chapter 15
Implementing Secure Machine Learning:
A Decision Tree Approach for Smart Grid Stability

J. Shanthi
Thiagarajar College of Engineering, India

M. Rajalakshmi
https://orcid.org/0000-0001-8532-2452
Thiagarajar College of Engineering, India

D.Gracia Nirmala Rani
https://orcid.org/0000-0001-7974-1183
Thiagarajar College of Engineering, India

S. Muthulakshmi
https://orcid.org/0009-0000-7293-2411
Velammal College of Engineering and Technology, India

ABSTRACT

This chapter investigates integrating a decision tree algorithm to boost smart grid stability within the 5G and fiber optics security framework. It stresses robust cyber defense for critical infrastructure, using machine learning, especially decision trees, to analyze smart grid stability data. Steps involve thorough data preprocessing, algorithm justification, model training, and evaluation. Visual aids aid interpretation. The chapter ends with future research suggestions, underlining machine learning's evolving role in enhancing security.

DOI: 10.4018/979-8-3693-2786-9.ch015

1. INTRODUCTION

1.1. Overview of Smart Grids

Smart grids are the modernized electricity distribution and management method, integrating advanced digital technologies, communication networks, and renewable energy sources to optimize grid performance and reliability. Smart grids, unlike conventional grids, support interactive communication between utilities and consumers. This capability enables real-time monitoring, management, and optimization of electricity distribution, as depicted in Figure 1. The main elements of smart grids consist of smart meters, sensors, automated control systems, and energy storage devices. These components are interconnected to enhance grid flexibility, efficiency, and resilience, as illustrated in Figure 2. Using data analytics and automation, smart grids enable stakeholders to make informed decisions, enhance energy efficiency, and seamlessly incorporate renewable energy sources into the grid infrastructure.

Figure 1. The structure of a smart grid

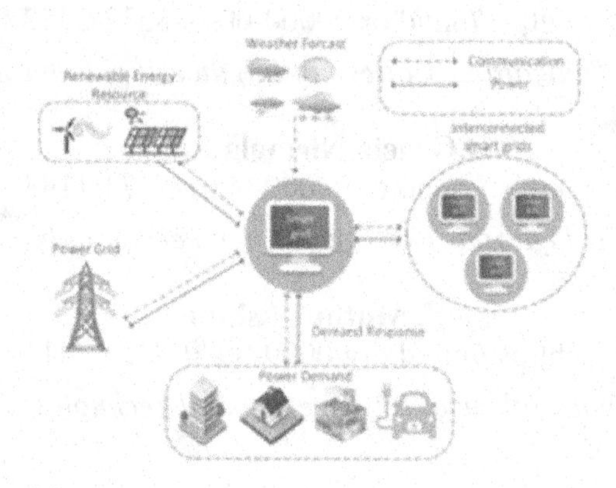

Figure 2. The components of a smart grid

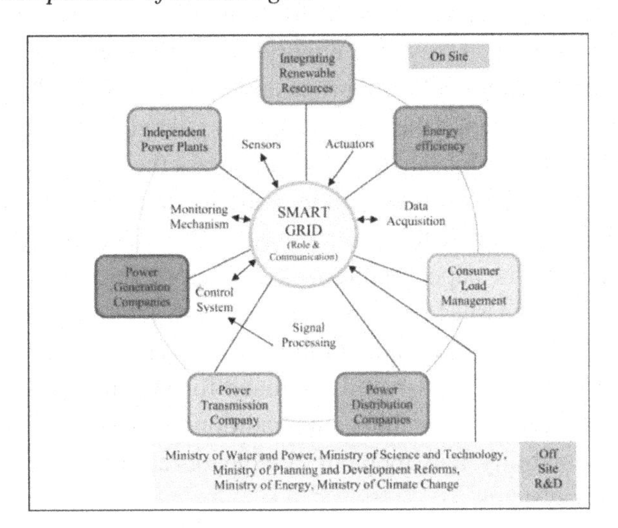

Several studies in the literature have focused on addressing the stability of smart grids. For example, one research project suggested a hierarchical transactive control strategy designed to uphold frequency regulation within an IEEE 30-bus system. The goal is to maintain an equilibrium between power generation and consumption by continually adapting power production to accommodate fluctuations in demand. In another study, a team proposed a robust nonlinear controller featuring an innovative method for compensating time delays to improve grid stability. Researchers also utilized an extreme gradient boosting approach to assess grid stability by analyzing the connection between power system properties and stability. Simulation results demonstrated a remarkable accuracy rate of 97.8% for this algorithm in power grid systems. In another study, the stability of power grids was evaluated using Artificial Neural Networks (ANNs) to assess voltage stability, with the model's validity confirmed through testing within a power grid system. A novel approach was developed (Shi et al., 2021) to address the issues of integrating renewable energy sources and demand response programs in smart-grid systems. The paper uses distributed model predictive control (DMPC) techniques to optimize power flow and coordinate real-time demand response actions to enhance grid stability, productivity, and consistency. The paper's main contribution lies in proposing a distributed model predictive control framework that integrates request response and ideal power distribution while considering the existence of renewable energy resources. The DMPC approach enables decentralized decision-making among multiple grid entities, such as consumers, generators, and storage systems, while ensuring coordination and coherence in achieving grid objectives. The paper

(Alsirhani et al.,2023) addresses the critical need for accurate and timely stability prediction in smart grid management, considering the increasing complexity and uncertainty introduced by renewable energy integration and demand variability. The paper's main contribution lies in proposing a hybrid MLP-ELM technique for stability prediction, leveraging the complementary strengths of both neural network architectures. The work (Mohsen et al.,2023) address the crucial need for accurate and efficient stability prediction methods in smart grid management, considering the increasing complexity and variability introduced by renewable energy integration and demand fluctuations. The paper emphasizes the importance of predictive analytics in anticipating stability issues and enabling proactive grid management strategies to mitigate risks and prevent cascading failures. The paper's main contribution lies in proposing an efficient ANN-based approach for stability prediction in smart grids. The paper concentrates on enhancing traditional ANN models' computational proficiency and scalability, making them suitable for real-time stability assessment in large-scale grid systems.

In one study (Cui et al.,2022), transient grid stability was estimated using transfer functions. A developed neural network served as a classifier for these transfer functions, and the proposed model's validity was assessed using the IEEE 39-bus test system. Another approach combined Feedforward Neural Networks (FNNs) with optimization algorithms like Particle Swarm Optimization (PSO) and Grey Wolf Optimizer (GWO) to create a resilient prediction way for determining the stability grade of the grid system. Furthermore, another study evaluated the grid system's stability status under significant disturbances using Support Vector Machines (SVM) (You et al., 2013). The survey (Ucar et al.,2023) provides an in-depth investigation into predicting smart grid stability, coupled with applying Explainable Artificial Intelligence (XAI) techniques. The study aims to enhance our understanding of smart grid stability while ensuring transparency and interpretability in the predictive models employed.

1.2. Importance of Stability in Smart Grids

Stability is paramount in smart grids to ensure reliable and uninterrupted electricity supply to consumers. Grid stability denotes the system's capability to maintain steadiness and withstand disturbances caused by fluctuations in demand, supply, or environmental conditions. Voltage fluctuations, frequency deviations, and line overloads can be reasons for power outages, equipment damage, and financial fatalities. Therefore, ensuring grid stability is critical for maintaining grid reliability, safety, and quality of service. In the framework of smart grids, stability becomes even more crucial due to the increased complexity and variability introduced by distributed energy resources, intermittent renewable generation, and dynamic load profiles. Effective grid stability management involves proactive monitoring, predictive analytics, and responsive control strategies to mitigate risks and maintain system equilibrium under varying operating conditions.

1.3. Role of Machine Learning in Smart Grid Management

Machine learning (ML) is vital in smart grid management as it enables decision-making based on data, analyzes predictions, and implements adaptive control strategies. ML algorithms analyze colossal datasets from smart grid sensors, meters, and control devices to classify patterns, forecast future trends, and optimize real-time grid operations. For instance, ML models can forecast electricity requests, detect anomalies, and augment energy dispatch to minimize costs and enhance grid stability. Moreover, ML algorithms such as reinforcement learning and autonomous agents enable self-healing capabilities in smart grids, allowing the system to adapt and respond autonomously to changing conditions and disturbances. By employing machine learning, smart grid operators can enhance grid reliability, optimize resource usage, and support the shift towards a more sustainable and resilient energy future.

2. DATASET FOR SMART GRID STABILITY

2.1. Introduction to the Dataset

The Smart Grid Stability Dataset offers crucial insights into the stability assessment of smart grid systems. With 60,000 data points, this dataset comprehensively views various factors affecting grid stability. The dataset includes features such as tau_1, tau_2, tau_3, tau_4, p_1, p_2, p_3, p_4, g_1, g_2, g_3, and g_4, each contributing to the understanding of grid behavior under different conditions as shown in Table 1. The parameter p_1 to p_4 denote the minimal power values of Gaussian Noise (GN) and Constant Noise (CN). tau_1 to tau_4 represents reaction time values of GN and CN. g_1 to g_4 Gamma coefficient, price elasticity for GN and CNs are. Moreover, the dataset includes labels indicating whether a particular instance represents a stable or unstable state, which is crucial for developing predictive models and understanding the dynamics of grid stability.

Table 1. Description of dataset features

Value	Description	Name of the feature
P	Minimal power values of GN and CNs	p_1 to p_4

continued on following page

Table 1. Continued

Value	Description	Name of the feature
τ	Reaction time values of GN and CNs	tau_1 to tau_4
γ	Gamma coefficient, price elasticity for GN and CNs	g_1 to g_4

2.2. Dataset Characteristics and Features

The Smart Grid Stability Dataset encompasses various characteristics and features for analyzing grid stability. The features include parameters like tau1 through tau4, representing time constants, and p1 through p4, indicating real power injections. Additionally, features g1 through g4 denote coefficients of the static energy function, providing insights into grid behavior. These features collectively capture the intricacies of smart grid systems, enabling researchers and practitioners to explore the relationships between various parameters and grid stability. Moreover, the dataset's substantial size, consisting of 60,000 data points, ensures ample data for robust analysis and model development.

We can identify two dependent variables: *stab* and *stabf.* The "stab" variable represents the root of a differential equation, indicating linear stability. A positive value signifies instability, while a negative value suggests stability. Conversely, stabf is a categorical variable that classifies the grid state as stable or unstable, acting as a binary label. These two variables are closely related, with *stabf* reflecting the classification of stability determined by *stab*. Consequently, we should select *"stab"* as the response variable for the regression approach and exclude *stabf.*

Conversely, for the classification approach, *stabf* should be chosen as the response variable while excluding *stab*. In this study, both situations are addressed separately, with one focusing on regression and the other on classification. It's important to note that neither dataset contains missing values.

2.3. Initial Data Exploration and Understanding

Before investigating in-depth analysis, performing initial exploration and understanding of the Smart Grid Stability Dataset is crucial. It involves analyzing summary statistics, including mean, median, standard deviation, and range, for each feature to understand their distributions and variability, as presented in Table 2. Visualizations, such as box plots, scatter plots, line plots, and histograms, can further help understand the different features and their impact on grid stability. Additionally, exploring the distribution of class labels (stable vs. unstable) provides an overview of the dataset's balance and potential challenges in modeling. By conducting this preliminary exploration, researchers can lay the groundwork for more advanced analyses and modeling techniques tailored to the unique characteristics of the smart grid stability dataset.

Table 2. Statistics of dataset

Features	tau1	tau2	tau3	tau4	p1	p2	p3	p4	g1	g2	g3	g4	stab
count	60000	60000	60000	60000	60000	60000	60000	60000	60000	60000	60000	60000	60000
mean	5.25	5.25	5.25	5.25	3.75	-1.25	-1.25	-1.25	0.525	0.525	0.525	0.525	
std						0.433	0.433	0.433					
min						-2	-2	-2	0.05	0.05	0.05	0.05	-0.081
25%		2.875	2.875	2.875									-0.016
50%	5.25	5.25	5.25	5.25	3.751	-1.25	-1.25	-1.25	0.525	0.525	0.525	0.525	
75%													
max						-0.5	-0.5	-0.5		1	1	1	

3. EXPLORATORY DATA ANALYSIS

3.1. Statistical Overview of the Dataset

An in-depth statistical overview of the dataset sheds light on its essential characteristics and distributions. Summary statistics provide a valuable understanding of each feature's central tendency, dispersion, and shape. Mean, median, standard deviation and range are the measures that offer a complete understanding of the data's variability. Additionally, exploring the distribution of class labels (stable vs. unstable) reveals the dataset's balance and potential challenges in classification tasks. Moreover, examining correlations between features can uncover potential relationships and dependencies, guiding further analysis and model development.

3.2. Visualization of Key Features

Visualizing key features is crucial for comprehending the dataset's structure and relationships. Histograms represent the distribution of each feature, highlighting patterns and outliers. Box plots concisely summarize feature distributions' central tendency, dispersion, and skewness, facilitating comparisons between stable and unstable instances. Scatter plots visualize the relationships between pairs of features, allowing for the identification of linear or nonlinear correlations. Additionally, heat maps of feature correlations provide a visual roadmap for understanding inter-feature relationships and identifying potential multicollinearity issues, as shown in Figure 3.

Figure 3. Correlation matrix of the raw dataset

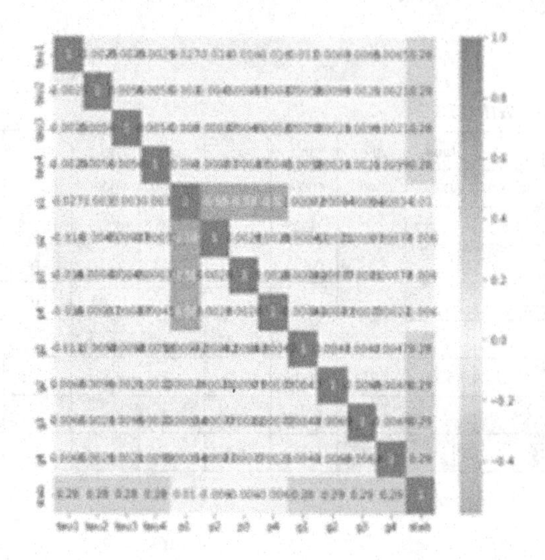

3.3. Identification of Patterns and Anomalies

Identifying patterns and anomalies within the dataset is critical for understanding underlying dynamics and potential areas of interest. Clustering techniques, such as k-means clustering, can uncover natural groupings or clusters within the data, revealing distinct patterns of grid stability behavior, as shown in the boxplot in Figure 4. Anomaly detection algorithms, such as isolation forests or autoencoders, help identify unusual or anomalous occurrences that diverge significantly from the standard. Exploring temporal patterns using time series analysis techniques can reveal trends, seasonality, and recurring patterns in grid stability dynamics. Researchers can gain deeper insights into the underlying mechanisms driving smart grid stability by systematically identifying patterns and anomalies and informing decision-making processes for system optimization and resilience enhancement. A scatter plot in Figure 5 shows the correlation between the principal components.

4. PREDICTIVE MODELING WITH DECISION TREES

Decision Trees (DT) offer a robust framework for predictive modeling, leveraging a hierarchical structure to classify or regress data based on feature values. In the context of smart grid stability analysis, decision trees prove invaluable for predicting whether a given configuration of grid parameters will result in a stable or unstable system state. Decision trees make sequential decisions by recursively partitioning the feature space, ultimately leading to a prediction at the leaf nodes. This approach provides predictive accuracy and interpretability, allowing stakeholders to understand the factors influencing stability predictions.

Figure 4. Representation of pattern in dataset

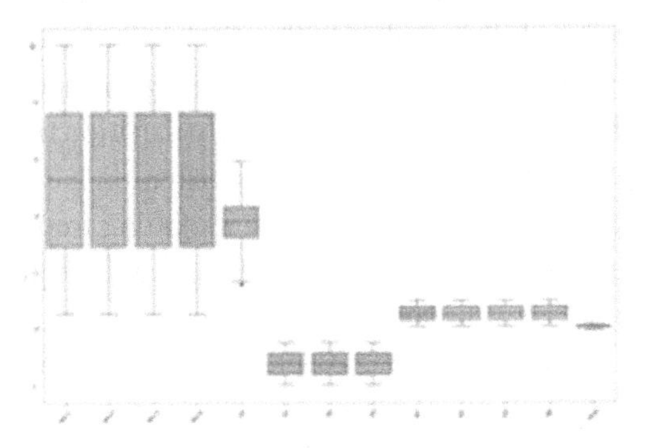

Figure 5. Correlation between the labels

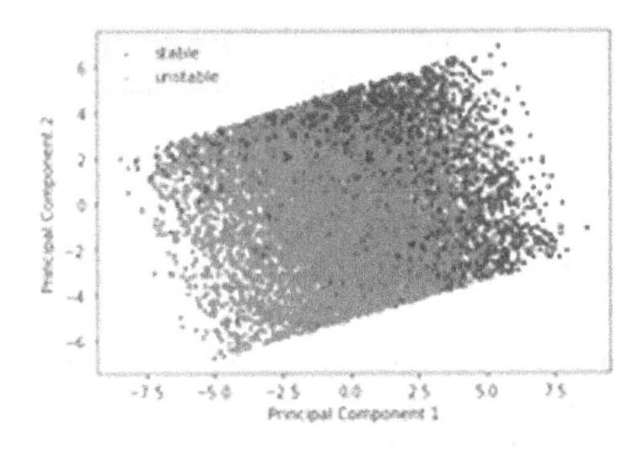

4.1. Application of Decision Tree Algorithm

The decision tree algorithm finds application in smart grid stability analysis by effectively capturing the complex relationships between grid parameters and stability outcomes. By examining features such as tau1 through tau4, p1 through p4, and g1 through g4, decision trees divide the dataset into subsets, iteratively refining predictions based on features. This process enables decision trees to discern patterns in the data and make informed predictions regarding grid stability. Additionally, decision trees can process both numerical and categorical data, making them highly versatile for analyzing the varied datasets typically encountered in smart grid stability analysis.

4.2. Model Training and Evaluation

Model training and evaluation are crucial in leveraging decision trees for predictive modeling in smart grid stability analysis. The dataset is divided into training and testing sets with an 80:20 ratio to facilitate model training and performance evaluation. During the training phase, decision trees recursively partition the training data based on feature values, optimizing criteria such as information gain or Gini impurity to enhance predictive accuracy. Hyperparameters, including tree depth and minimum samples per leaf, are adjusted to prevent overfitting and improve generalization. After training, the model is evaluated on the testing set using metrics like accuracy, precision, recall, and F1-score to gauge its predictive performance and robustness.

4.3. Interpretation of Decision Tree Results

Interpreting decision tree results provides valuable insights into the factors driving grid stability predictions. Decision trees offer transparency in decision-making, with each node representing a decision based on a specific feature threshold. By tracing decision paths from root to leaf nodes, stakeholders can understand how input features contribute to stability predictions. Furthermore, feature importances assigned by decision trees reveal the relative influence of each feature on stability outcomes, guiding stakeholders in prioritizing interventions or optimizations. Overall, interpretation of decision tree results empowers stakeholders to make well-versed decisions about smart grid management and optimization approaches. The classification report on DT is given in Table 3, which shows that an accuracy of 90% has been obtained using the DT algorithm. The confusion matrix in Figure 6. shows that the Decision tree classifier classifies the data well. Also, the Receiver Operating Characteristics (ROC) Area Under Curve (AUC) of 0.89 has been obtained, as shown in Figure 7. The performance of the DT classifier is determined using a k-fold cross-validation score with k = 5, as depicted in Figure 8.

Table 3. Classification report on DT

DT Classification Report				
	Precision	Recall	F1-score	Support
Stable	0.85	0.86	0.86	4322

continued on following page

Table 3. Continued

DT Classification Report				
	Precision	**Recall**	**F1-score**	**Support**
Unstable	0.92	0.92	0.92	7678
Accuracy			0.9	12000
Macro avg	0.89	0.89	0.89	12000
Weighted avg	0.9	0.9	0.9	12000

Figure 6. Confusion matrix of DT algorithm

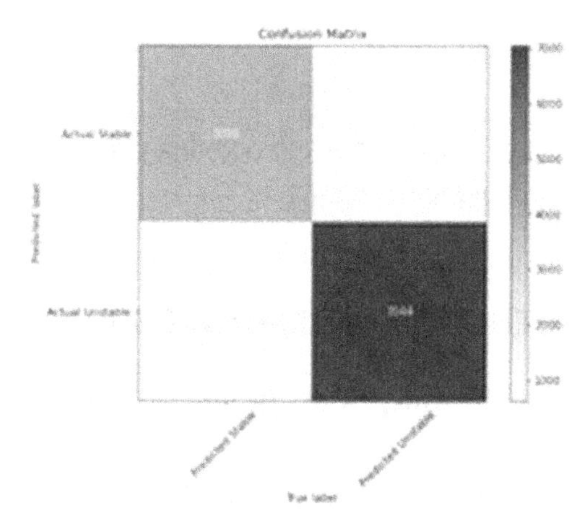

Figure 7. ROC curve of DT algorithm

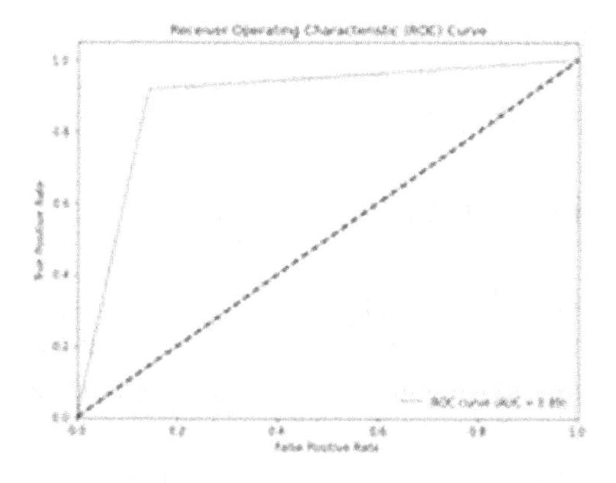

Figure 8. k-fold cross validation curve

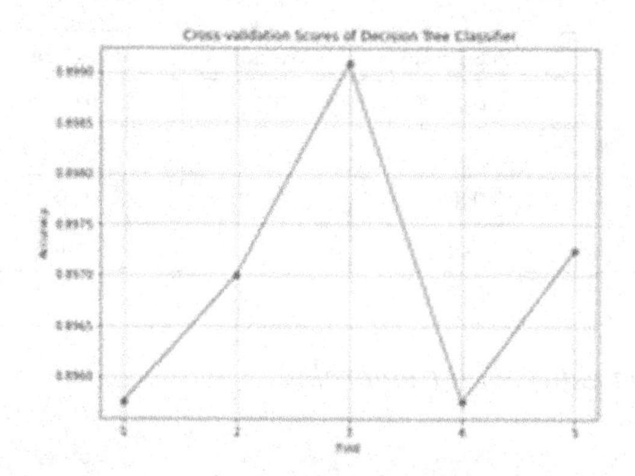

Cross-validation Scores: [0.89575 0.897 0.89908333 0.89575 0.89725]
Mean Accuracy: 0.8969666666666667

5. ENSEMBLE METHODS FOR IMPROVED ACCURACY

5.1. Introduction to Ensemble Methods

Ensemble approaches are influential ML methods that synergize several base models to increase prediction performance. The underlying principle is based on the concept that aggregating the predictions of various models leads to better results than any discrete model alone. Random Forest (RF) is the most famous ensemble method, which leverages a collection of decision trees. Ensemble methods are especially effective at managing complex datasets, minimizing overfitting, and improving model generalization and robustness. The collective perception of multiple models is harnessed into ensemble methods that offer greater predictive accuracy and stability than individual models.

5.2. Implementation of Random Forest

RF is a flexible ensemble learning algorithm for regression and classification tasks. It consists of multiple decision trees, each trained on a random subset of the training data and features. During the prediction process, the outputs from each tree are combined to produce a final estimate. The variability introduced during training diversifies the trees, which mitigates overfitting and enhances the model's ability to generalize to new data. RF is highly scalable and can effectively handle massive datasets with high dimensionality. Its user-friendly nature and reliability have made it widely adopted for a range of machine learning applications, including the analysis of smart grid stability. Table 4 lists the classification report of RF. It depicts that the accuracy has improved by 5% compared to the DT. Figure 9 represents the confusion matrix of RF, which shows that the True Positive is more in the left diagonal elements than the False Positive in the right diagonal elements. The ROC curve of the RF is displayed in Figure 10, which depicts that the AUC value has improved to 0.99, approximately equal to 1, the maximum value. So, it is said that the ensemble method has improved performance than the DT.

Table 4. Classification report of RF

Classification Report of RF				
	Precision	**Recall**	**F1-score**	**Support**
Stable	0.94	0.91	0.93	4322
Unstable	0.95	0.97	0.96	7678
Accuracy			0.95	12000
Macro avg	0.95	0.94	0.94	12000
Weighted avg	0.95	0.95	0.95	12000

Figure 9. Confusion matrix of RF

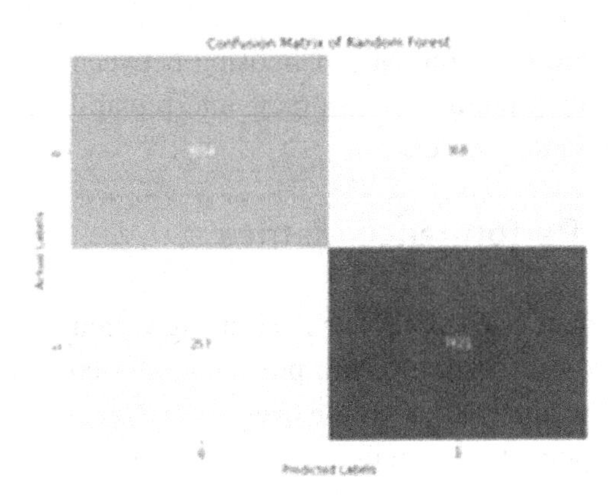

Figure 10. ROC curve of RF

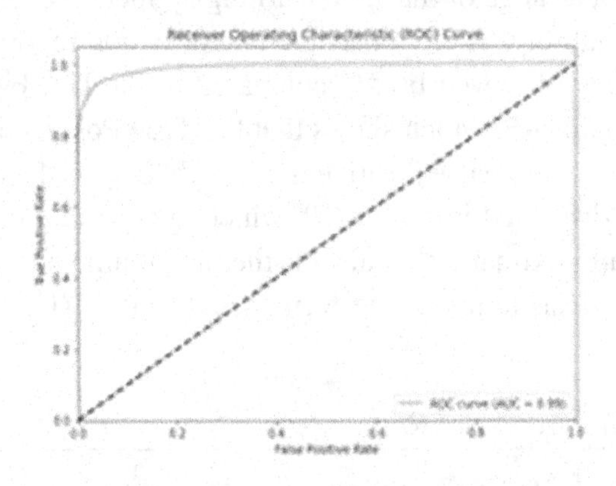

5.3. Comparative Analysis With Decision Trees

A comparative analysis between RF and DT delivers a remarkable understanding of their strengths and weaknesses. DT are valued for their simplicity and interpretability but are prone to overfitting, particularly on intricate datasets. In contrast, RF addresses overfitting by combining predictions from numerous trees, resulting in better generalization. Moreover, RF's ensemble approach enhances resilience against noisy data and outliers. However, Decision Trees often excel in interpretability, offering straightforward decision paths. Choosing between RF and DT hinges on specific needs, balancing predictive accuracy, interpretability, and computational efficiency for each unique problem.

5.4. Results and Performance Metrics

RF is applied to smart grid stability analysis using various performance metrics. The metrics encompass accuracy, recall, precision, F1-score, and area under the ROC curve (AUC). Accuracy assesses the overall correctness of predictions, while precision and recall gauge the model's accuracy in identifying stable and unstable instances. The F1-score balances precision and recall, comprehensively evaluating classification performance. Moreover, the AUC evaluates how well the model can discriminate between stable and unstable instances across a range of decision thresholds. By comprehensively evaluating these performance metrics, stakeholders can measure the efficiency of RF in predicting grid stability and make knowledgeable decisions regarding its distribution in real-world applications.

6. HYPERPARAMETER TUNING FOR OPTIMAL MODEL PERFORMANCE

6.1. Significance of Hyperparameter Tuning

Hyperparameter tuning is vital in improving ML algorithms for better performance and generalization. Hyperparameters dictate the learning process and model complexity, such as the maximum depth of a decision tree or the learning rate of a neural network. Choosing the right hyperparameters is crucial as it can greatly influence the model's ability to predict accurately and maintain stability. Through hyperparameter tuning, practitioners can systematically search for the best grouping of hyperparameters that maximize model performance on a test set. This process supports to mitigate overfitting, improve model robustness, and enhance predictive accuracy, ultimately leading to better outcomes in real-world applications.

6.2. Grid Search Approach

The grid search method is commonly employed for hyperparameter optimization in machine learning. This process entails establishing a range of hyperparameter values to investigate, often utilizing preset ranges or specific discrete values. The model undergoes training and evaluation using cross-validation for every set of hyperparameters in the grid. Each configuration's performance is subsequently assessed using evaluation metrics like accuracy or F1-score. Through exhaustive grid searching, practitioners can pinpoint the optimal combination of hyperparameters that achieves the highest performance on the testing set. Despite being computationally intensive, the grid search approach effectively finds optimal hyperparameters, especially in relatively small hyperparameter space scenarios.

6.3. Identification of Optimal Hyperparameters

Selecting the optimal hyperparameters involves identifying the combination that maximizes the model's performance on a test dataset. After conducting a grid search or other hyperparameter optimization techniques, practitioners analyze the performance metrics for each hyperparameter configuration. Optimal hyperparameters are usually selected based on the highest performance metric values, such as accuracy or F1-score. Considering both performance and computational complexity when selecting optimal hyperparameters is essential, as overly complex models may lead to overfitting or increased inference time. Once the optimal hyperparameters are identified, the final model can be trained using these values on the entire training dataset for deployment in production environments. By utilizing the best hyperparameters, the obtained accuracy is 94%.

6.4. Fine-Tuning the Model for Improved Stability Prediction

Fine-tuning the model involves refining the selected hyperparameters to improve stability prediction performance. This iterative process may include adjusting hyperparameters based on insights gained from initial model evaluations or domain knowledge. For example, in the context of smart grid stability analysis, fine-tuning may involve optimizing hyperparameters related to decision tree depth, regularization parameters, or ensemble size in Random Forest models. By fine-tuning the model, practitioners can report precise challenges or nuances in the data and further enhance the model's ability to predict grid stability accurately. Furthermore, continuous monitoring and validation of the model's performance are essential to verify its effectiveness in real-world scenarios.

7. MODEL EVALUATION AND VALIDATION

7.1. Cross-Validation Techniques

Cross-validation techniques are important for weighing ML models' performance and generalization ability. They involve partitioning the dataset into numerous subdivisions, or folds, to train and evaluate the model iteratively. One widely used technique is k-fold cross-validation, where the dataset is divided into k equal-sized folds. The model is trained k times, each time using k-1 folds for training and the remaining fold for validation. This method ensures that every data point is utilized for both training and validation, thereby minimizing bias in performance estimation. Cross-validation provides a more consistent analysis of the model's performance on unseen data compared to a single train-test split and helps classify issues such as overfitting or data leakage, as shown in Figure 11.

Figure 11. k-Fold cross-validation of RF

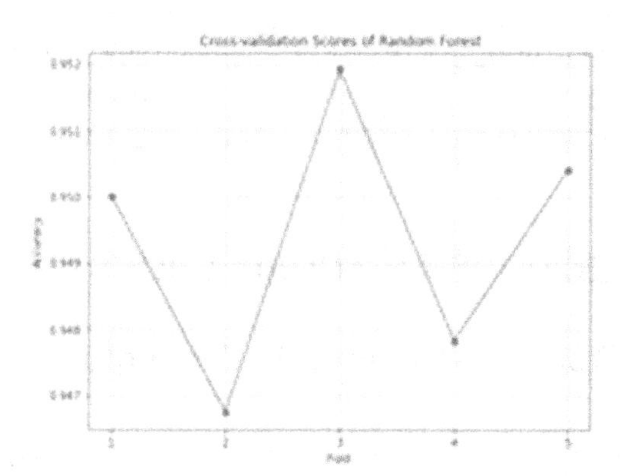

Cross-validation Scores: [0.95 0.94675 0.95191667 0.94783333 0.95041667]
Mean Accuracy: 0.9493833333333332

7.2. Robustness and Generalization of the Model

The robustness and generalization of the proposed ML model refer to its capability to execute well on unseen data and handle variations or noise in the dataset. A robust model exhibits consistent performance across different datasets or data distributions and is less susceptible to small changes in input data. Achieving robustness and generalization requires careful model design, appropriate feature selection, and hyperparameter tuning. Methods like regularization, ensemble learning, and data augmentation can bolster the model's robustness by mitigating overfitting and capturing intricate patterns within the data. Evaluating the model's proficiency using k-fold cross-validation and different datasets is crucial for assessing its robustness and generalization ability.

7.3. Addressing Overfitting and Underfitting

Overfitting and underfitting are prevalent challenges in machine learning that can undermine a model's performance and its ability to generalize. Overfitting happens when a model captures noise or irrelevant patterns from the training data, resulting in poor test data performance. Conversely, underfitting arises when a model is too essential to discern the underlying patterns in the data, leading to high bias and subpar performance on both training and test sets. To combat overfitting, strategies like regularization, dropout, and early stopping are employed to limit model complexity and prevent it from memorizing noise in the data. To mitigate underfitting, approaches might include boosting model complexity, incorporating additional features, or adopting more sophisticated model architectures. Achieving a balance between underfitting and overfitting demands meticulous experimentation and validation through cross-validation techniques. It ensures the model attains optimal performance and generalization across diverse datasets.

8. FUTURE DIRECTIONS AND CHALLENGES

8.1. Emerging Trends in Smart Grid Technology

Smart grid technology remains to evolve rapidly, driven by advancements in digitalization, renewable energy integration, and grid automation. Some emerging trends in smart grid technology include the widespread adoption of advanced metering infrastructure (AMI), the deployment of distributed energy resources (DERs) such as solar panels and energy storage systems, and the development of demand response programs to manage peak demand. Additionally, innovations in grid analytics, artificial intelligence, and Internet of Things (IoT) technologies are transforming grid operations and enabling more efficient, resilient, and sustainable energy systems. These trends are shaping the future of smart grids, ushering in an era of increased grid intelligence, flexibility, and reliability.

8.2. Challenges in Implementing Machine Learning for Grid Stability

While machine learning holds promise for enhancing grid stability analysis and management, several challenges hinder its widespread implementation. One of the major problems is the availability of massive and labeled data for training ML models. Smart grid data often suffer from incompleteness, noise, and imbalances, making building strong and precise predictive models crucial. Moreover, grid systems' complex and dynamic nature poses challenges for modeling and prediction, requiring sophisticated algorithms capable of capturing nonlinear relationships and temporal dependencies. Furthermore, regulatory constraints, data privacy concerns, and the need for domain expertise present additional barriers to adopting machine learning in grid stability applications.

8.3. Potential Solutions and Research Opportunities

Addressing the challenges of implementing machine learning for grid stability requires interdisciplinary research and innovation across multiple domains. One potential solution is the development of advanced data preprocessing techniques to clean, impute, and augment smart grid data, enhancing its quality and usability for machine learning applications. Moreover, researchers can explore novel machine learning algorithms tailored to the unique characteristics of smart grid data, such as deep learning architectures for time-series forecasting or ensemble methods for handling uncertainty. Collaborations between academia, industry, and government entities can facilitate data sharing, standardization, and validation efforts, enabling more robust and scalable machine learning solutions for grid stability. Additionally, research opportunities abound in areas such as explainable AI, federated learning, and online learning, which hold promise for addressing specific challenges and unlocking the full potential of machine learning in smart grid applications. By leveraging these solutions and opportunities, stakeholders can overcome barriers and harness the transformative power of machine learning to enhance grid stability, reliability, and resilience.

9. SUMMARY

In this book chapter, we analyzed smart grid stability prediction using machine learning algorithms, explicitly focusing on DT classification and RF ensemble methods. The key findings of our study revealed that while DT classification provided a straightforward and interpretable approach to predict grid stability, the RF ensemble method significantly improved accuracy by leveraging multiple decision trees. Our findings have significant implications for smart grid operations. Utility operators can enhance grid resilience and reliability by accurately predicting grid stability, leading to optimized energy distribution and reduced downtime. Implementing RF ensemble methods offers a robust solution for real-time monitoring and management of smart grid systems, ensuring efficient operation and minimizing disruptions.

Future research endeavors could further explore several avenues to enhance smart grid stability prediction. Firstly, investigating the integration of other advanced machine learning techniques, such as gradient boosting or neural networks, may offer additional insights into improving prediction accuracy. Additionally, incorporating real-time data streams and IoT sensors could enable dynamic modeling and forecasting of grid stability, thereby enhancing predictive capabilities. Moreover, exploring the influence of external factors, such as weather conditions and grid infrastructure, on stability prediction could provide a more comprehensive understanding of grid dynamics.

In summary, our study underscores the importance of accurate stability prediction in smart grid operations and highlights the efficacy of RF ensemble methods in improving prediction accuracy. By addressing the essential findings and implications outlined in this chapter, researchers and practitioners can pave the way for more resilient and efficient smart grid management in the future.

REFERENCES

Alsirhani, A., Mujib Alshahrani, M., Abukwaik, A., Taloba, A. I., Abd El-Aziz, R. M., & Salem, M. (2023). A novel approach to predicting the stability of the smart grid utilizing MLP-ELM technique. *Alexandria Engineering Journal*, 74, 495–508. 10.1016/j.aej.2023.05.063

Cui, H., Wang, Q., Ye, Y., Tang, Y., & Lin, Z. (2022). A combinational transfer learning framework for online transient stability predic- tion. *Sustain Energy Grids Netw*, 30, 1–10. 10.1016/j.segan.2022.100674

Mohsen, S., Bajaj, M., Kotb, H., Pushkarna, M., Alphonse, S., & Ghoneim, S. S. M. (2023). Efficient Artificial Neural Network for Smart Grid Stability Prediction. *International Transactions on Electrical Energy Systems*, 2023, 1–13. 10.1155/2023/9974409

Shi, Y., Tuan, H. D., Savkin, A. V., Lin, C. T., Zhu, J. G., & Poor, H. V. (2021). Distributed model predictive control for joint coordination of demand response and optimal power flow with renewables in smart grid. *Applied Energy*, 290(March), 116701. 10.1016/j.apenergy.2021.116701

Ucar, F. (2023). A Comprehensive Analysis of Smart Grid Stability Prediction along with Explainable Artificial Intelligence. *Symmetry*, 15(2), 289. 10.3390/sym15020289

You, D., Wang, K., Ye, L., Wu, J., & Huang, R. (2013). Transient stability assessment of power system using support vector machine with generator combinatorial trajectories inputs. *Int J Electr Power Energy Syst, 44*(1), 318–325. https://doi.org/ . 07.05710.1016/j.ijepes.2012

Chapter 16
Wireless Interference Identification in 5G Smart Networks

Rabia Bilal
UIT University, Pakistan

Bilal Muhammad Khan
National University of Sciences and Technology, Islamabad, Pakistan

ABSTRACT

In the rapidly evolving landscape of 5G communication, the identification and mitigation of wireless interference is paramount to maintaining the integrity and efficiency of data transmission. This chapter delves into the intricate process of wireless interference identification, emphasizing its critical role in the predictive analysis of modulation classification and the implementation of adaptive modulators. The discussion begins with a comprehensive overview of 5G architecture, and the inherent challenges posed by dense signal environments. Key techniques for interference identification are explored, including advanced machine learning algorithms and spectrum sensing methods that enable real-time detection and characterization of interference sources. The chapter then examines how these identification techniques inform the predictive analysis of modulation classification. By accurately predicting the modulation scheme of incoming signals, the system can adaptively adjust its modulator settings, thereby optimizing performance and minimizing error rates.

DOI: 10.4018/979-8-3693-2786-9.ch016

1. INTRODUCTION

Modulation classification was being initially used for military scenarios where electronic warfare, surveillance and threat analysis required the modulations to be classified in order to know adversary transceivers, to enhance jammers, and to recover the information from the received signal. Considerable research was carried out in the field of signal processing and communications to implement automatic modulation classification on FPGA and DSP. Modulation Classification is still being used for military purposes and more research in the field of Signal Processing and Communication Systems is being carried out.

Recently there have been many innovations in communications technology. Automatic Modulation Classification and Coding optimizes the reliability of the signal and data transmission rate by adaptively selecting modulation schemes according to channel conditions. The transmitter can select modulation scheme from the programmed modulations in RF Device, the receiver must have the knowledge of the modulation type to demodulate the signal so that the transmission can be successful. Automatic Modulation Classification System can identify the Modulation type of each frame when the Modulation type is changing rapidly. This Technique has become an integral part of intelligent radio systems, including cognitive radio and software-defined radio. In this chapter, Automatic modulation classification technique is used for classifying the modulation format of received signals and Adaptive Modulator has also been simulated for AWGN Channel with adjustments made in the Modulation Classifier for Reliable Communication System. It has multiple applications. In Military Application, the received signal can be demodulated by identifying the modulation type, and then appropriate signals can be generated. In Civilian Application, Modulation classification and Adaptive Modulator enables the optimization of the transmission reliability and data rate through the adaptive selection of modulation schemes according to channel conditions. The Objective of this research was to learn how we can classify the received modulated signals, and according to the channel conditions, required message signals can be adaptively modulated. Channelization can also be used for utilizing the maximum bandwidth of RF Signal by assigning channels of RF Signal to enhance Data Rate by splitting data in the form of Arrays. Then the data can be transmitted using maximum possible RF Bandwidth according to the channel conditions and also changing the modulations can also change data rate.

The Research has been motivated by the requirement of higher data rates that are in demand for good quality communication on devices. Most of the spectrum has already been utilized. Cognitive radio addresses the problems by detecting and using the spectrum holes. Cognitive Radios estimate the Signal to Interference plus Noise Ratio (SINR) of the channel, and also uses adaptive modulation (Mohammadi

& Kwasinski, 2018). Adaptive modulator changes the modulation format and maximizes throughput. Automatic Modulation Classification (AMC) is done by cyclic feature detection (Spooner & Mody, 2018). This Research is also applicable to National needs because Research is also being carried out to improve and produce Indigenously Manufactured Communication Equipments. AMC has an essential role in many military strategies. Electronic warfare (EW) consists of three components: electronic support (ES), electronic attack (EA) and electronic protect (EP). In ES, the information is gathered from radio frequency emissions. Automatic Modulation Classification is carried out after the signal detection is successfully completed. The resulting modulation information is useful for all the components in EW. It also has applications in Aircraft Communication using different Modulation Techniques for High Data Rate Communication for Aircraft to Aircraft Communication and Aircraft to Ground Communication and Drone Communication and Spectrum Sensing. Modulation Classification technique enables the transceiver to detect the Modulation type, the modulation type can be sent to Adaptive Modulator so that the signal can be modulated using the current modulation type, and the channel SNR can also be sensed and appropriate modulation can be utilized. BPSK is used for highly reliable communication, and higher order modulations like QAM can be used for high data rate communication.

2. LITERATURE REVIEW

2.1 Introduction

Opportunistic Spectrum Access is a modernized approach to tackle the spectrum congestion problem. Cognitive Radio (CR) is able to access spectrum hole, hence it can efficiently enhance electromagnetic spectrum usage. Cognitive radio (CR) is an innovation in communication technology, which is able to use different techniques for awareness of the surroundings. The main problem with the cognitive radio systems is to reduce interference. The Energy detection technique is a simple technique and it is efficient at higher SNR, but at lower SNR, its performance degrades. Adaptive modulator changes modulation type according to the SNR of the signal, and can also be used to maximize data rate. It is a part of modern systems. Dynamic Spectrum Access (DSA) is being utilized to overcome spectrum congestion. In a communication system, there is cooperation between the devices, so the receiver can demodulate a signal if the modulation format of the transmitted signal is known. The communication devices need knowledge of modulation type, nominal carrier frequency and modulation index in order to maintain stable communication systems. In digital communication systems (DCS), the receiver needs knowledge of modulation type, bits per symbol, deviation (for frequency modulated signals only), nominal carrier frequency, symbol constellation, frequency and the symbol rate.

In communication systems, the modulation classification was being done by using the banks of demodulator, where each of the banks was used for demodulating one modulation type. Demodulator output was the demodulated signal, corrupted by channel impairments. Implementation of modulation classification system is complicated, requiring storage. The modulated signals were being demodulated by demodulator's bank. With the advances in software-defined radios (SDRs), recently radio frequency (RF) communication devices can transmit in low power, alter the transmitting frequency, and change the modulation format during transmission. Adaptive modulation changes the data rate according to the channel conditions. If the modulation is being changed without sending the modulation information, then RF signals can be recognized by Automatic Modulation Classification. Automatic Modulation Classification (AMC) which is a widely used feature on receiver for adaptation. Link adaptation is also known as adaptive modulation and coding, develops an adaptive modulator scheme where a pool of different modulators have been installed in a system. It optimizes the transmission reliability and data rate by adaptively selecting modulation schemes according to the channel conditions. While transmitter is free to modulation format. The receiver needs to know the modulation format for demodulating the signal. For this purpose, modulation information is to be added in each frame, so that receiver can demodulate the modulated signal. This approach will affect the spectrum usage due to extra information about modulation in each frame. Automatic modulation classification is the solution.

Cyclic feature detection works well in noise, as well as carrier frequency offset (Federal Communications Commission, 2002). The techniques used for Modulation Classification include Energy Detection, Matched Filter Detection, Cyclostationary Feature Detection (Bhargavi & Murthy, 2010; Dobre et al., 2007). Out of these techniques Cyclostationary Feature Detection performs better. (FCC, 2015; Bhargavi & Murthy, 2010; Yucek, 2009). The automatic modulation classification (AMC) is performed between decoding of the signal carried by the RF wave and the demodulation process. AMC is a method for the classification of the modulation formats of the received signal on the receiver side. Automatic modulation classification has applications in the field of electronic warfare, military areas, electronic counter measures, public security, software defined radios and lately in cognitive radios. In military domains it is used for spectrum monitoring and communication intelligence, whereas in civil domain it includes data rate enhancement using Adaptive Modulation, spectrum management. Usage of Cognitive Radios in Communications is increasing because the Physical Layer research can be easily carried out by programming the physical layer using Software Defined Radio. With their ease of programmability and high bandwidth, they are ideal hardware platform choice for prototyping and research (Abid et al., 2022). In comparison to Application Specific Integrated Circuits (ASICs), Field Programmable Gate Array (FPGA) designs are cheaper and easier to develop due to their programmability, so they are also used in the development of low cost systems that have computing requirements beyond what a typical processor could offer. Historically, the FPGA in the Software Defined Radio was used for performing the detection of spectrum hole (Cabric et al., 2004)., spectrum sensing based on energy (Mitola & Maguire, 1999), and a system performing the role of Cognitive Radio (Frasch, 2017). Energy Detection Technique performs better in high SNR (Khan & Ali, 2011; Memon et al., 2021).

Cyclic feature detection was being used to determine the type of modulation being performed (Digham et al., 2003), but recently the Cognitive Radios are utilizing this type of classification method for modulation classification (Bhargavi & Murthy, 2010). It has portrayed the capability of classifying BPSK, QPSK, MSK, and FSK with an accuracy of at least 95% if the SINR was greater than 1 dB. Furthermore, when 4th order cummulants were combined, QAM signals were able to be detected. Compressed Sensing is also being used in Cognitive Radio (Gardner, 1988). The energy detection technique and power-spectral density do not have accuracy of classifying Modulation technique that Cyclic feature detection provides for Spectrum Monitoring, even at low Signal Strength (Ramkumar, 2009). By the use of Compressed Sensing, Cognitive Radios do not need to reconstruct the original signal from the Primary User precisely, Compressed Sensing provides good way to reduce the computing power* required, especially for spectrum sensing. When applied to AMC, the accuracy is greatly decreased (Tian et al., 2012). The Fuzzy Logic Approach for Cooperative Sensing performs better in Centralized Cognitive Radio Network as compared to local individual decision (Lagunas & Najar, 2015).

2.2 Cognitive Radios

The improvement in the radios systems is that, the radios can now be programmed to learn and autonomously make "intelligent" decisions, like altering the transmit power to avoid wasting energy, which gives it the new name Cognitive Radio (CR). Another key component of Cognitive Radios is Dynamic Spectrum Access, which determines how they use and reuse spectrum. Most bands of the spectrum are allocated by the FCC (Xie & Wan, 2017), meaning that only licensed bands can be used for transmission on those frequency bands. As the need for the radio spectrum increases, spectrum is becoming congested, in this situation limiting the bands to a single user becomes unrealistic. Cognitive Radios enable effective spectrum sharing without interfering with each other's communications. Normally, this function is performed by MAC Scheme, like Time Division Multiple Access or Frequency Division Multiple Access. The major goal of Cognitive Radio is to maintain interference free communication. Thus different access schemes for Cognitive Radio are needed, known as Dynamic Spectrum Access. There are three types of DSA: Underlay, Interweave, and Overlay (Zhang & Wang, 2011). Interweave involves finding spectrum holes, which are times when a channel not being used, and the Cognitive Radio can use the spectrum holes for communication. Overlay requires the Cognitive Radio to have knowledge about the Primary User's transmissions, and be able to decode them. Using this knowledge, the CR can transmit such that it does not interfere with the original message.

Finally, underlay DSA provides the CR opportunities to make use of extra power in the PU's transmission. While the Primary User's Signal Strength remains above a certain level, the Cognitive Radio is transmitting the data. Hence, Cognitive Radios are able to communicate for more time slots when comparing with interweave technique of DSA, and do not need to know as much about the PU's messages. However, in underlay technique, the Cognitive Radio needs a measure of the Signal to Interference plus Noise Ratio of the Primary User. These spectrum holes are known as black holes or white spaces (BTH, 2012a; Federal Communications Commission, 2018; Goldsmith et al., 2009). Performance of the spectrum sensing decreases due to the fading, and shadowing (BTH, 2012b). Link adaptation (LA), also known as adaptive modulation and coding (AM&C), creates an adaptive modulation scheme where specified modulations are employed by the same system (Bilal & Khan, 2019).

2.3 Modulation

For transmission of data over the air, radios modulate the information over a carrier wave. Modulation alters the spectrum of the data, thus improves the chance to recover the data at the receiver. The Modulation involves communication by using a High Frequency Carrier Wave, thereby reducing the size of antenna. The Examples include 2.4 GHz WiFi, or around 90 MHz for FM radio, TV Broadcast, Aircraft Communication, Drone Communication. Multiple Carrier Waves can be used for different uses as used in drones, the Control Frequency is 2.4 GHz, and Communication Frequency is 5.8 GHz. Modulation is the process in which a message signal is impressed on a reference electronic or optical carrier signal by varying any of the properties of the carrier signal according to message signal. Then the signal is modulated in-phase and quadrature components of a sinusoidal signal to represent the data which is being sent. In PSK, the amplitude or the frequency of the sinusoidal signal that is being transmitted does not change. The phase is changed according to the input bit stream.

2.4 Spectral Features

The spectrum provides the following parameters:

1. Amplitude.
2. Phase.
3. Frequency.

The amplitude, phase and frequency provide complete information about the modulated signal. These three parameters are obtained by obtaining Power Spectrum of the modulated signal. The spectrum is used for the classification of the modulation formats and demodulation and decoding of data.

2.4.1 Cyclo-Stationary Features

Modulated signals have some parameters which periodically change with time. Digital signal is modulated by periodic keying of amplitude, frequency and phase. The signals have the cyclo-stationary property that is used to classify modulation techniques. The Cyclic Domain Profile is extracted by using Cyclostationary Feature Detection Algorithm.

2.4.2 Time-Frequency Features

The Fourier transform (FT) displays the frequency components of the signal and cannot analyze the variations in time domain. For analyzing the frequency and time simultaneously, Short Time Fourier Transform can be utilized. Short Time Fourier Transform (STFT), is a time-frequency representation, in which a 2D representation is created by computing the Fourier Transform and using a sliding temporal window. By using the Short Time Fourier Transform (STFT) alteration of frequency with respect to time can be observed. The features that are extracted from the time-frequency analysis can help to classify the modulation formats. The Margenau-Hill distribution and Wigner- Vile distribution are also used in time-frequency analysis.

3. METHODOLOGY

3.1 1 Modulation Classification

Initially the Modulation Classification Algorithm was found to be capable of classifying the following Modulations:

1. BPSK.
2. QPSK

3. FSK
4. MSK

The Cyclic Domain Profiles were used to identify the modulation types.

The Parameters using which the Modulations have been simulated are given in Table 1.

Table 1. Modulation parameters

Modulations	BPSK, QPSK, FSK, MSK
Symbol Rates	1 kHz
Samples per Symbol	10
Intermediate Frequencies	2.5 kHz
TX Filter	Raised Cosine

3.1.1 1 Modulation Classification With Low Noise

In this process, initially the Message Signal is modulated into BPSK, QPSK, FSK, MSK Modulated Complex Baseband Waveform.

3.1.1.1 1 Modulation Classification With Noise

In this process, initially the Message Signal is modulated into BPSK, QPSK, FSK, MSK Modulated Complex Baseband Waveform.

The Parameters are given in Table 2.

Table 2. 1 modulation classification

TX Filter	Root Raised Cosine
Eb / No	0dB – 20dB

3.2 2 Modulations Classification

The Modulation Classifier has been adjusted to classify 2 Modulations, the details are given upcoming sections. The Cyclic Domain Profiles were used to identify the modulation types.

The Parameters using which the Modulations have been simulated are given in Table 3.

Table 3. Parameters and modulations

Modulations		**BPSK, QPSK, FSK, MSK**	
Symbol Rates		**1 kHz**	**10kHz**
Samples per Symbol		**8**	
Filter Parameters	Stopband start	1.5 kHz	15 kHz
	Stopband Gain	-100 dB	-120 Db
Intermediate Frequencies		2.5 kHz	25 kHz
TX Filter		Root Raised Cosine	

3.2.1 2 Modulations Classification Without Noise

In this process, initially the Message Signal is modulated into BPSK, QPSK, FSK, MSK Modulated Complex Baseband Waveform. The 2 Modulated Complex Baseband Waveforms are then Upconverted. The Waveforms are then Downconverted to Baseband, then Resampled by using Sampling Rate 10 times of Received Symbol Rates and then Upconverted back to Intermediate Frequencies.

3.2.2 2 Modulation Classification With Noise

As soon as the modulations were classified by the Modulation Classifier, further work was done to channelize 2 different signals on Intermediate Frequencies by using Upconversion. By using Channelization technique, we are able to transmit 2 signals on the RF Channel. The RF Bandwidth can be efficiently utilized by sending data on multiple Intermediate Frequency Channels to enhance data rate. Different Modulations can be applied to different channels as per requirement. Many Channels can be accessed simultaneously for more effective communication. Intermediate Frequency can be changed according to symbol rate. Then the Individual Signals at different Intermediate Frequencies can be separated by Downconverting to Baseband, then Resample the Signal, then again Upconvert to Intermediate Frequency.

The signals can be modulated independently on different symbol rates and modulations.

3.2.2.1 Adding Noise to Modulated Signal

A signal is being modulated with Digital Modulations, the Modulation can be selected during the simulation in the GUI at runtime. The AWGN Channel Impairments are added in the output complex waveform. The Eb/No can be changed to indicate change in channel condition as given in Table 4.

Table 4. Eb/No

Eb/No	0dB, 5dB, 10dB, 15dB, 20dB

3.2.2.2 Converting to Intermediate Frequencies

The Modulated Signals are upconverted to Intermediate Frequencies and then they are added. At Receiver they are then downconverted to their baseband signals and resampled and then again upconverted to the same Intermediate Frequency. This filters out the Required Signal. Then this filtered Signal on Intermediate Frequency is then sent Cyclostationary Feature Detector.

Table 5. Frequencies

Symbol Rates	1 kHz	10 kHz
Intermediate Frequencies	2.5 kHz	25 kHz
Resampling Rates	10 kHz	100 kHz

3.3 Modulation Classifiers

In Figure 1, a Modulation Classifier is shown. The Values of the Cyclic Domain Profile at specific points is used to classify different modulations. The values of the Array of CDP are being compared with fixed values for classification of Modulations.

Figure 1. Modulation classifiers

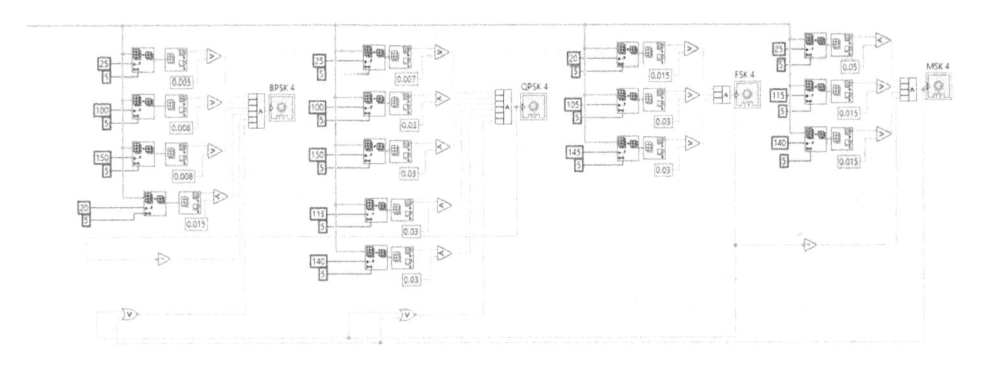

3.4 Adaptive Modulator

The Adaptive Modulation Scheme based on Eb/No. The functioning of Adaptive Modulator is shown in Figure 2. The Modulations change with respect to changing noise level.

Figure 2. Adaptive modulator

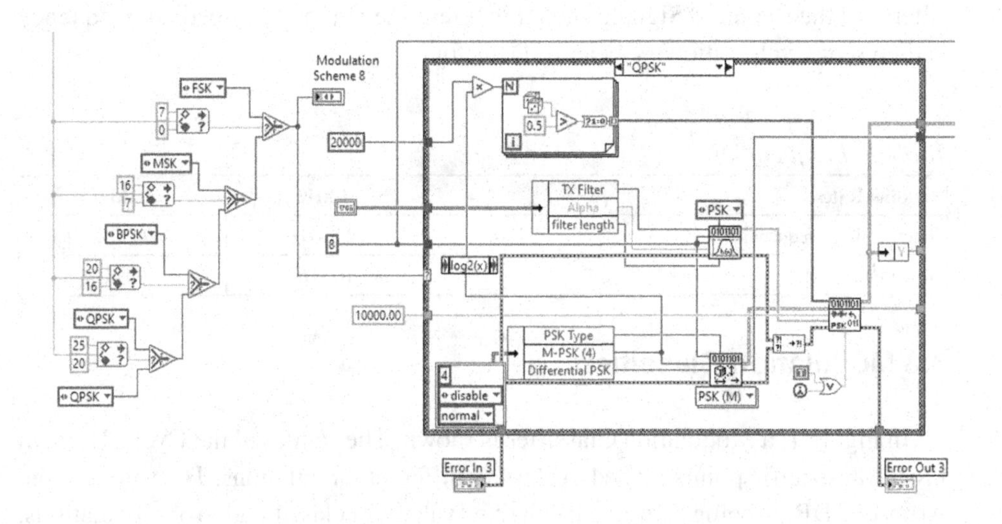

4. CELLULAR MODULATIONS CLASSIFICATION

The Cellular Modulations that can be classified are as follows:

1. GSM FSK
2. GSM MSK
3. EDGE 8PSK
4. WCDMA DL QPSK
5. WCDMA DL 16QAM
6. LTE DL 16QAM

In the LabVIEW Block Diagram given below, the TDMS waveform files of GSM FSK, GSM MSK, WCDMA DL QPSK, WCDMA DL 16QAM, and LTE DL 16QAM are in an Event Structure which will send the Cellular Modulated Waveform to Cyclostationary Feature Detector when the Cellular Modulation will be selected in the GUI during Simulation.

Figure 3. Cellular modulations input

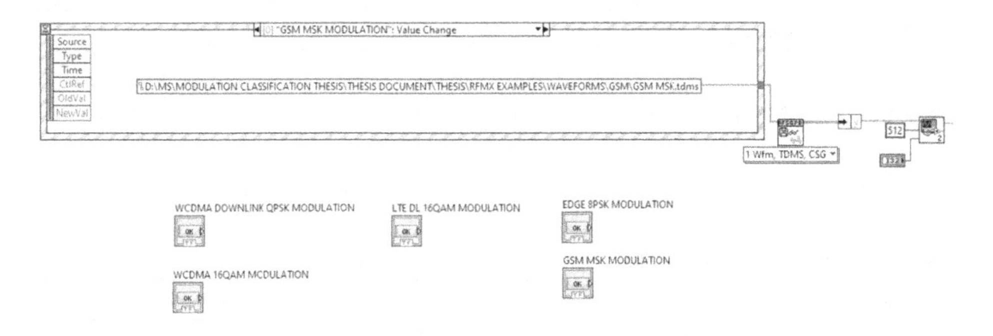

Table 6. Cellular standards

Cellular Standards	GSM, WCDMA, EDGE, LTE

4.1 GSM and EDGE

Table 7. GSM and EDGE specifications

Symbol Rate	**270.8 kHz**
Number of Symbols	1250

4.1.1 GSM MSK

4.1.1.1 Waveform Creation

Waveform of GSM MSK has been generated with the following Parameters

Table 8. GSM MSK waveform

Modulation	MSK

4.1.2 EDGE 8PSK

4.1.2.1 Waveform Creation

Waveform of EDGE 8PSK has been generated with the following Parameters

Table 9. EDGE 8PSK waveform

Modulation	8PSK
Oversampling Factor	6

4.2 WCDMA DL

Table 10. WCDMA downlink

Cellular Standard	WCDMA
Chip Rate	3.84 MHz
Number of Frames	1
Link Direction	Downlink

4.2.1 WCDMA DL QPSK

4.2.1.1 Waveform Creation

Waveform of WCDMA DL QPSK has been generated with the following Parameters

Table 11. WCDMA downlink QPSK waveform

Modulation	QPSK

4.2.2 WCDMA DL 16QAM

4.2.2.1 Waveform Creation

Waveform of WCDMA DL 16QAM has been generated with the following Parameters

Table 12. WCDMA downlink 16QAM waveform

Modulation	16QAM

4.3 LTE DL 16QAM

4.3.1 Waveform Creation

Waveform of LTE DL 16QAM has been generated with the following Parameters

Table 13. LTE downlink 16QAM waveform

Cellular Standard	LTE
Modulation	16QAM
Channel Bandwidth	20 MHz
Link Direction	Downlink

5. RESULTS AND DISCUSSION

Initially only 1 Modulation could be classified by the classifier.

5.1 BPSK Modulation Classification With Low Noise

In Figure 4, 1 Modulation has been classified with low noise, which has been shown in the GUI in LabVIEW.

Figure 4. BPSK classification

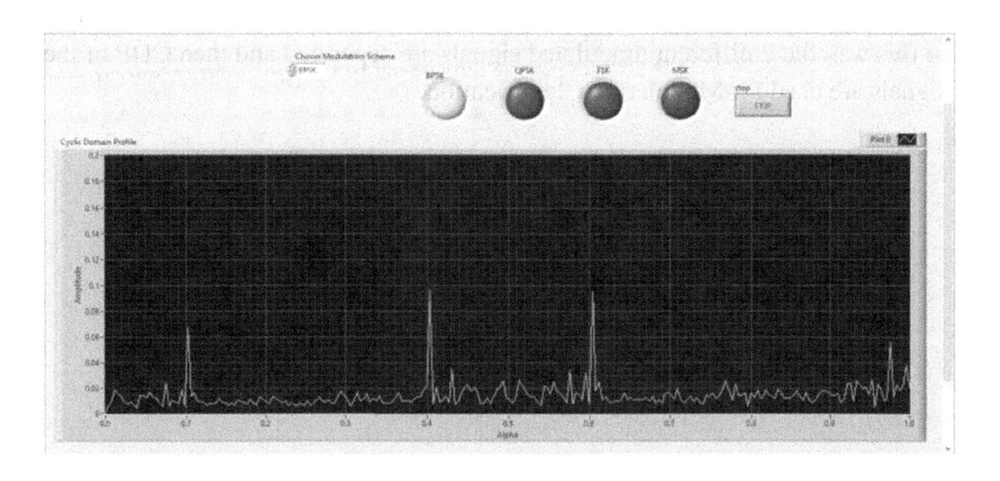

5.2 FSK Modulation Classification With Noise

In Figure 5, when AWGN Noise was added, 1 Modulation has been classified with noise, which has been shown in the GUI in LabVIEW.

Figure 5. FSK modulation classification with noise

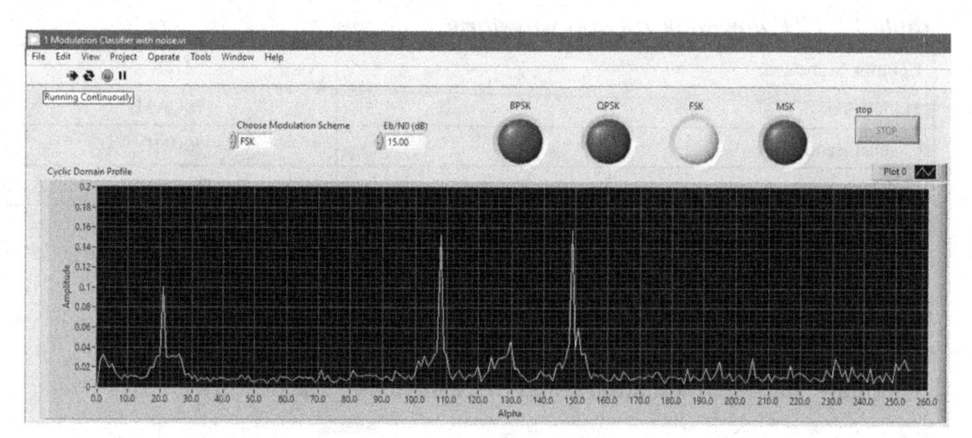

5.3 BPSK BPSK Modulations Classification Without Noise

The Modulation Classifiers have been tested without noise. There are 2 Modulations at different Intermediate Frequencies. The 2 Signals are first downconverted to their Baseband Frequency and then upconverted to their Intermediate Frequencies. In this way the 2 different modulated signals are separated and then CDP of these signals are used for Modulation Classification.

Figure 6. 2 modulations classification without noise

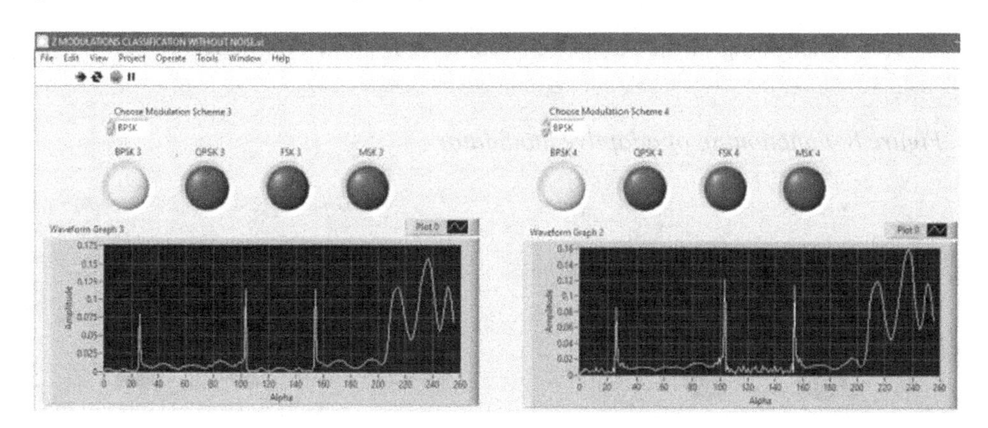

5.4 BPSK BPSK Modulations Classification in AWGN Channel

At 0Db in AWGN Channel, the modulation classification is shown in the figure below. The Modulations have been classified by the classifier based on the values of CDP. The CDP of BPSK at Symbol Rates of 10 kHz and 1 kHz have been displayed.

The Modulation Classifier has been tested with noise. There are 2 Modulations at different Intermediate Frequencies. The 2 Signals are first downconverted to their Baseband Frequency and then upconverted to their Intermediate Frequencies. In this way the 2 different modulated signals are separated and then CDP of these signals are used for Modulation Classification.

Figure 7. 2 Modulations classification without noise

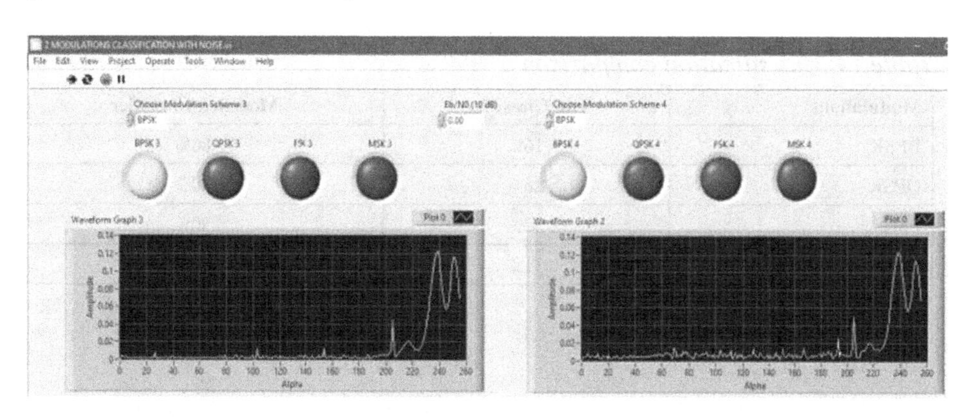

5.5 Adaptive Modulator

The Functioning of the Adaptive Modulator is shown in the figure below

Figure 8. Functioning of adaptive modulator

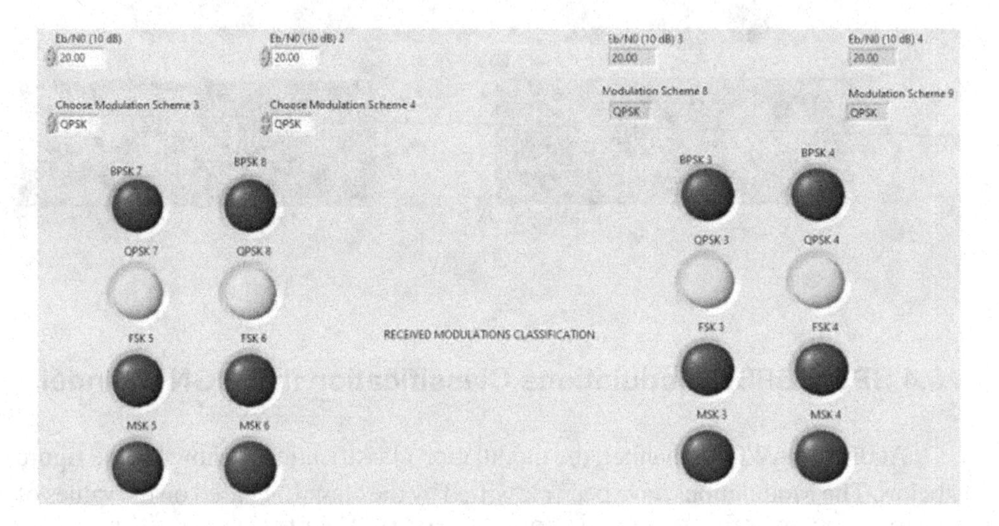

5.6 Comparison of the classifiers in noise

In Figure 9, a comparison of Average Classification Accuracy of different modulations at different noise levels from 0dB to 20dB between Old and Modified Classifiers have been shown.

Table 14. Classification comparison

Modulations	Old Classifier	Modified Classifier
BPSK	16%	80%
QPSK	26%	80%
FSK	32%	80%
MSK	31%	80%

Figure 9. Comparison of modulation classifiers

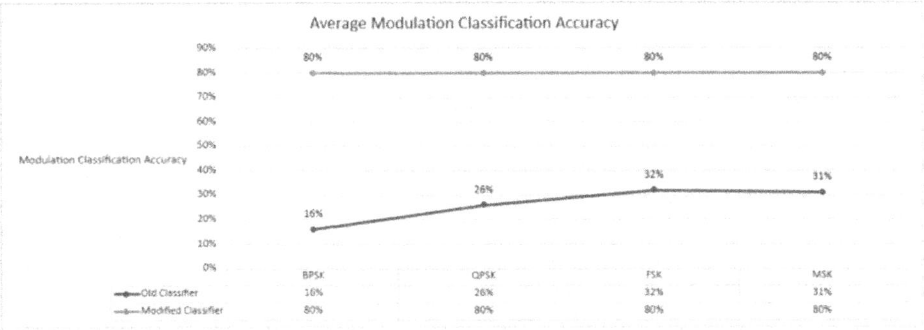

5.7 Cellular Modulation Classification

The output of the Cellular Modulations Classifier has been shown in the Figures, in which the Cellular Modulations have been classified using the classifier in the VI. The Indicators have been utilized to view which of the Modulation has been classified after Cyclostationary Processing. CDP Graph and IQ Vectors of the Cellular Modulations have also been shown in the Figures. In the Figure below, EDGE 8PSK has been classified, which has been displayed on LabVIEW GUI.

The Comparative Results of classification of Reference Cellular Modulations and Cellular Modulations in Noise have also been given in the table below and figures.

Table 15. Comparative results

	Reference Cellular Modulations Classification Accuracy	**Cellular Modulations Classification Accuracy with Noise**
GSM MSK	80%	70%
EDGE 8PSK	80%	70%
WCDMA QPSK DL	80%	70%
WCDMA 16QAM DL	80%	70%
LTE 16QAM DL	80%	70%

Figure 10. Cellular modulation classification accuracy

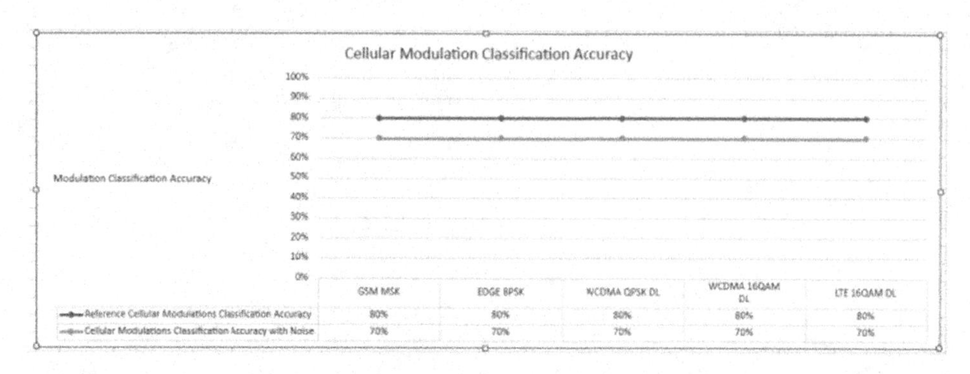

In the Figure 11 and Figure 12, the Classifications of Reference EDGE 8PSK and EDGE 8PSK in Noise has been given, which has been simulated using LabVIEW.

Figure 11. Reference EDGE 8PSK classification

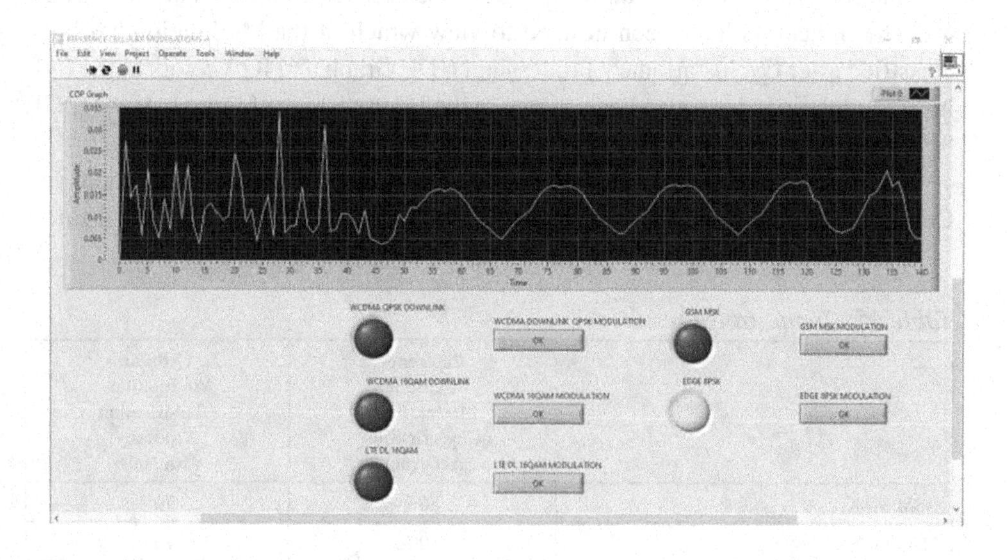

Figure 12. EDGE 8PSK classification in noise

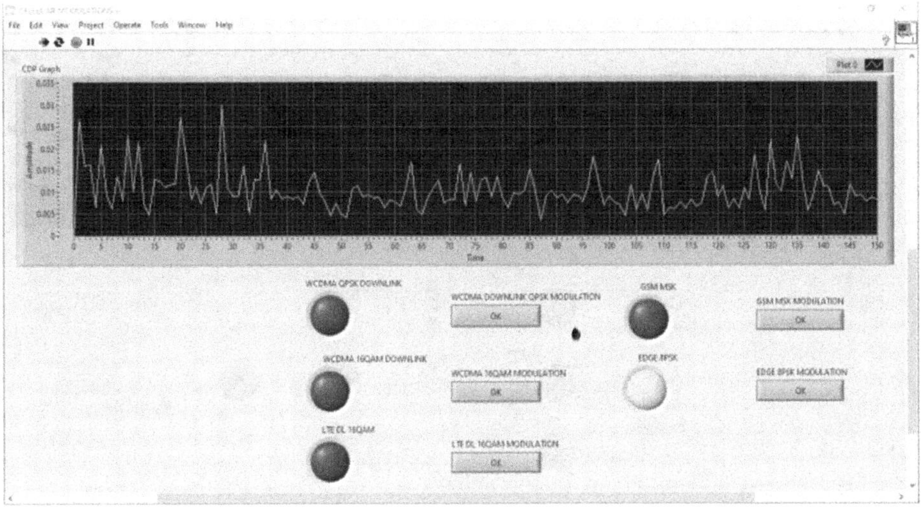

In the Figure 13 and Figure 14, the Classifications of Reference GSM MSK and GSM MSK in Noise has been given, which has been simulated using LabVIEW.

Figure 13. Reference GSM MSK classification

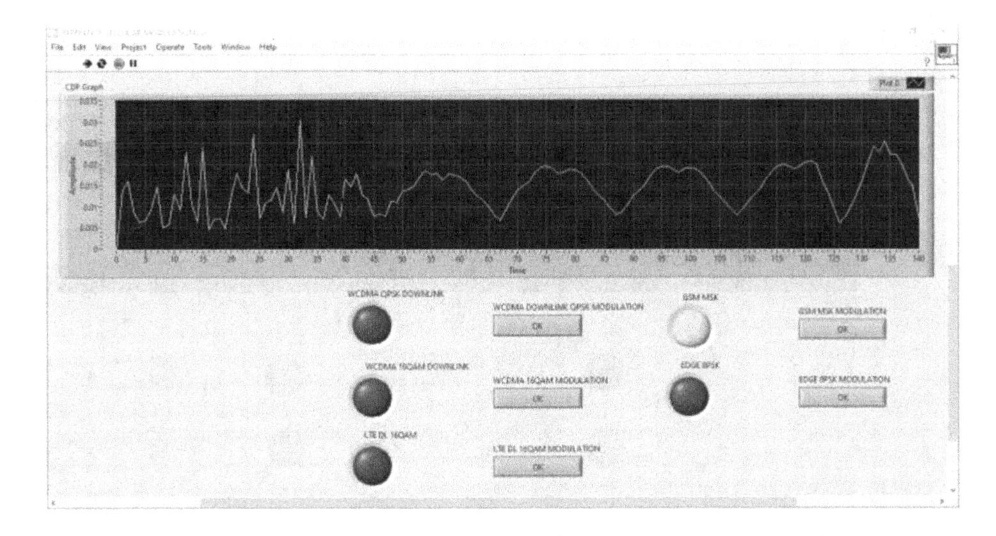

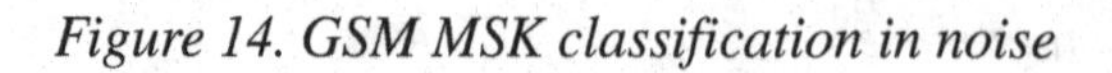

Figure 14. GSM MSK classification in noise

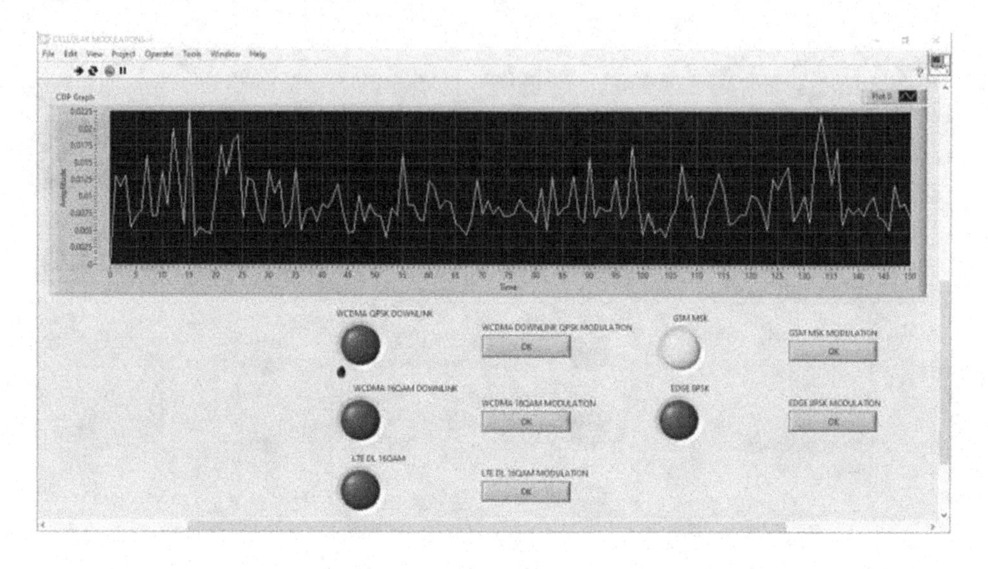

In the Figure 15 and Figure 16, the Classifications of Reference LTE DL 16QAM and LTE DL 16QAM in Noise has been given, which has been simulated using LabVIEW.

Figure 15. Reference LTE DL 16QAM classification

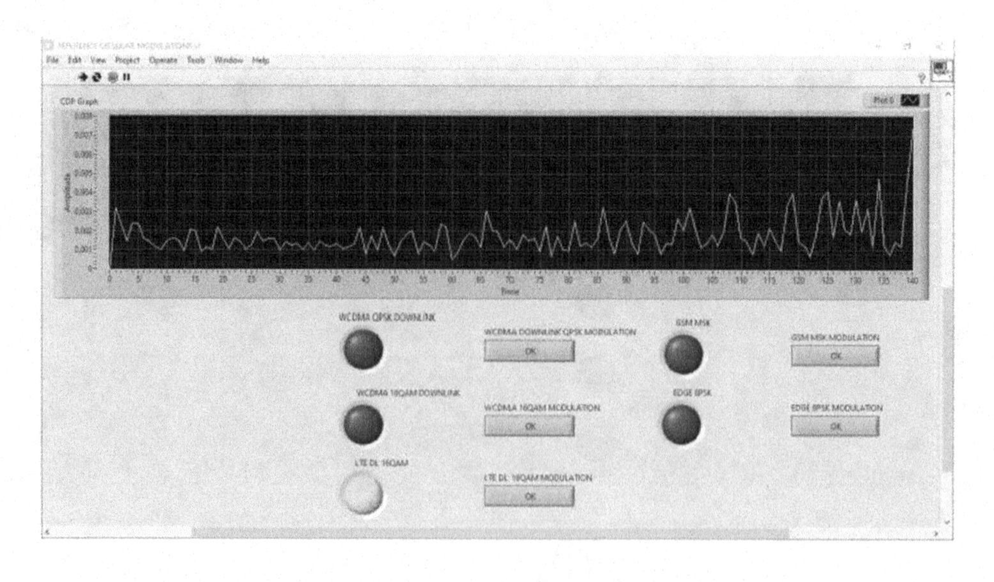

Figure 16. LTE 16QAM classification in noise

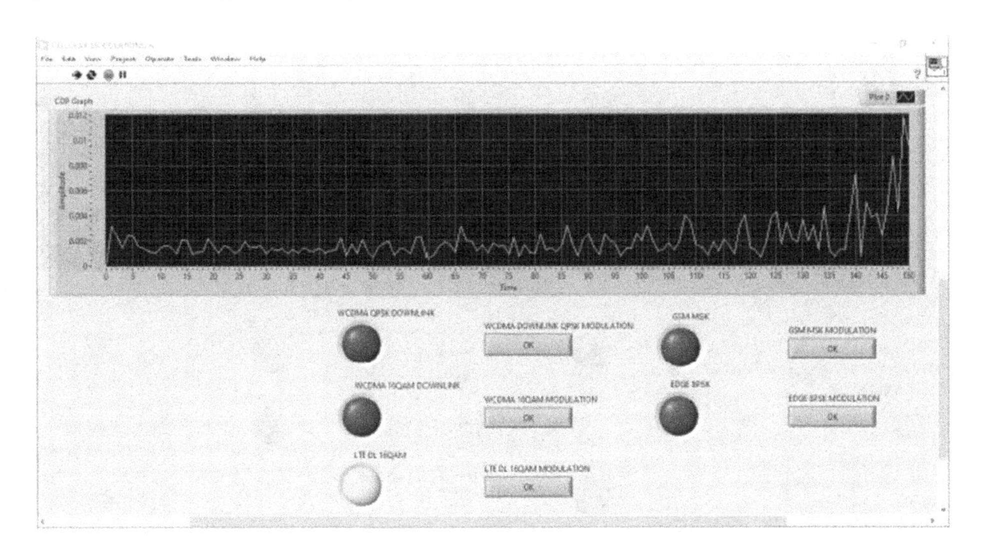

In the Figure 17 and Figure 18, the Classifications of Reference WCDMA 16QAM DL and WCDMA 16QAM DL in Noise has been given, which has been simulated using LabVIEW.

Figure 17. Reference WCDMA 16QAM DOWNLINK classification

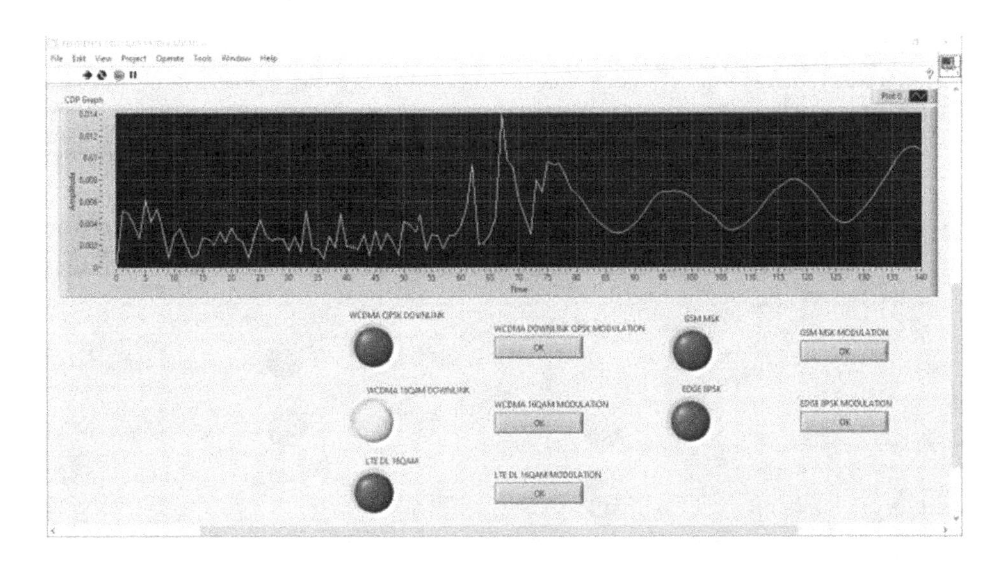

Figure 18. WCDMA 16QAM DOWNLINK classification in noise

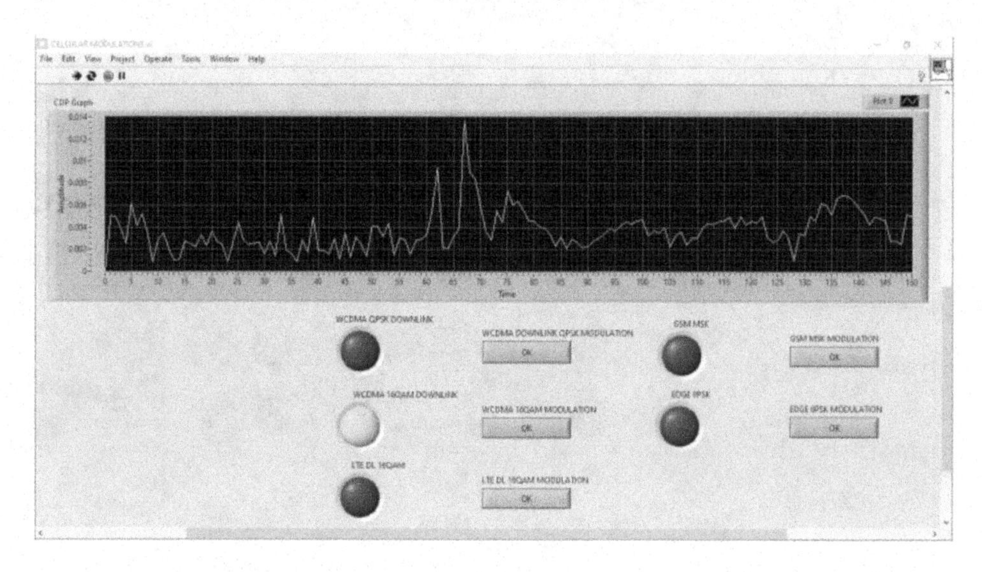

In the Figure 19 and Figure 20, the Classifications of Reference WCDMA QPSK DL and WCDMA QPSK DL in Noise has been given, which has been simulated using LabVIEW.

Figure 19. Reference WCDMA QPSK DOWNLINK classification

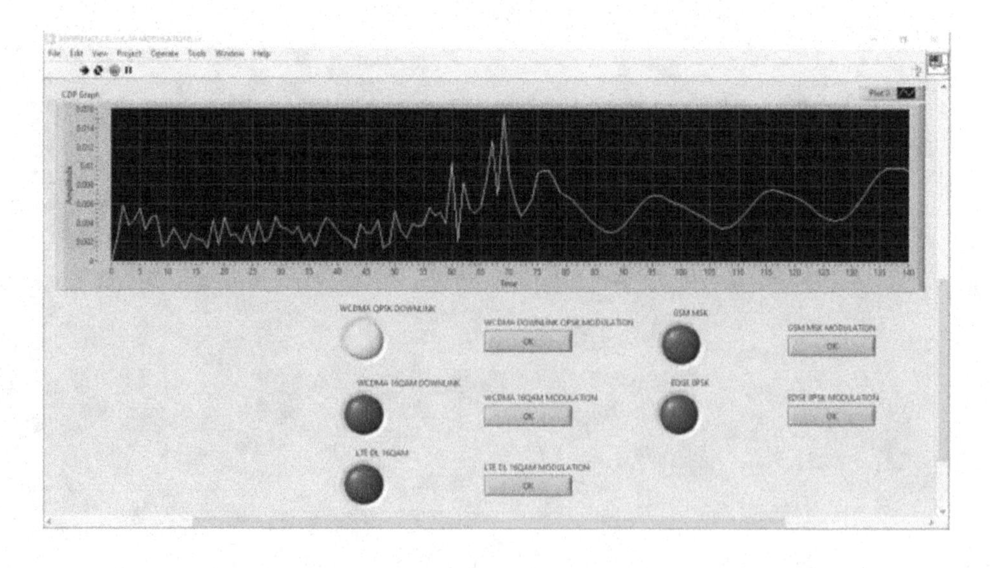

Figure 20. WCDMA QPSK DOWNLINK classification

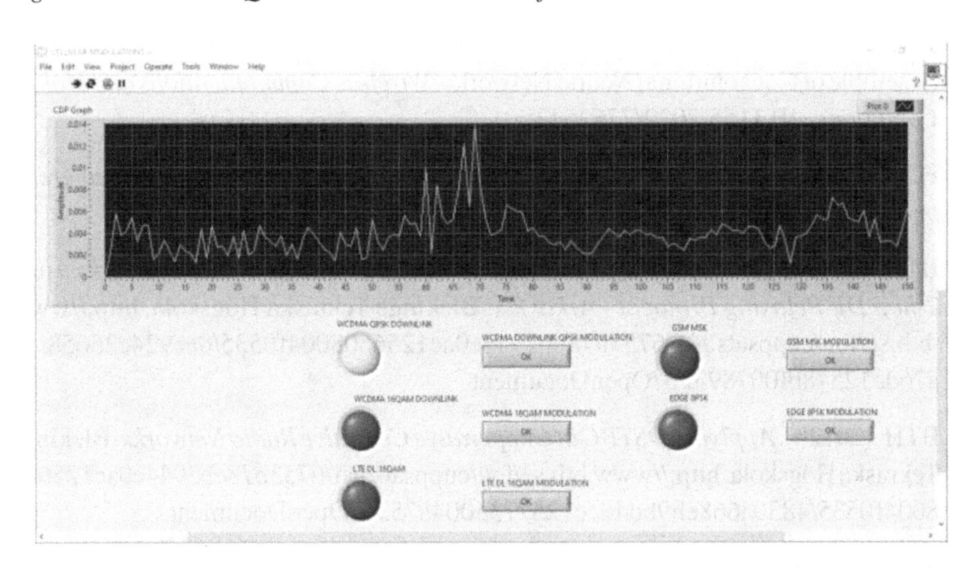

6. CONCLUSION

In this paper, Cyclostationary analysis of Digital Modulated Signals and its classification has been discussed with results and comparison, and based on these results an Adaptive Modulator has also been simulated. The Cyclostationary Features of the simulated signals are obtained using the Cyclostationary Feature Detection algorithm. The Adaptive Modulator for the simulated BPSK, QPSK, FSK and MSK modulated signals and is also capable of Adaptive Modulation based on Noise Level. The Modulation Classifier has also been modified for classifying Cellular Modulations including GSM FSK, GSM MSK, EDGE 8PSK, WCDMA Downlink QPSK, WCDMA Downlink 16QAM and LTE Downlink 16QAM. By utilizing the Modulation Classification and Adaptive Modulator we can achieve more reliable communication in different channel conditions and also adaptively modulate our signals as per requirements of channel to minimize the distortions caused by the noise.

REFERENCES

Abid, B., Khan, B. M., & Memon, R. A. (2022). Seismic Facies Segmentation Using Ensemble of Convolutional Neural Networks. *Wireless Communications and Mobile Computing*. 10.1155/2022/7762543

FCC. (2015). *Amendment of the commission's rules with regard to commercial operations in the 3550-3650 MHz band.* FCC.

BTH. (2012a). *Analysis of OSTBC in Cooperative Cognitive Radio Networks using 2-hop DF Relaying Protocol - Arkiv EX.* Blekinge Tekniska Högskola. http://www.bth.se/fou/cuppsats.nsf/6753b78eb2944e0ac1256608004f0535/0ae924c26658a76dc12578bf0078935b?OpenDocument

BTH. (2012b). *Applying OSTBC in Cooperative Cognitive Radio Networks*. Blekinge Tekniska Högskola. http://www.bth.se/fou/cuppsats.nsf/6753b78eb2944e0ac1256608004f0535/4830a668eb9bd48cc1257735004a7535?OpenDocument

Bhargavi, D., & Murthy, C. R. (2010). Performance comparison of energy, matched-filter and cyclostationarity-based spectrum sensing. *Signal Processing Advances in Wireless Communications (SPAWC), IEEE Eleventh International Workshop*. IEEE. 10.1109/SPAWC.2010.5670882

Bhargavi, D., & Murthy, C. R. (2010). Performance comparison of energy, matched-filter and cyclostationarity-based spectrum sensing. *Signal Processing Advances in Wireless Communications (SPAWC)*. IEEE. 10.1109/SPAWC.2010.5670882

Bilal, R., & Khan, B. M. (2019). Software-Defined Networks (SDN): A Survey. *Handbook of Research on Cloud Computing and Big Data Applications in IoT.* IGI Global. 10.4018/978-1-5225-8407-0.ch023

Cabric, D., Mishra, S. M., & Brodersen, R. W. (2004). Implementation issues in spectrum sensing for cognitive radios. *Conference Record of the Thirty-Eighth Asilomar Conference*. IEEE. 10.1109/ACSSC.2004.1399240

Digham, F. F., Alouini, M.-S., & Simon, M. K. (2003). On the energy detection of unknown signals over fading channels. *IEEE International Conference*. IEEE. 10.1109/ICC.2003.1204119

Dobre, O., Abdi, A., Bar-Ness, Y., & Su, W. (2007, April). Survey of automatic modulation classification techniques: Classical approaches and new trends. *IET Communications*, 1(2), 137–156. 10.1049/iet-com:20050176

Federal Communications Commission. (2002). *Revision of part 15 of the commission's rules regarding ultra-wideband transmission systems*. FCC.

Federal Communications Commission. (2018). *FCC online table of frequency allocations*. FCC. https://transition.fcc.gov/oet/spectrum/table/fcctable. pdf

Frasch, I. J. (2017). *Algorithmic framework and implementation of spectrum holes detection for cognitive radios*. [Master's thesis, Rochester Institute of Technology].

Gardner, W. (1988, August). Signal interception: A unifying theoretical framework for feature detection. *IEEE Transactions on Communications*, 36(8), 897–906. 10.1109/26.3769

Goldsmith, A., Jafar, S., Maric, I., & Srinivasa, S. (2009, May). Breaking spectrum gridlock with cognitive radios: An information theoretic perspective. *Proceedings of the IEEE*, 97(5), 894–914. 10.1109/JPROC.2009.2015717

Khan, B. M., & Ali, F. H. (2011). Mobility Adaptive Energy Efficient and Low Latency MAC for Wireless Sensor Networks.*2011 Fifth International Conference on Next Generation Mobile Applications, Services and Technologies*, Cardiff, UK. 10.1109/NGMAST.2011.46

Lagunas, E., & Najar, M. (2015, July). Spectral feature detection with sub-Nyquist sampling for wideband spectrum sensing. *IEEE Transactions on Wireless Communications*, 14(7), 3978–3990. 10.1109/TWC.2015.2415774

Memon, R. A., Qazi, S., & Khan, B. M. (2021). Design and Implementation of a Robust Convolutional Neural Network-Based Traffic Matrix Estimator for Cloud Networks. *Wireless Communications and Mobile Computing*. Wiley. 10.1155/2021/1039613

Mitola, J., & Maguire, G. (1999, August). Cognitive radio: Making software radios more personal. *IEEE Personal Communications*, 6(4), 13–18. 10.1109/98.788210

Mohammadi, F. S., & Kwasinski, A. (2018). *Neural network cognitive engine for autonomous and distributed underlay dynamic spectrum access*. arXiv preprint arXiv:1806.11038.

Ramkumar, B. (2009, June). Automatic modulation classification for cognitive radios using cyclic feature detection. *IEEE Circuits and Systems Magazine*, 9(2), 27–45. 10.1109/MCAS.2008.931739

Spooner, C. M., & Mody, A. N. (2018, June). Wideband cyclostationary signal processing using sparse subsets of narrowband subchannels. *IEEE Transactions on Cognitive Communications and Networking*, 4(2), 162–176. 10.1109/TCCN.2018.2790971

Tian, Z., Tafesse, Y., & Sadler, B. M. (2012, February). Cyclic feature detection with sub-Nyquist sampling for wideband spectrum sensing. *IEEE Journal of Selected Topics in Signal Processing*, 6(1), 58–69. 10.1109/JSTSP.2011.2181940

Xie, L., & Wan, Q. (2017, June). Cyclic feature-based modulation recognition using compressive sensing. *IEEE Wireless Communications Letters*, 6(3), 402–405. 10.1109/LWC.2017.2697853

Yucek, T. (2009). A survey of spectrum sensing algorithms for cognitive radio applications. *Communications Surveys Tutorials*. IEEE.

Zhang, H., & Wang, X. (2011). A Fuzzy Decision Scheme for Cooperative Spectrum Sensing in Cognitive Radio. *Vehicular Technology Conference (VTC Spring)*. IEEE. 10.1109/VETECS.2011.5956116

Chapter 17
Radar Cross Section Modelling and Analysis Using Various Estimation Techniques in FMCW Radar Frequencies

K. Hariharan
https://orcid.org/0000-0001-7540-6724
Thiagarajar College of Engineering, India

M. N. Suresh
Thiagarajar College of Engineering, India

B. Manimegalai
Thiagarajar College of Engineering, India

ABSTRACT

The radar cross-section (RCS) is a critical and crucial factor that determines presence of the target by radar signal, and it can be consistently detected so as to track the target by a radar system. The target's size can be fixed by changing the factor ratio of the RCS quantity. Raising the target's RCS improves its radar detestability. Reduced RCS, on the other hand, the chances for the target to miss from radars, which is frequently required in military settings. The equation for the RCS is depending upon various factor and constants, the surface area (A), the wavelength (λ), the Fresnel reflection coefficient (G) at normal incidence.

DOI: 10.4018/979-8-3693-2786-9.ch017

1. INTRODUCTION

Road infrastructure plays a fundamental role in facilitating transportation and commerce, serving as the lifeline of modern societies. However, maintaining road safety and integrity poses significant challenges, particularly in identifying and rectifying roadway defects such as potholes. Potholes, resulting from various factors including weathering, heavy traffic, and subpar construction materials, present substantial hazards to motorists, cyclists, and pedestrians. These roadway imperfections not only increase the risk of accidents but also contribute to vehicle damage, traffic congestion, and economic costs associated with repairs and medical expenses.

In recent years, advancements in radar technology have provided promising solutions for pothole detection and analysis. Among these technologies, Frequency Modulated Continuous Wave (FMCW) radar stands out for its ability to offer high-resolution range and velocity measurements, making it ideal for detecting small-scale surface irregularities like potholes. By leveraging the principles of radar cross-section (RCS) estimation, FMCW radar enables the detection and characterization of potholes with enhanced accuracy and efficiency.

This paper aims to explore and compare various RCS estimation techniques within the FMCW radar framework for pothole detection. Through a systematic analysis of these techniques, we seek to identify the most effective approach for detecting and analyzing potholes on roadways. By doing so, we contribute to the advancement of pothole detection methodologies, thereby facilitating proactive road maintenance practices and ultimately enhancing road safety for all users.

2. FUNDAMENTALS OF RADAR CROSS SECTION

Radar Cross Section (RCS) measures the electromagnetic energy an object reflects when illuminated by a radar signal. It is a fundamental concept in radar technology, crucial for target detection, tracking, and identification. Understanding the fundamentals of RCS involves several supporting concepts and equations:

2.1 Radar Equation

The radar equation connects the received power at the radar receiver to various parameters. Which includes the transmitted power, radar cross section (σ), distance to the target (R), wavelength of the radar signal (λ), and other factors such as antenna gains and losses. The basic form of the radar equation is:

$$P_R = \frac{P_T \bullet \sigma \bullet G_T \bullet G_R}{(4\pi)^2 \bullet R^4} \bullet \frac{\lambda^2}{4\pi}$$

Here;

- P_R is the received power.
- P_T is the transmitted power.
- σ the radar cross section.
- GT and GR represent the gains of the transmitting and receiving antennas, respectively.
- R represents the distance between the Radar and the target object.
  -lambda represents the radar signal's wavelength.

2.2 Radar Cross Section (RCS)

RCS measures how effectively an object reflects radar signals. It is defined as the ratio of the power scattered by the target toward the radar receiver to the power density of the incident radar wave. RCS is typically expressed in square meters. (m^2).

2.3 Monostatic and Bistatic RCS

Monostatic RCS: It's a radar system where the transmitter and receiver share one same antenna. The RCS is measured when the radar transmitter and receiver are positioned at the same location.

Bistatic RCS: It's a radar system where the transmitter and receiver have separate antennas. RCS is measured with the transmitter and receiver situated at different locations.

2.4 Polarization

The polarization of the incident radar signal and the reflected signal can influence RCS. Matching or mismatched polarizations can affect the amount of energy reflected back to the radar receiver.

2.5 Aspect Angle

The RCS of an object may change based on the aspect angle, which is the angle between the radar's line of sight and the object's orientation. Objects typically exhibit varying RCS values at different aspect angles.

2.6 Frequency Dependency

RCS can be frequency-dependent, meaning it may vary with the frequency of the incident radar signal. This frequency dependence is often exploited in radar stealth technology.

2.7 Scattering Mechanisms

RCS is influenced by various scattering mechanisms such as specular reflection, diffuse reflection, diffraction, and resonance phenomena. Each mechanism contributes differently to the overall RCS of an object.

Understanding these fundamental concepts and equations is essential for analyzing and interpreting RCS data in radar applications, including target detection, tracking, and radar system design.

3. MATHEMATICAL ANALYSIS OF RCS OF POTHOLES

3.1 Physical Optics

3.1.1 Physical Optics Method and RCS Analysis

Physical Optics (PO) is a widely employed method for analyzing Radar Cross Section (RCS), particularly for objects with smooth surfaces such as potholes on roadways. The fundamental concept of physical optics (PO) involves approximating the scattered electromagnetic field by accounting for the surface currents induced on the object by the incident electromagnetic wave. This approach simplifies the calculation of RCS by modeling the object as a collection of surface currents that interact with the incident field, leading to reflected and scattered waves.

When considering potholes, PO offers several advantages. Firstly, it provides a relatively simple yet effective means of estimating the RCS, making it suitable for initial assessments and quick evaluations. Secondly, PO is well-suited for analyzing smooth surfaces where the induced surface currents can be approximated more accurately. This is particularly relevant for potholes, which often have relatively uniform surface characteristics.

However, PO has limitations, especially when applied to objects with complex geometries or sharp edges, as encountered in certain types of road defects. The accuracy of PO diminishes in such cases, as the model struggles to capture the intricacies of electromagnetic interactions around irregular shapes. Additionally, PO neglects diffraction effects, which can be significant, especially for small structures or when dealing with high-frequency radar waves. These limitations highlight the need for complementary methods or advanced simulations to account for diffraction and complex geometries in RCS analysis for objects like potholes.

Mathematically, the scattered electric field $E_s(r)$ at an observation point **r** due to an incident electric field $E_i(r')$ can be expressed as:

$$E_s(r) = \frac{1}{4\pi}\int_S\left(\frac{k^2}{|r - r'|}\right)\left(n \times E_i(r')\right) \times n'ds'$$

Here, ***S*** represents the surface of the pothole, **n** and '**n**' are unit vectors normal to the surface at points **r** and '**r**' respectively, and k is the wave number.

3.2 PHYSICAL OPTICS METHOD

3.2.1 Methodology

1. Here, for efficient pothole analysis (Detecting the edges of the pothole), the pothole is modelled to be Dielectric Cylinders.
2. The RCS value of the pothole can be estimated by cumulating the backscattered fields from the cylinders that exhibit significant backscatter.
3. To assess the accuracy of the PO model in predicting the RCSs of surface potholes, a circular pothole is chosen as an illustrative example.
4. At millimeter-wave frequencies, the PO model tends to yield accurate results when targets are relatively large compared to the wavelength and are made of materials with significant losses (Li, 2003).

3.2.2 Assumptions for Applying Physical Optics Method

To utilize this technique for analyzing the backscatter response of the target, certain assumptions need to be made (Li, 2003).

1) The fields and equivalent currents are present solely within the region directly illuminated by the incident wave.
2) The surface's radii of curvature are assumed to be significantly larger than the wavelength, and the currents on the illuminated surface exhibit characteristics similar to those on an infinite plane tangent to the surface at the point of incidence.
3) The target's material is sufficiently lossy to prevent internal field scattering or penetration through the target.

4) We assume that in the non-illuminated area of the target, both the fields and equivalent currents are zero.

3.2.3 Challenges and Considerations in Po-Based Pothole RCS Analysis

While Physical Optics (PO) offers a straightforward approach to Radar Cross Section (RCS) analysis for smooth surfaces like potholes, it faces challenges when dealing with objects of irregular shapes or complex geometries. Potholes, being naturally occurring defects on road surfaces, often exhibit non-uniform shapes and sharp edges, posing difficulties for accurate RCS estimation using PO alone.

One of the key limitations of PO in pothole RCS analysis is its inability to account for diffraction effects. Diffraction plays a significant role, especially at higher frequencies or when dealing with small structures, influencing the scattering characteristics, and leading to potential inaccuracies in RCS predictions. This limitation underscores the importance of considering diffraction effects in conjunction with PO or utilizing advanced techniques such as the Physical Theory of Diffraction (PTD) for a thorough analysis of RCS.

Moreover, the accuracy of PO diminishes when applied to objects with intricate geometries, as encountered in certain types of potholes. The simplistic surface current approximation in PO may not adequately capture the electromagnetic interactions around irregular shapes, leading to deviations in RCS estimations. To address this, hybrid methods combining PO with geometric optics or other advanced techniques can be employed to enhance accuracy and capture the nuances of scattering from complex pothole geometries.

Figure 1. Comparing simulation outcomes with calculated values for the copolarized RCS of a ten cm diameter pothole

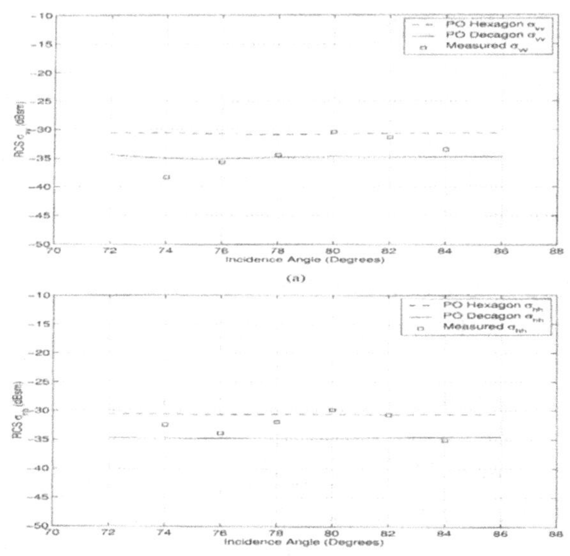

A) V-V response and B) H H response (Li, 2003)

3.2.4 Method of Moments (MOM)

The Method of Moments (MoM) is notable for its robustness and effectiveness as a numerical technique for analyzing the Radar Cross Section (RCS) of potholes on road surfaces. This method operates by discretizing the pothole's surface into small elements, treating each element as a separate entity for electromagnetic analysis. By doing so, MoM formulates integral equations that represent the electromagnetic interactions occurring between these discrete elements.

One of the notable advantages of employing MoM in pothole RCS analysis is its versatility. MoM demonstrates remarkable effectiveness and accuracy across a broad spectrum of frequencies, making it suitable for analyzing potholes regardless of the operating frequency of the radar system. This versatility ensures that the analysis remains comprehensive and accurate, capturing the electromagnetic behaviors of potholes across different frequency ranges.

However, it's essential to acknowledge that MoM can pose computational challenges, particularly when dealing with large structures. The computational intensity of MoM increases significantly for expansive road surfaces with numerous potholes. To mitigate this, careful attention must be given to meshing and discretization techniques. Optimal accuracy in pothole RCS analysis using MoM necessitates meticulous meshing strategies and discretization methods to manage computational resources effectively.

The scattered electric field *Es*(**r**) can be expressed as a sum of contributions from induced currents ***Ii*** on each element:

$$E_s(r) = \sum_{i=1}^{N} I_i e^{-jk|r-r_i|}$$

Here, ri represents the position vector of the i-th element, with N denoting the total number of elements.

3.2.5 Challenges and Considerations in Mom-Based Pothole RCS Analysis

The Method of Moments (MoM) is a powerful numerical technique designed specifically for analyzing Radar Cross Section (RCS) in potholes found on road surfaces. MoM achieves this by dividing the intricate geometry of a pothole into smaller, manageable elements. These elements are then used to create integral equations that precisely describe the electromagnetic interactions between them.

One of the primary strengths of MoM in pothole RCS analysis is its versatility and accuracy across a wide range of frequencies. Whether operating at lower or higher frequencies, MoM consistently delivers precise results, capturing the intricate electromagnetic phenomena associated with potholes. This capability ensures that the analysis remains reliable and applicable across various radar systems operating at different frequencies.

Despite its advantages, it's crucial to note that MoM can be computationally demanding, especially for large structures such as extensive road surfaces with multiple potholes. To achieve optimal accuracy, meticulous attention must be paid to meshing and discretization processes. Careful meshing ensures that the computational load is managed effectively, allowing MoM to deliver accurate and insightful pothole RCS analyses.

3.2.6 Physical Theory of Diffraction (PTD)

The Physical Theory of Diffraction (PTD) significantly enhances Radar Cross Section (RCS) analysis by incorporating diffraction effects, thereby elevating the accuracy and fidelity of pothole detection on road surfaces. PTD builds upon the foundation of Physical Optics (PO) but extends its capabilities by considering both specular and non-specular reflections, which is particularly valuable for objects characterized by sharp features and complex geometries.

In PTD, the scattered field is expressed as a comprehensive sum of contributions from various diffraction phenomena. This includes contributions from specular reflection, edge diffraction, and surface diffraction, each contributing significantly to the formation of the overall RCS pattern of an object, like a pothole on a road surface. By accounting for these diffraction effects, PTD provides a more nuanced and accurate representation of how electromagnetic waves interact with complex structures like potholes.

Using PTD for analyzing pothole RCS offers a key advantage in enhancing accuracy, particularly for structures with complex geometries and sharp features. PTD's inclusion of diffraction effects ensures that the analysis captures finer details and nuances in the scattering behavior, leading to more precise detection and characterization of potholes in radar imaging scenarios.

However, it's essential to note that PTD comes with its set of challenges. Its implementation complexity can be higher compared to simpler methods like Physical Optics (PO). Additionally, incorporating diffraction effects increases the computational demands of the analysis, potentially posing challenges in terms of computational resources and processing time. Despite these challenges, the improved accuracy and fidelity offered by PTD make it an essential resource for in-depth pothole RCS analysis.

3.2.7 Challenges and Considerations in PTD-Based Pothole RCS Analysis

The Physical Theory of Diffraction (PTD) method plays a crucial role in advancing Radar Cross Section (RCS) analysis, particularly when it comes to identifying and describing potholes on road surfaces. PTD enriches the analysis by integrating diffraction effects, which are essential for accurately modeling the scattering behavior of objects with complex geometries and sharp edges, such as potholes.

Unlike simpler methods like Physical Optics (PO), PTD goes beyond the approximation of only specular reflections and includes both distinctive and non-distinctive elements within the scattered field. This comprehensive approach involves considering edge diffraction and surface diffraction effects, providing a more detailed and accurate representation of how electromagnetic waves interact with the surfaces and edges of potholes.

The advantages of utilizing PTD in pothole RCS analysis are significant, primarily stemming from its ability to improve accuracy for complex structures. By capturing diffraction effects, PTD ensures that the RCS analysis accounts for subtle nuances in the scattering behavior, leading to more precise detection and characterization of potholes. This enhanced accuracy is particularly valuable in radar imaging applications where detailed information about the target's structure is essential.

However, it's important to acknowledge that implementing PTD comes with challenges. The method's complexity requires expertise in electromagnetic theory and numerical simulations. Additionally, incorporating diffraction effects increases the computational demands, requiring robust computational resources and efficient algorithms for accurate and timely analysis. Despite these challenges, the benefits of improved accuracy and detailed modeling offered by PTD make it a valuable tool in advanced pothole RCS analysis.

3.2.8 Geometric Theory of Diffraction (GTD)

The Geometric Theory of Diffraction (GTD) is a powerful mathematical framework that has gained prominence in Radar Cross Section (RCS) analysis, particularly in scenarios involving complex structures like potholes. GTD offers a robust approach by approximating the scattered field as a summation of diffracted rays originating from different edges and surfaces of the pothole. This methodology allows for efficient solutions to intricate scattering problems encountered in radar imaging and analysis.

One of the primary advantages of employing GTD in pothole RCS analysis is its efficiency in handling complex structures. By breaking down the scattering process into diffracted rays from specific edges and surfaces, GTD provides a streamlined and computationally efficient way to analyze the RCS of potholes. This efficiency is particularly valuable in scenarios where detailed analysis of complex structures is required without compromising computational resources or processing time.

However, the implementation of GTD may require careful consideration of diffraction mechanisms. While GTD offers efficient solutions, its accuracy and effectiveness heavily rely on correctly modeling diffraction phenomena at various edges and surfaces of the pothole. Incorrect modeling or oversimplification of diffraction effects can lead to inaccuracies in the RCS analysis, especially for highly detailed structures with intricate features.

3.2.9 Challenges and Considerations in GTD-Based Pothole RCS Analysis

Despite its efficiency, GTD may pose challenges when dealing with highly detailed structures. Ensuring an accurate representation of diffraction mechanisms and properly accounting for all relevant edges and surfaces can be complex and may require expertise in electromagnetic theory and numerical simulations. Additionally, GTD's effectiveness can be influenced by the complexity of the target structure and the specific radar imaging scenario, requiring careful validation and verification of results to ensure reliability.

3.2.10 Physical Geometrical Optics (PGO)

Physical Geometrical Optics (PGO) is a sophisticated methodology that combines aspects of Physical Optics (PO) and Geometrical Optics (GO) to create a comprehensive framework for analyzing the Radar Cross Section (RCS) of potholes, especially at high frequencies. PGO excels in accurately modeling both specular and non-specular reflections, offering valuable insights into the scattering properties of potholes and similar complex structures encountered in radar applications.

One significant benefit of employing PGO in pothole RCS analysis lies in its capability to offer accurate modeling of both specular and non-specular reflections. Specular reflections occur when electromagnetic waves bounce off smooth surfaces at predictable angles, while non-specular reflections involve scattering and diffraction effects from rough or irregular surfaces. PGO integrates these phenomena seamlessly, allowing for a more realistic representation of how potholes interact with radar signals.

PGO's accurate modeling of both specular and non-specular reflections leads to a deeper understanding of pothole scattering properties. By capturing the full range of electromagnetic interactions, PGO enables analysts to study various aspects of pothole behavior, such as backscattering characteristics, signal attenuation, and radar cross section variations across different frequencies and incident angles. This level of detail is crucial for designing effective radar systems and optimizing detection algorithms for road surface anomalies like potholes.

3.2.11 Challenges and Considerations in PGO-Based Pothole RCS Analysis

The implementation of PGO may pose challenges due to its complexity and computational demands. Specialized numerical techniques and advanced algorithms are often required to effectively apply PGO in RCS analysis, especially when dealing with intricate scattering scenarios or large-scale structures. Moreover, while PGO's precise modeling capabilities are impressive, they do demand higher computational resources including processing power and memory, which can limit its practical application in real-time or resource-constrained environments.

3.2.12 Finite Difference Time Domain Method

The Finite Difference Time Domain (FDTD) technique is a powerful numerical method widely used in computational electromagnetics for estimating the Radar Cross Section (RCS) of complex targets, including potholes. FDTD tackles Maxwell's equations by discretizing them on a grid, addressing both time and space domains, which allows for detailed modeling of electromagnetic wave interactions with intricate surface geometries and material properties. In the context of estimating the RCS of a pothole, FDTD is particularly advantageous because it can accurately capture transient scattering effects, near-field interactions, and the impact of varying incident waveforms. This method provides a time-domain solution, which is then transformed to the frequency domain to obtain the RCS. The ability of FDTD to handle arbitrary shapes and inhomogeneous materials makes it an essential tool for analyzing the complex scattering phenomena associated with potholes, contributing to a more comprehensive and accurate RCS estimation.

3.2.13 Challenges and Considerations in FDTD Method Based RCS Estimation

The Finite Difference Time Domain (FDTD) method for estimating the Radar Cross Section (RCS) of potholes faces several challenges. It requires significant computational resources due to the need to discretize the entire computational domain into small cells, which demands substantial memory and processing power. The computational domain must be large enough to accurately capture the scattering behavior, and improperly defined boundaries can lead to reflections that interfere with results. High grid resolution is necessary to model fine details, increasing computational demands, and the time step must be small to satisfy stability conditions, resulting in numerous time steps for large domains or high frequencies. Additionally, accurately modeling material properties, handling large amounts of data, and ensuring the accuracy and reliability of simulations through validation are complex tasks (Srivastava, Goyal and Ram, 2020).

To address these challenges, careful considerations and strategies are essential. Efficient use of parallel computing and hardware optimization can help manage computational load, while adaptive meshing can balance accuracy and efficiency by refining the grid in critical areas. Implementing absorbing boundary conditions like Perfectly Matched Layers (PML) minimizes artificial reflections. Advanced material modeling techniques that incorporate frequency-dependent properties and inhomogeneities improve simulation accuracy. Efficient data management and post-processing tools are necessary to handle the large data volumes generated by FDTD simulations, and techniques like Fast Fourier Transform (FFT) can aid in converting time-domain data to frequency-domain RCS values. Finally, comparing FDTD results with analytical solutions, experimental data, or other numerical methods, along with iterative testing and refinement, is crucial for validation and verification of the models.

Figure 2. Variation of RCS with respect to frequency at height (H), width (W), ground dielectric constant 10 S/m and ground conductivity 0.006 S/m [1]

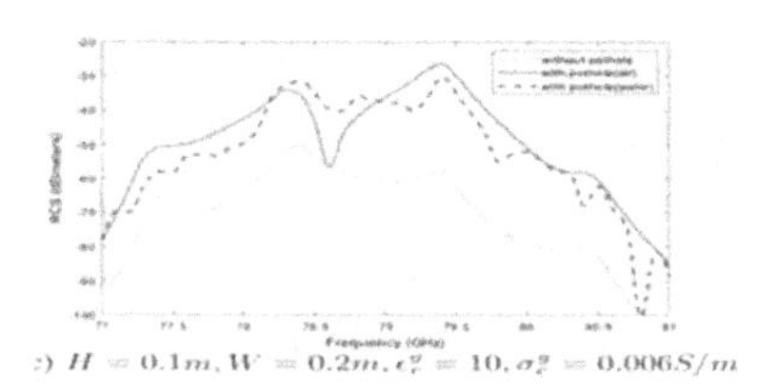

4. PROPOSED METHODOLOGY

Hybrid methodology for a pothole's Radar Cross Section (RCS) estimation

It integrates Physical Optics (PO), Geometrical Optics (GO), Physical Geometrical Optics (PGO), and Finite Difference Time Domain (FDTD) techniques to accurately calculate the Radar Cross Section (RCS) of a pothole. The proposed methodology harnesses the advantages of each technique to strike a balance between computational efficiency and accuracy.

4.1 Methodology Steps

1. Initial Geometrical Characterization Using Geometrical Optics (GO):

   Objective: Obtain a rapid estimation of the primary geometrical properties of the pothole.

   Method: Apply GO to model the large-scale shape of the pothole. Ray tracing techniques are used to determine primary reflections and shadowing regions.

   Output: Basic dimensions and orientation of the pothole surface, identifying large, smooth areas and basic structural features.
2. Detailed Surface Analysis Using Physical Optics (PO):

   Objective: Incorporate the effects of surface roughness and finer details that GO might miss.

Method: Utilize PO to account for surface currents and edge diffraction. PO is applied to regions identified by GO to refine the RCS by including physical surface properties.

Output: Detailed RCS estimate that includes contributions from surface roughness and finer structural details.

3. Integration with Physical Geometrical Optics (PGO):

    Objective: Integrate the advantages of PO and GO for a comprehensive analysis.

    Method: Apply PGO to combine PO-derived currents and GO-derived ray paths, considering multiple interactions and diffractions.

    Output: Enhanced RCS values that account for both large-scale geometrical features and detailed physical interactions.

4. Time-Domain Analysis Using Finite Difference Time Domain (FDTD):

    Objective: Include detailed time-domain characteristics and material properties.

    Method: Employ FDTD to analyze time-dependent scattering behavior, crucial for dynamic scenarios, detailed near-field interactions, and complex material properties.

    Output: Time-domain RCS data illustrating how the RCS varies with different incident waveforms and time instances.

5. Adaptive Meshing:

    Objective: Efficiently allocate computational resources by refining the mesh in regions with high complexity and coarsening it in simpler regions.

    Method:

    - Start with an initial coarse mesh based on GO.
    - Refine the mesh locally where PO and PGO indicate high gradients or rapid changes.
    - Use FDTD to further refine regions requiring detailed time-domain analysis.

    Output: An optimized mesh balancing computational efficiency and accuracy.

6. Hybrid Solver Integration:

    Objective: Seamlessly integrate different solvers (PO, GO, PGO, FDTD) based on region characteristics.

    Method:

- Classify regions of the pothole based on smoothness, roughness, and complexity.
- Assign GO to large, smooth regions; PO to detailed rough regions; PGO to areas needing combined effects; and FDTD to regions requiring detailed time-domain analysis.
- Use boundary conditions and interface matching to ensure smooth transitions between regions.

Output: Consistent and accurate RCS calculation across different regions of the pothole.

7. Iterative Solution and Validation:

Method:

- Solve the coupled system iteratively, allowing each solver to update based on adjacent regions' boundary conditions.
- Compare results with high-fidelity full-wave simulations or empirical data for validation.

Output: Validated RCS values with quantified error margins.

5. CONCLUSION

In this study, we have presented a comprehensive hybrid methodology for accurately estimating the Radar Cross Section (RCS) of potholes on road surfaces. Our approach integrates various electromagnetic modeling techniques including Physical Optics (PO), Geometrical Optics (GO), Physical Geometrical Optics (PGO), and Finite Difference Time Domain (FDTD) methods. Through a step-by-step methodology, we have leveraged the strengths of each technique to strike a balance between computational efficiency and accuracy in pothole RCS estimation.

5.1 Comparative Analysis

The comparison with established methods such as Physical Optics (PO), Method of Moments (MoM), Physical Theory of Diffraction (PTD), Geometric Diffraction, and Physical Geometrical Optics (PGO) reveals several advantages of our integrated methodology. While each traditional method has its merits, our hybrid approach offers distinct advantages:

Accuracy and Detail: Our methodology combines the accuracy of Physical Optics (PO) in capturing surface currents and edge diffraction with the detailed geometric analysis of Geometrical Optics (GO) and Physical Geometrical Optics (PGO). This results in a more comprehensive and detailed RCS estimation, especially for potholes with complex geometries and surface roughness.

Time-Domain Analysis: The inclusion of Finite Difference Time Domain (FDTD) analysis enables us to account for time-dependent scattering behavior, crucial for dynamic scenarios and complex material properties. This provides a more realistic and detailed understanding of the RCS variation over time and with different incident waveforms.

Computational Efficiency: Our methodology employs an adaptive meshing strategy, refining the mesh only in regions with high complexity while coarsening it in simpler areas. This optimizes computational resources and reduces computational overhead compared to full-wave simulations or brute-force methods.

Integration and Consistency: The seamless integration of different solvers (PO, GO, PGO, FDTD) based on region characteristics ensures a consistent and accurate RCS calculation across different parts of the pothole. This approach maintains continuity and avoids discontinuities that may arise in using disparate methods separately.

5.2 Advantages of Integrated Methodology

Enhanced Accuracy: Our integrated methodology provides a more accurate and detailed estimation of pothole RCS, considering both geometric features and physical interactions.

Time-Domain Analysis: The inclusion of FDTD allows for analyzing time-dependent scattering behavior, important for dynamic scenarios.

Computational Efficiency: Adaptive meshing and integration of solvers optimize computational resources, reducing computational burden without compromising accuracy.

Comprehensive Analysis: By combining the strengths of multiple techniques, our methodology offers a more comprehensive analysis of pothole RCS compared to individual methods alone.

In conclusion, our hybrid methodology stands out as a robust and effective approach for pothole RCS analysis, offering improved accuracy, detailed time-domain analysis, computational efficiency, and a comprehensive understanding of pothole characteristics compared to traditional methods.

REFERENCES

Leye, P. O., Khenchaf, A., & Pouliguen, P. (2016). Application of Gaussian beam summation method in high-frequency RCS of complex radar targets. *2016 IEEE Radar Conference (RadarConf)*, Philadelphia, PA, USA. 10.1109/RADAR.2016.7485208

Li, E. S. (2003, October). Physical optics models for the backscatter response of road-surface faults and roadside pebbles at millimeter-wave frequencies. *IEEE Transactions on Antennas and Propagation*, 51(10), 2862–2868. 10.1109/TAP.2003.818004

Reddy, C. J., Deshpande, M. D., Cockrell, C. R., & Beck, F. B. (1998, August). Fast RCS computation over a frequency band using method of moments in conjunction with asymptotic waveform evaluation technique. *IEEE Transactions on Antennas and Propagation*, 46(8), 1229–1233. 10.1109/8.718579

Shin, H., Yoon, D., Na, D.-Y., & Park, Y. B. (2022). Analysis of Radome Cross Section of an Aircraft Equipped With a FSS Radome. *IEEE Access : Practical Innovations, Open Solutions*, 10, 33704–33712. 10.1109/ACCESS.2022.3162262

Srivastava, A., Goyal, A., & Ram, S. S. (2020). Radar Cross-Section of Potholes at Automotive Radar Frequencies. *2020 IEEE International Radar Conference (RADAR)*, Washington, DC, USA. 10.1109/RADAR42522.2020.9114858

Usai, P., Borgese, M., Costa, F., & Monorchio, A. (2018). Hybrid physical optics-ray tracing method for the RCS calculation of electrically large objects covered with radar absorbing materials. *12th European Conference on Antennas and Propagation (EuCAP 2018)*, London, UK. 10.1049/cp.2018.0556

Chapter 18
Lattice-Aided Delay Phase Precoding for 6G THz Massive MIMO

Parthiban Ilango
https://orcid.org/0000-0002-7301-2147
Department of ECE, SRM Institute of Science and Technology, Chennai, India

V. Sudha
Department of ECE, National Institute of Technology, Tiruchirappalli, India

Hassan Pakarzadeh
https://orcid.org/0000-0002-7561-4088
Shiraz University of Technology, Iran

ABSTRACT

The future 6G wireless technology promotes applications like three-dimensional (3D) communication, extended virtual reality, digital twins, autonomous driven vehicles, etc. Such applications require tens of GHz bandwidth, such large bandwidth is promoted by Terahertz communication system. To compensate the effect of attenuation and to increase the coverage massive MIMO with traditional hybrid precoder is considered. The conventional hybrid precoder suffers from beam split effect, where the transmitted beams are oriented towards different direction leads to loss of array gain. To circumvent the array gain loss and to focus the orientation of all transmitted beams toward a particular direction a delay phase precoder (DPP) is considered. In delay phase precoding, the selected precoder is not redundant therefore consumes more power. To reduce the power consumption two lattice aided DPP namely: Boosted LLL aided DPP and Boosted KZ aided DPP were proposed. The performance in terms of array gain and data rate were plotted. The computa-

DOI: 10.4018/979-8-3693-2786-9.ch018

tional complexity analysis was also performed.

1. INTRODUCTION

The future 6G wireless technology delivers each user with Tbps of data rate to support high data applications like three dimensional (3D) communication, extended virtual reality, digital twins, autonomous driven vehicles, etc (Saad, 2020). Such applications require tenfold increase in bandwidth utilized for 5G and terahertz (THz) communication is a prominent candidate capable providing such wide bandwidth (Wang, 2020 & Gao, 2024). However similar to millimeterwave band (Ilango, 2021), THz signals suffer from severe attenuation due to very high frequency operation i.e., 300-1000 GHz. Massive multiple-input multiple-output (m-MIMO) uses a large antenna array for the generation directional beams can be employed to compensate severe attenuation in THz band (Akyildiz, 2018).

The large antenna arrays require special precoding architectures to reduce the utilization of power-hungry RF chains (Alkhateeb,2014). Various precoding architectures are designed to connect the large antenna arrays namely: (a) Analog Precoding, (b) Digital Precoding, (c) Hybrid Precoding (Ilango, 2023). These connections are performed by a group of circuits comprising of digital to analog converter, up-converter mixer, filters and power amplifiers are called as RF chains, hybrid precoding utilizes fewer RF chains and hence preferred usually (Gao, 2016). The RF chain is the prime factor that determines the complexity and cost of the system (Jiang, 2009). These tightly packed large antenna array forms MIMO channel which is sparse in nature (Ayach, 2014).

The maximal capacity of MIMO system can be achieved by deploying multiple beam propagation, but such systems are susceptible to inter beam interference (Hong, 2017). A phase shift network (PSN) was designed to align multiple beams over multiple RF chains (Xie, 2019). Instead of fully connected architecture in hybrid precoder design, a partial connected structure was proposed using alternating minimization algorithm to reduce the design complexity (Khalid, 2022). But in partial connected structure the Phase Shifter (PS) of infinite resolution is employed which consumes more power (Xianghao, 2016). Delay phase precoding was proposed for Terahertz communication system to alleviate the beam split effect (Dai, 2022). Also beamsplit effect was circumvented using beam management techniques (Kim, 2024)

From the literature it is inferred that 1-bit PSs are capable of dissipating low power requires adaptive connected structure (using special algorithm) to improve the data rate. In this paper, we propose a novel Lattice Aided Delay Phase Precoding for Terahertz Massive MIMO system.

The contributions made in this chapter are: (i) A Delay Phase Precoding aided Terahertz (THz) MIMO system using adaptive connected PSs were presented. (ii) Based on these adaptive connected PSs two lattice reduction methods namely DPP aided Boost-LLL Precoding Method and DPP aided Boosted KZ aided Precoding Method are proposed. (iii) The achievable rate for the proposed system is plotted by varying the signal to noise ratio and number of sub-carriers by comparing with existing methods. (iv) The analysis of computational complexity was also performed.

2. SYSTEM MODEL

2.1 Terahertz Channel Characteristics

A THz massive MIMO system with N_t transmitting antennas in uniform linear array for base station is considered. The number of RF chains is assumed as N_{RF} and each user assumed to be served by N_r receiving antennas. The operating frequency is divided by Orthogonal Frequency Multiple Access containing M subcarriers. The received signal at the mth subcarrier is represented as

$$\mathbf{y}_m = \sqrt{\rho}\,\mathbf{H}_m^H \mathbf{A}\mathbf{D}_m \mathbf{s}_m + \mathbf{n}_m, \tag{1}$$

Where, ρ represents the transmit power satisfying power constraint. $\mathbf{H}_m \in \mathscr{C}^{N_t \times N_r}$ corresponds to channel coefficient matrix of the mth sub-carrier. $\mathbf{A} \in \mathscr{C}^{N_t \times N_{RF}}$ corresponds to the analog beamforming vector. $\mathbf{D} \in \mathscr{C}^{N_{RF} \times N_s}$ refers to the Digital beamforming matrix, $\mathbf{s}_m \in \mathscr{C}^{N_s \times 1}$ represents the symbol vector of mth sub-carrier. $\mathbf{n}_m \sim \mathscr{C}\mathscr{N}\left(0, \sigma^2 \mathbf{I}_{N_s}\right)$ is the additive white Gaussian noise while σ^2 is noise power.

The channel matrix of the mth sub-carrier is further represented as

$$\mathbf{H}_m = \sum_{l=1}^{L} g_l e^{-j2\pi\tau_l f_m} \mathbf{f}_t\left(\bar{\theta}_{l,m}\right) \mathbf{f}_r\left(\bar{\phi}_{l,m}\right)^H, \tag{2}$$

Where, g_l refers to the path gain, τ_l refers to the path delay, f_m corresponds to the subcarrier frequency, $\mathbf{f}_t$ and $\mathbf{f}_r$ represents steering vector at transmitter and receiver respectively, $\bar{\theta}_{l,m}$ and $\bar{\phi}_{l,m}$ represents the angle of departure and angle of arrival of the lth path of mth subcarrier. The steering vector at transmitter can be represented as

$$\mathbf{f}_t\left(\bar{\theta}_{l,m}\right) = \frac{1}{\sqrt{N_t}}\left[1, e^{j\pi\bar{\theta}_{l,m}}, e^{j\pi 2\bar{\theta}_{l,m}}, \cdots, e^{j\pi(N_t-1)\bar{\theta}_{l,m}}\right]^H, \tag{3}$$

2.2 Problem Formulation

The normalized array gain is represented as

$$\eta(\mathbf{a}_l,\theta_l,f_c) = \left|\mathbf{f}_t\left(2d\frac{f_c}{c}\theta_l\right)^H \mathbf{a}_l\right| = \left|\mathbf{f}_t\left(\theta_l\right)^H \mathbf{f}_t\left(\theta_l\right)\right| = 1, \tag{4}$$

Where $\mathbf{a}_l$is the maximum array gain achievable at carrier frequency f_c.

The maximum value of array gain in each sub-carrier is considered as the maximization problem (Dai, 2022) and represented as $\mathscr{P}1 : \theta_{\text{opt}} = \arg\max_{\theta}\left|\eta\left(\mathbf{a}_{l,m},\theta,f_m\right)\right|$. Where θ_{opt} is the optimal beam angle.

The selection of optimal beam angle tends to increase in achievable rate (Xianghao, 2016), hence the above problem can be realized as achievable rate maximization problem

$$\begin{aligned} \mathscr{P}2 : \quad & \max_{\mathbf{A},\mathbf{D},\mathbf{W}_A,\mathbf{W}_D} \mathscr{R}_s\left(\mathbf{A},\mathbf{D},\mathbf{W}_A,\mathbf{W}_D\right), \\ & \text{s.t. } \|\mathbf{AD}\|_F^2 = N_S, \end{aligned} \tag{5}$$

Where, $\mathbf{W}_A, \mathbf{W}_D$corresponds to the analog and digital combining matrix respectively. The optimization problem $\mathscr{P}2$ involves four variables and the variables $\mathbf{A},\mathbf{D}$ follows non-convex constraints. The optimization problem $\mathscr{P}2$ can be formulated as minimization of distance vector problem given as (Yu, 2016)

$$\begin{aligned} \mathscr{P}3 : \quad & \min_{\mathbf{A},\mathbf{D}} \|\mathbf{F}_{opt} - \mathbf{AD}\|_F^2, \\ & \text{s.t. } \|\mathbf{AD}\|_F^2 = N_S, \end{aligned} \tag{6}$$

Where, $\mathbf{F}_{opt}$ corresponds to the optimal precoder comprising of first N_S columns of $\mathbf{V}$. where V is obtained by applying singular value decomposition (SVD) of $\mathbf{H}$. However, the optimization problem $\mathscr{P}3$ is difficult to solve because of the coupled nature of $\mathbf{A}$ and $\mathbf{D}$. To solve $\mathscr{P}3$ two stage iterative based Alternate minimization (Altmin) technique is employed. Where in first stage, the digital precoder $\mathbf{D}$ is optimized by keeping the Analog precoder $\mathbf{A}$ fixed as referred below.

$$\mathscr{P}4 : \quad \min_{\mathbf{D}} \|\mathbf{F}_{opt} - \mathbf{AD}\|_F^2 \tag{7}$$

In second stage, the Analog precoder **A** is optimized by keeping the digital precoder **D** fixed as described below

$$\mathscr{P}5: \min_{A} \left\| \mathbf{F}_{opt} - \mathbf{AD} \right\|_F^2 \tag{8}$$

2.3 Lattice Preliminaries and Closest Vector Problem

A n-dimension full rank lattice $\mathscr{L}$ is a additive distinct subgroup in $\mathbb{R}^n$. The Lattice generated by a basis $\mathbf{B} = \left[\mathbf{b}_1, \mathbf{b}_2, \cdots, \mathbf{b}_n\right] \in \mathbb{R}^{n\times n}$ is represented as $\mathscr{L}\left(\mathbf{B}\right) = \left\{\mathbf{v} \mid \mathbf{v} = \sum_{i\in[n]} c_i \mathbf{b}_i; c_i \in \mathbb{Z}\right\}$. For the lattice $\mathscr{L}(\mathbf{B})$, the fundamental region $\mathscr{P}(\mathbf{B})$ often defined by fundamental parallelotope is described as

$$\mathscr{P}\left(\mathbf{B}\right) = \left\{\mathbf{x} \mid \mathbf{x} = \sum_{l=1}^{m} \mathscr{V}_i \mathbf{b}_l; 0 \le \mathscr{V}_i \le 1\right\} \tag{9}$$

The fundamental region covers the entire span of **B** when shifted to all points in lattice. Different bases corresponds to different fundamental parallelotopes. The Voronoi region are the set of points closer to the origin to any point on the lattice is represented as

$$\mathscr{V}(\mathscr{L}(\mathbf{B})) = \{\mathbf{x} \mid \|\mathbf{x}\| \le \|\mathbf{x} - \mathbf{y}\| \forall\, \mathbf{y} \in \mathscr{L}(\mathbf{B})\} \tag{10}$$

The closest vector problem (CVP) of vector $\boldsymbol{y} \in \mathbb{R}^n$ in a lattice is obtained by finding a vector $\boldsymbol{v}$ such that

$$\|\mathbf{y} - \mathbf{v}\|^2 = \|\mathbf{y} - \mathbf{w}\|^2, \mathbf{w} \in \mathscr{L}(\mathbf{B}) \tag{11}$$

From optimization problem 5, the minimization function can be rewritten as (Lyu, 2021),

$$\left\|\mathbf{F}_{opt} - \boldsymbol{AD}\right\|_F^2 = \left\|\left[\mathbf{F}_{opt}^{\top}\right]_{1:N_s,n} - \boldsymbol{D}^{\top}\left[\boldsymbol{A}^{\top}\right]_{1:N_s,n}\right\|_F^2 \tag{12}$$

Therefore for $\mathscr{P}5$ applying CVP, the solution is obtained as

$$\mathscr{P}6: \min_{\mathbf{x}\in\mathscr{F}^{N_{rf}}} \left\|\tilde{\mathbf{y}} - \mathbf{Bx}\right\|^2 \tag{13}$$

where,

$$\mathbf{B} \triangleq \begin{bmatrix} R(\boldsymbol{D}^{\top}) \\ I(\boldsymbol{D}^{\top}) \end{bmatrix}, \tilde{\mathbf{y}} \triangleq \begin{bmatrix} R\left(\left[\mathbf{F}_{opt}^{\top} \right]_{1:N_s,n} \right) \\ I\left(\left[\mathbf{F}_{opt}^{\top} \right]_{1:N_s,n} \right) \end{bmatrix} \tag{14}$$

Here, $\mathscr{R}$ and $\mathscr{I}$ corresponds to real and imaginary part of the variable respectively. The optimized variable $\mathbf{x}$ is employed to construct the Analog precoder $\lfloor \boldsymbol{A}^{\top} \rfloor_{1:N_s,n}$

3. PROPOSED LATTICE AIDED MULTI BEAM HYBRID PRECODING

In this section we proposed lattice aided precoders (Lyu, 2017) for solving $\mathscr{P}6$ in DPP aided THz MIMO system having adaptive partial connected PSs as shown in Figure 1. Two algorithms based on lattice were proposed namely i) DPP aided Boosted Lenstra–Lenstra–Lovasz (DPP-BLLL) precoding method and ii) DPP aided Boosted Korkine Zolotarev (DPP-BKZ) precoding method. The detailed description are mentioned in the below subsections

Figure 1. Proposed lattice aided DPP structure

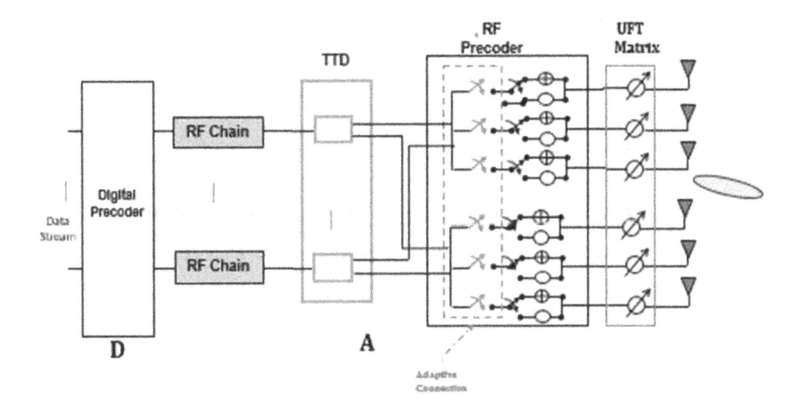

3.1 Delay Phase Precoding Aided Boosted LLL Method

In DPP aided Boosted LLL (DPP-BLLL) precoding method, an efficient solver for solving $\mathscr{P}6$ is proposed using Boosted LLL algorithm (Lyu, 2017) . In Boost-LLL method the length reduction is performed using parallel nearest plane (PNP) algorithm. If PNP is defined by a route number **L**. Here each route PNP corresponds individual active beams utilized in propagation from the Multiple beams, hence this Boost LLL methods are best suited for multiple beam system.

The application of PNP based length reduction leads to increase in complexity to a factor of **L** at the expense of accuracy in estimation. $\mathbf{L} = 1$ corresponds to LLL algorithm. The PNP route number $\mathbf{L} = \prod_{k=1}^{i-1} p_k$ indicates the total number of routes present in it. where, $(p_{i-1},\ldots,p_1) \in (\mathbb{Z}^+)^{i-1}$. Each route of PNP is marked as $(q_{i-1},\ldots,q_1) \in \{1,\ldots,p_{i-1}\} \times \cdots \times \{1,\ldots,p_1\}$.

At first QR factorization is performed for **D** i.e. $[\mathbf{Q},\mathbf{R}] = \mathrm{qr}(\mathbf{D})$. Where, $\mathbf{Q} \in \mathbb{C}^{N_s \times N_{RF}}$ is a matrix containing orthogonal vectors and $\mathbf{R} \in \mathbb{C}^{N_{RF} \times N_{RF}}$ is a upper triangular matrix. The First i[th] shortest vector among all the $\mathbf{L} + 1$ PNP routes are obtained and represented as $\mathbf{r}_i = \mathbf{r}_{i,(z_{i-1}^*,\ldots,z_1^*)}$. where,

$$\left(z_{i-1}^*,\ldots,z_1^*\right) = \arg\min_{(z_{i-1},\ldots,z_1)} \left\| \mathbf{r}_{i,(z_{i-1},\ldots,z_1)} \right\|. \tag{15}$$

Also, the unimodular transformation matrix **T** is simultaneously obtained for each PNP (as described in Algorithm 1), represented as, $\mathbf{t}_{i,(z_{i-1}^*,\ldots,z_1^*)} = \mathbf{T}_{\Gamma_{i+1}} \mathbf{c}_{(z_{i-1}^*,\ldots,z_1^*)}$, $\mathbf{T}_{1:n,1:i} = \mathbf{t}_{i,(z_{i-1}^*,\ldots,z_1^*)}$.

Since, the Siegel condition (Gama,2006) $|r_{k,i}/r_{k,k}| < 1/2$ for $k < 1$ is no longer satisfied, the diagonal reduction condition (Zhang, 2012) has to be maintained i.e. the upper triangular lattice basis **R** satisfies the diagonal reduction by satisfying the criterion for $\delta_l(1/2 < \delta_l < 1)$ and by following the condition

$$\delta_l r_{i-1,i-1}^2 \le r_{i,i}^2 + \left(r_{i-1,i} - \left\lfloor r_{i-1,i}/r_{i-1,i-1} \right\rceil r_{i-1,i-1}\right)^2 \tag{16}$$

for all $2 \le i \le n$, where δ_l is known as the Lovász constant.

Consider the sublattice $\mathscr{L}\left(\mathbf{R}_{\Gamma_{i+1}}\right)$ generated by the ith vectors and the potential of basis **R** is defined as

$$\mathrm{Pot}(\mathbf{R}) = \prod_{i=1}^{n} \det\left(\mathscr{L}\left(\mathbf{R}_{\Gamma_{i+1}}\right)\right)^2 = \prod_{i=1}^{n} r_{i,i}^{2(n-i+1)}. \tag{17}$$

If condition $\delta_l(1/2 < \delta_l < 1)$ fails then swapping of r_{i-1} and r is performed to reduce the potential of the basis. Now $\mathbf{R}_{i-1:i,i-1:i}$ becomes

$$\begin{bmatrix} r_{i-1,i} & r_{i-1,i-1} \\ r_{i,i} & 0 \end{bmatrix} \tag{18}$$

Let $\mathbf{G} \in \mathbb{C}^2 \times 2$ be a unitary matrix

$$\begin{bmatrix} \frac{r_{i-1,i}}{\sqrt{r_{i,i}^2 + r_{i-1,i}^2}} & \frac{r_{i,i}}{\sqrt{r_{i,i}^2 + r_{i-1,i}^2}} \\ -\frac{r_{i,i}}{\sqrt{r_{i,i}^2 + r_{i-1,i}^2}} & \frac{r_{i-1,i}}{\sqrt{r_{i,i}^2 + r_{i-1,i}^2}} \end{bmatrix} \tag{19}$$

The consecutive bases $\mathbf{R}' = \mathbf{G}\mathbf{R}_{i-1:i,1:n}$ forms a upper triangular matrix of form

$$\begin{bmatrix} \sqrt{r_{i,i}^2 + r_{i-1,i}^2} & \frac{r_{i-1,i} r_{i-1,i-1}}{\sqrt{r_{i,i}^2 + r_{i-1,i}^2}} \\ 0 & -\frac{r_{i,i} r_{i-1,i-1}}{\sqrt{r_{i,i}^2 + r_{i-1,i}^2}} \end{bmatrix} \tag{20}$$

From $\delta_l\left(1/2 < \delta_l < 1\right)$, the ratio between potential of two consecutive bases $\mathbf{R}'$ and $\mathbf{R}$ should satisfy the condition $\frac{\text{Pot}(\mathbf{R}')}{\text{Pot}(\mathbf{R})} \leq \delta_l$ i.e. only if $\left\lfloor r_{i-1,i}/r_{i-1,i-1} \right\rceil = 0$. Thus vector pairs $\mathbf{t}_{i,(z_{i-1}^*,\ldots,z_1^*)}$ and $\mathbf{r}_{i,(z_{i-1}^*,\ldots,z_1^*)}$ are obtained satisfying diagonal condition. From the value of $\mathbf{D}_r$, the estimated vector $\boldsymbol{x}$ is obtained as described in Algorithm 1. The DPP-BLLL analog precoder is obtained as

$$\lfloor \mathbf{A} \rfloor_{1:N_{RF},n}^{\mathsf{T}} = \hat{\mathbf{x}}^{\mathsf{T}} \tag{21}$$

3.2 Delay Phase Precoding Aided Boosted KZ Method

The system is configured similar to DPP-BoostLLL method. QR factorization is performed for $\mathbf{D}$ i.e. $[\mathbf{Q}, \mathbf{R}] = \text{qr}(\mathbf{D})$. The steps involved in boosted KZ are presented in Algorithm 2. Here, Schnorr and Euchner (SE) enumeration algorithm (Schnorr, 1994) is applied to solve Shortest Vector Problem (SVP) for the lattice $\mathscr{L}\left(\mathbf{R}_{i:n,i:n}\right)$. SE enumeration leads to stronger length reduction and generates accurate reduced bases. If $\mathbf{R}_{i:n,i}$ is the shortest vector (SV) of $\mathscr{L}\left(\mathbf{R}_{i:n,i:n}\right)$, then $\pi_{\mathbf{D}_{\Gamma_i}}^{\perp}\left(\mathbf{d}_i\right)$ is the shortest vector of the projected lattice $\pi_{\mathbf{D}_{\Gamma_i}}^{\perp}\left(\left[\mathbf{d}_i, \ldots, \mathbf{d}_n\right]\right)$

In order to retain the upper triangular property of $\mathbf{R}$ a unitary matrix is designed using Boosted LLL methods as shown in Algorithm 2. To obtain the $\mathbf{Q}$ and $\mathbf{R}$ matrices. Further SE enumeration is performed to obtain the reduced basis $\mathbf{D}_r$. From the value of $\mathbf{D}_r$, the estimated vector $\boldsymbol{x}$ is obtained. The DLL-BKZ analog precoder is obtained as

$$\lfloor \mathbf{A} \rfloor_{1:N_{RF},n}^{\mathsf{T}} = \hat{\mathbf{x}}^{\mathsf{T}} \tag{22}$$

4. RESULTS AND DISCUSSIONS

The DPP based THz MIMO system operating at f_c = 100GHz with number of transmitter antenna elements N_t =256, number of receiving antenna elements N_r =4, number of sub-carriers M=128, N_{RF} =4, K=8, is considered for simulation. We considered five simulation scenarios comprising of three existing methods (namely (a) Conventional hybrid precoding method, (b) Delay phase precoding method (DPP), (c) Optimal precoding method, along with two proposed methods namely (d) Proposed DPP aided Boost-LLL (DPP-BLLL) method and (e) Proposed DPP aided Boost-KZ (DPP-BKZ) method. The performance metrics in terms of achievable rate was plotted by changing values of signal to noise ratio (SNR).

Figure 2. Achievable rate vs. SNR for the proposed methods

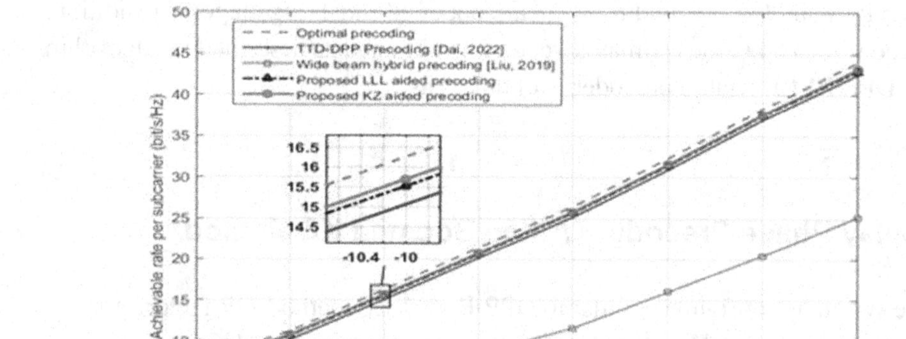

In Figure 2, the number of sub-carriers is kept at a constant value 4 for the DPP based THz MIMO system and other conventional systems mentioned in above scenarios. It can be inferred that the achievable rate of data transfer for the proposed DPP aided BoostLLL method and proposed DPP aided BKZ method is improved, while the later being the maximum. The reason for increase in rate in DPP-BLLL method is that the multiple terahertz beams are employed with PNP based length reduction is applied, therefore the accuracy of estimation of vector is improved. Whereas in DPP-BKZ method, the process of obtaining SV using SE enumeration algorithm leads to near perfect estimation of vector which increases the accuracy furthermore.

Figure 3. Achievable rate vs. number of sub-carriers for the proposed methods

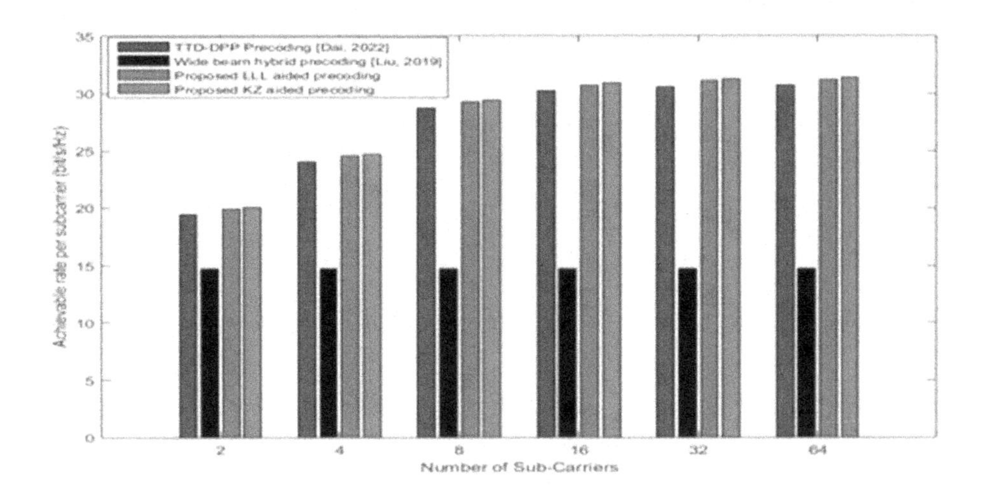

In Figure 3, the SNR is kept constant at 5 dB, the number of sub carriers are varied as 2, 4,8,16,32,64 and the corresponding achievable rate are plotted. It can be inferred that proposed DPP-BLLL method and DPP-BKZ method shows improved rate due to accuracy in estimation. As the number of sub-carriers are increased it can be inferred that the achievable rate reaches a saturation value, despite the increase in sub-carrier beyond certain value.

Figure 4. Normalized array gain vs number of sub-carriers for the proposed methods

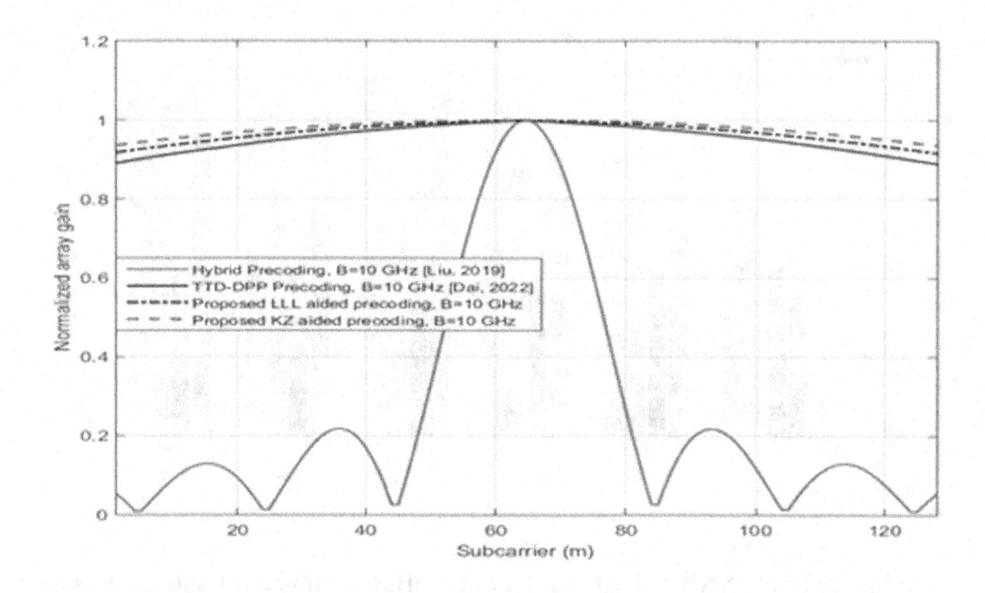

The array gain of (a) Conventional hybrid precoding method (Xianghao, 2016), (b) Delay phase precoding method (DPP), (c) Proposed DPP aided Boost-LLL (DPP-BLLL) method (Dai, 2022) and (d) Proposed DPP aided Boost-KZ (DPP-BKZ) method are shown in Figure 4. From the plot it can be inferred that the array gain for the proposed DPP-BLLL and proposed DPP-BKZ methods are uniform despite the increase in subcarriers. This shows that beam-split is avoided using the proposed DPP-BLLL and DPP-BKZ methods.

4.1 Complexity Analysis

In order to perform DPP-BLLL method $O(N_{iter} log\,(N_p)N_t\,(N_r^2 + N_r\,M)$ computations are required, where N_{iter} corresponds to the number of iteration, N_p refers to the number of PNP routes. Similarly for, the DPP-BKZ method $O(N_{iter}\,N_t\,(N_r^2 + N_r\,M)$ computations are required. The overall computation complexity as shown in Table 1. It can be inferred that the proposed DLL-BKZ method is proved to be less complex than other proposed methods.

Table 1. Computational complexity of proposed methods

Method	Complexity
Wideband Hybrid Method	$O(N_t (N_r + M))$
DPP precoding Method	$O(N_t (N_r^2 + M N_{RF}))$
Proposed DPP-BLLL method	$O(N_{iter} log (N_p) N_t (N_r^2 + N_r M))$
Proposed DPP-BKZ method	$O(N_{iter} N_t (N_r^2 + N_r M))$

5. CONCLUSION

In this chapter, a delay phase precoding aided THz MIMO system using adaptive connected PSs were proposed. To improve the array gain in the adaptive connected PSs system, two lattice methods namely DPP-BLLL method and DPP-BKZ method were proposed. From the achievable rate plot it can be inferred that proposed DPP-BLLL method and proposed DPP-BKZ method shows improved performance than all other methods due to the accurate estimation of vectors. The increase in the number of sub-carriers tend to increase in achievable rate until a maximal (saturation) rate is reached. In terms of computational complexity, the proposed DPP-BKZ method offers reduced complexity than other proposed lattice methods.

REFERENCES

Akyildiz, I. F., Han, C., & Nie, S. (2018). Combating the Distance Problem in the Millimeter Wave and Terahertz Frequency Bands. *IEEE Communications Magazine*, 56(6), 102–108. 10.1109/MCOM.2018.1700928

Alkhateeb, A., Mo, J., Gonzalez-Prelcic, N., & Heath, R. W. (2014). MIMO Precoding and Combining Solutions for Millimeter-Wave Systems. *IEEE Communications Magazine*, 52(12), 122–131. 10.1109/MCOM.2014.6979963

Ayach, O. E., Rajagopal, S., Abu-Surra, S., Pi, Z., & Heath, R. W. (2014). Spatially Sparse Precoding in Millimeter Wave MIMO Systems. *IEEE Transactions on Wireless Communications*, 13(3), 1499–1513. 10.1109/TWC.2014.011714.130846

Dai, L., Tan, J., Chen, Z., & Poor, H. V. (2022) Delay-Phase Precoding for Wideband THz Massive MIMO. *IEEE Transactions on Wireless Communications*, *21*(9). 7271-7286. 10.1109/TWC.2022.3157315

Gama, N., Howgrave-Graham, N., Koy, H., & Nguyen, P. Q. (2006) Rankin's Constant and Blockwise Lattice Reduction. In *Advances in Cryptology*. 10.1007/11818175_7

Gao, C., Zhong, C., Ye Li, G., Soriaga, J. B., & Behboodi, A. (2024). Spatially Sparse Precoding in Wideband Hybrid Terahertz Massive MIMO Systems. *IEEE Transactions on Wireless Communications*, 23(3), 1871–1885. 10.1109/TWC.2023.3292834

Gao, Z., Dai, L., Mi, D., Wang, Z., Imran, M. A., & Shakir, M. Z. (2015). MmWave massive-MIMO-based wireless backhaul for the 5G ultra-dense network. *IEEE Wireless Communications*, 22(5), 13–21. 10.1109/MWC.2015.7306533

Hong, W., Jiang, Z. H., Yu, C., Zhou, J., Chen, P., Yu, Z., Zhang, H., Yang, B., Pang, X., Jiang, M., Cheng, Y., Al-Nuaimi, M. K. T., Zhang, Y., Chen, J., & He, S. (2017). Multibeam Antenna Technologies for 5G Wireless Communications. *IEEE Transactions on Antennas and Propagation*, 65(12), 6231–6249. 10.1109/TAP.2017.2712819

Ilango, P., & Sudha, V. (2021). Energy efficient transmit antenna selection scheme for correlation relied beamspace mmWave MIMO system. *AEÜ. International Journal of Electronics and Communications*, 137, 153783. 10.1016/j.aeue.2021.153783

Ilango, P., & Sudha, V. (2023). Energy efficient degree of freedom aided multi beam transmit antenna precoding for mmWave MIMO system. *AEÜ. International Journal of Electronics and Communications*, 163, 154599. 10.1016/j.aeue.2023.154599

Jiang, Y., & Varanasi, M. K. (2009). The RF-chain limited MIMO system- part I: Optimum diversity-multiplexing tradeoff. *IEEE Transactions on Wireless Communications*, 8(10), 5238–5247. 10.1109/TWC.2009.081385

Khalid, S., Mehmood, R., Abbas, W., Khalid, F., & Naeem, M. (2022). Energy efficiency maximization of massive MIMO systems using RF chain selection and hybrid precoding. *Telecommunication Systems*, 80(2), 251–261. 10.1007/s11235-022-00900-7

Kim, J., Park, J., Moon, J., & Shim, B. (2024). Fast and Accurate Terahertz Beam Management via Frequency-Dependent Beamforming. *IEEE Transactions on Wireless Communications*, 23(3), 1699–1712. 10.1109/TWC.2023.3291440

Liu, X., & Qiao, D. (2019). Space-Time Block Coding-Based Beamforming for Beam Squint Compensation. *IEEE Wireless Communications Letters*, 8(1), 241–244. 10.1109/LWC.2018.2868636

Lyu, S., & Ling, C. (2017). Boosted KZ and LLL Algorithms. *IEEE Transactions on Signal Processing*, 65(18), 4784–4796. 10.1109/TSP.2017.2708020

Saad, W., Bennis, M., & Chen, M. (2020). A Vision of 6G Wireless Systems: Applications, Trends, Technologies, and Open Research Problems. *IEEE Network*, 34(3), 134–142. 10.1109/MNET.001.1900287

Schnorr, C. P., & Euchner, M. (1994). Lattice basis reduction: Improved practical algorithms and solving subset sum problems. *Mathematical Programming*, 66(1), 181–199. 10.1007/BF01581144

Wang, C.-X., Huang, J., Wang, H., Gao, X., You, X., & Hao, Y. (2020). 6G Wireless Channel Measurements and Models: Trends and Challenges. *IEEE Vehicular Technology Magazine*, 15(4), 22–32. 10.1109/MVT.2020.3018436

Xianghao, Y., Shen, J.-C., Zhang, J., & Letaief, K. B. (2016). Alternating Minimization Algorithms for Hybrid Precoding in Millimeter Wave MIMO Systems. *IEEE Journal of Selected Topics in Signal Processing*, 10(3), 485–500. 10.1109/JSTSP.2016.2523903

Xie, T., Dai, L., Ng, D. W. K., & Chae, C.-B. (2019). On the Power Leakage Problem in Millimeter-Wave Massive MIMO with Lens Antenna Arrays. *IEEE Transactions on Signal Processing*, 67(18), 4730–474. 10.1109/TSP.2019.2926019

Yu, X., Zhang, J., & Letaief, K. B. (2018). A hardware-efficient analog network structure for hybrid precoding in millimeter wave systems. *IEEE Journal of Selected Topics in Signal Processing*, 12(2), 282–297. 10.1109/JSTSP.2018.2814009

Zhang, W., Qiao, S., & Wei, Y. (2012). A Diagonal Lattice Reduction Algorithm for MIMO Detection. *IEEE Signal Processing Letters*, 19(5), 311–314. 10.1109/LSP.2012.2191614

Chapter 19
Case Studies on the Integration of 5G Technology Into Smart Grids With Emphasis on Cybersecurity Measures

M. Kumaran
https://orcid.org/0000-0002-1800-5972
Department of Mechanical Engineering, Faculty of Engineering and Technology, SRM Institute of Science and Technology (Deemed), Tiruchirappalli Campus, Tiruchirappalli, India

ABSTRACT

The incorporation of 5G technology into smart grids offers significant prospects for improving the efficiency, reliability, and flexibility of electricity distribution networks. This chapter covers numerous case studies demonstrating the adoption of 5G in smart grid systems, with a special emphasis on the cybersecurity measures used to protect these vital assets. The report illustrates the benefits of 5G by analyzing real-world applications in depth, including faster data transmission speeds, lower latency, and the capacity to handle a large number of connected devices. It also covers the new cybersecurity concerns posed by 5G integration, such as increased vulnerability to cyber-attacks and the complexity of safeguarding large, distributed networks. The case studies show a variety of tactics and technologies used to manage these risks, such as advanced encryption protocols, AI-powered threat detection, and robust network segmentation.

DOI: 10.4018/979-8-3693-2786-9.ch019

1. INTRODUCTION TO 5G TECHNOLOGY IN SMART GRIDS

Wireless communication has grown dramatically over the last three decades, transitioning from 1G to 4G technology. The primary objective of this investigation was to meet the criteria of large bandwidth and extremely minimal latency. 5G provides excellent data rates, higher quality of work, high coverage, reduced latency, cost-effective services, and high reliability. The services are divided into three types: (i) Extreme Mobile Broadband (eMBB): Utilizing a non-standalone architecture, it provides rapid internet connectivity. With acceptable latency and expanded bandwidth. This service supports virtual and augmented reality (AR/VR) media, videos streamed in Ultra-HD mode, and other advanced features. (ii) Ultra-Reliable Low Latency Communication (URLLC): Designed for minimum latency and extremely excellent accuracy, this service delivers a great quality of service compared to conventional mobile networks architectures cannot achieve. (iii). Massive Machine-Type Communication (mMTC): This service facilitates long-lasting and broadband machine-type communication at a low cost and with minimal electrical consumption. It is designed for on-demand real-time engagement, for example, communication between vehicles, remote surgery, smart grids, Industry 4.0, intelligent transportation processes, etc.

Recently, several authors have investigated the application of 5G technology in smart grids. (Ravishankar Borgaonkar et al., 2021) The integration of IoT devices can significantly accelerate digital transformation across various sectors, including the current power grid infrastructure. However, security remains a major concern for IoT in smart grid systems, particularly due to the cost-performance tradeoff. This article explores the communication needs of IoT devices in a low-voltage smart grid scenario and develops a corresponding threat model. The authors then investigate 5G specifications and analyze their application to security within this specific smart grid use case. While 5G networks cannot address all IoT security concerns, they offer a dependable communication channel with inherent security advantages for smart grid infrastructures. (Hongxun Hui et al., 2020) The rapid emergence and global deployment of 5G networks are impacting businesses across the board. These networks excel in managing large-scale machine-to-machine communication, high-speed data transfer, ultra-reliable connections, and low latency, making them ideal for implementing Demand Response (DR) programs in smart grids. To explore this potential, this study focuses on how 5G can enhance DR by examining: (1) Evolution of Network Technology: It analyzes the progression from 1G to 5G, highlighting key 5G features like millimeter wave (mmWave), massive MIMO, and ultra-dense cellular networks. (2) Smart Grid Disaster Recovery: It investigates recent advancements in communication technologies, cybersecurity measures, and dependability assessment tools relevant to disaster recovery in smart grids. (3) 5G Applications for DR: Finally, the research proposes three potential 5G application scenarios for DR: improved Mobile Broadband (eMBB), ultra-Reliable Low-Latency Communication (uRLLC), and massive Machine-Type Communication (mMTC). Sarmistha (De Dutta et al., 2019) This article dives into the exciting intersection of M2M communication and smart grid technology. It starts by unpacking the concepts of M2M communication, the abstract model of a smart grid, and the underlying M2M architecture. It then delves into the application of M2M in smart grids, emphasizing the importance of data analytics for optimizing performance. As 5G network deployment accelerates, the paper examines the evolution of wireless technologies, with a particular focus on 5G's suitability for smart grids due to its advanced capabilities. Recognizing 5G's significant advantages, the study explores the security challenges that need to be addressed for a secure and optimal power distribution system.

However, 5G provides high-speed, low-latency communication, which is critical for real-time monitoring and control of smart grids. It can connect millions of devices per square kilometer, enabling the widespread use of smart meters and sensors. 5G network slicing allows for specialized network resources for key smart grid applications, increasing reliability and security. Optimized power consumption in 5G networks aligns with smart grid sustainability goals by encouraging efficient energy use and management. 5G can deliver speeds up to 10 Gbps, vastly surpassing previous generations. Latency as low as 1 millisecond enables near-instantaneous data transmission, essential for applications requiring real-time responses. It is perfect for dense IoT environments, as it can accommodate up to one million devices per square kilometer. 5G provides faster internet speeds and improved user experiences in high-density areas. It ensures reliable, low-latency connections for critical applications, such as remote surgery and autonomous vehicles. Additionally, it facilitates extensive connectivity for IoT devices, enabling a wide range of applications, including smart cities and industrial automation. 5G allows multiple virtual networks on a single physical infrastructure, each optimized for different use cases, ensuring efficient resource allocation and management.

2. CYBERSECURITY CHALLENGES IN 5G ENABLED SMART GRIDS

Integrating 5G into smart grids introduces a range of cybersecurity challenges. Addressing these is crucial for guaranteeing reliable, secure, and confidential grid operations. 5G-enabled smart grids feature a large number of networked devices and systems, considerably increasing the potential attack surface. Every connected gadget might be a possible entry point for attackers. The integration of 5G technology with existing infrastructure adds complexity. Legacy systems may not have been developed with modern cybersecurity risks in mind, resulting in vulnerabilities when connected to newer, 5G-enabled components. Smart grids rely on real-time data to operate efficiently. Cyberattacks that cause latency or impair data flows can have serious repercussions, such as delayed responses to grid conditions, resulting in power outages or failures. Smart grids capture massive volumes of data, including sensitive information about energy consumption habits. Ensuring the privacy and confidentiality of this information is crucial. Cyberattacks aimed at data interception or breaches can result in serious privacy violations.

With so many devices connected to the grid, reliable procedures for authenticating and authorizing them are required. Weaknesses in authentication procedures can result in unauthorized access and control of grid components. 5G-enabled devices and components are frequently sourced from several manufacturers, making supply chain security a major problem. Compromised devices or components introduced through the supply chain can create vulnerabilities in the smart grid. Many 5G-enabled devices in smart grids are installed in remote or unattended areas, leaving them susceptible to physical tampering. Physical security of these devices is critical to preventing tampering and unauthorized access.

Smart grids are important infrastructure, rendering them vulnerable to sophisticated cyberattacks, especially advanced persistent threats. These threats include long-term and targeted attacks aimed at disrupting grid operations, frequently orchestrated by nation-states or well-funded adversaries. Denial-of-service attacks can overwhelm the network and interfere with communication between smart grid devices. Ensuring that the network can resist such attacks while remaining operational is a huge challenge. 5G networks are vulnerable to interference and jamming, which can impair communication channels for smart grid equipment. Protecting against physical layer threats is critical for ensuring reliable operations.

Figure 1. Challenges in cybersecurity for 5G-enabled smart grids

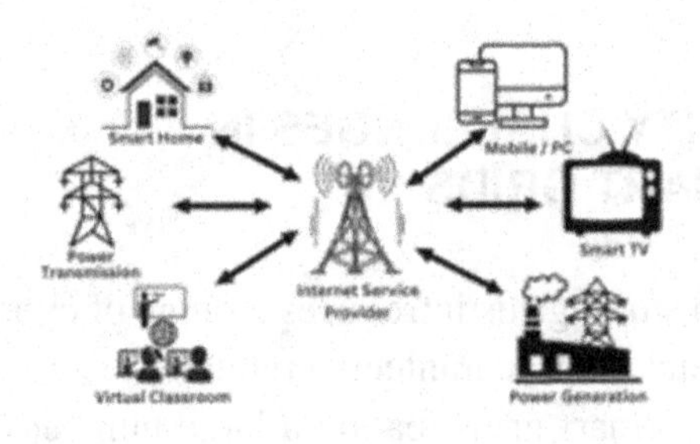

As 5G-enabled smart grids are deployed, they must meet a variety of regulatory criteria and standards. Ensuring compliance while maintaining strong cybersecurity procedures can be difficult, particularly as rules change. Insider threats occur when individuals with legitimate access misuse their privileges, posing a major danger. Employees, contractors, or third-party providers may purposefully or accidentally breach the grid's security (Javier Lopez et al., 2021).

3. CYBERSECURITY MEASURES AND SOLUTIONS

Implementing effective cybersecurity measures for 5G networks is critical for protecting them against numerous threats. This forms the foundation of all security strategies. It entails providing authorized users access to systems and data only at the level necessary for their work obligations. This can be accomplished using techniques such as user authentication (multi-factor authentication, passwords), role-based authorization (granting rights based on work duties), and data encryption. Firewalls act as the first line of defense, creating a secure barrier between your internal network and the external world. They meticulously screen incoming and outgoing traffic based on predefined rules, ensuring only authorized data passes through. Intrusion Detection and Prevention Systems (IDS/IPS) take security a step further. They constantly monitor network traffic for suspicious activity, and depending on the system, can even take action to prevent attacks before they happen. Network segmentation separates sensitive data into smaller zones with differing levels of access. This approach also involves safeguarding individual devices such as laptops, desktop computers, and mobile phones. Antivirus and anti-malware software detect and delete dangerous programs. Patch management ensures that software is updated with the most recent security fixes. Endpoint Detection and Response (EDR) systems constantly monitor devices for suspicious activity and can take action to mitigate attacks. Employees usually serve as the first line of defence against cyberattacks. Regular training programs can educate people about typical dangers such as phishing emails, social engineering techniques, and password hygiene practices, enabling them to identify and report suspicious conduct (Luiz Felipe Fernandes De Almeida et al., 2020).

Figure 2. Measures and solutions for cybersecurity

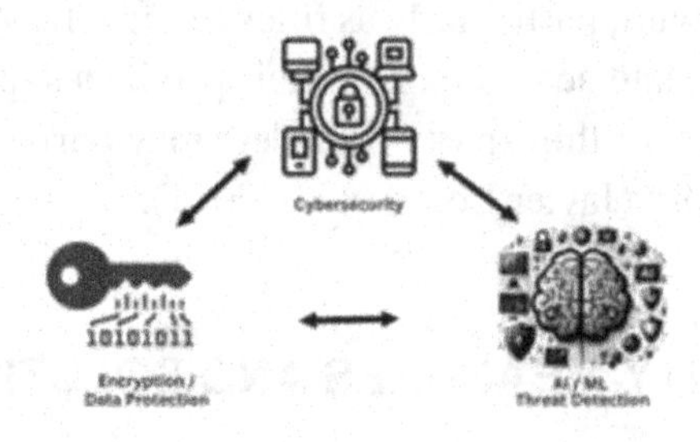

3.1 Encryption and Data Protection Strategies

Ever wondered how to keep your confidential information safe? Data encryption is the answer! It's a powerful tool that scrambles data into a secret code using a unique key. Only those who possess the authorized key can unlock the code and access the original information. Encryption safeguards your data in three ways: at rest (on your devices), in transit (traveling across networks), and even in use (while being processed). Common encryption algorithms like AES (Advanced Encryption Standard) and RSA (Rivest-Shamir-Adleman) act as the secret recipe for this data protection. Data Loss Prevention (DLP) solutions assist organizations in identifying and preventing sensitive data from being released or transferred without authorization. They can look for patterns in network traffic, emails, and endpoint behavior that may indicate a data breach.

Regular data backups are critical in the event of system failures, cyberattacks, or natural catastrophes. Backups should be securely stored, preferably offsite, to avoid loss or compromise in the same event that impacts primary data storage. Disaster recovery planning outlines the actions required to restore data and systems in the event of a significant outage, ensuring business continuity with minimal downtime (Ebenezer Esenogho et al., 2022).

3.2 AI and Machine Learning for Threat Detection

As 5G technology gets more widely adopted in numerous industries, the complexity and volume of network traffic grow, posing new cybersecurity challenges. Artificial Intelligence (AI) and Machine Learning (ML) provide increased capabilities for recognising and responding to such threats in real time. AI and ML algorithms can process massive amounts of data produced by 5G networks, such as network traffic, user behavior, and device logs. This analysis aids in identifying patterns and abnormalities that may signal a possible threat. AI can analyze sensor data from network equipment and detect probable errors before they happen. This preventive strategy can help prevent outages and reduce downtime. AI can constantly monitor for new and emerging threats by analyzing data from various sources, including threat intelligence feeds and security research publications. ML systems can learn typical network traffic patterns and identify deviations that may indicate suspicious activities. This can assist in detecting intrusions, malware infestations, and other harmful activity. Machine learning algorithms can analyze user behavior patterns to detect suspicious activity, such as unauthorized access attempts or anomalous data transfer patterns.

AI and ML systems can analyze data in real-time, allowing for faster detection and response to risks than previous approaches. With their ability to handle the colossal data volumes produced by 5G networks, these algorithms are perfectly suited for large-scale deployments. As new data and threats are introduced, ML models can continually learn and enhance their threat detection skills. AI and machine learning can automate many aspects of threat identification and response, freeing up security personnel's time to focus on more complex tasks (Yuanliang Li et al., 2023).

As 5G technology evolves, AI and machine learning will become increasingly crucial in network security. By tackling the problems and continuing to develop sophisticated algorithms, AI and ML can become the foundation of a comprehensive 5G security strategy. This will ensure that 5G networks are reliable, secure, and robust in the face of ever-increasing threats.

4. TECHNOLOGICAL INNOVATIONS AND FUTURE TRENDS

Although 5G technology is still in its early phases of development, it is already changing the way we connect and interact with the world around us. With significantly improved speed, capacity, and reliability compared to previous generations, 5G lays the groundwork for a plethora of technological advances and future trends that will transform many facets of our lives (Paola Vargas et al., 2023).

Figure 3. Technological innovations and future trends

5G network slicing unlocks the potential to create multiple virtual networks on a single physical infrastructure. This allows for application-specific customization, such as the creation of dedicated slices with guaranteed performance for mission-critical services like autonomous vehicles or remote surgery. Traditional cloud computing can cause latency concerns in 5G applications that require real-time response. Edge computing addresses this by placing computing capacity closer to the network edge, enabling faster data processing and decision-making in applications such as augmented reality and industrial automation. Millimeter Wave (mmWave) technology, a high-frequency band, provides ultra-fast speeds for future applications such as immersive entertainment experiences and fixed wireless access at speeds comparable to fiber optic connections. However, mmWave has a shorter range and requires more network infrastructure than lower frequency bands. By supporting a massive number of connected devices, 5G unlocks the full potential of the Internet of Things (IoT). Smart cities with interconnected infrastructure, connected factories with real-time machine monitoring, and large-scale environmental monitoring via sensor networks are all possible applications (A. V. Jha et al., 2021).

5G will be a major driver of smart cities, enabling real-time traffic management, intelligent garbage collection, and connected infrastructure to increase public safety and resource efficiency. Future factories will use 5G to provide real-time communication between machines, robotics, and control systems, resulting in enhanced automation, efficiency, and safety in industrial operations. 5G technology could revolutionize healthcare by facilitating remote patient monitoring, real-time diagnostics, and potentially remote surgery with low latency. Ultra-fast bandwidth and minimal latency will enable next-generation virtual reality (VR), education, transforming entertainment, and training simulations. Additionally, 5G will continue to enhance the mobile broadband experience, allowing for faster downloads, uninterrupted streaming, and lag-free online gaming (Charles Rajesh Kumar J et al., 2021).

5. CASE STUDIES OF 5G INTEGRATION IN SMART GRIDS

Integrating 5G technology into smart grids marks a significant leap forward in energy management, efficiency, and reliability. This comprehensive analysis delves into specific case studies where 5G has been successfully implemented in smart grids, highlighting the benefits, challenges, and lessons learned from these pioneering projects.

5.1 Case Study 1: Enel Distribuzione (Italy)

Enel Distribuzione, an Italian energy company, collaborated with Ericsson to build a private 5G network in a test area in Turin. This network integrates smart meters, sensors, and grid equipment, allowing for real-time monitoring, automated fault identification and isolation, and enhanced distribution automation. The initiative resulted in a significant decrease in outage duration (up to 70%) and increased grid resilience. Furthermore, real-time data collection enabled preemptive maintenance and optimal energy distribution. Collaboration between utilities and technology companies is critical for successful 5G integration in smart grids. Standardization and interoperability of equipment are essential for scaling.

5.2 Case Study 2: Sandia National Laboratories (US)

Sandia National Labs, along with partners, launched a project to assess 5G's role in modernizing the power grid. The study explored using 5G to manage distributed energy sources like solar panels and battery storage. Their research successfully demonstrated 5G's potential for real-time communication and control of these resources, paving the way for smoother integration of renewable energy into the grid. However, to fully harness this potential, regulatory frameworks need to evolve to accommodate new technologies like 5G in the energy sector. Additionally, careful consideration must be given to cybersecurity measures to ensure the continued safety of smart grids.

5.3 Case Study 3: Shanghai Electric (China)

Shanghai Electric, a Chinese power equipment maker, is developing a 5G-enabled smart grid platform. This technology uses 5G for real-time monitoring of electrical grids, including substations, transformers, and distribution lines. The platform aims to increase grid efficiency by optimizing power flow and reducing energy losses. It also seeks to enhance grid security by enabling real-time anomaly detection and automatic response to potential intrusions. Public-private collaborations can accelerate the development and deployment of 5G-based smart grid systems. User adoption and acceptance are critical to the success of such ventures.

5.4 Case Study 4: China Southern Power Grid (China)

China Southern Power Grid, a leading utility company in China, is at the forefront of grid modernization by integrating 5G technology into its operations. This initiative enhances power grid management by enabling real-time monitoring of distribution and transmission. 5G's capabilities facilitate swift responses to faults and optimize grid operations. As a result, the company has achieved improved grid reliability, reduced downtime, and enhanced data analytics capabilities for predictive maintenance.

5.5 Case Study 5: Southern Company (USA)

Southern Company, a leading utility company in the southeastern United States, is leveraging partnerships with technology providers to advance its smart grid initiatives through 5G integration. By focusing on real-time monitoring and automated fault detection systems, 5G is playing a key role in enhancing the reliability and security of power delivery. This translates to faster outage identification and resolution, improved grid resilience, and strengthened cybersecurity measures for critical infrastructure.

5.6 Case Study 6: ABB and Swisscom Partnership (Switzerland)

Industrial technology leader ABB joined forces with Swisscom, a leading telecom provider in Switzerland, to implement cutting-edge 5G-enabled smart grid solutions. This collaboration leverages 5G for advanced grid automation, including remote substation control and real-time analytics for predictive maintenance. The result? A more efficient grid with reduced operational costs and the ability to handle peak loads while seamlessly integrating renewable energy sources.

5.7 Case Study 7: EDF (Electricité de France) (France)

Electricité de France (EDF) is leading the way in smart grid innovation by taking advantage of 5G technology. Their ongoing trials focus on leveraging 5G to improve communication and control systems within the grid infrastructure. This initiative utilizes 5G for real-time data transmission from smart meters and sensors, enabling better load balancing and ultimately leading to more efficient energy management. As a result, EDF anticipates improved demand response, greater energy efficiency, and enhanced grid stability.

5.8 Case Study 8: KT Corporation and Korea Electric Power Corporation (South Korea)

KT Corporation and Korea Electric Power Corporation (KEPCO) are joining forces to revolutionize South Korea's energy landscape. Through a collaborative effort, they have implemented 5G technology to accelerate the development of smart grids nationwide. This groundbreaking system leverages 5G technology to enable real-time monitoring and management of power distribution, smart metering, and the seamless integration of renewable energy resources. The result? A more reliable grid, improved customer service with better outage management, and a significant boost in clean energy usage.

The future of smart grids is undoubtedly interconnected. By leveraging 5G technology, we can build a more efficient, intelligent, and sustainable energy ecosystem for future generations.

6. CONCLUSION

The integration of 5G technology into smart grids holds the potential to revolutionize the energy sector by boosting efficiency, reliability, and customer service. The case studies of Enel Distribuzione (Italy), Sandia National Laboratories (US), and Shanghai Electric (China) highlight the practical advantages and problems of integrating 5G in smart grids. Scalability, regulatory compliance, and stakeholder participation were among the most important lessons learnt. The growing adoption of 5G by utilities underscores the importance of these insights for guiding future deployments and unlocking the full potential of this transformative technology.

REFERENCES

Almeida, L. F. F. D. (2020). Control Networks and Smart Grid Teleprotection: Key Aspects, Technologies, Protocols, and Case-Studies. *IEEE Access : Practical Innovations, Open Solutions*, 8(8), 174049–174079. 10.1109/ACCESS.2020.3025235

Borgaonkar, R., Tøndel, I. A., Degefa, M. Z., & Jaatun, M. G. (2021). *Improving smart grid security through 5G enabled IoT and edge computing*. Wiley. 10.1002/cpe.6466

De Dutta, S., & Prasad, R. (2019). Security for Smart Grid in 5G and Beyond Networks. *Wireless Personal Communications*, 106(1), 261–273. 10.1007/s11277-019-06274-5

Esenogho, E., Djouani, K., & Kurien, A. M. (2022). Integrating Artificial Intelligence Internet of Things and 5G for Next-Generation Smartgrid: A Survey of Trends Challenges and Prospect. *IEEE Access : Practical Innovations, Open Solutions*, 10(10), 4794–4831. 10.1109/ACCESS.2022.3140595

Hui, H., Ding, Y., Shi, Q., Li, F., Song, Y., & Yan, J. (2020). 5G network-based Internet of Things for demand response in smart grid: A survey on application potential. *Applied Energy*, 257, 113972. 10.1016/j.apenergy.2019.113972

Jha, A. (2021). Smart grid cyber-physical systems: communication technologies, standards and challenges. *Wireless Networks, 27*, 2595–2613. 10.1007/s11276-021-02579-1

Li, Y. (2023). *Cybersecurity of Smart Inverters in the Smart Grid: A Survey*. IEEE. 10.1109/TPEL.2022.3206239

Lopez, J., Rubio, J. E., & Alcaraz, C. (2021). Digital Twins for Intelligent Authorization in the B5G-Enabled Smart Grid. *IEEE Wireless Communications*, 28(2), 2. 10.1109/MWC.001.2000336

Vargas, P. (2023). Impacts of 5G on cyber-physical risks for interdependent connected smart critical infrastructure systems. *International Journal of Critical Infrastructure Protection, 42*. 10.1016/j.ijcip.2023.100617

Compilation of References

5G Network Slicing, What is it? (2019). Viavi Solutions. Www.viavisolutions.com. https://www.viavisolutions.com/en-us/5g-network-slicing

Abdelghany, M. A., Fathy Abo Sree, M., Desai, A., & Ibrahim, A. A. (2022, July). Gain improvement of a dual-band CPW monopole antenna for sub-6 GHz 5G applications using AMC structures. *Electronics (Basel)*, 11(14), 2211. 10.3390/electronics11142211

Abdelwahab, S., Hamdaoui, B., Guizani, M., & Znati, T. (2016, April). Network function virtualization in 5G. *IEEE Communications Magazine*, 54(4), 84–91. 10.1109/MCOM.2016.7452271

Abdulla & Jameel. (2023). A Review on IoT Intrusion Detection Systems Using Supervised Machine Learning: Techniques, Datasets, and Algorithms. *UHD Journal of Science and Technology, 7*(1), 53–65. 10.21928/uhdjst.v7n1y2023.pp53-65

Abdullah, M., Altaf, A., Anjum, M. R., Arain, Z. A., Jamali, A. A., Alibakhshikenari, M., Falcone, F., & Limiti, E. (2021). Future Smartphone: MIMO Antenna System for 5G Mobile Terminals. *IEEE Access : Practical Innovations, Open Solutions*, 9, 91593–91603. 10.1109/ACCESS.2021.3091304

Abdulrahman, Y. (2023). Cybersecurity in Digital Transformation Era. From 5G to 6G: Technologies, Architecture, AI, and Security. IEEE. 10.1002/9781119883111.ch7

Abid, B., Khan, B. M., & Memon, R. A. (2022). Seismic Facies Segmentation Using Ensemble of Convolutional Neural Networks. *Wireless Communications and Mobile Computing*. 10.1155/2022/7762543

Abirami, B. S., & Sundarsingh, E. F. (2017, July). EBG-backed flexible printed Yagi– Uda antenna for on-body communication. *IEEE Transactions on Antennas and Propagation*, 65(7), 3762–3765. 10.1109/TAP.2017.2705224

Abubakar, H. S., Zhao, Z., Kiani, S. H., Rafique, U., Alabdulkreem, E., & Elmannai, H. (2024). Eight element dual-band MIMO array antenna for modern fifth generation mobile phones. *International Journal of Electronics and Communications*, 175, 155083. 10.1016/j.aeue.2023.155083

Afolabi, I., Taleb, T., Samdanis, K., Ksentini, A., & Flinck, H. (2018). Network Slicing and Softwarization: A Survey on Principles, Enabling Technologies, and Solutions. *IEEE Communications Surveys and Tutorials*, 20(3), 2429–2453. 10.1109/COMST.2018.2815638

Ahammed, M. T., & Khan, I. (2022). Ensuring power quality and demand-side management through IoT-based smart meters in a developing country. *Energy*, 250(Jul), 123747. 10.1016/j.energy.2022.123747

Ahmad, I., Kumar, T., Liyanage, M., Okwuibe, J., Ylianttila, M., & Gurtov, A. (2018). Overview of 5G Security Challenges and Solutions. *IEEE Communications Standards Magazine*, 2(1), 36–43. 10.1109/MCOMSTD.2018.1700063

Ahmed, I., Anisetti, M., Ahmad, A., & Jeon, G. (2022). A multilayer deep learning approach for malware classification in 5G-enabled IIoT. *IEEE Transactions on Industrial Informatics*, 19(2), 1495–1503. 10.1109/TII.2022.3205366

Aimasso, A., Vedova, M. D. L. D., & Maggiore, P. (2023). Sensitivity analysis of FBG sensors for detection of fast temperature changes. *Journal of Physics: Conference Series*, 2590(1), 012006. 10.1088/1742-6596/2590/1/012006

Akyildiz, I. F., Han, C., & Nie, S. (2018). Combating the Distance Problem in the Millimeter Wave and Terahertz Frequency Bands. *IEEE Communications Magazine*, 56(6), 102–108. 10.1109/MCOM.2018.1700928

Alamouti, S. M. (1998). A simple transmit diversity technique for wireless communications. *IEEE Journal on Selected Areas in Communications*, 16(8), 1451–1458. 10.1109/49.730453

Ali, K. S., Khan, A. A., Ur, T. P., Rehman, A., & Ouahada, K. (2023). Learned-SBL-GAMP based hybrid precoders/combiners in millimeter wave massive MIMO systems. *PLoS One*, 18(9), e0289868. 10.1371/journal.pone.028986837682816

Ali, K. S., Sampath, P., & Poongodi, C. (2019). Symbol Error Rate Performance of Hybrid DF/AF Relaying Protocol Using Particle Swarm Optimization Based Power Allocation. *2019 International Conference on Advances in Computing and Communication Engineering (ICACCE)*, Sathyamangalam, India. 10.1109/ICACCE46606.2019.9079970

Ali, U., Ullah, S., Shafi, M., Shah, S. A. A., Shah, I. A., & Flint, J. A. (2019, November). Design and comparative analysis of conventionaland metamaterial-based textile antennas for wearable applications. *International Journal of Numerical Modelling*, 32(6), 2567. 10.1002/jnm.2567

Aljanabi, M., Ismail, M. A., & Ali, A. H. (2021, January 1). Intrusion Detection Systems, Issues, Challenges, and Needs. *International Journal of Computational Intelligence Systems*.10.2991/ijcis.d.210105.001

Alkhateeb, A., Mo, J., Gonzalez-Prelcic, N., & Heath, R. W. (2014). MIMO Precoding and Combining Solutions for Millimeter-Wave Systems. *IEEE Communications Magazine*, 52(12), 122–131. 10.1109/MCOM.2014.6979963

Almeida, L. F. F. D. (2020). Control Networks and Smart Grid Teleprotection: Key Aspects, Technologies, Protocols, and Case-Studies. *IEEE Access : Practical Innovations, Open Solutions*, 8(8), 174049–174079. 10.1109/ACCESS.2020.3025235

Alotaibi, A., & Rassam, M. A. (2023). Adversarial Machine Learning Attacks against Intrusion Detection Systems: A Survey on Strategies and Defense. *Future Internet*, 15(2), 62. 10.3390/fi15020062

AlQahtani, S. A. (2023). Cooperative-Aware Radio Resource Allocation Scheme for 5G Network Slicing in Cloud Radio Access Networks. *Sensors (Basel)*, 23(11), 5111. 10.3390/s2311511137299838

Al-Shareeda, M. (2022). DDoS Attacks Detection Using Machine Learning and Deep Learning Techniques: Analysis and Comparison. *Bulletin of Electrical Engineering and Informatics, 12*(2). https://ssrn.com/abstract=4515135

Alsirhani, A., Mujib Alshahrani, M., Abukwaik, A., Taloba, A. I., Abd El-Aziz, R. M., & Salem, M. (2023). A novel approach to predicting the stability of the smart grid utilizing MLP-ELM technique. *Alexandria Engineering Journal*, 74, 495–508. 10.1016/j.aej.2023.05.063

Altulaihan, E., Almaiah, M. A., & Aljughaiman, A. (2022). Cybersecurity threats, countermeasures and mitigation techniques on the IoT: Future research directions. *Electronics (Basel)*, 11(20), 3330. 10.3390/electronics11203330

Amazon Web Services, Inc. (n.d.). *What is IoT? - Internet of Things Explained - AWS.*https://aws.amazon.com/what-is/iot/#:~:text=with%20AWS%20IoT-,What%20is%20the%20Internet%20of%20Things%20(IoT)%3F,as%20between%20the%20devices%20themselves

Anghel, P. A., Leus, G., & Kaveh, M. (2003). Multi-user space-time coding in cooperative networks. In *Proceedings of the IEEE International Conference on Acoustics, Speech and Signal Processing (ICASSP)*. IEEE.

Anghel, P., & Kaveh, M. (2004). Exact symbol error probability of a cooperative network in a Rayleigh-fading environment fading channels. *IEEE Transactions on Wireless Communications*, 3(5), 1416–1421. 10.1109/TWC.2004.833431

Ashourian, M., Salimian, R., & Nasab, H. M. (2013). A Low Complexity Resource Allocation Method for OFDMA System Based on Channel Gain. *Wireless Personal Communications*, 71(1), 519–529. 10.1007/s11277-012-0826-9

Ashraf, S. A., Aktas, I., Eriksson, E., Helmersson, K. W., & Ansari, J. (2016). Ultra-reliable and low-latency communication for wireless factory automation: From LTE to 5G. *2016 IEEE 21st International Conference on Emerging Technologies and Factory Automation (ETFA)*, (pp. 1–8). IEEE. https://doi.org/10.1109/ETFA.2016.7733543

Aslan, Ö., & Yilmaz, A. A. (2021). A new malware classification framework based on deep learning algorithms. *IEEE Access : Practical Innovations, Open Solutions*, 9, 87936–87951. 10.1109/ACCESS.2021.3089586

Atapattu, S., & Rajatheva, N. (2008). *'Analysis of Alamouti code transmission over TDMA-based Cooperative Protocol', Vehicular Technology. VTC Spring 2008*. IEEE.

Ayach, O. E., Rajagopal, S., Abu-Surra, S., Pi, Z., & Heath, R. W. (2014). Spatially Sparse Precoding in Millimeter Wave MIMO Systems. *IEEE Transactions on Wireless Communications*, 13(3), 1499–1513. 10.1109/TWC.2014.011714.130846

Ayomikun, E. (2020). Mokayef Mastaneh. Miniature microstrip antenna for IoT application. *Materials Today: Proceedings*, 29(43–47).

Azam, Z., Islam, M. M., & Huda, M. N. (2023). Comparative Analysis of Intrusion Detection Systems and Machine Learning-Based Model Analysis Through Decision Tree. *IEEE Access : Practical Innovations, Open Solutions*, 11, 80348–80391. 10.1109/ACCESS.2023.3296444

Aznabet, M., Mrabet, O. E., Floc'H, J. M., Falcone, F., & Drissi, M. (2014). A coplanar waveguide-fed printed antenna with complementary split ring resonator for wireless communication systems. *Waves in Random and Complex Media*, 25(1), 43–51. 10.1080/17455030.2014.956846

Bai, Z. (2013). Particle swarm optimization-based power allocation in DF cooperative communications. *Ubiquitous and Future Networks (ICUFN), 2013 Fifth International Conference*. IEEE.

Bai, Q. (2010). Analysis of Particle Swarm Optimization Algorithm. *Computer & Information Science*, 3(1), 3. 10.5539/cis.v3n1p180

Balamurugan, C., Singh, K., Ganesan, G., & Rajarajan, M. (2021). Post-quantum and code-based cryptography—Some prospective research directions. *Cryptography*, 5(4), 38. 10.3390/cryptography5040038

Bamy, C. L., Mbango, F. M., Dominic, B. O. K., & Mpele, P. M. (2021). A compact dual-band Dolly-shaped antenna with parasitic elements for automotive radar and 5G applications. *Heliyon*, 7(4), e06793. 10.1016/j.heliyon.2021.e0679333948514

Barrias, A., Casas, J. R., & Villalba, S. (2016). A Review of Distributed Optical Fiber Sensors for Civil Engineering Applications. *Sensors (Basel)*, 16(5), 748. 10.3390/s1605074827223289

Berahmand, K., Daneshfar, F., Salehi, E. S., Li, Y., & Xu, Y. (2024). Autoencoders and their applications in machine learning: A survey. *Artificial Intelligence Review*, 57(2), 28. 10.1007/s10462-023-10662-6

Bhargavi, D., & Murthy, C. R. (2010). Performance comparison of energy, matched-filter and cyclostationarity-based spectrum sensing. *Signal Processing Advances in Wireless Communications (SPAWC), IEEE Eleventh International Workshop*. IEEE. 10.1109/SPAWC.2010.5670882

Bhushan, B. (2023). *Synergizing AI, 5G, and Machine Learning: Unleashing Transformative Potential Across Industries with a Focus on Network Security*. 2023 3rd International Conference on Advancement in Electronics & Communication Engineering (AECE), Ghaziabad, India. 10.1109/AECE59614.2023.10428548

Bilal, R., & Khan, B. M. (2019). Software-Defined Networks (SDN): A Survey. *Handbook of Research on Cloud Computing and Big Data Applications in IoT.* IGI Global. 10.4018/978-1-5225-8407-0.ch023

Bjerkan, L. (2000). Application of fiber-optic Bragg grating sensors in monitoring environmental loads of overhead power transmission lines. *Applied Optics*, 39(4), 554–560. 10.1364/AO.39.00055418337925

Bletsas, A., Khisti, A., Reed, D. P., & Lippman, A. (2006). simple cooperative diversity method based on network path selection. *IEEE Journal on Selected Areas in Communications*, 24(3), 659–672. 10.1109/JSAC.2005.862417

Bodur, H., & Kara, R. (2015). Secure SMS Encryption Using RSA Encryption Algorithm on Android Message Application. In *3rd International Symposium On Innovative Technologies In Engineering And Science*. Research Gate.

Bondre, S., Sharma, A., & Bondre, V. (2023). 5G Technologies, Architecture and Protocols. In *Evolving Networking Technologies* (pp. 1–19). John Wiley & Sons, Ltd., 10.1002/9781119836667.ch1

Borgaonkar, R., Tøndel, I. A., Degefa, M. Z., & Jaatun, M. G. (2021). *Improving smart grid security through 5G enabled IoT and edge computing*. Wiley. 10.1002/cpe.6466

Boualouache, A., & Engel, T. (2023). A survey on machine learning-based misbehavior detection systems for 5g and beyond vehicular networks. *IEEE Communications Surveys and Tutorials*, 25(2), 1128–1172. 10.1109/COMST.2023.3236448

Brennan, D. G. (2003). Linear Diversity Combining Techniques. *Proceedings of the IEEE*, 91(2), 331–356. 10.1109/JPROC.2002.808163

BTH. (2012a). *Analysis of OSTBC in Cooperative Cognitive Radio Networks using 2-hop DF Relaying Protocol - Arkiv EX.* Blekinge Tekniska Högskola. http://www.bth.se/fou/cuppsats.nsf/6753b78eb2944e0ac1256608004f0535/0ae924c26658a76dc12578bf0078935b?OpenDocument

BTH. (2012b). *Applying OSTBC in Cooperative Cognitive Radio Networks.* Blekinge Tekniska Högskola. http://www.bth.se/fou/cuppsats.nsf/6753b78eb2944e0ac1256608004f0535/4830a668eb9bd48cc1257735004a7535?OpenDocument

Buell, D. (2021). Simple Ciphers. In *Fundamentals of Cryptography: Introducing Mathematical and Algorithmic Foundations* (pp. 11–26). Springer International Publishing. 10.1007/978-3-030-73492-3_2

Cabric, D., Mishra, S. M., & Brodersen, R. W. (2004). Implementation issues in spectrum sensing for cognitive radios. *Conference Record of the Thirty-Eighth Asilomar Conference*. IEEE. 10.1109/ACSSC.2004.1399240

Cai, T., Jia, T., Adepu, S., Li, Y., & Yang, Z. (2023). ADAM: An Adaptive DDoS Attack Mitigation Scheme in Software-Defined Cyber-Physical System. *IEEE Transactions on Industrial Informatics*, 19(6), 7802–7813. 10.1109/TII.2023.3240586

Cetin, B. K., & Pratas, N. K. (2021). Resource sharing and scheduling in device-to-device communication underlying cellular networks. *Pamukkale Üniversitesi Mühendislik Bilimleri Dergisi*, 27(5), 604–609.

Chai, Q., Luo, Y., Ren, J., Zhang, J., Yang, J., Yuan, L., & Peng, G. (2019). Review on fiber-optic sensing in health monitoring of power grids. *Optical Engineering (Redondo Beach, Calif.)*, 58(7), 072007. 10.1117/1.OE.58.7.072007

Chandranegara, D. R., Djawas, J. S., Nurfaizi, F. A., & Sari, Z. (2023). Malware Image Classification Using Deep Learning InceptionResNet-V2 and VGG-16 Method. *Jurnal Online Informatika*, 8(1), 61–71. 10.15575/join.v8i1.1051

Chen, J. (2020). Smart Grid Technology and 5G Integration: Advantages and Challenges. *Journal of Energy Technology*.

Chen, C. (2023). A Single-Layer Single-Patch Dual-Polarized High-Gain Cross-Shaped Microstrip Patch Antenna. *IEEE Antennas and Wireless Propagation Letters*, 22(10), 2417–2421. 10.1109/LAWP.2023.3289861

Chen, X., Grzegorczyk, T. M., Wu, B., Pacheco, J., & Kong, J. A. (2004). Robust method to retrieve the constitutive effective parameters of metamaterials. *Physical Review. E*, 70(1), 016608. 10.1103/PhysRevE.70.01660815324190

Chen, Y. (2024). Collaborative Approach of NFV and SDN for Cybersecurity in Smart Grids. *IEEE Transactions on Smart Grid*.

Chitra, S., Ramesh, S., & Beulah Jackson, S. (2020). Performance enhancement of generalized frequency division multiplexing with RF impairments compensation for efficient 5G wireless access. *AEÜ. International Journal of Electronics and Communications*, 127, 153467. 10.1016/j.aeue.2020.153467

Chuan-long, Yue-fei, Jin-long, & Xin-zheng. (2017). A Deep Learning Approach for Intrusion Detection Using Recurrent Neural Networks. *IEEE Access*. IEEE. . 10.1109/ACCESS.2017.2762418

Chukunda, C.D., Matthias, & Bennett. E.O. (2022) Malware detection classification system using random forest. *Journal of Software Engineering and Simulation, 8(5)*, 16-25

Cloudflare. (n.d.). *What is IoT security?* https://www.cloudflare.com/learning/security/glossary/iot-security/#:~:text=Internet%20of%20Things%20(IoT)%20devices%20are%20computerized%20Internet%2Dconnected,introduce%20threats%20into%20a%20network

Cui, H., Wang, Q., Ye, Y., Tang, Y., & Lin, Z. (2022). A combinational transfer learning framework for online transient stability predic- tion. *Sustain Energy Grids Netw*, 30, 1–10. 10.1016/j.segan.2022.100674

Cybrosys. (n.d.). *An Overview of Network-based Intrusion Detection & Prevention Systems*. https://www.cybrosys.com/blog/an-overview-of-network-based-intrusion-detection-and-prevention-systems

Dai, L., Tan, J., Chen, Z., & Poor, H. V. (2022) Delay-Phase Precoding for Wideband THz Massive MIMO. *IEEE Transactions on Wireless Communications*, *21*(9). 7271-7286. 10.1109/TWC.2022.3157315

Dangi, R., Lalwani, P., Choudhary, G., You, I., & Pau, G. (2022). Study and Investigation on 5G Technology: A Systematic Review. *Sensors (Basel)*, 22(1), 1. 10.3390/s2201002635009569

Das, U., & Namboodiri, V. (2019, February). A quality-aware multi-level data aggregation approach to manage smart grid AMI traffic. *IEEE Transactions on Parallel and Distributed Systems*, 30(2), 245–256. 10.1109/TPDS.2018.2865937

De Dutta, S., & Prasad, R. (2019). Security for Smart Grid in 5G and Beyond Networks. *Wireless Personal Communications*, 106(1), 261–273. 10.1007/s11277-019-06274-5

Deavours, C. A., & Reeds, J. (1977). The Enigma part I historical perspectives. *Cryptologia*, 1(4), 381–391. 10.1080/0161-117791833183

Debicha, I., Cochez, B., Kenaza, T., Debatty, T., Dricot, J.-M., & Mees, W. (2023). Adv-Bot: Realistic adversarial botnet attacks against network intrusion detection systems. *Computers & Security, 129.* 10.1016/j.cose.2023.103176

Dellaoui, S., Kaabal, A., El Halaoui, M., & Asselman, A. (2018). Patch array antenna with high gain using EBG superstrate for future 5G cellular networks. *Procedia Manufacturing*, 22, 463–467. 10.1016/j.promfg.2018.03.071

Deng, X., & Haimovich, A. M. (2005). Cooperative relaying in wireless networks with local channel state information. *Proc.IEEE Veh. Tech. Conf. (VTC).* IEEE.

Desai, P., Sonawane, A., Mane, T., & Jaiswal, R. (2023). Network based intrusion detection system. *International Research Journal of Modernization in Engineering Technology and Science*, 5, 3851–3857. 10.56726/IRJMETS35232

Dhirani, L. L., Mukhtiar, N., Chowdhry, B. S., & Newe, T. (2023). Ethical Dilemmas and Privacy Issues in Emerging Technologies: A Review. *Sensors (Basel)*, 23(3), 1151. 10.3390/s2303115136772190

Digham, F. F., Alouini, M.-S., & Simon, M. K. (2003). On the energy detection of unknown signals over fading channels. *IEEE International Conference.* IEEE. 10.1109/ICC.2003.1204119

Dixit, S., Jain, R., & Patel, H. B. (2024). Impact of 5G Wireless Technologies on Cloud Computing and Internet of Things (IOT). *Advances in Robotic Technology.* https://ssrn.com/abstract=4700149

Dobre, O., Abdi, A., Bar-Ness, Y., & Su, W. (2007, April). Survey of automatic modulation classification techniques: Classical approaches and new trends. *IET Communications*, 1(2), 137–156. 10.1049/iet-com:20050176

Do, D.-T., Nguyen, M.-S. V., Jameel, F., Jäntti, R., & Ansari, I. S. (2020). Performance Evaluation of Relay-Aided CR-NOMA for Beyond 5G Communications. *IEEE Access : Practical Innovations, Open Solutions*, 8, 134838–134855. 10.1109/ACCESS.2020.3010842

Dohler, M., & Li, Y. (2010). *Cooperative Communications: Hardware, Channel and PHY*. Wiley. 10.1002/9780470740071

Dong, Y., Toyao, H., & Itoh, T. (2012). Design and characterization of miniaturized patch antennas loaded with complementary Split-Ring resonators. *IEEE Transactions on Antennas and Propagation*, 60(2), 772–785. 10.1109/TAP.2011.2173120

Dorsch, N., Kurtz, F., Georg, H., Hägerling, C., & Wietfeld, C. (2014). Software-defined networking for Smart Grid communications: Applications, challenges, and advantages. *2014 IEEE International Conference on Smart Grid Communications (SmartGridComm)*, Venice, Italy. 10.1109/SmartGridComm.2014.7007683

Dragičević, T., Siano, P., & Prabaharan, S. R. (2019). Future generation 5G wireless networks for smart grid: A comprehensive review. *Energies*, 12(11), 2140. 10.3390/en12112140

Dutt, T. A. (2014). PSO based Power Allocation for Single and Multi Relay AF Cooperative Network. *IEEE Sponsored International Conference on Green Computing, Communication and Electrical Engineering (ICGCCEE 2014)*. IEEE.

El Atrash, M., Abdalla, M. A., & Elhennawy, H. M. (2019, October). A wearable dual-band low profile high gain low SAR antenna AMC-backed for WBAN applications. *IEEE Transactions on Antennas and Propagation*, 67(10), 6378–6388. 10.1109/TAP.2019.2923058

El-Gammal, H. M., El-Badawy, E.-S. A., Rizk, M. R. M., & Aly, M. H. (2020). A new hybrid FBG with a π-shift for temperature sensing in overhead high voltage transmission lines. *Optical and Quantum Electronics*, 52(1), 53. 10.1007/s11082-019-2171-7

Elrawy, M., Awad, A., & Hamed, H. (2018). Intrusion Detection Systems for IoT-based smart environments: A survey. *Journal of Cloud Computing (Heidelberg, Germany)*, 7(1), 21. 10.1186/s13677-018-0123-6

Elsayed, Y., & Gabbar, H. A. (2022). FBG Sensing Technology for an Enhanced Microgrid Performance. *Energies*, 15(24), 9273. Advance online publication. 10.3390/en15249273

Equihua, C., Anides, E., García, J. L., Vázquez, E., Sánchez, G., Avalos, J.-G., & Sánchez, G. (2021). A low-cost and highly compact FPGA-based encryption/decryption architecture for AES algorithm. *Revista IEEE América Latina*, 19(9), 1443–1450. 10.1109/TLA.2021.9468436

Esenogho, E., Djouani, K., & Kurien, A. M. (2022). Integrating Artificial Intelligence Internet of Things and 5G for Next-Generation Smartgrid: A Survey of Trends Challenges and Prospect. *IEEE Access : Practical Innovations, Open Solutions*, 10(10), 4794–4831. 10.1109/ACCESS.2022.3140595

Eslami, Y. (2006). An area-efficient universal cryptography processor for smart cards. *IEEE transactions on very large scale integration (VLSI) systems*. IEEE.

Esmaeily, A., & Kralevska, K. (2021). Small-Scale 5G Testbeds for Network Slicing Deployment: A Systematic Review. *Wireless Communications and Mobile Computing*, 2021, e6655216. 10.1155/2021/6655216

European Telecommunications Standards Institute (ETSI). (2021). *Standardization in 5G-Enabled Smart Grids*. ETSI.

Fadhil, T. Z., Murad, N. A., Rahim, M. K. A., Hamid, M. R., & Nur, L. O. (2022). A Beam-Split metasurface antenna for 5G applications. *IEEE Access : Practical Innovations, Open Solutions*, 10, 1162–1174. 10.1109/ACCESS.2021.3137324

Fan, C., Cui, J., Jin, H., Zhong, H., Bolodurina, I., & He, D. (2024). Auto-Updating Intrusion Detection Systems for Vehicular Network: A Deep Learning Approach Based on Cloud-Edge-Vehicle Collaboration. *IEEE Transactions on Vehicular Technology*. IEEE. 10.1109/TVT.2024.3399219

Fang, X., Xiong, F., & Chen, J. (2019). An Experimental Study on Fiber Bragg Grating-Point Heat Source Integration System for Seepage Monitoring. *IEEE Sensors Journal*, 19(24), 12346–12352. 10.1109/JSEN.2019.2937155

Farhadi, G. (2007). Selective Decode-and-Forward Relaying Scheme for Multi-Hop Diversity Transmission Systems. *Global Telecommunications Conference, GLOBECOM '07*. IEEE.

Farhangi, H. (2010, January-February). The path of the smart grid. *IEEE Power & Energy Magazine*, 8(1), 18–28. 10.1109/MPE.2009.934876

Faruk, M. J. H., Tahora, S., Tasnim, M., Shahriar, H., & Sakib, N. (2022). A review of quantum cybersecurity: threats, risks and opportunities. In *2022 1st International Conference on AI in Cybersecurity (ICAIC)*, (pp. 1-8). IEEE.

FCC. (2015). *Amendment of the commission's rules with regard to commercial operations in the 3550-3650 MHz band*. FCC.

Federal Communications Commission. (2002). *Revision of part 15 of the commission's rules regarding ultra-wideband transmission systems*. FCC.

Federal Communications Commission. (2018). *FCC online table of frequency allocations*. FCC. https://transition.fcc.gov/oet/spectrum/table/fcctable. pdf

Fei, L., Xi-mei, L., Tao, L., & Guang-xin, Y. (2008). Optimal Power Allocation to minimize SER for multimode amplify-and-forward cooperative communication systems. *Journal of China Universities of Posts and Telecommunications*, 15(4), 14–23. 10.1016/S1005-8885(08)60395-7

Flinders, M., & Flinders, M. (2024, March 14). The future of 5G: What to expect from this transformational technology. *IBM Blog*. IBM. https://www.ibm.com/blog/5g-future/

Floridia, C., Rosolem, J. B., Leonardi, A. A., Hortencio, C. A., Fonseca, R. F., Moreira, R. O. C., Souza, G. C. L., Melo, A. L., & Nascimento, C. A. M. (2013). Temperature sensing in high voltage transmission lines using fiber Bragg grating and free-space-optics. *Proceedings of the Society for Photo-Instrumentation Engineers*, 8722, 87220N. 10.1117/12.2017934

Frasch, I. J. (2017). *Algorithmic framework and implementation of spectrum holes detection for cognitive radios*. [Master's thesis, Rochester Institute of Technology].

Gadallah, W., Ibrahim, H., & Omar, N. (2024). A deep learning technique to detect distributed denial of service attacks in software-defined networks. *Computers & Security*, 137, 103588. 10.1016/j.cose.2023.103588

Gala, Y., Vanjari, N., Doshi, D., & Radhanpurwala, I. (2023). *AI based Techniques for Network-based Intrusion Detection System: A Review.* 10th International Conference on Computing for Sustainable Global Development (INDIACom), New Delhi, India.

Gama, N., Howgrave-Graham, N., Koy, H., & Nguyen, P. Q. (2006) Rankin's Constant and Blockwise Lattice Reduction. In *Advances in Cryptology*. 10.1007/11818175_7

Gangopadhyay, T. K., Paul, M. C., & Bjerkan, L. (2009). Fiber-optic sensor for real-time monitoring of temperature on high voltage (400KV) power transmission lines. *Proceedings of the Society for Photo-Instrumentation Engineers*, 7503, 75034M. 10.1117/12.835447

Gao, C., Zhong, C., Ye Li, G., Soriaga, J. B., & Behboodi, A. (2024). Spatially Sparse Precoding in Wideband Hybrid Terahertz Massive MIMO Systems. *IEEE Transactions on Wireless Communications*, 23(3), 1871–1885. 10.1109/TWC.2023.3292834

Gao, Z., Dai, L., Mi, D., Wang, Z., Imran, M. A., & Shakir, M. Z. (2015). MmWave massive-MIMO-based wireless backhaul for the 5G ultra-dense network. *IEEE Wireless Communications*, 22(5), 13–21. 10.1109/MWC.2015.7306533

Garcia, F. C. C., & Muga, F. P., II. (2016). Random forest for malware classification. *arXiv preprint arXiv:1609.07770.*

Gardner, W. (1988, August). Signal interception: A unifying theoretical framework for feature detection. *IEEE Transactions on Communications*, 36(8), 897–906. 10.1109/26.3769

General International Group. (n.d.). *Advantages & Disadvantages of Intrusion Detection Systems (IDS) Types.*https://generalintlgroup.com/en/blog/advantages-and-disadvantages-of-intrusion-detection-system-ids types#:~:text=Limited%20Zero%2DDay%20Threat%20Detection,and%20deploy%2C%20leaving%20networks%20vulnerable

Genovesi, S., Costa, F., Fanciulli, F., & Monorchio, A. (2016). Wearable inkjet-printed wideband antenna by using miniaturized AMC for subGHz applications. *IEEE Antennas and Wireless Propagation Letters*, 15, 1927–1930. 10.1109/LAWP.2015.2513962

Ghazal, T. M., Hasan, M. K., Alzoubi, H. M., Alshurideh, M., Ahmad, M., & Akbar, S. S. (2023). Internet of Things connected wireless sensor networks for smart cities. In Alzoubi, H. M., & Salloum, S. (Eds.), *The Effect of Information Technology on Business and Marketing Intelligence Systems* (pp. 1953–1968). Springer. 10.1007/978-3-031-12382-5_107

Ghosh, S. (2023). Design of a highly-isolated, high-gain, compact 4-port MIMO antenna loaded with CSRR and DGS for millimeter wave 5G communications. *AEU - International Journal of Electronics and Communications, 169*, 154721.

Goldena, N. J. (2022, January 4). *Essentials of the Internet of Things (IoT).* Auerbach Publications eBooks. 10.1201/9781003119784-3

Goldsmith, A., Jafar, S., Maric, I., & Srinivasa, S. (2009, May). Breaking spectrum gridlock with cognitive radios: An information theoretic perspective. *Proceedings of the IEEE*, 97(5), 894–914. 10.1109/JPROC.2009.2015717

Goran, S. (2008). Demand side management: Benefits and challenges. *Energy Policy, 36*(12). 10.1016/j.enpol.2008.09.030

Gray, I. I. I., & James, W. (1992). Toward a mathematical foundation for information flow security. *Journal of Computer Security*, 1(3-4), 255–294. 10.3233/JCS-1992-13-405

Gui, X., Li, Z., Fu, X., Guo, H., Wang, Y., Wang, C., Wang, J., & Jiang, D. (2023). Distributed Optical Fiber Sensing and Applications Based on Large-Scale Fiber Bragg Grating Array [Review]. *Journal of Lightwave Technology*, 41(13), 4187–4200. 10.1109/JLT.2022.3233707

Guo, H., et al. (2024). Interoperability in 5G-Enabled Smart Grids: Challenges and Solutions. *IEEE Transactions on Industrial Informatics*. IEEE.

Guo, L., Zhu, Z., Lau, F. C. M., Zhao, Y., & Yu, H. (2022). Joint Security and Energy-Efficient Cooperative Architecture for 5G Underlaying Cellular Networks. *Symmetry*, 14(6), 1160. 10.3390/sym14061160

Guo, S., Hou, X., & Wang, H. (2018). Dynamic TDD and interference management towards 5G. *2018 IEEE Wireless Communications and Networking Conference (WCNC)*, (pp. 1–6). IEEE. 10.1109/WCNC.2018.8377314

Guo, Y., Mao, X., Tong, X., & Wu, H. (2023). A real time digital vibration acceleration fiber sensing system based on a multi-carrier modulation/demodulation technique. *Optics & Laser Technology*, 167, 109724. 10.1016/j.optlastec.2023.109724

Gupta, A., & Jha, R. K. (2015). A Survey of 5G Network: Architecture and Emerging Technologies. *IEEE Access : Practical Innovations, Open Solutions*, 3, 1206–1232. 10.1109/ACCESS.2015.2461602

Gutierrez-Garcia, J. L., Sanchez-DelaCruz, E., & Pozos-Parra, M. P. (2023). A Review of Intrusion Detection Systems Using Machine Learning: Attacks, Algorithms and Challenges. In Arai, K. (Ed.), *Advances in Information and Communication. FICC 2023. Lecture Notes in Networks and Systems* (Vol. 652). Springer. 10.1007/978-3-031-28073-3_5

Hamed, R. (2012). Cooperative Subcarrier and Power Allocation in OFDM Based Relaying Systems. [Unpublished Dissertation, Ryerson University, Canada].

Hamza, W. S., Ibrahim, H. M., Shyaa, M. A., & Stephan, J. J. (2020). Iot botnet detection: Challenges and issues. *Test Eng. Manag*, 83, 15092–15097.

Hao, Y., Huang, L., Wei, J., Liang, W., Pan, R., & Yang, L. (2023). The Detecting System and Method of Quasi-Distributed Fiber Bragg Grating for Overhead Transmission Line Conductor Ice and Composite Insulator Icing Load. *IEEE Transactions on Power Delivery*, 38(3), 1799–1809. 10.1109/TPWRD.2022.3222774

Hao, Y., Huang, L., Wei, J., Pan, R., Zhang, W., & Yang, L. (2022). Interface quasi-distributed fibre Bragg grating positioning detection of glaze icing load on composite insulators. *IET Science, Measurement & Technology*, 16(5), 316–325. 10.1049/smt2.12106

Haq, I. U., Khan, T. A., & Akhunzada, A. (2021). A dynamic robust DL-based model for android malware detection. *IEEE Access : Practical Innovations, Open Solutions*, 9, 74510–74521. 10.1109/ACCESS.2021.3079370

He, K., Kim, D. D., & Asghar, M. R. (2023). Adversarial Machine Learning for Network Intrusion Detection Systems: A Comprehensive Survey. *IEEE Communications Surveys & Tutorials*. IEEE. 10.1109/COMST.2022.3233793

Henry, A., Gautam, S., Khanna, S., Rabie, K., Shongwe, T., Bhattacharya, P., Sharma, B., & Chowdhury, S. (2023). Composition of Hybrid Deep Learning Model and Feature Optimization for Intrusion Detection System. *Sensors (Basel)*, 23(2), 890. 10.3390/s2302089036679684

Himsoon, T., Su, W., & Liu, K. J. R. (2005). Differential transmission for amplify-and-forward cooperative communications. *IEEE Signal Processing Letters*, 12(9), 597–600. 10.1109/LSP.2005.853067

Hnamte, V., & Hussain, J. (2023). DCNNBiLSTM: An Efficient Hybrid Deep Learning-Based Intrusion Detection System. *Telematics and Informatics Reports, 10*. 10.1016/j.teler.2023.100053

Hnamte, V., & Hussain, J. (2023). Dependable Intrusion Detection Systems using deep convolutional neural network: A Novel framework and performance evaluation approach. *Telematics and Informatics Reports, 11*. 10.1016/j.teler.2023.100077

Huang, Q., Zhang, C., Liu, Q., Ning, Y., & Cao, Y. (2010). New type of fiber optic sensor network for smart grid interface of transmission system. *IEEE PES General Meeting*, (pp. 1–5). IEEE. 10.1109/PES.2010.5589596

Huang, H., & Han, Z. (2024). Computational ghost imaging encryption using RSA algorithm and discrete wavelet transform. *Results in Physics*, 56, 107282. 10.1016/j.rinp.2023.107282

Huang, M., Gong, F., Zhang, N., Li, G., & Qian, F. (2021). Reliability and Security Performance Analysis of Hybrid Satellite-Terrestrial Multi-Relay Systems with Artificial Noise. *IEEE Access : Practical Innovations, Open Solutions*, 9, 34708–34721. 10.1109/ACCESS.2021.3058734

Huang, Q. (2019). Real-time Cyber Threat Monitoring with NFV and SDN. *IEEE Transactions on Dependable and Secure Computing*.

Hui, H., Ding, Y., Shi, Q., Li, F., Song, Y., & Yan, J. (2020). 5G network-based Internet of Things for demand response in smart grid: A survey on application potential. *Applied Energy*, 257, 113972. 10.1016/j.apenergy.2019.113972

Ibrahim, A., Sadek, A., Su, W., & Liu, K. J. R. (2005). Cooperative communications with partial channel state information: when to cooperate? *In Proceedings of the IEEE Global Telecommunications Conference, GLOBECOM*. IEEE. 10.1109/GLOCOM.2005.1578321

Ibrahim, M., & Elhafiz, R. (2023). Modeling an intrusion detection using recurrent neural networks. *Journal of Engineering Research, 11*(1). 10.1016/j.jer.2023.100013

Ikki, S., & Ahmed, M. (2007). Performance analysis of cooperative diversity wireless networks over Nakagami-m fading channel. *IEEE Communications Letters*, 11(4), 334–336. 10.1109/LCOM.2007.348292

Ikki, S., & Ahmed, M. (2008). Performance of multiple-relay cooperative diversity systems with best relay selection over Rayleigh fading channels. *EURASIP Journal on Advances in Signal Processing*, 2008(1), 580368. 10.1155/2008/580368

Ilango, P., & Sudha, V. (2021). Energy efficient transmit antenna selection scheme for correlation relied beamspace mmWave MIMO system. *AEÜ. International Journal of Electronics and Communications*, 137, 153783. 10.1016/j.aeue.2021.153783

Ilango, P., & Sudha, V. (2023). Energy efficient degree of freedom aided multi beam transmit antenna precoding for mmWave MIMO system. *AEÜ. International Journal of Electronics and Communications*, 163, 154599. 10.1016/j.aeue.2023.154599

Inga, E., Inga, J., & Hincapié, R. (2023). Maximizing resource efficiency in wireless networks through virtualization and opportunistic channel allocation. *Sensors (Basel)*, 23(8), 3949. 10.3390/s2308394937112290

International Electrotechnical Commission (IEC). (2022). *IEC Standards for Cybersecurity in Smart Grids*. IEC.

Islam, M. S., Ullah, M. A., Beng, G. K., Amin, N., & Misran, N. (2019). A modified Meander Line microstrip patch antenna with enhanced bandwidth for 2.4 GHz ISM-Band internet of things (IoT) applications. *IEEE Access : Practical Innovations, Open Solutions*, 7, 127850–127861. 10.1109/ACCESS.2019.2940049

Ismail, H., Utomo, R. G., & Bawono, M. W. A. (2024). Comparison of Support Vector Machine and Random Forest Method on Static Analysis Windows Portable Executable (PE) Malware Detection. *JURNAL MEDIA INFORMATIKA BUDIDARMA*, 8(1), 154–162.

Jain, K., & Kushwah, V. S. (2021). Design and development of dual band antenna for sub-6 frequency band application. *Materials Today: Proceedings*, 47, 6795–6798. 10.1016/j.matpr.2021.05.133

Jha, A. (2021). Smart grid cyber-physical systems: communication technologies, standards and challenges. *Wireless Networks, 27*, 2595–2613. 10.1007/s11276-021-02579-1

Jiang, Y., & Varanasi, M. K. (2009). The RF-chain limited MIMO system- part I: Optimum diversity-multiplexing tradeoff. *IEEE Transactions on Wireless Communications*, 8(10), 5238–5247. 10.1109/TWC.2009.081385

Jin, L., Zhang, W., Zhang, H., Liu, B., Zhao, J., Tu, Q., Kai, G., & Dong, X. (2006). An embedded FBG sensor for simultaneous measurement of stress and temperature. *IEEE Photonics Technology Letters*, 18(1), 154–156. 10.1109/LPT.2005.860046

Juraszek, J. (2020). Fiber Bragg Sensors on Strain Analysis of Power Transmission Lines. *Materials (Basel)*, 13(7), 1559. 10.3390/ma13071559 32230998

Kabalci, Y., & Kabalci, E. (2017). Modeling and analysis of a smart grid monitoring system for renewable energy sources. *Solar Energy*, 153, 262–275. https://doi.org/https://doi.org/10.1016/j.solener.2017.05.063. 10.1016/j.solener.2017.05.063

Kaknjo, A., Rao, M., Omerdic, E., Newe, T., & Toal, D. (2019). Real-Time Secure/Unsecure Video Latency Measurement/Analysis with FPGA-Based Bump-in-the-Wire Security. *Sensors (Basel)*, 19(13), 2984. 10.3390/s19132984 31284580

Kalizhanova, A., Kunelbayev, M., Wojcik, W., & Kozbakova, A. (2024). Experimental study of a temperature measurement system for an overhead power line using sensors based on TFBG. *International Journal of Innovative Research and Scientific Studies*, 7(1), 180–188. 10.53894/ijirss.v7i1.2596

Kaltenberger, F., Silva, A. P., Gosain, A., Wang, L., & Nguyen, T.-T. (2020). OpenAirInterface: Democratizing innovation in the 5G Era. *Computer Networks*, 176, 107284. 10.1016/j.comnet.2020.107284

Kasongo, A. (2023). A Deep Learning technique for Intrusion Detection Systems using a Recurrent Neural Networks based framework. *Computer Communications, 199,* 113-125. ,10.1016/j.comcom.2022.12.010

Kennedy, J. (2010). *'Particle Swarm Optimization', Encyclopedia of Machine Learning*. Springer US.

Kerr, O. S. (2000). The fourth amendment in cyberspace: Can encryption create a reasonable expectation of privacy. *Connecticut Law Review*, 33, 503.

Khalid, S., Mehmood, R., Abbas, W., Khalid, F., & Naeem, M. (2022). Energy efficiency maximization of massive MIMO systems using RF chain selection and hybrid precoding. *Telecommunication Systems*, 80(2), 251–261. 10.1007/s11235-022-00900-7

Khan, A., Umar, A., Munir, A., Shirazi, S., Khan, M., & Adnan, M. (2021, December). A QoS-aware machine learning-based framework for AMI applications in smart grids. *Energies*, 14(23), 1–22. 10.3390/en14238171

Khan, B. M., & Ali, F. H. (2011). Mobility Adaptive Energy Efficient and Low Latency MAC for Wireless Sensor Networks.*2011 Fifth International Conference on Next Generation Mobile Applications, Services and Technologies*, Cardiff, UK. 10.1109/NGMAST.2011.46

Khuntia, M., & Singh, D., & Sahoo, S. (2021). *Impact of Internet of Things (IoT) on 5G.* Springer. 10.1007/978-981-15-6202-0_14

Kiani, S. H., Ibrahim, I. M., Savcı, H. Ş., Rafique, U., Alsunaydih, F. N., Alsaleem, F., Alhassoon, K., & Mostafa, H. (2024). Side-edge dual-band MIMO antenna system for 5G cellular devices. *AEÜ. International Journal of Electronics and Communications*, 173, 154992. 10.1016/j.aeue.2023.154992

Kim, H.-K., & Cho, Y., ahiagbe, E. E., & Jo, H.-S. (2018). Adjacent Channel Interference from Maritime Earth Station in Motion to 5G Mobile Service. *2018 International Conference on Information and Communication Technology Convergence (ICTC)*, (pp. 1164–1169). IEEE. https://doi.org/10.1109/ICTC.2018.8539401

Kim, S. (2020). Cybersecurity Challenges in 5G-Enabled Smart Grids. *Journal of Information Security and Applications.*

Kim, D., & Lee, J. (2023). Cross-Sector Collaboration for Interoperable Cybersecurity Solutions. *International Journal of Critical Infrastructure Protection.*

Kim, J., Park, J., Moon, J., & Shim, B. (2024). Fast and Accurate Terahertz Beam Management via Frequency-Dependent Beamforming. *IEEE Transactions on Wireless Communications*, 23(3), 1699–1712. 10.1109/TWC.2023.3291440

Kim, K. C., Ko, E., Kim, J., & Yi, J. H. (2019). Intelligent Malware Detection Based on Hybrid Learning of API and ACG on Android. *Journal of Internet Services and Information Security.*, 9(4), 39–48.

Kim, S., & Nam, S. (2022). Wideband and Ultrathin 2×2 Dipole Array Antenna for 5G mm Wave Applications. *IEEE Antennas and Wireless Propagation Letters*, 21(12), 2517–2521. 10.1109/LAWP.2022.3199695

Kim, S., Ren, Y., Lee, H., Rida, A., Nikolaou, S., & Tentzeris, M. M. (2012). Monopole antenna with inkjet-printed EBG array on paper substrate for wearable applications. *IEEE Antennas and Wireless Propagation Letters*, 11, 663–666. 10.1109/LAWP.2012.2203291

Kim, Y.-S., Basir, A., Herbert, R., Kim, J., Yoo, H., & Yeo, W.-H. (2020, January). Soft materials, stretchable mechanics, and optimized designs for bodywearable compliant antennas. *ACS Applied Materials & Interfaces*, 12(2), 3059–3067. 10.1021/acsami.9b2023331842536

Kuhn, G. G., de Morais Sousa, K., & da Silva, J. C. C. (2018). Dynamic Strain Analysis of Transformer Iron Core with Fiber Bragg Gratings. *Advanced Photonics 2018 (BGPP, IPR, NP, NOMA, Sensors, Networks, SPPCom, SOF)*, JTu2A.74. https://opg.optica.org/abstract.cfm?URI=Sensors-2018-JTu2A.74

Kuhn, G. G., Sousa, K. M., Martelli, C., Bavastri, C. A., & da Silva, J. C. C. (2020). Embedded FBG Sensors in Carbon Fiber for Vibration and Temperature Measurement in Power Transformer Iron Core. *IEEE Sensors Journal*, 20(22), 13403–13410. 10.1109/JSEN.2020.3005884

Kumar, S., Dixit, A. S., Malekar, R. R., Raut, H. D., & Shevada, L. K. (2020). Fifth Generation Antennas: A comprehensive review of design and performance enhancement techniques. *IEEE Access : Practical Innovations, Open Solutions*, 8, 163568–163593. 10.1109/ACCESS.2020.3020952

Lad, S. S., & Adamuthe, A. C. (2020). Malware classification with improved convolutional neural network model. *Int. J. Comput. Netw. Inf. Secur*, 12(6), 30–43.

Lagunas, E., & Najar, M. (2015, July). Spectral feature detection with sub-Nyquist sampling for wideband spectrum sensing. *IEEE Transactions on Wireless Communications*, 14(7), 3978–3990. 10.1109/TWC.2015.2415774

Lakshmi Satya Nagasri, D., & Marimuthu, R. (2023). Review on advanced control techniques for microgrids. *Energy Reports, 10*.10.1016/j.egyr.2023.09.162

Lalem, F., Laouid, A., Kara, M., Al-Khalidi, M., & Eleyan, A. (2023). A novel digital signature scheme for advanced asymmetric encryption techniques. *Applied Sciences (Basel, Switzerland)*, 13(8), 5172. 10.3390/app13085172

Laneman, J., Tse, D., & Wornell, G. (2004). Cooperative diversity in wireless networks: Efficient protocols and outage behavior. *IEEE Transactions on Information Theory*, 50(12), 3062–3080. 10.1109/TIT.2004.838089

Leligou, H. C., Zahariadis, T., Sarakis, L., Tsampasis, E., Voulkidis, A., & Velivassaki, T. E. (2018). Smart grid: a demanding use case for 5G technologies. *2018 IEEE International Conference on Pervasive Computing and Communications Workshops (Percom Workshops)*, (pp. 215–220). IEEE. 10.1109/PERCOMW.2018.8480296

Leones Sherwin Vimalraj, S. (2023). Power allocation algorithm for capacity maximization in 5G MIMO systems. *Measurement: Sensors*, *30*, 100919.

Lessig, L. (2009). *Code: And other laws of cyberspace*. ReadHowYouWant. com.

Letaief, K. B., Chen, W., Shi, Y., Zhang, J., & Zhang, Y. J. A. (2019). The roadmap to 6G: AI empowered wireless networks. *IEEE Communications Magazine*, 57(8), 84–90. 10.1109/MCOM.2019.1900271

Levin, G. (2012). *Amplify-and-Forward Versus Decode-and-Forward Relaying: Which is Better*? ETH.

Levinsky, J. (2022). Encryption: The History and Implementation.

Leye, P. O., Khenchaf, A., & Pouliguen, P. (2016). Application of Gaussian beam summation method in high-frequency RCS of complex radar targets. *2016 IEEE Radar Conference (RadarConf)*, Philadelphia, PA, USA. 10.1109/RADAR.2016.7485208

Li, H. (2021). Centralized Control for Cybersecurity in 5G-Enabled Smart Grids: The Role of SDN. *Journal of Cybersecurity and Energy Systems*.

Li, H. (2021). Securing the Fusion: Vulnerabilities and Security Challenges in the Integration of 5G into Smart Grids. *Journal of Energy Cybersecurity*.

Li, H. (2022). Low-Profile All-Textile Multiband Microstrip Circular Patch Antenna for WBAN Applications. *IEEE Antennas And Wireless Propagation Letters*, (*vol. 21,* no. 4).

Li, M., & Zhang, S. (2024). Emerging Technologies in 5G-Integrated Smart Grids: A Comprehensive Review. *IEEE Transactions on Power Systems*. IEEE.

Li, N. (2010). Research on Diffie-Hellman key exchange protocol. In *2010 2nd International Conference on Computer Engineering and Technology*. IEEE.

Li, W. (2024). Ensuring Data Privacy in Smart Grids: A Comprehensive Approach. *Journal of Information Security and Privacy.*

Li, X. (2022). Collaborative Security Solutions for 5G-Enabled Smart Grids. *International Journal of Critical Infrastructure Protection.*

Li, Y. (2023). *Cybersecurity of Smart Inverters in the Smart Grid: A Survey*. IEEE. 10.1109/TPEL.2022.3206239

Liang, Y. (2018). NFV and SDN for Cybersecurity in Smart Grids: Dynamic Security Measures." . *International Journal of Critical Infrastructure Protection.*

Li, E. S. (2003, October). Physical optics models for the backscatter response of road-surface faults and roadside pebbles at millimeter-wave frequencies. *IEEE Transactions on Antennas and Propagation*, 51(10), 2862–2868. 10.1109/TAP.2003.818004

Li, K., Li, H., Chen, X., & Li, Y. (2020). A survey of data-driven methods for fault diagnosis in smart grids. *IEEE Access : Practical Innovations, Open Solutions*, 8, 159437–159453.

Limniotis, K. (2021). Cryptography as the Means to Protect Fundamental Human Rights. *Cryptography*, 5(4), 34. 10.3390/cryptography5040034

Lin, W. (2009). SER Performance Analysis and Optimal Relay Location of Cooperative Communications with Distributed Alamouti Code. *43rd Annual Conference on Information Sciences and Systems,* (pp. 646-651). IEEE. 10.1109/CISS.2009.5054798

Li, S., Chen, Y., Chen, L., Jing, L., Kuang, C., Li, K., Liang, W., & Xiong, N. (2023). Post-Quantum Security: Opportunities and Challenges. *Sensors (Basel)*, 23(21), 8744. 10.3390/s2321874437960442

Li, S.-H., & Li, J. (2018). Smart patch wearable antenna on Jeans textile for body wireless communication. *12th International Symposium on Antennas, Propagation and EM Theory (ISAPE).* IEEE. 10.1109/ISAPE.2018.8634084

Liu, J., and Wu, H. (2022). Privacy Concerns and Data Protection in 5G-Enabled Smart Grids. *International Journal of Information Management.*

Liu, M. (2024). Enhancing Cyber-Resiliency of DER-Based Smart Grid: A Survey. *IEEE Transactions on Smart Grid.* IEEE. 10.1109/TSG.2024.3373008

Liu, X. (2022). Dynamic Deployment of Network Functions in 5G: A Focus on NFV. *Journal of Communication Networks.*

Liu, Y., & Chen, Z. (2023). Service Chains in 5G Networks: Enabling End-to-End Network Services. *Wireless Communications and Mobile Computing.*

Liu, F., Wei, S., Li, B., Tan, Y., Guo, X., & Fu, X. (2023). A novel fast response and high precision water temperature sensor based on Fiber Bragg Grating. *Optik (Stuttgart)*, 289, 171257. 10.1016/j.ijleo.2023.171257

Liu, L., Liu, T., Zheng, Y., Jin, Z., & Sun, Z. (2022). Archimedean spiral antenna based on metamaterial structure with wideband circular polarization. *AEÜ. International Journal of Electronics and Communications*, 152, 154257. 10.1016/j.aeue.2022.154257

Liu, P., Shen, C., Liu, C., Cintrón, F. J., Zhang, L., Cao, L., Rouil, R., & Roy, S. (2023). Towards 5G new radio sidelink communications: A versatile link-level simulator and performance evaluation. *Computer Communications*, 208, 231–243. 10.1016/j.comcom.2023.06.005

Liu, S. (2020). Anomaly Detection Techniques for Cyber Defense in Smart Grids." . *IEEE Transactions on Smart Grid.*

Liu, X., Chen, X., & Chen, Y. (2021). A survey of integration of big data analytics with smart grid. *IEEE Access : Practical Innovations, Open Solutions*, 9, 14310–14326.

Liu, X., & Qiao, D. (2019). Space-Time Block Coding-Based Beamforming for Beam Squint Compensation. *IEEE Wireless Communications Letters*, 8(1), 241–244. 10.1109/LWC.2018.2868636

Liu, Y., Li, L., Zhao, L., Wang, J., & Liu, T. (2017). Research on a new fiber-optic axial pressure sensor of transformer winding based on fiber Bragg grating. *Photonic Sensors*, 7(4), 365–371. 10.1007/s13320-017-0427-z

Liu, Y., Ren, A., Liu, H., Wang, H., & Sim, C. (2019). Eight-Port MIMO array using characteristic mode theory for 5G smartphone applications. *IEEE Access : Practical Innovations, Open Solutions*, 7, 45679–45692. 10.1109/ACCESS.2019.2909070

Li, Y., & Chen, J. (2022). Design of miniaturized high gain Bow-Tie antenna. *IEEE Transactions on Antennas and Propagation*, 70(1), 738–743. 10.1109/TAP.2021.3098595

Li, Y., & Chen, M. (2015). Software-Defined Network Function Virtualization: A Survey. *IEEE Access : Practical Innovations, Open Solutions*, 3, 2542–2553. 10.1109/ACCESS.2015.2499271

Li, Y., & Liu, Q. (2021). A comprehensive review study of cyber-attacks and cyber security; Emerging trends and recent developments. *Energy Reports*, 7, 8176–8186. 10.1016/j.egyr.2021.08.126

Liyanage, M., Ahmad, I., Abro, A. B., Gurtov, A., & Ylianttila, M. (2018). *A comprehensive guide to 5G security.* Wiley Online Library. 10.1002/9781119293071

Li, Z., Hu, H., Hu, H., Huang, B., Ge, J., & Chang, V. (2021). Security and energy-aware collaborative task offloading in D2D communication. *Future Generation Computer Systems*, 1, 1–28. 10.1016/j.future.2021.06.009

Logeshwaran, J. (2024). *Next-Generation Tele-Communication Networks and IoT: Advancements and Challenges.* 2024 11th International Conference on Reliability, Infocom Technologies and Optimization (Trends and Future Directions) (ICRITO), Noida, India. 10.1109/ICRITO61523.2024.10522190

Lopez, J., Rubio, J. E., & Alcaraz, C. (2021). Digital Twins for Intelligent Authorization in the B5G-Enabled Smart Grid. *IEEE Wireless Communications*, 28(2), 2. 10.1109/MWC.001.2000336

Lu, H., & Nikookar, H. (2009). *A thresholding strategy for DF-AF hybrid cooperative wireless networks and its performance*. Proc IEEE SCVT '09, UCL, Louvain.

Luo, J., Hao, Y., Ye, Q., Hao, Y., & Li, L. (2013). Development of Optical Fiber Sensors Based on Brillouin Scattering and FBG for On-Line Monitoring in Overhead Transmission Lines. *Journal of Lightwave Technology*, 31(10), 1559–1565. 10.1109/JLT.2013.2252882

Lyu, S., & Ling, C. (2017). Boosted KZ and LLL Algorithms. *IEEE Transactions on Signal Processing*, 65(18), 4784–4796. 10.1109/TSP.2017.2708020

Madhu, B., Chari, M. V. G., Vankdothu, R., Silivery, A. K., & Aerranagula, V. (2023, February 1). Intrusion detection models for IOT networks via Deep Learning approaches. *Measurement. Sensors*, 25, 100641. 10.1016/j.measen.2022.100641

Ma, G., Li, C., Quan, J., Jiang, J., & Cheng, Y. (2011). A Fiber Bragg Grating Tension and Tilt Sensor Applied to Icing Monitoring on Overhead Transmission Lines. *IEEE Transactions on Power Delivery*, 26(4), 2163–2170. 10.1109/TPWRD.2011.2157947

Ma, G.-M., Zhou, H.-Y., Shi, C., Li, Y.-B., Zhang, Q., Li, C.-R., & Zheng, Q. (2018). Distributed Partial Discharge Detection in a Power Transformer Based on Phase-Shifted FBG. *IEEE Sensors Journal*, 18(7), 2788–2795. 10.1109/JSEN.2018.2803056

Mahin, A. U., Islam, S. N., Ahmed, F., & Hossain, Md. F. (2022). Measurement and monitoring of overhead transmission line sag in smart grid: A review. *IET Generation, Transmission & Distribution, 16*(1), 1–18.

Mahto, D., & Yadav, D. K. (2017). RSA and ECC: A comparative analysis. *International Journal of Applied Engineering Research: IJAER*, 12(19), 9053–9061.

Majeed, A., & Lee, S. (2020). Anonymization Techniques for Privacy Preserving Data Publishing: A Comprehensive Survey. *IEEE Access : Practical Innovations, Open Solutions*, 9, 8512–8595. 10.1109/ACCESS.2020.3045700

Malviya, L. (2020). *Highly isolated inset-feed 28 GHz MIMO-antenna array for 5G wireless application*. Research Gate.

Malviya, L., Parmar, A., Solanki, D., Gupta, P., & Malviya, P. (2020). Highly isolated inset-feed 28 GHz MIMO-antenna array for 5G wireless application. *Procedia Computer Science*, 171, 1286–1292. 10.1016/j.procs.2020.04.137

Mangla, C., Rani, S., Faseeh Qureshi, N. M., & Singh, A. (2023). Mitigating 5G security challenges for next-gen industry using quantum computing. *Journal of King Saud University. Computer and Information Sciences*, 35(6), 101334. 10.1016/j.jksuci.2022.07.009

Mao, C. (2022). Design of Computer Storage System Based on Cloud Computing. In *International Conference on Communication, Devices and Networking*. Singapore: Springer Nature Singapore.

Mao, C. X., Zhang, L., Khalily, M., Gao, Y., & Xiao, P. (2021). A Multiplexing Filtering Antenna. *IEEE Transactions on Antennas and Propagation*, 69(8), 5066–5071. 10.1109/TAP.2020.3048589

Mao, S., Zhang, H., Wu, W., Liu, J., Li, S., & Wang, H. (2014). A resistant quantum key exchange protocol and its corresponding encryption scheme. *China Communications*, 11(9), 124–134. 10.1109/CC.2014.6969777

McDaniel, P., & McLaughlin, S. (2009, May-June). Security and Privacy Challenges in the Smart Grid. *IEEE Security and Privacy*, 7(3), 75–77. 10.1109/MSP.2009.76

Memon, R. A., Qazi, S., & Khan, B. M. (2021). Design and Implementation of a Robust Convolutional Neural Network-Based Traffic Matrix Estimator for Cloud Networks. *Wireless Communications and Mobile Computing*. Wiley. 10.1155/2021/1039613

Mendonça, S., Damásio, B., Charlita de Freitas, L., Oliveira, L., Cichy, M., & Nicita, A. (2022). The rise of 5G technologies and systems: A quantitative analysis of knowledge production. *Telecommunications Policy*, 46(4), 102327. 10.1016/j.telpol.2022.102327

Meng, S., Wang, Z., Tang, M., Wu, S., & Li, X. (2019). Integration application of 5g and smart grid. *2019 11th International Conference on Wireless Communications and Signal Processing (WCSP)*, 1–7.

Mersani, A. (2018). Improved Radiation Performance of Textile Antenna Using AMC Surface. *15th International Multi-Conference on Systems, Signals & Devices (SSD)*. AMC.

Mijumbi, R., Serrat, J., Gorricho, J.-L., Bouten, N., De Turck, F., & Boutaba, R. (2016). Network Function Virtualization: State-of-the-Art and Research Challenges. *IEEE Communications Surveys and Tutorials*, 18(1), 236–262. 10.1109/COMST.2015.2477041

Milias, C., Andersen, R. B., Lazaridis, P. I., Zaharis, Z. D., Muhammad, B., Kristensen, J. T. B., Mihovska, A., & Hermansen, D. D. S. (2022). Miniaturized Multiband Metamaterial Antennas With Dual-Band Isolation Enhancement. *IEEE Access : Practical Innovations, Open Solutions*, 10, 64952–64964. 10.1109/ACCESS.2022.3183800

Min, L., Li, S., Zhang, X., Zhang, F., Sun, Z., Wang, M., Zhao, Q., Yang, Y., & Ma, L. (2018). The Research of Vibration Monitoring System for Transformer Based on Optical Fiber Sensing. *2018 IEEE 3rd Optoelectronics Global Conference (OGC)*, (pp. 126–129). IEEE. 10.1109/OGC.2018.8529978

Mishra, G. P., & Mangaraj, B. B. (2019). Miniaturised microstrip patch design based on highly capacitive defected ground structure with fractal boundary for X-band microwave communications. *IET Microwaves, Antennas & Propagation*, 13(10), 1593–1601. 10.1049/iet-map.2018.5778

Mishra, V., Lohar, M., & Amphawan, A. (2016). Improvement in temperature sensitivity of FBG by coating of different materials. *Optik (Stuttgart)*, 127(2), 825–828. 10.1016/j.ijleo.2015.10.014

Mitola, J., & Maguire, G. (1999, August). Cognitive radio: Making software radios more personal. *IEEE Personal Communications*, 6(4), 13–18. 10.1109/98.788210

Mohammadi, F. S., & Kwasinski, A. (2018). *Neural network cognitive engine for autonomous and distributed underlay dynamic spectrum access.* arXiv preprint arXiv:1806.11038.

Mohammed. (2022). Vulnerabilities and Strategies of Cybersecurity in Smart Grid - Evaluation and Review. *2022 3rd International Conference on Smart Grid and Renewable Energy (SGRE)*, Doha, Qatar. 10.1109/SGRE53517.2022.9774038

Mohsen, S., Bajaj, M., Kotb, H., Pushkarna, M., Alphonse, S., & Ghoneim, S. S. M. (2023). Efficient Artificial Neural Network for Smart Grid Stability Prediction. *International Transactions on Electrical Energy Systems*, 2023, 1–13. 10.1155/2023/9974409

Moro, R., Agneessens, S., Rogier, H., & Bozzi, M. (2018). Circularly-polarised cavity-backed wearable antenna in SIW technology. *IET Microwaves, Antennas & Propagation*, 12(1), 127–131. 10.1049/iet-map.2017.0271

Mozumder, M. (2023). The Metaverse for Intelligent Healthcare using XAI, Blockchain, and Immersive Technology. *2023 IEEE International Conference on Metaverse Computing, Networking and Applications (MetaCom)*, Kyoto, Japan. 10.1109/MetaCom57706.2023.00107

Nahas, M. (2022). Design of a high-gain dual-band LI-slotted microstrip patch antenna for 5G mobile communication systems. *Journal of Radiation Research and Applied Sciences*, 15(4), 100483. 10.1016/j.jrras.2022.100483

Nan, Y., Xie, W., Min, L., Cai, S., Ni, J., Yi, J., Luo, X., Wang, K., Nie, M., Wang, C., Peng, G.-D., & Guo, T. (2020). Real-Time Monitoring of Wind-Induced Vibration of High-Voltage Transmission Tower Using an Optical Fiber Sensing System. *IEEE Transactions on Instrumentation and Measurement*, 69(1), 268–274. 10.1109/TIM.2019.2893034

Nawshin, F., Unal, D., Hammoudeh, M., & Suganthan, P. N. (2024). AI-powered malwaredetection with Differential Privacy for zero trust security in Internet of Things networks. *Ad Hoc Networks*, 161, 103523. 10.1016/j.adhoc.2024.103523

Nguyen, V. M., & Nur, A. Y. "Major CyberSecurity Threats in Healthcare During Covid-19 Pandemic," *2023 International Symposium on Networks, Computers and Communications (ISNCC)*, Doha, Qatar. 10.1109/ISNCC58260.2023.10323723

Nicholls, M., & Nicholls, M. (2024, March 20). *The key challenges of intrusion detection and how to overcome them.* Redscan. https://www.redscan.com/news/the-key-challenges-of-intrusion-detection-and-how-to-overcome-them/

Nissanov, U., & Singh, G. (2023). Grounded coplanar waveguide microstrip array antenna for 6G wireless networks. *Sensors International*, 4, 100228. 10.1016/j.sintl.2023.100228

Nostratinia, A., Hunter, T. E., & Hedayat, A. (2004). Cooperative communication in wirelesss networks. *IEEE Communications Magazine*, 42(10), 74–80. 10.1109/MCOM.2004.1341264

Oktaviana, B., & Utama Siahaan, A. P. (2016). Three-pass protocol implementation in caesar cipher classic cryptography. *IOSR Journal of Computer Engineering*, 18(04), 26–29. 10.9790/0661-1804032629

Oprea, S.-V., & Bâra, A. (2023). An Edge-Fog-Cloud computing architecture for IoT and smart metering data. *Peer-to-Peer Networking and Applications*, 16(no. 2), 818–845. 10.1007/s12083-022-01436-y

Osama, M., El Ramly, S., & Abdelhamid, B. (2021). Interference Mitigation and Power Minimization in 5G Heterogeneous Networks. *Electronics (Basel)*, 10(14), 14. 10.3390/electronics10141723

Ourahou, M., Ayrir, W., Hassouni, B. E., & Haddi, A. (2020). Review on smart grid control and reliability in presence of renewable energies: Challenges and prospects. *Mathematics and Computers in Simulation*, 167, 19–31. 10.1016/j.matcom.2018.11.009

Padhan, A. K., Kumar Sahu, H., Sahu, P. R., & Samantaray, S. R. (2023). Performance analysis of smart grid wide area network with RIS assisted three hop system. *IEEE Transactions on Signal and Information Processing Over Networks*, 9, 48–59. 10.1109/TSIPN.2023.3239652

Pant, D., & Bista, R. (2021). Image-based malware classification using deep convolutional neural network and transfer learning. *Proceedings of the 3rd International Conference on Advanced Information Science and System* (pp. 1-6). ACM. 10.1145/3503047.3503081

Parchin, N. O., Basherlou, H. J., Al-Yasir, Y. I. A., & Abd-Alhameed, R. A. (2020). A broadband multiple-input multiple-output loop antenna array for 5G cellular communications. *AEÜ. International Journal of Electronics and Communications*, 127, 153476. 10.1016/j.aeue.2020.153476

Parveez Shariff, B. (2023). Planar MIMO antenna for mmWave applications: Evolution, present status & future scope. *Heliyon, 9*(2), e13362.

Pathmudi, V. R., Khatri, N., Kumar, S., Antar, S. H. A.-Q., & Vyas, A. K. (2023). A systematic review of IoT technologies and their constituents for smart and sustainable agriculture applications. *Scientific African, 19*.10.1016/j.sciaf.2023.e01577

Pavitra, A. R. R., Lawrence, I. D., & Maheswari, P. U. (2023). To Identify the Accessibility and Performance of Smart Healthcare Systems in IoT-Based Environments. In *Using Multimedia Systems, Tools, and Technologies for Smart Healthcare Services* (pp. 229–245). IGI Global. 10.4018/978-1-6684-5741-2.ch014

Paya, A., Arroni, S., García-Díaz, V., & Gómez, A. (2024). Apollon: A robust defense system against Adversarial Machine Learning attacks in Intrusion Detection Systems. *Computers & Security, 136*.10.1016/j.cose.2023.103546

Petroulakis, N. E., Fysarakis, K., Askoxylakis, I. G., & Spanoudakis, G. (2017). Reactive Security for SDN/NFV-enabled Industrial Networks leveraging Service Function Chaining. *Transactions on Emerging Telecommunications Technologies*.

Pharkkavi, D., & Maruthanayagam, D. (2018). Time complexity analysis of RSA and ECC based security algorithms in cloud data. *International Journal of Advanced Research in Computer Science*, 9(3), 201–208. 10.26483/ijarcs.v9i3.6104

Pinto, A., Herrera, L.-C., Donoso, Y., & Gutierrez, J. A. (2023). Survey on Intrusion Detection Systems Based on Machine Learning Techniques for the Protection of Critical Infrastructure. *Sensors (Basel)*, 23(5), 2415. 10.3390/s2305241536904618

Politi, A. (2010). Quantum information science with photonic chips. In *36th European Conference and Exhibition on Optical Communication*, (pp. 1-3). IEEE.

Popović, I., Rakić, A., & Petruševski, I. D. (2022, January). Multi-agent real-time advanced metering infrastructure based on fog computing. *Energies*, 15(1), 373. 10.3390/en15010373

Preneel, B. (1994). Cryptographic hash functions. *European Transactions on Telecommunications*, 5(4), 431–448. 10.1002/ett.4460050406

Priyadarshini, I., Kumar, R., Sharma, R., Singh, P. K., & Satapathy, S. C. (2021). Pradeep Kumar Singh, Suresh Chandra Satapathy, Identifying cyber insecurities in trustworthy space and energy sector for smart grids. *Computers & Electrical Engineering*, 93, 107204. 10.1016/j.compeleceng.2021.107204

Priyanka, M. (2013). BER analysis of Alamouti space time block coded 2×2 MIMO systems using Rayleigh dent mobile radio channel. *Advance Computing Conference (IACC), 2013 IEEE 3rd International*. IEEE.

Qamar, F., Hindia, M. H. D. N., Dimyati, K., Noordin, K. A., & Amiri, I. S. (2019). Interference management issues for the future 5G network: A review. *Telecommunication Systems*, 71(4), 627–643. 10.1007/s11235-019-00578-4

Qualcomm. (2017). *Everything You Need to Know About 5G*. Qualcomm. https://www.qualcomm.com/5g/what-is-5g

Qualcomm. (2023). *What is 5G | Everything You Need to Know About 5G*. Qualcomm. https://www.qualcomm.com/5g/what-is-5g (accessed: October 25, 2023).

Qureshi, S., He, J., Tunio, S., Zhu, N., Akhtar, F., Ullah, F., Nazir, A., & Wajahat, A. (2021). A Hybrid DL-Based Detection Mechanism for Cyber Threats in Secure Networks. *IEEE Access : Practical Innovations, Open Solutions*, 9, 73938–73947. 10.1109/ACCESS.2021.3081069

Rafique, M. F., Ali, M., Qureshi, A. S., Khan, A., & Mirza, A. M. (2019). Malware classification using deep learning based feature extraction and wrapper based feature selection technique. *arXiv preprint arXiv:1910.10958*.

Raheman, F. (2022). The future of cybersecurity in the age of quantum computers. *Future Internet*, 14(11), 335. 10.3390/fi14110335

Rahman, A.U., Mahmud, M., Iqbal, T., Saraireh, L., Kholidy, H., Gollapalli, M., Musleh, D., Alhaidari, F., Almoqbil, D. and Ahmed, M.I.B. (2022). Network Anomaly Detection in 5G Networks. *Mathematical Modelling of Engineering Problems, 9*(2).

Ramkumar, B. (2009, June). Automatic modulation classification for cognitive radios using cyclic feature detection. *IEEE Circuits and Systems Magazine*, 9(2), 27–45. 10.1109/MCAS.2008.931739

Ramnarine, V., Peesapati, V., & Djurović, S. (2023). Fibre Bragg Grating Sensors for Condition Monitoring of High-Voltage Assets: A Review. *Energies*, 16(18), 6709. Advance online publication. 10.3390/en16186709

Raviteja, V., Kumar, S. A., & Shanmuganantham, T. (2019). *CPW-fed inverted six shaped antenna design for internet of things (IoT) applications.* In: *International Conference on Microwave Integrated Circuits, Photonics and Wireless Networks (IMICPW)*, Tiruchirappalli, India. 10.1109/IMICPW.2019.8933248

Ray Liu, K. J., & Ahmed, K. (2009). *Cooperative Communication and Networking.* Cambridge University Press, the Edinburgh Building, Cambridge.

Reddy, C. J., Deshpande, M. D., Cockrell, C. R., & Beck, F. B. (1998, August). Fast RCS computation over a frequency band using method of moments in conjunction with asymptotic waveform evaluation technique. *IEEE Transactions on Antennas and Propagation*, 46(8), 1229–1233. 10.1109/8.718579

Rehmani, M. H., Davy, A., Jennings, B., & Assi, C. (2019). Software Defined Networks-Based Smart Grid Communication: A Comprehensive Survey. *IEEE Communications Surveys and Tutorials*, 21(3), 2637–2670. 10.1109/COMST.2019.2908266

Reply. (2024). *Low Latency: what makes 5G different.* Reply. https://www.reply.com/en/telco-and-media/low-latency-what-makes-5g-different

Ribeiro, L. E. (2019). High-profile data breaches: Designing the right data protection architecture based on the law, ethics and trust. *Applied Marketing Analytics*, 5(2), 146–158.

Rochman, M. I., Sathya, V., Fernandez, D., Nunez, N., Ibrahim, A. S., Payne, W., & Ghosh, M. (2023). A comprehensive analysis of the coverage and performance of 4G and 5G deployments. *Computer Networks*, 237, 10060. 10.1016/j.comnet.2023.110060

Roseline, S. A., Geetha, S., Kadry, S., & Nam, Y. (2020). Intelligent vision-based malware detection and classification using deep random forest paradigm. *IEEE Access : Practical Innovations, Open Solutions*, 8, 206303–206324. 10.1109/ACCESS.2020.3036491

Rubinstein-Salzedo, S., & Rubinstein-Salzedo, S. (2018). *The vigenere cipher.* Cryptography. 10.1007/978-3-319-94818-8_5

S. A. K. (2021). Approximate Message Passing for mmWave Massive MIMO Architecture using Optimal Hybrid Precoder/Combiner. 2021 Smart Technologies, Communication and Robotics. STCR. 10.1109/STCR51658.2021.9588908

S. A. K. (2022). GM-LAMP with Residual Learning Network for Millimetre Wave MIMO Architectures. 2022 Smart Technologies, Communication and Robotics. STCR. 10.1109/STCR55312.2022.10009163

Saad, W., Bennis, M., & Chen, M. (2020). A Vision of 6G Wireless Systems: Applications, Trends, Technologies, and Open Research Problems. *IEEE Network*, 34(3), 134–142. 10.1109/MNET.001.1900287

Sabban, A. (2019). Small New Wearable Antennas for IOT, Medical and Sport Applications. 13th European Conference on Antennas and Propagation (EuCAP). Research Gate.

Sadek, A. K., Su, W., & Liu, K. J. R. (2005). Performance analysis for multimode decode-and-forward relaying in collaborative wireless networks. Proc. IEEE Int. Conf. Acoust., Speech, Signal Process. (ICASSP), Philadelphia, PA.

Sadr, S., Anpalagan, A., & Raahemifar, K. (2009). A novel subcarrier allocation algorithm for multiuser OFDM system with fairness: User's perspective. *Vehicular Technology Conference*, (pp. 1772-1776). Research Gate.

Sadr, S., Anpalagan, A., & Raahemifar, K. (2009). Radio Resource Allocation Algorithms for the Downlink of Multiuser OFDM Communication Systems. *IEEE Communications Surveys and Tutorials*, 11(3), 92–106. 10.1109/SURV.2009.090307

Saeed, M. M., Saeed, R. A., Abdelhaq, M., Alsaqour, R., Hasan, M. K., & Mokhtar, R. A. (2023). Anomaly detection in 6G networks using machine learning methods. *Electronics, 12(15),* 3300. 37188858

Saheed, Y. K., Abiodun, A. I., Misra, S., Holone, M. K., & Colomo-Palacios, R. (2022, December 1). A machine learning-based intrusion detection for detecting internet of things network attacks. Alexandria Engineering Journal. *Alexandria Engineering Journal*, 61(12), 9395–9409. 10.1016/j.aej.2022.02.063

Santamargarita, D., Molinero, D., Bueno, E., Marrón, M., & Vasić, M. (2023). On-Line Monitoring of Maximum Temperature and Loss Distribution of a Medium Frequency Transformer Using Artificial Neural Networks. *IEEE Transactions on Power Electronics*, 38(12), 15818–15828. 10.1109/TPEL.2023.3308613

Santos, V. F., Albuquerque, C., Passos, D., Quincozes, S. E., & Mossé, D. (2023). Assessing Machine Learning Techniques for Intrusion Detection in Cyber-Physical Systems. *Energies*, 16(16), 6058. 10.3390/en16166058

Sanyal, J., & Samanta, T. (2021). Game Theoretic Approach to Enhancing D2D Communications in 5G Wireless Networks. *International Journal of Wireless Information Networks*, 28(4), 421–436. 10.1007/s10776-021-00531-w

Sargam, S., Gupta, R., Sharma, R., & Jain, K. (2023). A Comprehensive review on 5G-based Smart Healthcare Network Security: Taxonomy, Issues, Solutions and Future research directions. *Array (New York, N.Y.)*, 18, 100290. 10.1016/j.array.2023.100290

Sarkar, B., Koley, C., Roy, N. K., & Kumbhakar, P. (2015). Condition monitoring of high voltage transformers using Fiber Bragg Grating Sensor. *Measurement*, 74, 255–267. 10.1016/j.measurement.2015.07.014

Scalise, P., Boeding, M., Hempel, M., Sharif, H., Delloiacovo, J., & Reed, J. (2024). A Systematic Survey on 5G and 6G Security Considerations, Challenges, Trends, and Research Areas. *Future Internet*, 16(3), 67. 10.3390/fi16030067

Schnorr, C. P., & Euchner, M. (1994). Lattice basis reduction: Improved practical algorithms and solving subset sum problems. *Mathematical Programming*, 66(1), 181–199. 10.1007/BF01581144

Sendonaris, A., Erkip, E., & Aazhang, B. (2003). User cooperation diversity - Part I: System description. *IEEE Transactions on Communications*, 51(11), 1927–1938. 10.1109/TCOMM.2003.818096

Serghiou, D., Khalily, M., Singh, V., Araghi, A., & Tafazolli, R. (2020). Sub-6 GHz Dual-Band 8 × 8 MIMO Antenna for 5G Smartphones. *IEEE Antennas and Wireless Propagation Letters*, 19(9), 1546–1550. 10.1109/LAWP.2020.3008962

Shahzad, A., Paracha, K. N., Naseer, S., Ahmad, S., Malik, M., Farhan, M., Ghaffar, A., Hussien, M., & Sharif, A. B. (2021, November). An artificial magnetic conductor-backed compact wearable antenna for smart watch IoT applications. *Electronics (Basel)*, 10(23), 2908. 10.3390/electronics10232908

Shanbhag, J. (2023). Layered Division Multiplexing in 5G NR. *International Conference for Advancement in Technology (ICONAT)*, Goa, India.

Sharma, R. (2021). Securing Critical Energy Infrastructure: A Study on 5G-Enabled Smart Grids. *IEEE Transactions on Industrial Informatics*. IEEE.

Shen, T., Yuan, Y., Chen, Q., Xu, S., Zheng, W., Xiao, H., Liu, D., Liu, D., & Zhao, S. (2024). Fiber optic sensor for transformer temperature detection. *Microwave and Optical Technology Letters*, 66(1), e33813. 10.1002/mop.33813

Shen, X., Liu, Y., Zhao, L., Huang, G.-L., Shi, X., & Huang, Q. (2019). A Miniaturized Microstrip Antenna Array at 5G Millimeter-Wave Band. *IEEE Antennas and Wireless Propagation Letters*, 18(8), 1671–1675. 10.1109/LAWP.2019.2927460

Shetu, S. F., Saifuzzaman, M., Moon, N. N., & Nur, F. N. (2019). A survey of botnet in cyber security. In *2019 2nd International Conference on Intelligent Communication and Computational Techniques (ICCT)* (pp. 174-177). IEEE. 10.1109/ICCT46177.2019.8969048

Shi, C., Ma, G., Mao, N., Zhang, Q., Zheng, Q., Li, C., & Zhao, S. (2017). Ultrasonic detection coherence of fiber Bragg grating for partial discharge in transformers. *2017 IEEE 19th International Conference on Dielectric Liquids (ICDL)*, (pp. 1–4). IEEE. 10.1109/ICDL.2017.8124639

Shin, H., Yoon, D., Na, D.-Y., & Park, Y. B. (2022). Analysis of Radome Cross Section of an Aircraft Equipped With a FSS Radome. *IEEE Access : Practical Innovations, Open Solutions*, 10, 33704–33712. 10.1109/ACCESS.2022.3162262

Shi, Y., Tuan, H. D., Savkin, A. V., Lin, C. T., Zhu, J. G., & Poor, H. V. (2021). Distributed model predictive control for joint coordination of demand response and optimal power flow with renewables in smart grid. *Applied Energy*, 290(March), 116701. 10.1016/j.apenergy.2021.116701

Shoukath Ali, K. (2019). Sampath Palaniswami, Particle swarm optimization-based power allocation for Alamouti amplify and forward relaying protocol. *International Journal of Communication Systems*, 32(10).

Shoukath Ali, K., & Sampath, P. (2021). Time Domain Channel Estimation for Time and Frequency Selective Millimeter Wave MIMO Hybrid Architectures: Sparse Bayesian Learning-Based Kalman Filter. *Wireless Personal Communications*, 117(3), 2453–2473. 10.1007/s11277-020-07986-9

Shoukath Ali, K., & Sampath, P. (2023). Sparse Bayesian Learning Kalman Filter-based Channel Estimation for Hybrid Millimeter Wave MIMO Systems: A Frequency Domain Approach. *Journal of the Institution of Electronics and Telecommunication Engineers*, 69(7), 4243–4253. 10.1080/03772063.2021.1951367

Sim, C., Liu, H., & Huang, C. (2020). Wideband MIMO Antenna Array Design for Future Mobile Devices Operating in the 5G NR Frequency Bands n77/n78/n79 and LTE Band 46. *IEEE Antennas and Wireless Propagation Letters*, 19(1), 74–78. 10.1109/LAWP.2019.2953334

Siriwongpairat, W. P., Himsoon, T., & Su, W. (2006). Optimum threshold-selection relaying for decode-and-forward cooperation protocol. In *Proc. IEEE Wireless Communications and Networking Conference*. Las Vegas, NV.

Sisavath, C., & Yu, L. (2021). Design and implementation of security system for smart home based on IOT technology. *Procedia Computer Science, 183*. https://www.sciencedirect.com/science/article/pii/S187705092100487710.1016/j.procs.2021.02.023

Siva Shankar, S., Hung, B. T., Chakrabarti, P., Chakrabarti, T., & Parasa, G. (2024). A novel optimization based Deep Learning with artificial intelligence approach to detect intrusion attack in network system. *Education and Information Technologies*, 29(4), 3859–3883. 10.1007/s10639-023-11885-4

Skorupski, K., Harasim, D., Panas, P., Cięszczyk, S., Kisała, P., Kacejko, P., Mroczka, J., & Wydra, M. (2020). Overhead Transmission Line Sag Estimation Using the Simple Opto-Mechanical System with Fiber Bragg Gratings—Part 2: Interrogation System. *Sensors (Basel)*, 20(9), 2652. Advance online publication. 10.3390/s20092652323847l5

Smida, A., Iqbal, A., Alazemi, A. J., Waly, M. I., Ghayoula, R., & Kim, S. (2020). Wideband wearable antenna for biomedical telemetry applications. *IEEE Access : Practical Innovations, Open Solutions*, 8, 15687–15694. 10.1109/ACCESS.2020.2967413

Smith, D. R., Schultz, S., Markos, P., & Soukoulis, C. M. (2002). Determination of effective permittivity and permeability of metamaterials from reflection and transmission coefficients. *Physical Review*, 65(19).

Soh, P. J., Vandenbosch, G. A. E., Wee, F. H., Zoinol, M., Abdul, A., & Campus, P. (2013). Bending investigation of broadband wearable all-textile antennas. *Australian Journal of Basic and Applied Sciences*, 7(5), 91–94.

Song, H., & Wang, Y. (2024). Industry Partnerships for Cybersecurity in Smart Grids: A Collaborative Approach. *Journal of Cybersecurity and Energy Systems*.

Song, Q. (2019). Application of NFV and SDN in Enhancing Cybersecurity in Smart Grids. *Smart Grids: Fundamentals and Technologies*.

Spooner, C. M., & Mody, A. N. (2018, June). Wideband cyclostationary signal processing using sparse subsets of narrowband subchannels. *IEEE Transactions on Cognitive Communications and Networking*, 4(2), 162–176. 10.1109/TCCN.2018.2790971

Srinivasan, S., & Deepalakshmi, P. (2023). Enhancing the security in cyber-world by detecting the botnets using ensemble classification based machine learning. *Measurement. Sensors*, 25, 100624. 10.1016/j.measen.2022.100624

Srivastava, A., Goyal, A., & Ram, S. S. (2020). Radar Cross-Section of Potholes at Automotive Radar Frequencies.*2020 IEEE International Radar Conference (RADAR)*, Washington, DC, USA. 10.1109/RADAR42522.2020.9114858

Standard, Data Encryption. (1999). Data encryption standard. *Federal Information Processing Standards Publication*, 112, 3.

Stefanov, A., & Erkip, E. (2004). Cooperative coding for wireless networks. *IEEE Transactions on Communications*, 52(9), 14701476. 10.1109/TCOMM.2004.833070

Sullivan, S., Brighente, A., Kumar, S. A. P., & Conti, M. (2021). 5G Security Challenges and Solutions: A Review by OSI Layers. *IEEE Access : Practical Innovations, Open Solutions*, 9, 116294–116314. 10.1109/ACCESS.2021.3105396

Sung, T.-W., Xu, Y., Hu, X., Lee, C.-Y., & Fang, Q. (2022, March). Optimizing data aggregation point location with grid-based model for smart grids. *Journal of Intelligent & Fuzzy Systems*, 42(4), 3189–3201. 10.3233/JIFS-210881

Susilo, B., & Sari, R. F. (2020, May 21). Intrusion Detection in IoT Networks Using Deep Learning Algorithm. *Information (Basel)*, 11(5), 279. 10.3390/info11050279

Su, W., Sadek, A. K., & Liu, K. J. R. (2005). SER performance analysis and optimum power allocation for decode-and-forward cooperation protocol in wireless networks. Proc. *IEEE Wireless Commun. Netw. Conf. (WCNC'05)*, New Orleans, LA.

Su, W., Sadek, A. K., & Ray Liu, K. J. (2008). Cooperative Communication Protocols in Wireless Networks: Performance Analysis and Optimum Power Allocation. *Wireless Personal Communications*, 44(2), 181–217. 10.1007/s11277-007-9359-z

Swasdio, W., & Pirak, C. (2010). *A Novel Alamouti-Coded Decode-and-Forward Protocol for Cooperative Communications.* In *TENCON 2010-2010 IEEE Region 10 Conference*, (pp. 2091-2095). IEEE. 10.1109/TENCON.2010.5686616

Swasdio, W., Pirak, C., & Ascheid, G. (2011). Alamouti-Coded Decode-and-Forward Protocol with Optimum Relay Selection for Cooperative Communications. In *Ultra Modern Telecommunications and Control Systems and Workshops (ICUMT), 3rd International Congress*. IEEE.

Swasdio, W., Pirak, C., Jitapunkul, S., & al Gerd, A. (2014). Alamouti-coded decode-and-forward protocol with optimum relay selection and Power Allocation for cooperative communications. *EURASIP Journal on Wireless Communications and Networking*, 1(1), 1–13. 10.1186/1687-1499-2014-112

Tariq, U., Ahmed, I., Bashir, A. K., & Shaukat, K. (2023). A Critical Cybersecurity Analysis and Future Research Directions for the Internet of Things: A Comprehensive Review. *Sensors (Basel)*, 23(8), 4117. 10.3390/s2308411737112457

Thantharate, A., & Beard, C. (2022). ADAPTIVE6G: Adaptive Resource Management for Network Slicing Architectures in Current 5G and Future 6G Systems. *Journal of Network and Systems Management*, 31(1), 9. 10.1007/s10922-022-09693-1

Tian, Z., Tafesse, Y., & Sadler, B. M. (2012, February). Cyclic feature detection with sub-Nyquist sampling for wideband spectrum sensing. *IEEE Journal of Selected Topics in Signal Processing*, 6(1), 58–69. 10.1109/JSTSP.2011.2181940

Tiwari, R. Sharma, R., & Dubey, R. (2023). 2X2 & 4X4 dumbbell shape microstrip patch antenna array design for 5G Wi-Fi communication application. *Materials Today.*

Trabelsi, N., Fourati, L. C., & Chen, C. S. (2024). Interference management in 5G and beyond networks: A comprehensive survey. *Computer Networks*, 239, 110159. 10.1016/j.comnet.2023.110159

Trzebiatowski, K., Rzymowski, M., Kulas, L., & Nyka, K. (2021). Simple 60 GHz Switched Beam Antenna for 5G Millimeter-Wave Applications. *IEEE Antennas and Wireless Propagation Letters*, 20(1), 38–42. 10.1109/LAWP.2020.3038260

Tuballa, M. L., & Abundo, M. L. (2016). A review of the development of Smart Grid technologies. *Renewable & Sustainable Energy Reviews*, 59, 710–725. 10.1016/j.rser.2016.01.011

Ucar, F. (2023). A Comprehensive Analysis of Smart Grid Stability Prediction along with Explainable Artificial Intelligence. *Symmetry*, 15(2), 289. 10.3390/sym15020289

Usai, P., Borgese, M., Costa, F., & Monorchio, A. (2018). Hybrid physical optics-ray tracing method for the RCS calculation of electrically large objects covered with radar absorbing materials. *12th European Conference on Antennas and Propagation (EuCAP 2018)*, London, UK. 10.1049/cp.2018.0556

Vaczi, M., Ding, Z., & Poor, H. V. (2018). *Multiple Access Techniques for 5G Wireless Networks and Beyond.* Springer Cham publications.

Vargas, P. (2023). Impacts of 5G on cyber-physical risks for interdependent connected smart critical infrastructure systems. *International Journal of Critical Infrastructure Protection, 42.* 10.1016/j.ijcip.2023.100617

Vaudenay, S. (2006). Prehistory of Cryptography. *A Classical Introduction to Cryptography: Applications for Communications Security*. Research Gate.

Verma, A., & Ranga, V. (2019). On evaluation of Network Intrusion Detection Systems: Statistical analysis of CIDDS-001 dataset using Machine Learning Techniques. *TechRxiv.*

Veselago, V. G. (1968). The electrodynamics of substances with simultaneously negative values of e and μ. *Soviet Physics - Uspekhi*, 10(4), 509–514. 10.1070/PU1968v010n04ABEH003699

Victor-Mgbachi, T. O. Y. I. N. (2024). *Navigating Cybersecurity Beyond Compliance: Understanding Your Threat Landscape and Vulnerabilities*. Research Gate.

Wang, J. (2019). Vulnerabilities in Communication Protocols of Smart Grids: A Comprehensive Analysis. *Journal of Cybersecurity and Energy Systems*.

Wang, L. (2020). SDN in 5G Networks: Optimizing Performance and Resource Utilization. *IEEE Transactions on Network and Service Management*. IEEE.

Wang, Q. (2023). AI-Based Strategies for Detecting and Responding to Cyber Threats in 5G-Enabled Smart Grids. *IEEE Transactions on Smart Grid*. IEEE.

Wang, B., Zhao, Z., Sun, K., Du, C., Yang, X., & Yang, D. (2023). Wideband Series-Fed Microstrip Patch Antenna Array With Flat Gain Based on Magnetic Current Feeding Technology. *IEEE Antennas and Wireless Propagation Letters*, 22(4), 834–838. 10.1109/LAWP.2022.3226461

Wang, C.-X., Huang, J., Wang, H., Gao, X., You, X., & Hao, Y. (2020). 6G Wireless Channel Measurements and Models: Trends and Challenges. *IEEE Vehicular Technology Magazine*, 15(4), 22–32. 10.1109/MVT.2020.3018436

Wang, H., Jiang, Z. H., Yu, C., Zhou, J., Chen, P., Yu, Z., Zhang, H., Yang, B., Pang, X., Jiang, M., Cheng, Y. J., Al-Nuaimi, M. K. T., Zhang, Y., Chen, J., & He, S. (2017). Multibeam antenna technologies for 5G wireless communications. *IEEE Transactions on Antennas and Propagation*, 65(12), 6231–6249. 10.1109/TAP.2017.2712819

Wang, M., Yang, Z., Wu, J., Bao, J., Liu, J., Cai, L., Dang, T., Zheng, H., & Li, E. (2018, June). Investigation of SAR reduction usingcflexible antenna with metamaterial structure in wireless body area network. *IEEE Transactions on Antennas and Propagation*, 66(6), 3076–3086. 10.1109/TAP.2018.2820733

Wang, Z., & Dong, Y. (2022, November). Metamaterial-Based, Vertically Polarized, Miniaturized Beam-Steering Antenna for Reconfigurable Sub-6 GHz Applications. *IEEE Antennas and Wireless Propagation Letters*, 21(11), 2239–2243. 10.1109/LAWP.2022.3188548

Wang, Z., Ning, Y., & Dong, Y. (2022). Hybrid Metamaterial-TL-Based, Low-Profile, Dual-Polarized omnidirectional antenna for 5G indoor application. *IEEE Transactions on Antennas and Propagation*, 70(4), 2561–2570. 10.1109/TAP.2021.3137242

Wang, Z., Zhao, S., & Dong, Y. (2022). Pattern reconfigurable, Low-Profile, vertically polarized, ZOR-Metasurface antenna for 5G application. *IEEE Transactions on Antennas and Propagation*, 70(8), 6581–6591. 10.1109/TAP.2022.3162332

Wassan, S., Dongyan, H., Suhail, B., Jhanjhi, N. Z., Xiao, G., Ahmed, S., & Murugesan, R. K. (2024). Deep convolutional neural network and IoT technology for healthcare. *Digital Health*, 10. 10.1177/2055207623122012338250147

Wen, L., Yu, Z., Zhu, L., & Zhou, J. (2021). High-Gain Dual-Band Resonant Cavity Antenna for 5G Millimeter-Wave Communications. *IEEE Antennas and Wireless Propagation Letters*, 20(10), 1878–1882. 10.1109/LAWP.2021.3098390

What Are 5G Speeds? (2024). Cisco. https://www.cisco.com/c/en/us/solutions/what-is-5g/what-are-5g-speeds.html

Wong, C. Y., & Cheng, R. S. (1999). Multiuser OFDM with adaptive subcarrier, bit and power allocation. *IEEE Journal on Selected Areas in Communications*, 17(10), 1747–1758. 10.1109/49.793310

Wu, S. & Zhang, J. (2023). Energy-Aware Security Measures in 5G-Enabled Smart Grids. *IEEE Transactions on Sustainable Energy*. IEEE.

Wu, Z. (2023). Integration of NFV and SDN for Enhanced Security in 5G-Enabled Smart Grids. *Journal of Network and Computer Applications*.

Wydra, M., Kisala, P., Harasim, D., & Kacejko, P. (2018). Overhead Transmission Line Sag Estimation Using a Simple Optomechanical System with Chirped Fiber Bragg Gratings. Part 1: Preliminary Measurements. *Sensors (Basel)*, 18(1), 309. 10.3390/s1801030929361714

Xianghao, Y., Shen, J.-C., Zhang, J., & Letaief, K. B. (2016). Alternating Minimization Algorithms for Hybrid Precoding in Millimeter Wave MIMO Systems. *IEEE Journal of Selected Topics in Signal Processing*, 10(3), 485–500. 10.1109/JSTSP.2016.2523903

Xie, L., & Wan, Q. (2017, June). Cyclic feature-based modulation recognition using compressive sensing. *IEEE Wireless Communications Letters*, 6(3), 402–405. 10.1109/LWC.2017.2697853

Xie, T., Dai, L., Ng, D. W. K., & Chae, C.-B. (2019). On the Power Leakage Problem in Millimeter-Wave Massive MIMO with Lens Antenna Arrays. *IEEE Transactions on Signal Processing*, 67(18), 4730–474. 10.1109/TSP.2019.2926019

Xu, L. (2023). Resilience and Redundancy in Smart Grid Cybersecurity. *Journal of Energy Security and Resilience*.

Xu, L., Hu, H., & Liu, Y. (2023). SFCSim: A network function virtualization resource allocation simulation platform. *Cluster Computing*, 26(1), 423–436. 10.1007/s10586-022-03670-8

Yadav, S., & Nanivadekar, S. (2023). Hybrid optimization assisted green power allocation model for QoS-driven energy-efficiency in 5G networks. *Cybernetics and Systems*, (Feb), 1–16. 10.1080/01969722.2023.2175147

Yalduz, H., Tabaru, T. E., Kilic, V. T., & Turkmen, M. (2020). Design and analysis of low profile and low SAR full-textile UWB wearable antenna with metamaterial for WBAN applications. *AEÜ. International Journal of Electronics and Communications*, 126(Nov), 153465. 10.1016/j.aeue.2020.153465

Yamamoto, S., Nuimura, S., & Takikawa, M. (2021). A Design Concept of Grid-loaded Step Reflector Antenna with Coaxial-Mode Excitation. *2020 International Symposium on Antennas and Propagation (ISAP)*, Osaka, Japan. 10.23919/ISAP47053.2021.9391172

Yang, J. (2022). Dynamic Resource Allocation in NFV: Adapting to Varying Network Demands." . *Journal of Network and Systems Management*.

Yang, M., Tjuawinata, I., & Lam, K.-Y. (2022). K-Means Clustering With Local d -Privacy for Privacy-Preserving Data Analysis. *IEEE Transactions on Information Forensics and Security*, 17(1), 2524–2537. 10.1109/TIFS.2022.3189532

Yan, Q., Zhou, C., Feng, X., Deng, C., Hu, W., & Xu, Y. (2022). Galloping Vibration Monitoring of Overhead Transmission Lines by Chirped FBG Array. *Photonic Sensors*, 12(3), 220310. 10.1007/s13320-021-0651-4

Yassein, M. B., Aljawarneh, S., Qawasmeh, E., Mardini, W., & Khamayseh, Y. (2017). Comprehensive study of symmetric key and asymmetric key encryption algorithms. In *2017 international conference on engineering and technology (ICET)* (pp. 1-7). IEEE. 10.1109/ICEngTechnol.2017.8308215

Yazdinejad, A., Parizi, R. M., Dehghantanha, A., & Choo, K.-K. R. (2019). Blockchain-enabled authentication handover with efficient privacy protection in sdn-based 5g networks. *IEEE Transactions on Network Science and Engineering*, 8(2), 1120–1132. 10.1109/TNSE.2019.2937481

Yeboah-Akowuah, B., Tchao, E. T., Ur-Rehman, M., Khan, M. M., & Ahmad, S. (2021, September). Study of a printed split-ring monopole for dual-spectrum communications. *Heliyon*, 7(9), e07928. 10.1016/j.heliyon.2021.e0792834589621

You, D., Wang, K., Ye, L., Wu, J., & Huang, R. (2013). Transient stability assessment of power system using support vector machine with generator combinatorial trajectories inputs. *Int J Electr Power Energy Syst, 44*(1), 318–325. https://doi.org/. 07.05710.1016/j.ijepes.2012

Yucek, T. (2009). A survey of spectrum sensing algorithms for cognitive radio applications. *Communications Surveys Tutorials*. IEEE.

Yucel, M., Ozturk, N. F., & Gemci, C. (2016). Design of a Fiber Bragg Grating multiple temperature sensor. *2016 Sixth International Conference on Digital Information and Communication Technology and Its Applications (DICTAP)*, (pp. 6–11). IEEE. 10.1109/DICTAP.2016.7543992

Yu, X., Zhang, J., & Letaief, K. B. (2018). A hardware-efficient analog network structure for hybrid precoding in millimeter wave systems. *IEEE Journal of Selected Topics in Signal Processing*, 12(2), 282–297. 10.1109/JSTSP.2018.2814009

Zahran, R., Abdalla, M. A., & Gaafar, A. (2019, July). New thin wide-band braceletlike antenna with low SAR for on-arm WBAN applications. *IET Microwaves, Antennas & Propagation*, 13(8), 1219–1225. 10.1049/iet-map.2018.5801

Zeb, J., Hassan, A., & Nisar, M. D. (2021). Joint power and spectrum allocation for D2D communication overlaying cellular networks. *Computer Networks*, 184, 1–13. 10.1016/j.comnet.2020.107683

Zhang, H., & Wang, X. (2011). A Fuzzy Decision Scheme for Cooperative Spectrum Sensing in Cognitive Radio. *Vehicular Technology Conference (VTC Spring)*. IEEE. 10.1109/VETECS.2011.5956116

Zhang, L., & Chen, W. (2023). Machine Learning and Artificial Intelligence in Cybersecurity for Smart Grids. *Journal of Cybersecurity Research*.

Zhang, Q. (2021). Integration of 5G into Smart Grids: Cybersecurity Measures. *International Journal of Smart Grids and Communications.*

Zhang, Q. (2022). Machine Learning Algorithms for Cyber Defense in 5G-Integrated Smart Grids. *International Journal of Smart Grid and Clean Energy.*

Zhang, H., & Wang, G. (2021). Unified Network View and Centralized Control: SDN's Contribution to Network Management . *Computer Networks.*

Zhang, J., Xie, W., & Yang, F. (2015). An architecture for 5g mobile network based on sdn and nfv, *International Conference on Wireless, Mobile and Multi-Media.* Wiley. 10.1002/spy2.271

Zhang, K., Soh, P. J., & Yan, S. (2020, December). Meta-wearable antennas—A review of metamaterial based antennas in wireless body area networks. *Materials (Basel)*, 14(1), 149. 10.3390/ma14010149 33396333

Zhang, L., Ruan, J., Du, Z., Huang, D., & Deng, Y. (2023). Transmission line tower failure warning based on FBG strain monitoring and prediction model. *Electric Power Systems Research*, 214, 108827. 10.1016/j.epsr.2022.108827

Zhang, R. (2013). Application of optical fiber sensors in Smart Grid. *Proceedings of the Society for Photo-Instrumentation Engineers*, 9044, 90440J. 10.1117/12.2037572

Zhang, W., Qiao, S., & Wei, Y. (2012). A Diagonal Lattice Reduction Algorithm for MIMO Detection. *IEEE Signal Processing Letters*, 19(5), 311–314. 10.1109/LSP.2012.2191614

Zhang, X., Zhou, Z., & Niu, Y. (2018). An image encryption method based on the feistel network and dynamic DNA encoding. *IEEE Photonics Journal*, 10(4), 1–14. 10.1109/JPHOT.2018.2858823

Zhang, Y., Huang, T., & Bompard, E. F. (2018). Big data analytics in smart grids: A review. *Energy Informatics*, 1(1), 8. 10.1186/s42162-018-0007-5

Zhao, X., & Liu, W. (2023). Future Trends in 5G, NFV, and SDN Integration for Cybersecurity in Smart Grids. *Journal of Energy Engineering.*

Zhao, Y. (2019). NFV in 5G Networks: Flexibility, Scalability, and Programmability. *IEEE Transactions on Network and Service Management.* IEEE.

Zhao, J., Dong, W., Hinds, T., Li, Y., Splain, Z., Zhong, S., Wang, Q., Bajaj, N., To, A., Ahmed, M., Petrie, C. M., & Chen, K. P. (2023). Embedded Fiber Bragg Grating (FBG) Sensors Fabricated by Ultrasonic Additive Manufacturing for High-Frequency Dynamic Strain Measurements. *IEEE Sensors Journal*, 1, 1. 10.1109/JSEN.2023.3343604

Zheng, Q., Ma, G., Jiang, J., Li, C., & Zhan, H. (2015). A comparative study on partial discharge ultrasonic detection using fiber Bragg grating sensor and piezoelectric transducer. *2015 IEEE Conference on Electrical Insulation and Dielectric Phenomena (CEIDP)*, (pp. 282–285). IEEE. 10.1109/CEIDP.2015.7352071

Zhou, X., He, D., Khan, M. K., Wu, W., & Choo, K.-K. R. (2023). An Efficient Blockchain-Based Conditional Privacy-Preserving Authentication Protocol for VANETs. *IEEE Transactions on Vehicular Technology*, 72(1), 81–92. 10.1109/TVT.2022.3204582

Zhu, M., Zhang, Y., Zhao, W., Hu, Y., Wan, H., Li, K., & Zhou, A. (2023). Temperature Measurement of Pulsed Inductive Coil Continuous Discharge Based on FBG. *IEEE Sensors Journal*, 23(19), 22524–22532. 10.1109/JSEN.2023.3305092

Zhu, T., Li, J., Hu, X., Xiong, P., & Zhou, W. (2020). The Dynamic Privacy-Preserving Mechanisms for Online Dynamic Social Networks. *IEEE Transactions on Knowledge and Data Engineering*, 34(6), 2962–2974. 10.1109/TKDE.2020.3015835

Zicheng, Z., Keyan, T., & Yang, N. (2020). An FSS integrated with reflector antenna. *2020 9th Asia-Pacific Conference on Antennas and Propagation (APCAP)*, Xiamen, China. 10.1109/APCAP50217.2020.9246126

Zou, H., & Lu, M. (2021). Developing High-Frequency Fiber Bragg Grating Acceleration Sensors to Monitor Transmission Line Galloping. *IEEE Access : Practical Innovations, Open Solutions*, 9, 30893–30897. 10.1109/ACCESS.2021.3055820

Zu, B., Wu, B., Yang, P., Li, W., & Liu, J. (2021, August). Wideband and high-gain wearable antenna array with specific absorption rate suppression. *Electronics (Basel)*, 10(17), 2056. 10.3390/electronics10172056

About the Contributors

Prabhakar Gunasekaran is currently working as an Assistant Professor in the Department of Electronics and Communication Engineering, Thiagarajar College of Engineering, Madurai – 15 (A Govt. Aided Autonomous Institution Affiliated to Anna University) Tamilnadu, India. He obtained his B.E degree in Electronics & Communication Engineering from Arulmigu Kalasalingam College of Engineering, Krishnankoil, under Anna University, Chennai in the year of2009, and his M.Tech. Degree in the specialization of Embedded Systems from Hindustan University, Chennai in the year 2011. He obtained his Ph.D. degree in the year 2018 under the faculty of Electrical Engineering, at Anna University, Chennai. He is a recognized Ph. D Supervisor of Anna University, Chennai, under the Faculty of Electrical Engineering, and also guiding 5 Ph.D. scholars. He has published more than 47 research articles around the world including reputed journal transactions like IET, Springer, Taylor & Francis, and Elsevier.

N. Ayyanar received the B.E. degree in electronics and communication from the Narasu's Sarathy Institute of Technology, Salem, India, in 2013, the M.E. degree in optical communication from the Alagappa Chettiar Government College of Engineering and Technology, Karaikudi, India, in 2015, and the Ph.D. degree in electronics and communication engineering from the National Institute of Technology, Tiruchirappalli, India, in 2020. He is currently working as an Assistant Professor with the Department of Electronics and Communication Engineering, Thiagarajar College of Engineering, Madurai, Tamil Nadu, India. He has published 35 articles in refereed international journals and over 21 papers in conferences. His research interests include PCF-based optical fiber sensor, few mode fibers, fiber laser, plasmonics, and few mode amplifier system designs.

S. Rajaram is currently working as a Professor and Head in the Department of Electronics and Communication Engineering, Thiagarajar College of Engineering, Madurai – 15 (A Govt. Aided Autonomous Institution Affiliated to Anna University) Tamilnadu, India. He obtained his B.E degree in Electronics & Communication Engineering from Thiagarajar College of Engineering, Madurai, in the year of 1994, and his M.E. degree in the specialization of Microwave and Optical Engineering from Alagappa Chettiar College of Engineering and Technology in the year of 1996. He obtained his Ph.D. degree in the year 2008 from Anna University, Chennai. He did his Post-Doctoral Research at the Georgia Institute of Technology, Atlanta, USA. He is a recognized Ph. D Supervisor at Anna University, Chennai, and guided more than 15 Ph.D. scholars. He has published more than 250 research articles around the world including reputed journal transactions like IEEE, IET, Springer, Taylor & Francis, and Elsevier.

***

About the Contributors

B. Manimegalai is an esteemed professor and researcher in the field of Electronics and Communication Engineering. She is currently working as a Professor at Department of Electronics and Communication Engineering at Thiagarajar College of Engineering (TCE) in Madurai, Tamil Nadu, India. She completed her Bachelor's degree in Electronics and Communication Engineering from the Government College of Engineering in Salem, Tamil Nadu, her Master's degree in Digital Communication and Networking from the PSG College of Technology in Coimbatore, Tamil Nadu. She later earned her Ph.D. in Wireless Sensor Networks from Anna University, Chennai. She has over 20 years of teaching and research experience and has authored over 100 research publications in various international journals and conferences

Meenambal Bose received her B.E degree in Electronics and Communication Engineering (2010) and M.E degree in Optical Communication (2013). She is currently pursuing the Ph.D. degree in Information and Communication Engineering, specializing in the field of antennas. Her research focuses on the development and optimization of antennas for various wireless communication applications. Her work involves exploring innovative designs and methodologies to enhance the performance and efficiency of antennas in modern communication systems. Her current research interests include metamaterial inspired antennas, mobile terminal antennas and multiple-input multiple-output (MIMO) antennas.

Akashdeep Bhardwaj is working as Professor (Cyber Security & Digital Forensics) and Head of Cybersecurity Center of Excellence at University of Petroleum & Energy Studies (UPES), Dehradun, India. An eminent IT Industry expert with over 28 years of experience in areas such as Cybersecurity, Digital Forensics and IT Operations, Dr. Akashdeep mentors graduate, masters and doctoral students and leads several projects. Dr. Akashdeep is a Post-Doctoral from Majmaah University, Saudi Arabia, Ph.D. in Computer Science, Post Graduate Diploma in Management (equivalent to MBA), and holds an Engineering Degree in Computer Science. Dr. Akashdeep has published over 120 research works (including copyrights, patent, papers, authored & edited books) in international journals. Dr. Akashdeep worked as Technology Leader for several multinational organizations during his time in the IT industry. Dr. Akashdeep is certified in multiple technologies including Compliance Audits, Cybersecurity, and industry certifications in Microsoft, Cisco, and VMware technologies.

Rabia Bilal holds an Assistant professor position in the Department of Electrical Engineering at Usman Institute, Karachi, Pakistan. She has an MPhil in Engineering from the University of Sussex, UK, and a BS degree in Electronic Engineering and MS degree in Electronic with specialization in Telecommunication from Sir Syed University of Engineering and Technology, Karachi, Pakistan. She has decade of experience in engineering universities and industry where she was involved in teaching engineering undergraduates, research and publications. Her publications include refereed journal, conference papers, book chapters and books. She is a lifetime member of Pakistan Engineering Council (PEC).

D. Gracia Nirmala Rani received BE degree in Electronics and Communication Engineering from Madurai Kamarajar University, Madurai, India, in 2004, and ME degree in VLSI Design from Karunya University, Tamilnadu, India in 2007. She has been awarded PhD degree in VLSI Design from Anna University, Tamilnadu, India in 2014. She is working as an Assistant Professor in Thiagarajar College of Engineering, Madurai since 2007. She has authored or co-authored 25 international journal and conference papers. In 2018, she was the technical program chair of the 22nd International Symposium on VLSI Design and Test. She is serving as a reviewer in IEEE Transaction on Nanotechnology, Elsevier and Inderscience Journals, respectively. Her research interests include RFIC Design, Physical Design Automation, Optimization Algorithms for IC and mixed signal circuits and systems for Bio-medical Devices

D. Vijayalakshmi received her Bachelor of Engineering degree in Electronics and Communication Engineering from Anna University, Chennai in 2009 and a Master of Engineering degree in Applied Electronics from Anna University, Chennai in 2012. Presently, she is working as an Assistant Professor in the Department of Electronics and Communication Engineering at SRM Institute of Science and Technology, Kattankulathur, Kancheepuram, India, working here since 2013. She is currently pursuing her PhD degree in the Department of Electronics and Communication Engineering at SRMIST, Chennai, India. Her research interests include Optical Sensors, Optical networking and the Internet of Things.

E. Murugavalli received B.E. degree in Electronics and Communication Engineering from Bharathidasan University, India in 1996, M.E. degree in Optical Communication from Anna University, India in 2005 and Ph.D. degree from Anna University, Chennai, India in 2018. She is now an Assistant Professor in the Department of Electronics and Communication Engineering, Thiagarajar College of Engineering, Madurai, India. Her research interest includes Wireless communication networks, Millimeter wave communication, Internet of Things and Security.

G. Prabhakar is currently working as an Assistant Professor in the Department of Electronics and Communication Engineering, Thiagarajar College of Engineering, Madurai – 15 (A Govt. Aided Autonomous Institution Affiliated to Anna University) Tamilnadu, India. he started his career as a DSP Design Engineer in Venmsol Technologies Pvt. Ltd. Chennai. There, he carried out iCanTek Korean Client projects in the domain of H.264 High definition standard video streaming for Network IP camera and involved in the design & development of PCM based face recognition algorithms. After two years, he change-over his career and started to work as an Assistant Professor in the Department of Electrical & Electronics Engineering, Syed Ammal Engineering College, Ramanathapuram. He obtained his Ph.D. degree in the year of 2018 under the faculty of Electrical Engineering, Anna University, Chennai. After his Ph.D, he worked as an Associate Professor in the department of EEE, VSB Engineering College, Karur, for a period 3 years. His areas of interest are Embedded Systems, Cyber Physical Systems, Robotics & Control, Nonlinear Control Systems, Mechatronics and Image & Signal processing.

Josep M. Guerrero is with AAU Energy, Aalborg University, 9220 Aalborg East, Denmark.His research interests mainly include power electronics, distributed energy storage, Wireless Networks, IoT, and microgrids.

J. Shanthi, has completed her B.E Degree in Electronics and Communication Engineering from Mepco Schlenk Engineering College, Tamilnadu, India in 2004 and M.E Degree in Embedded Systems from Anna University in 2010. She has completed her research in the field of VLSI physical design at Thiagarajar College of Engineering and obtained her PhD Degree from Anna University, Chennai, India. She has published her research articles in reputed SCI journals of Q2. Currently, she is working in Fault detection in through Silicon Via of 3D IC, FinFET modelling and bio sensor applications. Her research interests also include VLSI Physical Design of 2D and 3D ICs,Through Silicon Vias, Machine Learning Algorithms, Optimization Algorithms, Processor Architectures, and Device Modelling.

Sherene Jacob is a dedicated research scholar at Thiagarajar College of Engineering, specializing in the field of antennas. With a strong background in electrical engineering, Ms. Jacob's research focuses on the development and optimization of antennas for various wireless communication applications. Her work involves exploring innovative designs and methodologies to enhance the performance and efficiency of antennas in modern communication systems. Ms. Jacob's passion for advancing technology and her commitment to academic excellence drive her pursuit of knowledge and innovation in the field of antennas.

K. Hariharan was born in Tamil Nadu, India. He awarded Ph.D. in 2011 and currently working as an Associate Professor in Department of Electronics and Communication Engineering in Thiagarajar College of Engineering, Tamilnadu, India. He published many research papers in optimisation techniques and analog devices testing applications. He was awarded best project guide Renesas-Japan in 2008.

K. Rajeswari (Senior Member, IEEE) received the Ph.D. degree in wireless communication from Anna University, Chennai, in 2015. She has a working experience as a Lecturer for 11 years and as an Assistant Professor for 12 years. She is currently an Associate Professor with the Department of Electronics and Communication Engineering, Thiagarajar College of Engineering, Madurai. Her research interests include digital signal processing, 5G wireless communication and biomedical signal processing.

K Shoukath Ali received his B.E. degree in Electronics and Communication Engineering, M.E. degree in Communication Systems and Ph.D. degree from Anna University, Chennai, Tamil Nadu in the year 2009, 2012 and 2019 respectively. He is currently an Assistant Professor at Presidency University, Itgalpura, Rajanukunte, Yelahanka, Bengaluru, Karnataka 560064, India. His main research interests include wireless communication, millimeter wave communication and 5G communication.

About the Contributors

K. Vijayakumar was born in Pollachi, Tamilnadu, India in 1977. He received his M.E. (Process Control and Instrumentaion) from Annamali Univeristy during 2004 and Ph.D (Electrical Enigneering) from Anna University during 2015. He has a teaching and research experience for nearly 22 years at various levels in engineeirng institutions and has published 15 research articles in reputed international and national journals 10 research articles in international conferences. Presently he is associated with Dr. N.G.P. Institute of Technology, Coimbatore in the department of Biomedical Enigneerng. His research interest is Process control, Machine learning algorithms, Sensors and Instrumentation.

Arfat Ahmad Khan received the B.Eng. degree in electrical engineering from the University of Lahore, Pakistan, in 2013, the M.Eng. degree in electrical engineering from the Government College University Lahore, Pakistan, in 2015, and the Ph. D. degree in telecommunication and computer engineering from the Suranaree University of Technology, Thailand, in 2018. From 2014–2016, he was an RF Engineer with Etisalat, UAE. From 2018–2022, he worked as a lecturer and senior researcher with the Suranaree University of Technology. Currently, he is working as a senior lecturer and researcher at Khon Kaen University, Thailand. His research interests include optimization and stochastic processes, channel and the mathematical modeling, wireless sensor networks, ZigBee, green communications, massive MIMO, OFDM, wireless technologies, signal processing, and the advance wireless communications.

M. Kumaran received his B.E. degree in Mechanical Engineering and M.E. degree in Product Design and Development from Anna University, Chennai, India, in 2008 and 2011, respectively. After completing his post-graduation, he worked as an Assistant Professor at various engineering colleges from 2011 to 2017. He registered for a Ph.D. in the Department of Production Engineering at NIT Tiruchirappalli and completed it in October 2022. Following his full-time Ph.D., he served as an Assistant Professor at several deemed universities. Currently, he is an Assistant Professor in the Department of Mechanical Engineering at SRM Institute of Science and Technology (Deemed to be University) - Tiruchirappalli Campus, Tamil Nadu, India. He has published five SCI-indexed research articles, six conference proceedings, and six book chapters in various reputed international journals. His research interests include Additive Manufacturing, Material Joining, and Product Design and Development.

M. N. Suresh was born in Tamil Nadu, India. He awarded Ph.D. in 2011 and currently working as an Associate Professor in Department of Electronics and Communication Engineering in Thiagarajar College of Engineering, Tamilnadu, India. He published many research papers in signal processing

M. Isaivani is a Full-time Faculty at the Department of Electrical and Electronics Engineering, Vaigai College of Engineering, Madurai. Her Research Interests Include VLSI, Smart grid, Embedded Systems & IoT.

Rajalakshmi Murugesan was born in Madurai, Tamil Nadu, India in 1988. She graduated from Electronics and Instrumentation Engineering of Kamaraj College of Engineering and Technology in 2010. She completed post-graduation in Degree in the Faculty of Instrumentation and Control Engineering from Kalasalingam Academy of Research and Education (KARE), in 2012 and completed her Ph.D. in the faculty of Electrical Engineering from Anna University, Chennai, 2020. She is currently employed as an Assistant Professor in the Thiagarajar College of Engineering, Mechatronics Department, Madurai. She has academic background, with Ten years of teaching experience as an Assistant/Associate professor at various institutions. Besides a research background, she has published several international journals (SCI/Scopus) and conferences (Scopus) from 2012 to till date. Her professional interests focus on Machine learning, Artificial Intelligence, linear and nonlinear control systems, system identification, and her current projects include modelling and controlling nonlinear processes (machine learning algorithms for Biomedical & Robotics).

Hassan Pakarzadeh was born in Dezful, Iran, in 1980. He received the B.Sc., M.Sc., and Ph.D. degrees in physics from Shiraz University, Shiraz, Iran, in 2002, 2005, and 2011, respectively. During 2011-2016, he was an Assistant Professor with Physics Department, Shiraz University of Technology, Shiraz, Iran, and since 2017, he has been an Associate Professor of physics. He is the first author or co-author of more than 160 peer-reviewed scientific publications in journals and conferences. His research interests include optics and photonics, fiber optical parametric amplifiers (FOPAs), fiber sensors, THz photonics, and nonlinear effects in optical fibers and devices. He is a Member of the Optics and Photonics Society of Iran, Physics Society of Iran, and IEEE Photonics Society. He is a Reviewer for a number of journals, which include the Optics Communications, Optics Express, Optical Fiber Technology, Applied Optics, JOSA B, Optik, Optical Review, International Journal of Optics and Photonics, Frontiers in Energy,...From 2008 to 2009, he was the recipient of the Scholarship of Visiting Ph.D. Student withDTU FOTONIK, the Department of Photonics Engineering, Denmark.

P. Nedumal Pugazhenthi was born in Ramanathapuram, Tamilnadu, India in 1979. He received his B.E (Electrical and Electronics Engineering) from Madurai kamaraj University during 2000, M.E. (Process Control and Instrumentaion) from Annamali Univeristy during 2004 and Ph.D (Electrical Enigneering) from Anna University during 2023. He has a teaching and research experience for nearly 20 years at various levels in engineeirng institutions and has published 15 research articles in reputed international and national journals 8 research articles in international conferences. He has two design patent granted and three copyrights registered to his credit. Presently he is associated with Dr. N.G.P. Institute of Technology, Coimbatore in the department of Biomedical Enigneerng. His research interest is Nonlinear system modeling and control, Machine learning algorithms and Regualtory aspects and Intellectual Property rights etc.,

Aravinda Raj R, a Full-time Faculty at the Department of Electrical and Electronics Engineering, Vaigai College of Engineering, Madurai, Tamilnadu, India. His research interest includes Smart Grid, Electric Vehicles Design, Power System Protection.

A. Rehash Rushmi Pavitra is a Full-time Faculty at the Department of Data Science and Business Systems, SRM Institute of Science and Technology, Faculty of Engineering and Technology, Kattankulathur, Chengalpattu district, Tamilnadu - 603203, India. Her research interest includes IoT, Optimization, AI, Machine learning, Wireless Sensor Networks, and Embedded systems.

Krithiga S received her Ph.D in the field of wireless communication by SRM Institute of Science and Technology, Kattankulathur, India in Jan 2019. Currently, she is an Associate Professor in the Department of Electronics and Communication Engineering at SRMIST, Kattankulathur, India working here since 2009. Her research interests include Wireless MIMO, OFDM, Digital Image Processing and Internet of things. She has got many international level journal publications under SCI and scopus indexing. Additionally she has got international conference publications indexed in ieeexplore and published book chapters. She is currently guiding her research scholars in the areas of IoT and deep learning techniques.

S.A. Sadik is a researcher in the field of computer science and telecommunication. He holds a MSc degree in Electrical and Electronics Engineering from Kutahya Dumlupinar University and a PhD degree in Electrical and Electronics Engineering from Kutahya Dumlupinar University. Sadik's research focuses on optical fiber technology and applying artificial intelligence techniques to solve real-world problems, with a particular interest in machine learning applications in optical fiber systems.

S. Muthulakshmi working as an Assistant professor II in the Department of Electrical and Electronics Engineering at Velammal college of Engineering and Technology viraganoor, Madurai She graduated in Engineering at Madurai kamarajar University. She secured Master of Engineering in Applied Electronics at Anna University.She secured Ph.D. in the department of Information and Communication Engineering , Anna University, India. She is in teaching profession for more than 20 years. She has presented 22 papers in National and International Journals, Conference and Symposiums. Her main area of interest includes power electronics, Renewable energy system and Special Electrical Machines.

About the Contributors

Selvaperumal Sundaramoorthy completed his PhD in Electrical Engineering from Anna University, Chennai 2013. Dr. Selvaperumal Sundaramoorthy has been serving as a Professor & Head of the EEE Department and Dean of Research at Mohamed Sathak Engineering College, Kilakarai, Ramanathapuram, Tamilnadu, India for the past one year and three months. With over 17 years of teaching experience and 8 years in the industry, Dr. Selvaperumal has made significant contributions to academia. He has published over 86 articles in prestigious journals, including IET and IEEE Transactions, and authored over 40 textbooks in Electrical Engineering to support rural students' education. He has filed 20 utility patents, with 17 published and 2 granted, and 9 design patents, with 8 granted. His Google Scholar citation indices include 827 citations and an H-index of 15. As a recognized supervisor at Anna University, he is currently guiding 8 PhD scholars and has successfully produced 12 PhD graduates in Electrical Engineering and Information Communication Engineering. Additionally, he has secured Rs. 40,00,000 in funding from AICTE and TNSCST.

Subaselvi Sundarraj received her B.E. degree in Electronics and Communication Engineering from Kumaraguru College of Technology, Coimbatore, India in 2015 and received her M.E. degree in Communication Engineering from Coimbatore Institute of Technology, Coimbatore, India in 2017. She received her Ph.D. degree from College of Engineering, Guindy (CEG), Anna University, Chennai in 2022. Presently she is working as an Assistant Professor in the Department of Electronics and Communication Engineering, M.Kumarasamy College of Engineering, Karur, India. His research interests include Software Defined Wireless Sensor Networks, Security and wireless Communication systems.

Kavitha V, a Full-time Faculty at the Department of Electronics and Communication Engineering,Sethu Institute of Technology,Virudhunagar, Tamilnadu, India. Her research interest includes IoT, Energy Management, Optimization, Microgrid, Wireless Networks, and Embedded systems.

V. Vinoth Thyagarajan received B.E. degree in Electronics and Communication Engineering from Thiagarajar College of Engineering, India in 2003, M.E. degree in Wireless Technologies from Anna University, India in 2005 and Ph.D. degree from Anna University, Chennai, India in 2016. He is now an Associate Professor in the Department of Electronics and Communication Engineering, Thiagarajar College of Engineering, Madurai, India. His research interest includes Low power VLSI circuits, FPGA implementation of signal processing and image processing algorithms.

Kavitha V is a Full-time Faculty at the Department of Electronics and Communication Engineering, Sethu Institute of Technology, Virudhunagar, Tamilnadu, India. Her research interests include IoT, Energy Management, Optimization, Microgrid, Wireless Networks, and Embedded Systems.

Index

Symbols

A

C

D

E

F

S

T

V

W

Printed in the United States
by Baker & Taylor Publisher Services